VAPOUR–LIQUID EQUILIBRIUM DATA AT NORMAL PRESSURES

EDUARD HÁLA, IVAN WICHTERLE, JIŘI POLÁK
AND TOMÁŠ BOUBLIK

Institute of Chemical Process Fundamentals
Czechoslovak Academy of Sciences
Prague, Czechoslovakia

THE QUEEN'S AWARD
TO INDUSTRY 1966

PERGAMON PRESS

OXFORD · LONDON · EDINBURGH · NEW YORK

TORONTO · SYDNEY · PARIS · BRAUNSCHWEIG

Pergamon Press Ltd., Headington Hill Hall, Oxford
4 & 5 Fitzroy Square, London W.1.
Pergamon Press (Scotland) Ltd., 2 & 3 Teviot Place, Edinburgh 1
Pergamon Press Inc., 44–01 21st Street, Long Island City, New York 11101
Pergamon of Canada Ltd., 207 Queen's Quay West, Toronto 1
Pergamon Press (Aust.) Pty. Ltd., 19a Boundary Street, Rushcutters Bay, N.S.W. 2011
Pergamon Press S.A.R.L., 24 rue des Écoles, Paris 5^e
Vieweg & Sohn GmbH, Burgplatz 1, Braunschweig

First edition 1968

Library of Congress Catalog Card No. 67-31076

PRINTED IN GREAT BRITAIN BY A. WHEATON AND CO., EXETER
08 012652 9

VAPOUR–LIQUID EQUILIBRIUM DATA
AT NORMAL PRESSURES

CONTENTS

INTRODUCTION

DISTILLATION and rectification are the most important separation processes used in the chemical industry. For the design of rectification columns and for the determination of their optimal operating conditions, quantitative information is necessary on vapour–liquid equilibrium for systems whose components are to be separated.

In the literature many articles have been published dealing with vapour–liquid equilibrium of mixtures both of nonelectrolytes and of electrolytes (summary review see [13]). In this book the direct experimental data of a set of selected systems are collected and correlated with the aid of equations expressing the dependence of the activity coefficients or separation functions on the composition of the liquid phase. The quantitative characterization of the systems is thus condensed into the constants of the correlation equations evaluated using the weighted least squares method (by using an automatic computer National Elliott 803 B). In the last columns of the tables the deviations of the calculated from the direct experimental data are presented which give information on the quality of the data and on the flexibility of the correlation relations used.

The choice of systems which are analysed in the tables was made from the point of view partly of the systems' practical importance, partly of the quality of the data, and partly of the suitability of the constituent combinations in the light of Ewell classification of liquids to permit the estimation of the behaviour of similar systems.

We wish to thank Prof. V. Bažant for suggesting the writing of this book and Prof. G. Standart for the critical reading of the manuscript.

Prague   THE AUTHORS
July 1966

PART I

CORRELATION EQUATIONS

CHAPTER 1

CORRELATION OF DATA IN TWO-COMPONENT SYSTEMS

1.1. Nonelectrolytes

The mutual dependence of the equilibrium compositions of the liquid and vapour phases in a two-component system of nonelectrolytes is expressed by means of the equations

$$y_1 = \frac{a_{12}(x_1/x_2)}{1 + a_{12}(x_1/x_2)}, \qquad (1)$$

$$y_2 = \frac{1}{1 + a_{12}(x_1/x_2)}, \qquad (2)$$

where x_1, x_2 are the mole fractions of the components 1, 2 in the liquid phase and y_1, y_2 their mole fractions in the vapour phase. The relative volatility a_{12}, defined by the relation

$$a_{12} = \frac{y_1 x_2}{y_2 x_1} \qquad (3)$$

is given in the region of normal and low pressures by the equation

$$a_{12} = \frac{\gamma_1 P_1^0}{\gamma_2 P_2^0}, \quad [* g], \qquad (4)$$

where P_1^0, P_2^0 denote the vapour pressures of the pure components 1, 2 at given temperature (in mmHg), γ_1, γ_2 their activity coefficients and the symbol $[* g]$ indicates that we assume ideal behaviour of vapour phase.

The dependence of the vapour pressure of the pure component i on temperature is expressed by means of the Antoine equation [1]

$$\log P^0 = A_i - \frac{B_i}{t + C_i}, \qquad (5)$$

where t is the temperature (in °C) and A_i, B_i, C_i are constants which are characteristic for the component i at the given temperature range. Their values were taken from Dreisbach's collection [6] or evaluated from the direct experimental data using the weighted least squares method [5, 7, 15].

3

The activity coefficients γ_1, γ_2 are generally functions of the temperature, pressure, and composition of the liquid phase. The temperature and pressure dependence was not considered, as is usual in chemical engineering practice. The concentration dependence was expressed by means of one of the equations:

van Laar 3rd order [21, 22, 25, 26]:

$$\log \gamma_1 = \frac{A_{12}\, x_2^2}{[x_1\, (A_{12}/A_{21}) + x_2]^2},$$

$$\log \gamma_2 = \frac{A_{21}\, x_1^2}{[x_2\, (A_{21}/A_{12}) + x_1]^2}.$$

(6)

Margules 3rd order [16, 25, 26]:

$$\log \frac{\gamma_1}{\gamma_2} = x_2^2 A_{12} - x_1^2 A_{21} - 2x_1 x_2 \, (A_{12} - A_{21}).$$

(7)

Margules 4th order:

$$\log \frac{\gamma_1}{\gamma_2} = x_2^2 A_{12} - x_1^2 A_{21} - 2x_1 x_2 \, [A_{12} - A_{21} + (x_2 - x_1)D_{12}],$$

(8)

where A_{12}, A_{21}, D_{12} are constants which are characteristic for the given mixture. Their values were determined from the direct experimental data of the binary system under consideration.

In the region of normal and higher pressures the adjusted series expansion of the relative volatility [8, 13] was also used for the correlation which in the two-component case has the form:

2nd order:

$$a_{12} = \frac{1 + a_{12}x_2}{1 + a_{21}x_1};$$

(9)

3rd order:

$$a_{12} = \frac{1 + a_{12}x_2 + a_{122}x_2^2}{1 + a_{21}x_1 + a_{211}x_1^2},$$

(10)

where a_{12}, a_{21}, a_{122}, a_{211} are constants characteristic of the given binary system. Their values were determined from the direct vapour–liquid equilibrium data of the given binary system.

Note: It can be shown [13] that the empirical equations proposed by Clark [4], Prahl [18], Kretschmer and Wiebe [14], Spinner, Lu, and Graydon [19] are special cases of equations (9) or (10).

Experimental data were adjusted by means of the weighted least squares method (see Appendix) for the correlation equations presented above when expressed in the form

$$\log \frac{\gamma_1}{\gamma_2} = F(x_1). \tag{11}$$

The weighting factor for this equation (11) is

$$w = \left[\frac{1}{2{,}303^2} \left(\frac{1}{x_1^2 x_2^2} + \frac{1}{y_1^2 y_2^2} \right) + \left(\frac{\partial F(x_1)}{\partial x_1} \right)^2 \right]^{-1} \tag{12}$$

1.2. Electrolytes

The mutual dependence of the equilibrium compositions of the vapour and liquid phases in binary systems containing electrolytic components is expressed [10, 13] by means of the equations

$$y_1 = \frac{\beta_{12}(x_{\pm}^{\nu 1}/x_{2\pm})}{1 + \beta_{12}(x_{1\pm}^{\nu 1}/x_{2\pm})} , \tag{13}$$

$$y_2 = \frac{1}{1 + \beta_{12}(x_{1\pm}^{\nu 1}/x_{2\pm})} ,$$

where $x_{1\pm}$ is the mean mole fraction of the electrolytic component 1 defined by the relations

$$x_{1\pm} = (x_{1+}^{\nu 1+} x_{1-}^{\nu 1-})^{1/\nu 1}, \tag{14}$$

$$x_{1+} = \frac{\nu_{1+} n_1}{\nu_1 n_1 + n_2} , \tag{15a}$$

$$x_{1-} = \frac{\nu_{1-} n_1}{\nu_1 n_1 + n_2} , \tag{15b}$$

$$\nu_1 = \nu_{1+} + \nu_{1-} \tag{16}$$

where n_1, n_2 denote the moles of the components 1, 2 and ν_{1+}, ν_{1-} the number of positive and negative ions respectively, formed by the complete dissociation of one molecule of the component 1; $x_{2\pm}$ is the mole fraction of the liquid solvent given by relation

$$x_{2\pm} = \frac{n_2}{\nu_1 n_1 + n_2} . \tag{17}$$

The separation function β_{12} is defined by the equation

$$\beta_{12} = \frac{y_1 \, x_{2\pm}}{y_2 \, x_{1\pm}^{\nu 1}}, \tag{18}$$

where y_1, y_2 are the mole fractions of the components 1, 2 in the vapour phase.

The dependence of the separation function β_{12} on composition of the liquid phase was expressed using the equations:

the 1st order:

$$\log \beta_{12} = a_{12} + b_{12} I^{1/2}; \tag{19}$$

or the 2nd order:

$$\log \beta_{12} = a_{12} + b_{12} I^{1/2} + c_{12} (x_2 - x_1), \tag{20}$$

where x_1, x_2 are the (analytical) mole fractions of the components 1, 2

$$x_1 = \frac{n_1}{n_1 + n_2}, \tag{21}$$

$$x_2 = \frac{n_2}{n_1 + n_2}, $$

and the ionic strength I is defined by the relation

$$I = \tfrac{1}{2} (\nu_{1+} z_{1+}^2 + \nu_{1-} z_{1-}^2) x_1, \tag{22}$$

where z_{1+} is the valence of the cation and z_{1-} the valence of the anion formed from the electrolytic component 1.

The constants a_{12}, b_{12}, c_{12} are characteristic of the given system and are evaluated from the direct experimental binary data. The weighting factor was derived by a method similar to that used for nonelectrolytical systems (see Appendix).

CHAPTER 2

CORRELATION OF DATA IN THREE-COMPONENT SYSTEMS

2.1. Nonelectrolytes

The mutual dependence of the equilibrium compositions of the liquid and vapour phases in a three-component system of nonelectrolytes is expressed by means of the equations

$$y_1 = \frac{a_{13}\,(x_1/x_3)}{1 + a_{13}\,(x_1/x_3) + a_{23}\,(x_2/x_3)},$$

$$y_2 = \frac{a_{23}\,(x_2/x_3)}{1 + a_{13}\,(x_1/x_3) + a_{23}\,(x_2/x_3)}, \tag{23}$$

$$y_3 = \frac{1}{1 + a_{13}\,(x_1/x_3) + a_{23}\,(x_2/x_3)},$$

where x_1, x_2, x_3 are the mole fractions of the components 1, 2, 3 in the liquid phase and y_1, y_2, y_3 are their mole fractions in the vapour phase. The independent relative volatilities a_{13}, a_{23} defined by the relations

$$a_{13} = \frac{y_1\,x_3}{y_3\,x_1},$$

$$a_{23} = \frac{y_2\,x_3}{y_3\,x_2}, \tag{24}$$

were calculated in the region of normal and low pressures according to the equations

$$a_{13} = \frac{\gamma_1\,P_1^0}{\gamma_3\,P_3^0} \quad [*g],$$

$$a_{23} = \frac{\gamma_2\,P_2^0}{\gamma_3\,P_3^0} \quad [*g], \tag{25}$$

where P_1^0, P_2^0, P_3^0 denote the vapour pressure of the pure components 1, 2, 3 at the given temperature and γ_1, γ_2, γ_3 their activity coefficients.

The dependence of the vapour pressure of the pure components i ($i = 1, 2, 3$) on temperature is again expressed by means of the Antoine equation (5). The concentration dependence of the activity coefficients is expressed by means of the equations:

Margules 3rd order [16, 25, 26]:

$$\log \frac{\gamma_1}{\gamma_3} = x_3^2 A_{13} - x_1^2 A_{31} - 2x_1 x_3 (A_{13} - A_{31}) - x_2 [2x_3 A_{23} -$$
$$- 2x_1 A_{21} + x_2 (A_{32} - A_{12}) + (x_1 - x_3)(A_{21} + A_{13} + A_{32} - C_{123})]$$

$$\log \frac{\gamma_2}{\gamma_3} = x_3^2 A_{23} - x_2^2 A_{32} - 2x_2 x_3 (A_{23} - A_{32}) - x_1 [2x_3 A_{13} -$$
$$- 2x_2 A_{12} + x_1 (A_{31} - A_{21}) + (x_2 - x_3)(A_{21} + A_{13} + A_{32} - C_{123})]; \tag{26}$$

Margules 4th order:

$$\log \frac{\gamma_1}{\gamma_3} = x_2^2 [A_{12} + 2x_1 (A_{21} - A_{12} - D_{12}) + 3x_1^2 D_{12}] +$$
$$+ x_3^2 [A_{13} + 2x_1 (A_{31} - A_{13} - D_{13}) + 3x_1^2 - D_{13}] +$$
$$+ x_2 x_3 [A_{21} + A_{13} - A_{32} + 2x_1 (A_{31} - A_{13}) + 2x_3 (A_{32} - A_{23}) +$$
$$+ 3x_2 x_3 D_{23} - C_{1123} x_1 (2 - 3x_1) - C_{1223} x_2 (1 - 3x_1) -$$
$$- C_{1233} x_3 (1 - 3x_1)] -$$
$$- x_1^2 [A_{31} + 2x_3 (A_{13} - A_{31} - D_{13}) + 3x_3^2 D_{13}] -$$
$$- x_2^2 [A_{32} + 2x_3 (A_{23} - A_{32} - D_{23}) + 3x_3^2 D_{23}] -$$
$$- x_1 x_2 [A_{13} + A_{32} - A_{21} + 2x_3 (A_{23} - A_{32}) + 2x_2 (A_{21} - A_{12}) +$$
$$+ 3x_1 x_2 D_{12} - C_{1233} x_3 (2 - 3x_3) - C_{1123} x_1 (1 - 3x_3) -$$
$$- C_{1223} x_2 (1 - 3x_3)],$$

$$\log \frac{\gamma_2}{\gamma_3} = x_1^2 [A_{21} + 2x_2 (A_{12} - A_{21} - D_{12}) + 3x_2^2 D_{12}] +$$
$$+ x_3^2 [A_{23} + 2x_2 (A_{32} - A_{23} - D_{23}) + 3x_2^2 D_{23}] +$$
$$+ x_1 x_3 [A_{32} + A_{21} - A_{13} + 2x_2 (A_{12} - A_{21}) + 2x_1 (A_{13} - A_{31}) +$$
$$+ 3x_1 x_3 D_{13} - C_{1223} x_2 (2 - 3x_2) - C_{1233} x_3 (1 - 3x_2) -$$
$$- C_{1123} x_1 (1 - 3x_2)] -$$
$$- x_1^2 [A_{31} + 2x_3 (A_{13} - A_{31} - D_{13}) + 3x_3^2 D_{13}] -$$
$$- x_2^2 [A_{32} + 2x_3 (A_{23} - A_{32} - D_{23}) + 3x_3^2 D_{23}] -$$
$$- x_1 x_2 [A_{13} + A_{32} - A_{21} + 2x_3 (A_{23} - A_{32}) + 2x_2 (A_{21} - A_{12}) +$$
$$+ 3x_1 x_2 D_{12} - C_{1233} x_3 (2 - 3x_3) - C_{1123} x_1 (1 - 3x_3) -$$
$$- C_{1223} x_2 (1 - 3x_3)], \tag{27}$$

where A_{12}, A_{21}, D_{12}, A_{13}, A_{31}, D_{13}, A_{23}, A_{32}, D_{23} are constants which are characteristic of the binary systems indicated by the indices; C_{123}, C_{1123}, C_{1223}, C_{1233} are constants which are characteristic of the given ternary system. They were evaluated (determined) from the direct experimental data of the binary and ternary system under consideration.

Note: Ternary constants in the Margules equations were calculated in the following way: the individual constants C_{123} *or* C_{1123}, C_{1223}, C_{1233} were independently evaluated by means of the weighted least squares method from each of the equations for $\log \frac{\gamma_1}{\gamma_2}$, $\log \frac{\gamma_1}{\gamma_3}$ and $\log \frac{\gamma_2}{\gamma_3}$. The reported constants were then obtained as the mean of these individual values.

In the region of normal and higher pressures the adjusted series expansions of the relative volatilities [9, 13] are also used which in the three-component system have the form:

the 2nd order:

$$a_{13} = \frac{1 + a_{13}x_3 + a_{12}x_2}{1 + a_{31}x_1 + a_{32}x_2},$$

$$a_{23} = \frac{1 + a_{23}x_3 + a_{21}x_1}{1 + a_{32}x_2 + a_{31}x_1};$$

(28)

the 3rd order:

$$a_{13} = \frac{1 + a_{13}x_3 + a_{12}x_2 + a_{133}x_3^2 + 2a_{123}x_2x_3 + a_{122}x_2^2}{1 + a_{31}x_1 + a_{32}x_2 + a_{311}x_1^2 + 2a_{312}x_1x_2 + a_{322}x_2^2},$$

$$a_{23} = \frac{1 + a_{23}x_3 + a_{21}x_1 + a_{233}x_3^2 + 2a_{213}x_1x_3 + a_{211}x_1^2}{1 + a_{32}x_2 + a_{31}x_1 + a_{322}x_2^2 + 2a_{312}x_1x_2 + a_{311}x_1^2},$$

(29)

where a_{12}, a_{21}, a_{122}, a_{211}, a_{13}, a_{31}, a_{133}, a_{311}, a_{23}, a_{32}, a_{233}, a_{322} are constants characteristic of the given binary systems indicated by indices, and a_{123}, a_{312}, a_{213} are constants which are characteristic for the given ternary system. Their values were determined from the direct experimental data on the binary and ternary systems under consideration.

Note: The Margules type equations were chosen for correlations because their binary constants are (in distinction to the van Laar constants) free from restricting conditions—for details see the critical discussion [23].

2.2. Electrolytes

The mutual dependence of the equilibrium compositions of the vapour and liquid phases in ternary systems of electrolytic components [two electrolytes (1, 2), one nonelectrolytic solvent (3)] are expressed by means of the equations [12, 13]:

$$y_1 = \frac{\beta_{13}\,(x_{1\pm}^{v_1}/x_{3\pm})}{1 + \beta_{13}\,(x_{1\pm}^{v_1}/x_{3\pm}) + \beta_{23}\,(x_{2\pm}^{v_2}/x_{3\pm})},$$

$$y_2 = \frac{\beta_{23}\,(x_{2\pm}^{v_2}/x_{3\pm})}{1 + \beta_{13}\,(x_{1\pm}^{v_1}/x_{3\pm}) + \beta_{23}\,(x_{2\pm}^{v_2}/x_{3\pm})},$$

(30)

$$y_3 = \frac{1}{1 + \beta_{13}\,(x_{1\pm}^{v_1}/x_{3\pm}) + \beta_{23}\,(x_{2\pm}^{v_2}/x_{3\pm})},$$

B

where $x_{1\pm}$, $x_{2\pm}$ are the mean mole fractions of the electrolytic components 1, 2 defined by the relations

$$x_{1\pm} = (x_{1+}^{\nu_{1+}} x_{1-}^{\nu_{1-}})^{1/\nu_1},$$

$$x_{2\pm} = (x_{2+}^{\nu_{2+}} x_{2-}^{\nu_{2-}})^{1/\nu_2}, \tag{31}$$

where $\nu_1 = \nu_{1+} + \nu_{1-}$, $\nu_2 = \nu_{2+} + \nu_{2-}$.

When the components 1 and 2 have no common ion then

$$x_{1+} = \frac{\nu_{1+} n_1}{\nu_1 n_1 + \nu_2 n_2 + n_3},$$

$$x_{2+} = \frac{\nu_{2+} n_2}{\nu_1 n_1 + \nu_2 n_2 + n_3}, \tag{32}$$

$$x_{1-} = \frac{\nu_{1-} n_1}{\nu_1 n_1 + \nu_2 n_2 + n_3},$$

$$x_{2-} = \frac{\nu_{2-} n_2}{\nu_1 n_1 + \nu_2 n_2 + n_3}. \tag{33}$$

If the components 1 and 2 have a common cation, then

$$x_{1+} = x_{2+} = \frac{\nu_{1+} n_1 + \nu_{2+} n_2}{\nu_1 n_1 + \nu_2 n_2 + n_3} \tag{34}$$

while x_{1-}, x_{2-} are given by equations (33).

For components 1 and 2 with the common anion x_{1+} and x_{2+} are given by equations (32) while now

$$x_{1-} = x_{2-} = \frac{\nu_{1-} n_1 + \nu_{2-} n_2}{\nu_1 n_1 + \nu_2 n_2 + n_3}. \tag{35}$$

In equations (31)–(35), n_1, n_2, n_3 denote the analytically determined moles of the (undissociated) components 1, 2, 3 and ν_{1+}, ν_{1-}, ν_{2+}, ν_{2-}, the number of positive and negative ions respectively, formed from the complete dissociation of one molecule of the component 1 resp. 2; $x_{3\pm}$ is the mole fraction of the liquid solvent given by the relation

$$x_{3\pm} = \frac{n_3}{\nu_1 n_1 + \nu_2 n_2 + n_3}. \tag{36}$$

The separation functions β_{13}, β_{23} are defined by the equations

$$\beta_{13} = \frac{y_1\, x_{3\pm}}{y_3\, x_{1\pm}^{\nu_1}},$$

$$\beta_{23} = \frac{y_2\, x_{3\pm}}{y_3\, x_{2\pm}^{\nu_2}},$$

(37)

where y_1, y_2, y_3 are the mole fractions of the (undissociated) components 1, 2, 3 in the vapour phase.

The dependences of the separation functions β_{13}, β_{23} on the composition of the liquid phase were expressed using the equations:

the 1st order:

$$\log \beta_{13} = a_{13} + b_{13}I^{1/2} + (A_{12} - A_{32})\, x_2,$$

$$\log \beta_{23} = a_{23} + b_{23}I^{1/2} + (A_{21} - A_{31})\, x_1;$$

(38)

2nd order:

$$\log \beta_{13} = a_{13} + b_{13}I^{1/2} + c_{13}\,(x_3 - x_1) + (A_{12} - A_{32})\, x_2,$$

$$\log \beta_{23} = a_{23} + b_{23}I^{1/2} + c_{23}\,(x_3 - x_2) + (A_{21} - A_{31})\, x_1,$$

(39)

where x_1, x_2, x_3 are the (analytical) mole fractions of the components 1, 2, 3.

$$x_1 = \frac{n_1}{n_1 + n_2 + n_3},$$

$$x_2 = \frac{n_2}{n_1 + n_2 + n_3},$$

(40)

$$x_3 = \frac{n_3}{n_1 + n_2 + n_3}.$$

The concentration variable (modified ionic strength) I is defined by the relation

$$I = \tfrac{1}{2}\,[(\nu_{1+}z_{1+}^2 + \nu_{1-}z_{1-}^2)x_1 + (\nu_{2+}z_{2+}^2 + \nu_{2-}z_{2-}^2)x_2];$$

(41)

z_{1+}, z_{1-}, z_{2+}, z_{2-} in equation (41) are the valencies of the cations and anions formed from the electrolytic components 1, 2; a_{13}, a_{23}, b_{13}, b_{23}, c_{13}, c_{23} are constants characteristic of the indicated systems which were determined from the vapour–liquid equilibrium data on the binary systems. The constants $(A_{12} - A_{32})$ and $(A_{21} - A_{31})$ are expressed as

$$D_{13} = A_{12} - A_{32}$$

$$D_{23} = A_{21} - A_{31}$$

(42)

and these values were determined from the direct ternary data.

Note: If the component 2 is a nonvolatile electrolyte then $y_2 = 0$ and thus $\beta_{23} = 0$.

CHAPTER 3

CORRELATION OF DATA IN FOUR- AND MORE-THAN-FOUR-COMPONENT SYSTEMS

THE mutual dependence of the equilibrium compositions of the liquid and vapour phases in four- and more-than-four-component systems of nonelectrolytes was expressed by means of the equation

$$y_i = \frac{a_{ik}\,(x_i/x_k)}{\sum\limits_{j=1}^{k} a_{jk}\,[x_j\,/\,x_k]} \quad (i, j = 1, 2, \ldots, k), \tag{43}$$

where x_i, x_j, x_k are the mole fractions of the components i, j, k in the liquid phase and y_i the mole fraction of the component i in the vapour phase.

The relative volatilities, defined by the relations

$$a_{ik} = \frac{y_i\,x_k}{y_k\,x_i},$$

$$\tag{44}$$

$$a_{jk} = \frac{y_j\,x_k}{y_k\,x_j},$$

are expressed in the region of normal and low pressures by the equations

$$a_{ik} = \frac{\gamma_i\,P^o_i}{\gamma_k\,P^o_k} \quad [*g],$$

$$\tag{45}$$

$$a_{jk} = \frac{\gamma_j\,P^o_j}{\gamma_k\,P^o_k} \quad [*g],$$

where P^o_i, P^o_j, P^o_k, denote the vapour pressures of the pure components i, j, k at given temperatures, $\gamma_i, \gamma_j, \gamma_k$ their activity coefficients.

The dependence of the vapour pressure of the pure component i on temperature was again expressed by means of the Antoine equation (5). The concentration dependence of the activity coefficients in the k–component system is expressed by means of the equations:

Margules–Wohl 3rd order [11, 13]:

$$\log \frac{\gamma_i}{\gamma_k} = \sum_{\substack{i \neq r}}^{k} x_r \left(2x_i A_{ri} + x_r A_{ir}\right) + \sum_{\substack{r < s \\ i \neq r,s}}^{k} x_r x_s \left(R_{(i)rs} - C_{(i)rs}\right) -$$

(46)

$$- \sum_{\substack{k \neq r}}^{k} x_r \left(2x_k A_{rk} + x_r A_{kr}\right) - \sum_{\substack{r < s \\ k \neq r,s}}^{k} x_r x_s \left(R_{(k)rs} - C_{(k)rs}\right);$$

Margules–Wohl 4th order:

$$\log \frac{\gamma_i}{\gamma_k} = \sum_{\substack{i \neq r}}^{k} x_r \left(2x_i A_{ri} + x_r A_{ir} - 2x_i x_r D_{ir}\right) +$$

$$+ \sum_{\substack{r < s \\ i \neq r,s}} x_r x_s \left(R_{(i)rs} - 2x_i C_{(ii)rs} - x_r C_{(i)rrs} - x_s C_{(i)rss}\right)$$

$$+ \sum_{\substack{r < s < t \\ i \; r,s\,t}}^{k} x_r x_s x_t E_{(i)rst} -$$

$$- \sum_{\substack{k \neq r}}^{k} x_r \left(2x_r A_{rk} + x_r A_{kr} - 2x_k x_r D_{kr}\right) -$$

$$- \sum_{\substack{r < s \\ k \neq r,s}}^{k} x_r x_s \left(R_{(k)rs} - 2x_k C_{(kk)rs} - x_r C_{(k)rrs} - x_s C_{(K)rss}\right) -$$

$$- \sum_{\substack{r < s < t \\ k \neq r,s,t}}^{k} x_r x_s x_t E_{(k)rst}$$

(47)

$$R_{rst} \equiv A_{sr} + A_{rt} + A_{ts}$$

$$R_{(i)rs} \begin{array}{l} \nearrow R_{irs} \quad \text{for} \quad i < r < s \\ \rightarrow R_{ris} \quad \text{for} \quad r < i < s \\ \searrow R_{rsi} \quad \text{for} \quad r < s < i \end{array}$$

$$C_{(i)rs} \begin{array}{l} \nearrow C_{irs} \quad \text{for} \quad i < r < s \\ \rightarrow C_{ris} \quad \text{for} \quad r < i < s \\ \searrow C_{rsi} \quad \text{for} \quad r < s < i, \quad \text{etc.} \end{array}$$

where A_{ir}, A_{ri}, A_{rk}, A_{kr}, D_{ir}, D_{kr} are constants which are characteristic of the binary systems indicated by the indices and R, C are constants which are characteristic of the given ternary systems. Their values were evaluated from the direct experimental data of the binary and ternary systems under consideration.

Note: In the calculations the quarternary constants E in the 4th order Margules–Wohl equation were taken equal to zero as the accuracy of vapour–liquid equilibrium data is rarely sufficient to show any effect of quarternary interactions.

In the region of normal and higher pressures the adjusted series expansions of the relative volatilities [9, 13] are also used which in the k-component system have the form:

2nd order:

$$a_{ik} = \frac{1 + \sum\limits_{\substack{i \neq r}}^{k} x_r a_{ir}}{1 + \sum\limits_{\substack{k \neq r}}^{k} x_r a_{kr}} \; ; \tag{48}$$

3rd order:

$$a_{ik} = \frac{1 + \sum\limits_{\substack{i \neq r}}^{k} x_r (a_{ir} + x_r a_{irr}) + \sum\limits_{\substack{r < s \\ i \neq r,s}}^{k} x_r x_s 2 a_{irs}}{1 + \sum\limits_{\substack{k \neq r}}^{k} x_r (a_{kr} + x_r a_{krr}) + \sum\limits_{\substack{r < s \\ k \neq r,s}}^{k} x_r x_s 2 a_{krs}} \; , \tag{49}$$

where a_{ir}, a_{kr}, a_{irr}, a_{krr} are constants characteristic of the binary systems indicated by the indices and a_{irs}, a_{krs} are constants which characteristic of the given ternary systems. Their values were evaluated from the direct experimental data of the binary and ternary systems under consideration.

APPENDIX

Vapour–liquid equilibrium correlation relations for a k-component system can be formally expressed by the equation

$$G(x_1, x_2, \ldots, x_{k-1}, y_1, y_2, \ldots, y_{k-1}, T, P) = 0$$

with

$$\sum_{i}^{k} x_i = 1; \quad \sum_{i}^{k} y_i = 1$$

where $x_1, x_2, \ldots, x_k$ are the mole fractions of the constituents in the liquid, $y_1, y_2, \ldots, y_k$ are those in the vapour, P is the total pressure, and T is the absolute temperature of the system. Then the statistical weight can be generally expressed as

$$w = \left[\sum_{i=1}^{k-1} \left(\frac{\partial G}{\partial x_i}\right)^2_{x_{j \neq i}, y_j, T, P} \sigma_{xi}^2 + \sum_{i=1}^{k-1} \left(\frac{\partial G}{\partial y_i}\right)^2_{x_j, y_{j \neq i}, P, T} \sigma_{yi}^2 + \left(\frac{\partial G}{\partial T}\right)^2_{x_j, y_j, P} \sigma_T^2 + \left(\frac{\partial G}{\partial P}\right)^2_{x_j, y_j, T} \sigma_p^2 \right]^{-1},$$

where $j = 1, 2, \ldots, k$ and σ^2 is the variance.

This expression was simplified using the assumptions:

(1) The error in measuring pressure and temperature considered negligible compared to that for concentrations so that $\sigma_T^2 = 0; \sigma_P^2 = 0$.
(2) The variances σ_x^2, σ_y^2 are constant over the whole concentration range.
(3) The variance for vapour and liquid concentrations are equal $\sigma_{xi}^2 = \sigma_{yi}^2$. This assumption is justified if both phases are analysed in the same way.

As the weight is a relative quantity, we can absorb the variance into it and thus obtain finally

$$w = \left[\sum_{i=1}^{k-1} \left[\left(\frac{\partial G}{\partial x_i}\right)^2_{x_{j \neq i}, y_j} + \left(\frac{\partial G}{\partial y_i}\right)^2_{x_j, y_{j \neq i}} \right] \right]^{-1},$$

where $j = 1, 2, \ldots, k$.

PART II

TABLES

CHAPTER 1

GUIDE TO TABLES

BOTH binary and multicomponent systems are listed in these tables. In looking for the calculated values for a given system, we find (any) one of its constituents in the column headed "Name" and the other constituent(s) in the adjoining column (Chapter 5). The substances are listed in the tables according to the number of carbon atoms in the molecule. The number in the last column denotes the page where it is possible to find the desired system.

The vapour pressures were taken from the Dreisbach compilation [6]. For substances which are not cited in this work we used Antoine constants taken from the original literature or calculated from the experimental data using the method employed by the National Bureau of Standards [24]. This case concerns the substances (reference in parentheses): acetone [2], ethyl alcohol [2], ethyl ether [20], ethyl acetate [17], butyl alcohol [3], methyl acetate [17], methyl alcohol [2], propyl alcohol [3], *iso*propyl alcohol [3].

The total pressure of an isobaric system is denoted as P and the boiling point of an isothermal system as T. The pressures are expressed in mmHg and temperatures in °C.

In the adjusted series expansion of the relative volatility, the constants are printed as capitals in the tables but as small letters in the text.

Constants of the type -0.123×10^5 are printed as $-0.123/05$.

Deviations are defined as $\triangle Y = y_{\text{calc}} - y_{\text{exp}}$. "MEAN" is defined as $\Sigma \mid \triangle y \mid /\text{number}$ of points.

For binary systems we write $X = x_1$ and $Y = y_1$.

Binary systems of electrolytes are correlated according to equations (19) and (20). The constants a_{12}, b_{12} and c_{12} are printed in the tables as A12, B12 and C12. Only two kinds of electrolytic systems are tabulated:

(a) 1-1 Electrolytes with water. For these systems we find from equation (18) that the separation function is given by the expression

$$\beta = \frac{y_1}{y_2} \cdot \frac{x_2 (2x_1 + x_2)}{x_1^2}$$

and modified ionic strength $I = x_1$.

(b) 2-1 Electrolytes with water, for which

$$\beta = \frac{y_1}{y_2} \frac{x_2 (3x_1 + x_2)^2}{4x_1^3}$$

and $I = 3x_1$.

Ternary systems of electrolytes are systems of the type: volatile electrolyte (1)–nonvolatile electrolyte (2)–water (3) which are correlated according to equations (38) and (39). Since constituent (2) is non-volatile, $y_2 = 0$ for the whole range of concentrations and thus $\beta_{23} = 0$ also.

TWO-COMPONENT SYSTEMS

2.1. Nonelectrolytes

BROMINE(1)-CARBON TETRACHLORIDE(2)

KORZHEV P.P.,ROSSINSKAYA I.M.:ZH.KHIM.PROM.12,610(1935).

ANTOINE VAPOR PRESSURE CONSTANTS

	A	B	C
(1)	6.95897	1169.300	228.000
(2)	6.93390	1242.430	230.000

P =   737.00

EQUATION	ORDER	A12	A21	D	A122	A211
VAN LAAR	3	0.2900	0.3424	-	-	-
MARGULES	3	0.2906	0.3376	-	-	-
MARGULES	4	0.3529	0.4146	0.2578	-	-
ALPHA	2	2.3453	0.3171	-	-	-
ALPHA	3	6.2267	2.7575	-	-3.1053	-2.2893

			DEVIATION IN VAPOR PHASE COMPOSITION				
X	Y	TEMP.	LAAR 3	MARG 3	MARG 4	ALPHA2	ALPHA3
0.0700	0.1990	71.70	-0.0079	-0.0077	0.0020	-0.0092	0.0083
0.1510	0.3530	68.60	-0.0157	-0.0153	-0.0149	-0.0163	-0.0075
0.3420	0.5360	63.10	0.0047	0.0045	-0.0024	0.0079	0.0019
0.4410	0.6060	61.20	0.0043	0.0037	0.0020	0.0093	0.0040
0.5580	0.6900	59.90	-0.0103	-0.0110	-0.0059	-0.0040	-0.0050
0.7710	0.8100	58.10	-0.0072	-0.0075	-0.0031	-0.0039	0.0019
0.8650	0.8740	57.90	-0.0054	-0.0052	-0.0076	-0.0052	-0.0016
0.9400	0.9270	57.70	0.0070	0.0074	0.0015	0.0053	0.0045
0.9690	0.9660	58.00	-0.0020	-0.0017	-0.0065	-0.0035	-0.0053
		MEAN	0.0072	0.0071	0.0051	0.0072	0.0045

BROMINE(1)-TRIFLUOROTOLUENE(2)

SPICER W.M.,MEYER L.N.:J.AM.CHEM.SOC. 73,934(1951)

ANTOINE VAPOR PRESSURE CONSTANTS
```
          A          B          C
(1) 6.95897    1169.300    228.000
(2) 7.00708    1331.300    220.580
```

P =   760.00

EQUATION	ORDER	A12	A21	D	A122	A211
VAN LAAR	3	0.3130	0.7267	-	-	-
MARGULES	3	0.2706	0.6067	-	-	-
MARGULES	4	0.3470	0.7140	0.3486	-	-
ALPHA	2	7.3852	0.0527	-	-	-
ALPHA	3	9.0295	-0.6425	-	-3.1454	1.1054

			DEVIATION IN VAPOR PHASE COMPOSITION				
X	Y	TEMP.	LAAR 3	MARG 3	MARG 4	ALPHA2	ALPHA3
0.0400	0.2490	96.50	-0.0217	-0.0326	-0.0133	0.0027	-0.0248
0.0800	0.3860	90.20	-0.0046	-0.0128	0.0010	0.0170	-0.0092
0.1750	0.5780	80.10	0.0178	0.0206	0.0163	0.0205	0.0142
0.2980	0.7360	71.10	-0.0019	0.0042	-0.0038	-0.0150	-0.0055
0.6300	0.8650	61.00	0.0085	0.0030	0.0093	-0.0048	0.0006
0.9340	0.9420	58.40	0.0061	0.0108	0.0051	0.0105	0.0006
0.9670	0.9670	58.10	0.0014	0.0064	0.0014	0.0050	-0.0029
0.9880	0.9780	58.50	0.0087	0.0115	0.0089	0.0104	0.0064
		MEAN	0.0088	0.0127	0.0074	0.0107	0.0080

CARBON TETRACHLORIDE(1)-ACETONITRILE(2)

BROWN I.,SMITH F.:AUSTR.J.CHEM.7,264(1954).

ANTOINE VAPOR PRESSURE CONSTANTS

	A	B	C
(1)	6.93390	1242.430	230.000
(2)	7.24299	1397.929	238.894

T =   45.00

EQUATION	ORDER	A12	A21	D	A122	A211
VAN LAAR	3	0.6900	0.8258	-	-	-
MARGULES	3	0.6872	0.8151	-	-	-
MARGULES	4	0.7935	0.9302	0.3754	-	-
ALPHA	2	8.5309	6.0039	-	-	-
ALPHA	3	11.3819	3.9589	-	-5.2107	4.1321

			DEVIATION IN VAPOR PHASE COMPOSITION				
X	Y	PRESS.	LAAR 3	MARG 3	MARG 4	ALPHA2	ALPHA3
0.0347	0.1801	247.96	-0.0130	-0.0135	0.0122	0.0354	0.0032
0.0892	0.3309	291.58	-0.0108	-0.0110	0.0086	0.0278	0.0014
0.1914	0.4603	335.99	-0.0007	-0.0003	-0.0024	0.0049	-0.0030
0.2887	0.5129	355.85	0.0086	0.0088	-0.0012	-0.0008	0.0003
0.3752	0.5429	364.57	0.0084	0.0082	-0.0001	-0.0040	0.0005
0.4567	0.5633	368.82	0.0049	0.0042	0.0017	-0.0046	0.0006
0.4790	0.5684	369.62	0.0033	0.0025	0.0019	-0.0048	0.0004
0.5060	0.5740	370.23	0.0015	0.0005	0.0023	-0.0045	0.0004
0.6049	0.5936	371.09	-0.0054	-0.0069	0.0022	-0.0027	0.0001
0.7164	0.6181	368.95	-0.0101	-0.0113	-0.0009	0.0016	-0.0005
0.8069	0.6470	362.78	-0.0052	-0.0054	-0.0035	0.0073	-0.0012
0.8959	0.7004	346.97	0.0175	0.0191	0.0029	0.0177	-0.0008
0.9609	0.8001	314.43	0.0454	0.0476	0.0225	0.0287	0.0037
		MEAN	0.0104	0.0107	0.0048	0.0111	0.0012

CARBON TETRACHLORIDE(1)-BENZENE(2)

FOWLER R.T.,LIM S.C.:J.APLL.CHEM.,6,74(1956)

ANTOINE VAPOR PRESSURE CONSTANTS
	A	B	C
(1)	6.93390	1242.430	230.000
(2)	6.90565	1211.033	220.790

T =   40.00

EQUATION	ORDER	A12	A21	D	A122	A211
VAN LAAR	3	0.0621	0.0430	-	-	-
MARGULES	3	0.0595	0.0412	-	-	-
MARGULES	4	0.0598	0.0415	0.0009	-	-
ALPHA	2	0.2985	-0.0294	-	-	-
ALPHA	3	0.3227	0.0417	-	0.0003	-0.0890

DEVIATION IN VAPOR PHASE COMPOSITION

X	Y	PRESS.	LAAR 3	MARG 3	MARG 4	ALPHA2	ALPHA3
0.0994	0.1244	185.00	0.0015	0.0013	0.0014	-0.0012	-0.0001
0.1858	0.2233	189.00	0.0010	0.0010	0.0010	-0.0014	-0.0004
0.3401	0.3824	195.00	0.0018	0.0020	0.0019	0.0015	0.0014
0.4393	0.4815	198.00	-0.0011	-0.0010	-0.0010	-0.0006	-0.0012
0.5324	0.5681	201.00	-0.0000	-0.0001	-0.0001	0.0005	-0.0002
0.6063	0.6363	203.00	0.0003	0.0002	0.0002	0.0004	-0.0001
0.6815	0.7043	205.00	0.0014	0.0013	0.0013	0.0008	0.0008
0.7812	0.7964	207.00	0.0008	0.0008	0.0008	-0.0008	-0.0002
0.8926	0.8990	209.00	0.0009	0.0010	0.0010	-0.0010	-0.0000
0.9529	0.9555	210.00	0.0004	0.0005	0.0005	-0.0007	-0.0001
0.9632	0.9653	210.00	0.0002	0.0003	0.0003	-0.0007	-0.0002
		MEAN	0.0009	0.0009	0.0009	0.0009	0.0004

24

CARBON TETRACHLORIDE(1)-BENZENE(2)

FOWLER R.T.,LIM S.C.:J.APPL.CHEM.,6,74(1956)

ANTOINE VAPOR PRESSURE CONSTANTS
```
           A         B          C
(1) 6.93390   1242.430   230.000
(2) 6.90565   1211.033   220.790
```

P =   760.00

EQUATION	ORDER	A12	A21	D	A122	A211
VAN LAAR	3	0.0692	0.0322	-	-	-
MARGULES	3	0.0581	0.0263	-	-	-
MARGULES	4	0.0660	0.0341	0.0245	-	-
ALPHA	2	0.2028	0.0004	-	-	-
ALPHA	3	1.5399	1.3167	-	-1.2713	-1.3324

| | | | DEVIATION IN VAPOR PHASE COMPOSITION | | | | |
X	Y	TEMP.	LAAR 3	MARG 3	MARG 4	ALPHA2	ALPHA3
0.0883	0.1056	79.50	0.0016	0.0008	0.0015	-0.0027	-0.0001
0.2043	0.2308	78.80	0.0017	0.0020	0.0020	-0.0011	0.0003
0.3288	0.3577	78.20	-0.0002	0.0005	-0.0000	-0.0002	-0.0007
0.4183	0.4439	77.80	0.0006	0.0010	0.0007	0.0017	0.0005
0.5207	0.5425	77.45	0.0001	0.0000	0.0001	0.0012	0.0001
0.6231	0.6397	77.15	0.0004	0.0000	0.0005	0.0005	0.0001
0.7284	0.7399	76.90	0.0003	0.0001	0.0004	-0.0011	-0.0005
0.8196	0.8254	76.70	0.0016	0.0019	0.0017	-0.0006	0.0005
0.9092	0.9119	76.65	0.0008	0.0013	0.0008	-0.0012	-0.0002
0.9579	0.9587	76.65	0.0008	0.0012	0.0007	-0.0005	0.0001
		MEAN	0.0008	0.0009	0.0008	0.0011	0.0003

c

CARBON TETRACHLORIDE(1)-BENZENE(2)

SCATCHARD G.,WOOD S.E.,MOCHEL J.M.:J.AM.CHEM.SOC.62,712(1940)

ANTOINE VAPOR PRESSURE CONSTANTS
           A   B   C
(1) 6.93390   1242.430   230.000
(2) 6.90565   1211.033   220.790

T =   40.00

EQUATION	ORDER	A12	A21	D	A122	A211
VAN LAAR	3	0.0673	0.0413	-	-	-
MARGULES	3	0.0623	0.0381	-	-	-
MARGULES	4	0.0619	0.0377	-0.0013	-	-
ALPHA	2	0.2957	-0.0214	-	-	-
ALPHA	3	0.2332	-0.0633	-	0.0669	0.0361

			DEVIATION IN VAPOR PHASE COMPOSITION				
X	Y	PRESS.	LAAR 3	MARG 3	MARG 4	ALPHA2	ALPHA3
0.1398	0.1703	190.18	0.0030	0.0028	0.0028	-0.0005	-0.0003
0.2378	0.2774	194.70	0.0024	0.0026	0.0026	0.0002	0.0004
0.3735	0.4159	200.07	0.0003	0.0006	0.0006	0.0001	0.0000
0.4919	0.5295	204.02	-0.0002	-0.0001	-0.0001	0.0000	-0.0001
0.4986	0.5359	204.20	-0.0003	-0.0003	-0.0003	-0.0001	-0.0003
0.6201	0.6475	207.44	0.0012	0.0011	0.0010	0.0004	0.0003
0.7585	0.7739	210.37	0.0024	0.0024	0.0023	-0.0001	0.0000
0.8718	0.8783	211.97	0.0025	0.0027	0.0028	-0.0004	-0.0001
		MEAN	0.0015	0.0016	0.0016	0.0002	0.0002

CARBON TETRACHLORIDE(1)-BENZENE(2)

SCATCHARD G.,WOOD S.E.,MOCHEL J.M.:J.AM.CHEM.SOC.62,712(1940)

ANTOINE VAPOR PRESSURE CONSTANTS
```
           A           B          C
(1) 6.93390   1242.430   230.000
(2) 6.90565   1211.033   220.790
```

T = 70.00

EQUATION	ORDER	A12	A21	D	A122	A211
VAN LAAR	3	0.0575	0.0378	-	-	-
MARGULES	3	0.0543	0.0356	-	-	-
MARGULES	4	0.0556	0.0370	0.0039	-	-
ALPHA	2	0.2312	0.0025	-	-	-
ALPHA	3	0.2878	0.0631	-	-0.0510	-0.0636

DEVIATION IN VAPOR PHASE COMPOSITION

X	Y	PRESS.	LAAR 3	MARG 3	MARG 4	ALPHA2	ALPHA3
0.1428	0.1666	568.89	0.0024	0.0022	0.0023	-0.0003	-0.0000
0.2394	0.2702	579.13	0.0015	0.0016	0.0016	-0.0002	-0.0001
0.3791	0.4105	591.62	0.0006	0.0008	0.0007	0.0004	0.0003
0.4930	0.5204	600.77	-0.0002	-0.0001	-0.0001	-0.0000	-0.0001
0.4939	0.5215	599.67	-0.0004	-0.0004	-0.0004	-0.0003	-0.0004
0.6224	0.6411	607.22	0.0010	0.0009	0.0010	0.0004	0.0004
0.7624	0.7719	613.08	0.0016	0.0016	0.0016	-0.0003	-0.0002
0.8750	0.8780	616.02	0.0020	0.0022	0.0021	-0.0001	0.0000
		MEAN	0.0012	0.0012	0.0012	0.0002	0.0002

CARBON TETRACHLORIDE(1)-BENZENE(2)

ZAWIDZKI J.:Z.PHYS.CHEM.35,129(1900)

ANTOINE VAPOR PRESSURE CONSTANTS
```
            A          B          C
(1) 6.93390    1242.430   230.000
(2) 6.90565    1211.033   220.790
```

T =   49.99

EQUATION	ORDER	A12	A21	D	A122	A211
VAN LAAR	3	0.0726	0.0496	-	-	-
MARGULES	3	0.0694	0.0467	-	-	-
MARGULES	4	0.0674	0.0444	-0.0061	-	-
ALPHA	2	0.3126	0.0139	-	-	-
ALPHA	3	1.6686	1.2696	-	-1.2957	-1.2558

			DEVIATION IN VAPOR PHASE COMPOSITION				
X	Y	PRESS.	LAAR 3	MARG 3	MARG 4	ALPHA2	ALPHA3
0.0507	0.0693	272.20	-0.0028	-0.0030	-0.0032	-0.0046	-0.0028
0.0507	0.0670	271.60	-0.0005	-0.0007	-0.0009	-0.0023	-0.0005
0.1170	0.1457	277.60	0.0009	0.0007	0.0005	-0.0013	0.0006
0.1745	0.2114	281.30	-0.0001	-0.0002	-0.0003	-0.0017	-0.0007
0.1772	0.2126	281.70	0.0016	0.0015	0.0014	0.0001	0.0011
0.2506	0.2883	285.60	0.0035	0.0035	0.0036	0.0031	0.0029
0.2525	0.2922	285.20	0.0015	0.0016	0.0016	0.0012	0.0010
0.2947	0.3365	288.30	-0.0001	0.0001	0.0001	0.0003	-0.0005
0.3947	0.4357	295.20	-0.0017	-0.0016	-0.0016	-0.0003	-0.0017
0.3959	0.4385	293.90	-0.0033	-0.0033	-0.0032	-0.0019	-0.0033
0.5561	0.5860	301.30	-0.0009	-0.0010	-0.0011	0.0000	-0.0006
0.5600	0.5860	301.00	0.0027	0.0026	0.0025	0.0036	0.0030
0.6737	0.6926	305.50	0.0009	0.0007	0.0006	0.0001	0.0006
0.6774	0.6960	305.00	0.0009	0.0008	0.0006	0.0000	0.0006
0.7646	0.7768	306.70	0.0006	0.0006	0.0006	-0.0015	-0.0005
0.7658	0.7783	306.90	0.0002	0.0003	0.0002	-0.0019	-0.0009
		MEAN	0.0014	0.0014	0.0014	0.0015	0.0013

CARBON TETRACHLORIDE(1)-BUTYL ALCOHOL(2)

SMITH C.P.,ENGEL E.W.:J.AM.CHEM.SOC.51,2660(1929).

ANTOINE VAPOR PRESSURE CONSTANTS
	A	B	C
(1)	6.93390	1242.430	230.000
(2)	7.54472	1405.873	183.908

T = 50.00

EQUATION	ORDER	A12	A21	D	A122	A211
VAN LAAR	3	0.5481	1.1653	-	-	-
MARGULES	3	0.4988	0.9752	-	-	-
MARGULES	4	0.5738	1.0987	0.3571	-	-
ALPHA	2	36.0941	0.9061	-	-	-
ALPHA	3	12.6301	0.0640	-	22.6609	-0.1491

			DEVIATION IN VAPOR PHASE COMPOSITION				
X	Y	PRESS.	LAAR 3	MARG 3	MARG 4	ALPHA2	ALPHA3
0.0782	0.7231	86.30	-0.0097	-0.0186	-0.0060	0.0077	0.0063
0.1352	0.8157	121.60	-0.0068	-0.0081	-0.0066	0.0020	0.0020
0.1766	0.8559	146.40	-0.0110	-0.0099	-0.0117	-0.0056	-0.0051
0.2592	0.8887	181.50	-0.0035	-0.0011	-0.0045	-0.0016	-0.0008
0.3161	0.9024	210.00	-0.0013	0.0006	-0.0021	-0.0002	0.0005
0.3847	0.9156	233.30	-0.0018	-0.0012	-0.0023	-0.0006	-0.0002
0.4759	0.9269	254.60	-0.0029	-0.0046	-0.0033	-0.0002	-0.0004
0.4949	0.9247	260.30	0.0008	-0.0015	0.0003	0.0039	0.0037
0.5836	0.9338	274.90	-0.0037	-0.0082	-0.0046	0.0025	0.0016
0.6714	0.9454	285.60	-0.0136	-0.0194	-0.0152	-0.0031	-0.0045
0.7764	0.9470	296.10	-0.0155	-0.0197	-0.0175	0.0017	0.0002
0.8872	0.9570	304.60	-0.0216	-0.0174	-0.0216	-0.0003	0.0008
0.9755	0.9834	309.00	-0.0131	-0.0049	-0.0103	-0.0079	-0.0006
		MEAN	0.0081	0.0089	0.0081	0.0029	0.0021

CARBON TETRACHLORIDE(1)-1,2-DICHLOROETHANE(2)

KIREEV V.A.,MONAKHOVA Z.D.:ZH.FIZ.KHIM.7,71(1936)

ANTOINE VAPOR PRESSURE CONSTANTS
```
          A          B          C
(1) 6.93390    1242.430    230.000
(2) 7.18431    1358.500    232.000
```

P =   760.00

EQUATION	ORDER	A12	A21	D	A122	A211
VAN LAAR	3	0.2320	0.2182	-	-	-
MARGULES	3	0.2317	0.2179	-	-	-
MARGULES	4	0.2395	0.2260	0.0244	-	-
ALPHA	2	1.0441	0.3786	-	-	-
ALPHA	3	0.9468	0.8166	-	0.3791	-0.6219

DEVIATION IN VAPOR PHASE COMPOSITION

X	Y	TEMP.	LAAR 3	MARG 3	MARG 4	ALPHA2	ALPHA3
0.0500	0.0900	82.17	0.0053	0.0052	0.0062	0.0033	0.0119
0.1000	0.1800	81.08	-0.0060	-0.0060	-0.0051	-0.0080	0.0024
0.1500	0.2500	80.10	-0.0091	-0.0091	-0.0086	-0.0103	-0.0014
0.2000	0.3100	79.27	-0.0111	-0.0111	-0.0111	-0.0110	-0.0048
0.2500	0.3600	78.55	-0.0097	-0.0096	-0.0100	-0.0081	-0.0051
0.3000	0.4000	77.91	-0.0032	-0.0031	-0.0036	-0.0002	-0.0001
0.3500	0.4400	77.37	-0.0003	-0.0003	-0.0008	0.0039	0.0015
0.4000	0.4800	76.88	-0.0002	-0.0001	-0.0005	0.0050	0.0008
0.4500	0.5200	76.44	-0.0018	-0.0018	-0.0020	0.0039	-0.0013
0.5000	0.5500	76.18	0.0053	0.0053	0.0053	0.0114	0.0059
0.5500	0.5900	75.93	0.0019	0.0019	0.0022	0.0079	0.0029
0.6000	0.6300	75.68	-0.0014	-0.0014	-0.0010	0.0041	0.0004
0.6500	0.6700	75.50	-0.0041	-0.0041	-0.0036	0.0005	-0.0014
0.7000	0.7100	75.37	-0.0056	-0.0057	-0.0052	-0.0022	-0.0017
0.7500	0.7500	75.25	-0.0054	-0.0054	-0.0051	-0.0034	-0.0003
0.8000	0.8000	75.35	-0.0128	-0.0128	-0.0128	-0.0123	-0.0067
0.8500	0.8400	75.40	-0.0069	-0.0069	-0.0073	-0.0078	-0.0003
0.9000	0.8900	75.54	-0.0069	-0.0069	-0.0076	-0.0089	-0.0007
0.9500	0.9300	75.76	0.0082	0.0082	0.0076	0.0063	0.0125
		MEAN	0.0055	0.0055	0.0056	0.0062	0.0033

CARBON TETRACHLORIDE(1)-ETHYL ALCOHOL(2)

BARKER J.A.,BROWN J.,SMITH F.:DISC.FARAD.SOC.15,142(1953).

ANTOINE VAPOR PRESSURE CONSTANTS

	A	B	C
(1)	6.93390	1242.430	230.000
(2)	8.16290	1623.220	228.980

T = 45.00

EQUATION	ORDER	A12	A21	D	A122	A211
VAN LAAR	3	0.5993	0.9913	-	-	-
MARGULES	3	0.5638	0.9180	-	-	-
MARGULES	4	0.6813	1.0342	0.4100	-	-
ALPHA	2	9.0673	4.4251	-	-	-
ALPHA	3	28.0003	3.8269	-	-22.6127	8.9023

			DEVIATION IN VAPOR PHASE COMPOSITION				
X	Y	PRESS.	LAAR 3	MARG 3	MARG 4	ALPHA2	ALPHA3
0.0212	0.1210	192.73	-0.0099	-0.0164	0.0063	0.0426	-0.0024
0.0356	0.1863	205.65	-0.0142	-0.0224	0.0055	0.0508	-0.0031
0.1016	0.3756	253.77	-0.0146	-0.0202	-0.0023	0.0408	0.0008
0.1638	0.4731	286.53	-0.0089	-0.0097	-0.0079	0.0205	0.0015
0.2736	0.5665	321.69	0.0005	0.0039	-0.0080	-0.0027	-0.0002
0.3905	0.6177	339.65	0.0054	0.0076	-0.0019	-0.0125	-0.0012
0.4845	0.6423	346.88	0.0053	0.0041	0.0025	-0.0132	-0.0005
0.5931	0.6631	350.51	-0.0005	-0.0059	0.0018	-0.0096	0.0001
0.6050	0.6644	350.47	-0.0007	-0.0065	0.0019	-0.0082	0.0009
0.7088	0.6822	350.97	-0.0106	-0.0177	-0.0066	-0.0005	0.0005
0.8076	0.7019	348.98	-0.0164	-0.0192	-0.0159	0.0139	-0.0001
0.8985	0.7351	342.23	-0.0033	0.0048	-0.0102	0.0385	-0.0019
0.9541	0.7822	328.31	0.0343	0.0489	0.0240	0.0671	0.0026
		MEAN	0.0096	0.0144	0.0073	0.0247	0.0012

CARBON TETRACHLORIDE(1)-ETHYL ALCOHOL(2)

BARKER J.A.,BROWN J.,SMITH F.:DISC.FARAD.SOC.15,142(1953).

ANTOINE VAPOR PRESSURE CONSTANTS
	A	B	C
(1)	6.93390	1242.430	230.000
(2)	8.16290	1623.220	228.890

T =   65.00

EQUATION	ORDER	A12	A21	D	A122	A211
VAN LAAR	3	0.5863	0.8968	-	-	-
MARGULES	3	0.5620	0.8436	-	-	-
MARGULES	4	0.6650	0.9558	0.3454	-	-
ALPHA	2	6.8600	4.6027	-	-	-
ALPHA	3	17.3309	3.7697	-	-13.1158	6.5900

			DEVIATION IN VAPOR PHASE COMPOSITION				
X	Y	PRESS.	LAAR 3	MARG 3	MARG 4	ALPHA2	ALPHA3
0.0237	0.1075	481.24	-0.0096	-0.0135	0.0041	0.0367	-0.0001
0.0404	0.1673	508.51	-0.0129	-0.0178	0.0041	0.0448	0.0001
0.1061	0.3270	595.71	-0.0146	-0.0182	-0.0027	0.0356	-0.0008
0.1745	0.4206	657.86	-0.0078	-0.0083	-0.0057	0.0179	-0.0005
0.2877	0.5058	716.91	0.0038	0.0055	-0.0022	-0.0001	0.0010
0.2925	0.5083	716.86	0.0042	0.0059	-0.0019	-0.0006	0.0011
0.3001	0.5122	720.63	0.0048	0.0065	-0.0015	-0.0013	0.0012
0.4005	0.5576	745.46	0.0046	0.0050	-0.0004	-0.0119	-0.0020
0.4924	0.5841	756.18	0.0036	0.0016	0.0024	-0.0131	-0.0012
0.5993	0.6092	760.82	-0.0026	-0.0071	0.0004	-0.0105	-0.0010
0.6087	0.6097	759.79	-0.0017	-0.0064	0.0015	-0.0085	0.0005
0.7040	0.6300	757.36	-0.0079	-0.0128	-0.0042	-0.0004	0.0013
0.8023	0.6573	748.32	-0.0110	-0.0124	-0.0118	0.0135	0.0004
0.8961	0.7013	721.28	0.0067	0.0137	-0.0036	0.0411	0.0021
0.9483	0.7688	678.54	0.0250	0.0365	0.0102	0.0536	-0.0041
		MEAN	0.0080	0.0114	0.0038	0.0193	0.0012

CARBON TETRACHLORIDE(1)-ETHYL ALCOHOL(2)

TYRER D.:J.CHEM.SOC.101,1104(1912).

ANTOINE VAPOR PRESSURE CONSTANTS
```
        A          B          C
(1) 6.93390   1242.430   230.000
(2) 8.16290   1623.220   228.980
```

P = 745.00

EQUATION	ORDER	A12	A21	D	A122	A211
VAN LAAR	3	0.6484	0.9896	-	-	-
MARGULES	3	0.6376	0.9097	-	-	-
MARGULES	4	0.7975	1.1624	0.6664	-	-
ALPHA	2	6.0778	3.5169	-	-	-
ALPHA	3	-2.9972	0.0557	-	7.7399	-0.0625

			DEVIATION IN VAPOR PHASE COMPOSITION				
X	Y	TEMP.	LAAR 3	MARG 3	MARG 4	ALPHA2	ALPHA3
0.0322	0.1664	74.82	-0.0347	-0.0367	-0.0033	0.0042	-0.0155
0.0696	0.2647	72.44	-0.0205	-0.0224	0.0107	0.0210	0.0033
0.1137	0.3544	70.25	-0.0145	-0.0155	0.0040	0.0148	0.0064
0.1664	0.4350	68.35	-0.0149	-0.0152	-0.0115	-0.0019	-0.0004
0.2304	0.4978	66.64	-0.0114	-0.0118	-0.0194	-0.0136	-0.0049
0.3100	0.5358	65.32	0.0032	0.0013	-0.0083	-0.0083	0.0018
0.4113	0.5688	64.42	0.0085	0.0036	0.0038	-0.0022	0.0008
0.5567	0.5972	63.88	0.0041	-0.0047	0.0153	0.0135	-0.0004
0.7294	0.6691	64.30	-0.0581	-0.0643	-0.0454	-0.0025	0.0000
		MEAN	0.0189	0.0195	0.0135	0.0091	0.0037

CARBON TETRACHLORIDE(1)-HEPTANE(2)

SMITH C.P.,ENGEL E.W.:J.AM.CHEM.SOC.,51,2646(1929)

ANTOINE VAPOR PRESSURE CONSTANTS

	A	B	C
(1)	6.93390	1242.430	230.000
(2)	6.90240	1268.115	216.900

T =   50.00

EQUATION	ORDER	A12	A21	D	A122	A211
VAN LAAR	3	0.0618	0.2164	-	-	-
MARGULES	3	0.0340	0.1403	-	-	-
MARGULES	4	0.0220	0.1258	-0.0410	-	-
ALPHA	2	1.6543	-0.4515	-	-	-
ALPHA	3	-2.0885	-1.9870	-	3.6882	1.4641

			DEVIATION IN VAPOR PHASE COMPOSITION				
X	Y	PRESS.	LAAR 3	MARG 3	MARG 4	ALPHA2	ALPHA3
0.0332	0.0798	146.70	0.0006	-0.0031	-0.0045	0.0033	0.0020
0.0983	0.2159	159.80	0.0004	-0.0045	-0.0062	0.0054	0.0039
0.1714	0.3437	174.00	-0.0018	-0.0044	-0.0048	0.0034	0.0029
0.3024	0.5279	198.70	-0.0113	-0.0100	-0.0093	-0.0083	-0.0068
0.3569	0.5805	207.40	-0.0052	-0.0035	-0.0029	-0.0033	-0.0016
0.4324	0.6452	220.70	0.0013	0.0024	0.0026	0.0021	0.0030
0.5012	0.7000	232.30	0.0030	0.0029	0.0026	0.0033	0.0030
0.5700	0.7515	244.70	0.0015	0.0001	-0.0006	0.0019	0.0003
0.6496	0.8053	256.90	-0.0009	-0.0035	-0.0043	0.0003	-0.0019
0.7323	0.8554	271.10	-0.0033	-0.0060	-0.0064	-0.0004	-0.0015
0.8126	0.9000	282.40	-0.0054	-0.0065	-0.0064	-0.0003	0.0009
0.8992	0.9468	295.40	-0.0076	-0.0059	-0.0052	-0.0008	0.0021
0.9649	0.9914	304.10	-0.0151	-0.0127	-0.0122	-0.0104	-0.0085
		MEAN	0.0044	0.0050	0.0052	0.0033	0.0030

CARBON TETRACHLORIDE(1)-CYCLOHEXANE(2)

BROWN I.,EWALD A.H.:AUSTR.J.SCI.RES.A3,306(1950).

ANTOINE VAPOR PRESSURE CONSTANTS

	A	B	C
(1)	6.93390	1242.430	230.000
(2)	6.84498	1203.526	222.863

T = 70.00

EQUATION	ORDER	A12	A21	D	A122	A211
VAN LAAR	3	0.0357	0.0297	-	-	-
MARGULES	3	0.0352	0.0295	-	-	-
MARGULES	4	0.0385	0.0329	0.0101	-	-
ALPHA	2	0.2197	-0.0433	-	-	-
ALPHA	3	0.4231	0.1099	-	-0.2117	-0.1414

			DEVIATION IN VAPOR PHASE COMPOSITION				
X	Y	PRESS.	LAAR 3	MARG 3	MARG 4	ALPHA2	ALPHA3
0.0538	0.0641	551.10	0.0010	0.0010	0.0013	0.0003	-0.0000
0.1396	0.1622	560.58	0.0014	0.0014	0.0016	0.0003	-0.0001
0.2327	0.2629	570.28	0.0016	0.0016	0.0016	0.0007	0.0005
0.2641	0.2968	573.20	0.0006	0.0007	0.0005	-0.0002	-0.0003
0.3135	0.3476	577.94	0.0006	0.0007	0.0005	-0.0000	-0.0000
0.4041	0.4388	585.90	0.0000	0.0001	-0.0001	-0.0004	-0.0001
0.5153	0.5464	594.31	0.0002	0.0002	0.0002	-0.0003	0.0001
0.5994	0.6259	600.32	0.0004	0.0004	0.0006	-0.0002	0.0001
0.7012	0.7209	606.27	0.0007	0.0007	0.0009	-0.0003	-0.0003
0.8009	0.8125	611.19	0.0019	0.0019	0.0018	0.0006	0.0002
0.8995	0.9045	614.95	0.0015	0.0016	0.0013	0.0004	-0.0001
0.9479	0.9500	616.17	0.0012	0.0012	0.0010	0.0005	0.0000
		MEAN	0.0009	0.0009	0.0010	0.0004	0.0002

CARBON TETRACHLORIDE(1)-CYCLOHEXANE(2)

DVORAK K.,BOUBLIK T.:COLLECTION CZECH.CHEM.COMMUN.28,1249(1963).

ANTOINE VAPOR PRESSURE CONSTANTS
	A	B	C
(1)	6.93390	1242.430	230.000
(2)	6.84498	1203.526	222.863

T =   10.00

EQUATION	ORDER	A12	A21	D	A122	A211
VAN LAAR	3	0.0376	0.1184	-	-	-
MARGULES	3	0.0219	0.0842	-	-	-
MARGULES	4	0.0306	0.0920	0.0268	-	-
ALPHA	2	0.3967	-0.0790	-	-	-
ALPHA	3	1.5497	0.7930	-	-1.1577	-0.8346

			DEVIATION IN VAPOR PHASE COMPOSITION				
X	Y	PRESS.	LAAR 3	MARG 3	MARG 4	ALPHA2	ALPHA3
0.0706	0.0937	46.66	-0.0034	-0.0051	-0.0043	0.0010	0.0005
0.1668	0.2129	48.17	-0.0066	-0.0076	-0.0074	-0.0004	-0.0012
0.2450	0.2990	49.10	-0.0040	-0.0037	-0.0042	0.0017	0.0009
0.4251	0.4858	51.59	-0.0012	0.0000	-0.0005	0.0006	0.0008
0.5043	0.5610	52.39	-0.0023	-0.0019	-0.0020	-0.0019	-0.0013
0.5577	0.6083	53.09	-0.0003	-0.0006	-0.0004	-0.0004	0.0004
0.6196	0.6644	53.68	-0.0009	-0.0021	-0.0016	-0.0010	-0.0002
0.7393	0.7683	54.48	-0.0015	-0.0033	-0.0029	0.0004	0.0005
0.7957	0.8176	55.25	-0.0032	-0.0044	-0.0043	0.0003	-0.0001
0.8840	0.8944	55.42	-0.0048	-0.0040	-0.0044	0.0011	-0.0001
0.9434	0.9472	55.91	-0.0041	-0.0023	-0.0028	0.0013	0.0001
0.9614	0.9640	56.01	-0.0037	-0.0020	-0.0024	0.0007	-0.0002
		MEAN	0.0030	0.0031	0.0031	0.0009	0.0005

CARBON TETRACHLORIDE(1)-CYCLOHEXANE(2)

DVORAK.K.,BOUBLIK T.:COLLECTION CZECH.CHEM.COMMUN.28,1249(1963).

ANTOINE VAPOR PRESSURE CONSTANTS

	A	B	C
(1)	6.93390	1242.430	230.000
(2)	6.84498	1203.526	222.863

T =   60.00

EQUATION	ORDER	A12	A21	D	A122	A211
VAN LAAR	3	0.0473	0.0272	-	-	-
MARGULES	3	0.0437	0.0237	-	-	-
MARGULES	4	0.0272	0.0080	-0.0501	-	-
ALPHA	2	0.2268	-0.0436	-	-	-
ALPHA	3	-.3983/ 05	-.3551/ 05	-	.3925/ 05	.3551/ 05

			DEVIATION IN VAPOR PHASE COMPOSITION				
X	Y	PRESS.	LAAR 3	MARG 3	MARG 4	ALPHA2	ALPHA3
0.1035	0.1236	399.00	0.0012	0.0009	-0.0005	-0.0011	0.0037
0.1718	0.2003	404.07	0.0005	0.0004	-0.0000	-0.0014	-0.0011
0.2327	0.2640	408.95	0.0016	0.0004	0.0017	0.0005	-0.0015
0.3732	0.4084	419.18	-0.0002	-0.0001	0.0010	0.0003	-0.0029
0.4817	0.5150	425.49	-0.0011	-0.0012	-0.0009	-0.0002	-0.0020
0.5617	0.5904	429.81	0.0000	-0.0003	-0.0008	0.0005	0.0003
0.6402	0.6637	432.49	0.0009	0.0006	-0.0003	0.0007	0.0021
0.7283	0.7451	436.98	0.0021	0.0020	0.0013	0.0010	0.0035
0.8278	0.8368	440.78	0.0032	0.0034	0.0037	0.0014	0.0036
0.9010	0.9124	443.14	-0.0043	-0.0040	-0.0030	-0.0060	-0.0057
		MEAN	0.0015	0.0015	0.0014	0.0013	0.0026

CARBON TETRACHLORIDE(1)-CYCLOHEXANE(2)

SCATCHARD G.,WOOD S.E.,MOCHEL J.M.:J.AM.CHEM.SOC.61,3206(1939)

ANTOINE VAPOR PRESSURE CONSTANTS

	A	B	C
(1)	6.93390	1242.430	230.000
(2)	6.84498	1203.526	222.863

T = 40.00

EQUATION	ORDER	A12	A21	D	A122	A211
VAN LAAR	3	0.0585	0.0351	-	-	-
MARGULES	3	0.0540	0.0318	-	-	-
MARGULES	4	0.0513	0.0291	-0.0082	-	-
ALPHA	2	0.2639	-0.0308	-	-	-
ALPHA	3	-0.0980	-0.3313	-	0.3626	0.2892

			DEVIATION IN VAPOR PHASE COMPOSITION				
X	Y	PRESS.	LAAR 3	MARG 3	MARG 4	ALPHA2	ALPHA3
0.1262	0.1515	190.62	0.0031	0.0029	0.0027	-0.0001	0.0001
0.2453	0.2822	195.62	0.0016	0.0018	0.0019	-0.0002	0.0000
0.3669	0.4066	200.08	-0.0001	0.0001	0.0002	-0.0004	-0.0003
0.4739	0.5103	203.40	-0.0004	-0.0004	-0.0004	-0.0002	-0.0004
0.5151	0.5474	204.59	0.0016	0.0015	0.0015	0.0016	0.0015
0.4753	0.5116	203.45	-0.0004	-0.0004	-0.0003	-0.0002	-0.0003
0.6061	0.6341	206.97	0.0003	0.0002	0.0000	-0.0003	-0.0005
0.7542	0.7702	210.15	0.0018	0.0018	0.0017	-0.0003	-0.0002
0.8756	0.8822	212.04	0.0023	0.0025	0.0027	-0.0002	0.0002
		MEAN	0.0013	0.0013	0.0013	0.0004	0.0004

CARBON TETRACHLORIDE(1)-CYCLOHEXANE(2)

SCATCHARD G.,WOOD S.E.,MOCHEL J.M.:J.AM.CHEM.SOC.61,3206(1939)

ANTOINE VAPOR PRESSURE CONSTANTS
```
          A           B          C
(1) 6.93390     1242.430    230.000
(2) 6.84498     1203.526    222.863
```

T =   70.00

EQUATION	ORDER	A12	A21	D	A122	A211
VAN LAAR	3	0.0420	0.0306	-	-	-
MARGULES	3	0.0406	0.0296	-	-	-
MARGULES	4	0.0407	0.0296	0.0001	-	-
ALPHA	2	0.2236	-0.0361	-	-	-
ALPHA	3	0.1967	-0.0751	-	0.0193	0.0449

			DEVIATION IN VAPOR PHASE COMPOSITION				
X	Y	PRESS.	LAAR 3	MARG 3	MARG 4	ALPHA2	ALPHA3
0.1248	0.1458	558.78	0.0022	0.0021	0.0021	0.0004	0.0000
0.2468	0.2787	571.59	0.0011	0.0012	0.0012	-0.0000	-0.0003
0.3640	0.3981	582.59	0.0008	0.0009	0.0009	0.0004	0.0004
0.4836	0.5150	591.97	0.0005	0.0005	0.0005	0.0003	0.0005
0.5153	0.5473	594.42	-0.0015	-0.0015	-0.0015	-0.0017	-0.0015
0.4796	0.5113	591.73	0.0003	0.0003	0.0003	0.0002	0.0003
0.6074	0.6320	600.52	0.0010	0.0009	0.0009	0.0004	0.0005
0.7535	0.7680	608.79	0.0016	0.0016	0.0016	0.0003	0.0001
0.8757	0.8818	613.97	0.0018	0.0018	0.0018	0.0002	-0.0001
		MEAN	0.0012	0.0012	0.0012	0.0004	0.0004

CARBON TETRACHLORIDE(1)-CYCLOHEXANE(2)

YUAN K.S.,LU B.C.Y.:J.CHEM.ENG.DATA 8(4),549(1963).

ANTOINE VAPOR PRESSURE CONSTANTS
```
             A          B          C
(1) 6.93390    1242.430    230.000
(2) 6.84498    1203.526    222.863
```

P = 760.00

EQUATION	ORDER	A12	A21	D	A122	A211
VAN LAAR	3	0.0183	-0.0014	-	-	-
MARGULES	3	0.0612	-0.0076	-	-	-
MARGULES	4	0.1054	0.0428	0.1434	-	-
ALPHA	2	0.2134	-0.0041	-	-	-
ALPHA	3	.1623/ 04	.1509/ 04	-	-.1588/ 04	-.1508/ 04

			DEVIATION IN VAPOR PHASE COMPOSITION				
X	Y	TEMP.	LAAR 3	MARG 3	MARG 4	ALPHA2	ALPHA3
0.0450	0.0680	80.34	-0.0023	-0.0114	-0.0077	-0.0143	0.0009
0.1130	0.1350	79.73	0.0063	0.0003	0.0039	-0.0034	0.0034
0.1240	0.1550	79.60	-0.0066	-0.0075	-0.0043	-0.0111	-0.0055
0.1340	0.1630	79.66	-0.0061	-0.0046	-0.0017	-0.0080	-0.0033
0.1460	0.1710	79.54	-0.0028	0.0004	0.0028	-0.0028	0.0008
0.1980	0.2300	79.25	-0.0083	-0.0040	-0.0036	-0.0056	-0.0057
0.2010	0.2200	79.10	0.0048	0.0091	0.0094	0.0076	0.0073
0.2010	0.2300	79.17	-0.0052	-0.0009	-0.0006	-0.0024	-0.0027
0.2560	0.2730	78.88	0.0098	0.0117	0.0103	0.0123	0.0095
0.2600	0.2850	78.80	0.0020	0.0037	0.0022	0.0044	0.0015
0.3140	0.3450	78.46	-0.0015	-0.0030	-0.0053	-0.0006	-0.0047
0.4040	0.4270	78.07	0.0092	0.0023	0.0007	0.0065	0.0024
0.4120	0.4360	78.06	0.0083	0.0010	-0.0004	0.0053	0.0013
0.5190	0.5450	77.63	0.0067	-0.0046	-0.0032	-0.0011	-0.0030
0.5400	0.5690	77.53	0.0035	-0.0083	-0.0063	-0.0053	-0.0065
0.5790	0.6000	77.40	0.0107	-0.0013	0.0014	0.0004	0.0003
0.6750	0.6930	77.06	0.0101	-0.0005	0.0027	-0.0029	-0.0002
0.6850	0.6960	77.07	0.0166	0.0063	0.0094	0.0035	0.0064
0.7910	0.8050	76.74	0.0069	0.0009	0.0009	-0.0064	-0.0016
		MEAN	0.0067	0.0043	0.0040	0.0055	0.0035

CARBON TETRACHLORIDE(1)-PROPYL ALCOHOL(2)

CARLEY J.F.,BERTELSEN L.W.:IND.ENG.CHEM.41,2806(1949).

ANTOINE VAPOR PRESSURE CONSTANTS

	A	B	C
(1)	6.93390	1242.430	230.000
(2)	7.85418	1497.910	204.112

P = 760.00

EQUATION	ORDER	A12	A21	D	A122	A211
VAN LAAR	3	0.4744	0.7590	-	-	-
MARGULES	3	0.4537	0.7076	-	-	-
MARGULES	4	0.5391	0.8073	0.3317	-	-
ALPHA	2	6.4478	1.2874	-	-	-
ALPHA	3	13.3331	1.2081	-	-8.6936	1.6929

DEVIATION IN VAPOR PHASE COMPOSITION

X	Y	TEMP.	LAAR 3	MARG 3	MARG 4	ALPHA2	ALPHA3
0.0580	0.2550	90.80	-0.0067	-0.0113	0.0081	0.0354	-0.0023
0.1410	0.4500	84.50	0.0030	0.0022	0.0061	0.0260	0.0015
0.3000	0.6330	78.30	0.0055	0.0076	-0.0014	-0.0027	-0.0009
0.4410	0.7070	75.80	0.0060	0.0057	0.0033	-0.0085	0.0015
0.5220	0.7390	74.70	0.0016	-0.0007	0.0023	-0.0118	-0.0013
0.6830	0.7820	73.90	-0.0023	-0.0066	0.0017	-0.0047	-0.0000
0.8180	0.8180	73.40	0.0008	-0.0002	0.0009	0.0083	0.0008
0.8510	0.8310	73.60	0.0018	0.0026	0.0002	0.0114	-0.0004
0.9500	0.8950	74.20	0.0166	0.0220	0.0108	0.0237	-0.0002
		MEAN	0.0049	0.0065	0.0039	0.0147	0.0010

D

41

CARBON TETRACHLORIDE(1)-PROPYL ALCOHOL(2)

PAPOUSEK D.,PAPOUSKOVA Z.,PAGO L.:Z.PHYS.CHEM.(LEIPZIG) 211,231(1959).

ANTOINE VAPOR PRESSURE CONSTANTS

	A	B	C
(1)	6.93390	1242.430	230.000
(2)	7.85418	1497.910	204.112

T =   70.00

EQUATION	ORDER	A12	A21	D	A122	A211
VAN LAAR	3	0.4567	0.8243	-	-	-
MARGULES	3	0.4226	0.7477	-	-	-
MARGULES	4	0.5398	0.8586	0.4206	-	-
ALPHA	2	7.9421	1.4624	-	-	-
ALPHA	3	19.4566	2.6087	-	-12.4444	0.9405

			DEVIATION IN VAPOR PHASE COMPOSITION				
X	Y	PRESS.	LAAR 3	MARG 3	MARG 4	ALPHA2	ALPHA3
0.0825	0.3850	368.90	-0.0148	-0.0215	0.0021	0.0144	-0.0036
0.1260	0.4760	417.25	-0.0054	-0.0080	0.0018	0.0155	0.0010
0.2125	0.5920	505.80	0.0022	0.0053	-0.0033	0.0069	0.0005
0.2775	0.6460	551.80	0.0059	0.0102	-0.0029	0.0020	0.0007
0.4600	0.7350	633.60	0.0072	0.0074	0.0034	-0.0058	0.0017
0.5700	0.7730	664.95	-0.0012	-0.0050	-0.0004	-0.0116	-0.0027
0.6320	0.7880	676.80	-0.0033	-0.0088	-0.0009	-0.0102	-0.0018
0.8080	0.8260	689.80	-0.0040	-0.0071	-0.0036	0.0037	0.0026
0.8775	0.8470	686.55	0.0022	0.0045	-0.0004	0.0139	0.0039
0.9120	0.8700	683.80	0.0013	0.0067	-0.0027	0.0130	-0.0031
0.9630	0.9130	666.10	0.0133	0.0207	0.0091	0.0203	-0.0030
		MEAN	0.0055	0.0096	0.0028	0.0106	0.0022

CARBON TETRACHLORIDE(1)-ISO-PROPYL ALCOHOL(2)

NAGATA I.:J.CHEM.ENG.DATA 10,106(1965).

ANTOINE VAPOR PRESSURE CONSTANTS

	A	B	C
(1)	6.93390	1242.430	230.000
(2)	7.75634	1366.142	197.970

P = 760.00

EQUATION	ORDER	A12	A21	D	A122	A211
VAN LAAR	3	0.4669	0.7441	-	-	-
MARGULES	3	0.4436	0.6957	-	-	-
MARGULES	4	0.5216	0.7785	0.2802	-	-
ALPHA	2	4.0891	2.1167	-	-	-
ALPHA	3	11.5822	2.4773	-	-8.9691	2.5976

DEVIATION IN VAPOR PHASE COMPOSITION

X	Y	TEMP.	LAAR 3	MARG 3	MARG 4	ALPHA2	ALPHA3
0.0340	0.1140	79.90	-0.0071	-0.0108	0.0018	0.0258	-0.0040
0.0620	0.1850	78.50	-0.0051	-0.0093	0.0046	0.0353	-0.0009
0.0920	0.2520	76.80	-0.0051	-0.0086	0.0027	0.0336	-0.0015
0.1730	0.3760	74.00	0.0049	0.0049	0.0048	0.0255	0.0056
0.2240	0.4420	72.80	-0.0008	0.0007	-0.0045	0.0077	-0.0020
0.3420	0.5340	70.60	0.0047	0.0068	-0.0010	-0.0073	-0.0002
0.4120	0.5750	69.70	0.0033	0.0042	-0.0007	-0.0147	-0.0023
0.4860	0.6040	69.10	0.0066	0.0056	0.0053	-0.0129	0.0020
0.5780	0.6400	69.00	0.0012	-0.0022	0.0030	-0.0133	0.0004
0.6470	0.6650	68.80	-0.0031	-0.0077	-0.0000	-0.0110	-0.0014
0.7300	0.6920	68.90	-0.0049	-0.0092	-0.0021	-0.0011	-0.0007
0.8040	0.7190	69.10	-0.0026	-0.0043	-0.0021	0.0133	0.0008
0.8800	0.7600	70.20	0.0044	0.0078	0.0007	0.0325	0.0015
0.9430	0.8210	72.20	0.0208	0.0284	0.0143	0.0509	0.0060
0.9700	0.8930	74.10	0.0057	0.0129	-0.0001	0.0294	-0.0120
		MEAN	0.0054	0.0082	0.0032	0.0210	0.0028

CARBON TETRACHLORIDE(1)-ISO-PRPYL ALCOHOL(2)

PAPOUSEK D.,PAPOUSKOVA Z.,PAGO L.:Z.PHYS.CHEM.(LEIPZIG) 211,231(1959).

ANTOINE VAPOR PRESSURE CONSTANTS
	A	B	C
(1)	6.93390	1242.430	230.000
(2)	7.75634	1366.142	197.970

T = 70.00

EQUATION	ORDER	A12	A21	D	A122	A211
VAN LAAR	3	0.4637	0.7493	-	-	-
MARGULES	3	0.4369	0.6979	-	-	-
MARGULES	4	0.5027	0.7647	0.2237	-	-
ALPHA	2	4.3201	2.1823	-	-	-
ALPHA	3	7.3372	1.7074	-	-4.0699	1.8093

			DEVIATION IN VAPOR PHASE COMPOSITION				
X	Y	PRESS.	LAAR 3	MARG 3	MARG 4	ALPHA2	ALPHA3
0.0860	0.2550	579.60	-0.0064	-0.0111	-0.0001	0.0266	0.0038
0.1660	0.3920	659.00	-0.0097	-0.0105	-0.0090	0.0101	-0.0017
0.2380	0.4690	707.80	-0.0064	-0.0048	-0.0092	-0.0003	-0.0023
0.3840	0.5610	766.30	0.0033	0.0050	-0.0001	-0.0071	0.0021
0.5590	0.6340	798.20	-0.0003	-0.0031	-0.0002	-0.0101	0.0000
0.7110	0.6860	804.30	-0.0071	-0.0116	-0.0061	-0.0017	0.0002
0.8150	0.7330	793.80	-0.0128	-0.0139	-0.0135	0.0074	-0.0019
0.9450	0.8460	723.50	0.0021	0.0101	-0.0012	0.0281	0.0037
0.9750	0.9160	675.10	-0.0004	0.0064	-0.0028	0.0165	-0.0029
		MEAN	0.0054	0.0085	0.0047	0.0120	0.0021

CARBON TETRACHLORIDE(1)-TRICHLOROETHYLENE(2)

ACHARYA M.V.R.,VENKATA RAO C.:TRANS.INDIAN INST.CHEM.ENGRS.
6,129(1953-1954).

ANTOINE VAPOR PRESSURE CONSTANTS
         A   B   C
(1) 6.93390   1242.430   230.000
(2) 7.02808   1315.000   230.000

P =   760.00

EQUATION   ORDER   A12   A21   D   A122   A211

VAN LAAR   3   0.0010   -0.0016   -   -   -
MARGULES   3   -0.0302   0.0436   -   -   -
MARGULES   4   -0.0107   0.0655   0.0630   -   -
ALPHA   2   0.4297   -0.3675   -   -   -
ALPHA   3   2.2079   1.2890   -   -1.5492   -1.8296

| | | | DEVIATION IN VAPOR PHASE COMPOSITION | | | | |
X	Y	TEMP.	LAAR 3	MARG 3	MARG 4	ALPHA2	ALPHA3
0.0420	0.0700	86.20	-0.0134	-0.0161	-0.0145	-0.0109	-0.0040
0.1250	0.1820	85.60	-0.0182	-0.0210	-0.0195	-0.0112	-0.0015
0.1500	0.2120	85.20	-0.0171	-0.0190	-0.0179	-0.0088	0.0001
0.2000	0.2720	84.70	-0.0164	-0.0159	-0.0158	-0.0059	0.0005
0.2350	0.3150	84.10	-0.0108	-0.0156	-0.0161	-0.0062	-0.0020
0.2890	0.3650	83.60	-0.0060	-0.0010	-0.0020	0.0075	0.0082
0.3350	0.4250	83.00	-0.0147	-0.0080	-0.0091	-0.0002	-0.0022
0.4000	0.4920	82.50	-0.0117	-0.0038	-0.0046.	0.0037	-0.0013
0.4430	0.5400	81.80	-0.0143	-0.0069	-0.0072	0.0007	-0.0056
0.5100	0.5900	81.20	0.0069	0.0092	0.0097	0.0179	0.0110
0.5550	0.6500	80.80	0.0046	-0.0087	-0.0078	0.0011	-0.0053
0.6650	0.7500	79.90	0.0075	-0.0128	-0.0115	0.0003	-0.0022
0.7000	0.7800	79.30	-0.0083	-0.0139	-0.0128	0.0000	-0.0006
0.7700	0.8400	78.90	-0.0154	-0.0181	-0.0178	-0.0031	0.0002
0.8200	0.8800	78.50	-0.0158	-0.0193	-0.0197	-0.0046	0.0013
0.8800	0.9270	78.00	-0.0161	-0.0202	-0.0214	-0.0077	0.0000
0.9180	0.9550	77.50	-0.0153	-0.0190	-0.0204	-0.0091	-0.0017
0.9400	0.9650	77.00	-0.0088	-0.0120	-0.0133	-0.0041	0.0024
		MEAN	0.0127	0.0134	0.0134	0.0057	0.0028

CHLOROFORM(1)-BENZENE(2)

BUKALA M.,MAJEWSKI J.:PRZEM.CHEM-.,8,551(1952)

ANTOINE VAPOR PRESSURE CONSTANTS

	A	B	C
(1)	6.90328	1163.000	227.000
(2)	6.90565	1211.000	220.790

P =   760.00

EQUATION	ORDER	A12	A21	D	A122	A211
VAN LAAR	3	-0.2863	-0.0849	-	-	-
MARGULES	3	-0.1997	-0.0316	-	-	-
MARGULES	4	-0.2221	-0.0611	-0.0817	-	-
ALPHA	2	0.4276	-0.6827	-	-	-
ALPHA	3	5.0731	1.3857	-	-5.0917	-2.0171

| | | | DEVIATION IN VAPOR PHASE COMPOSITION |||||
X	Y	TEMP.	LAAR 3	MARG 3	MARG 4	ALPHA2	ALPHA3
0.0500	0.0600	80.00	-0.0065	-0.0016	-0.0033	0.0112	-0.0029
0.0800	0.1000	79.80	-0.0075	-0.0030	-0.0050	0.0136	-0.0029
0.1200	0.1500	79.20	-0.0005	0.0016	0.0001	0.0197	0.0043
0.2000	0.2700	78.50	0.0002	-0.0026	-0.0025	0.0098	0.0033
0.3200	0.4450	77.00	-0.0021	-0.0050	-0.0036	-0.0077	-0.0014
0.3800	0.5200	76.30	0.0000	-0.0006	0.0003	-0.0085	0.0005
0.4500	0.6100	75.30	-0.0085	-0.0065	-0.0066	-0.0167	-0.0076
0.5600	0.7100	73.50	0.0018	0.0058	0.0042	-0.0000	0.0043
0.6400	0.7800	72.30	0.0002	0.0033	0.0015	0.0046	0.0041
0.7800	0.8900	69.00	-0.0094	-0.0102	-0.0105	0.0024	-0.0041
0.8200	0.9100	68.00	-0.0046	-0.0065	-0.0061	0.0077	0.0008
		MEAN	0.0038	0.0043	0.0040	0.0093	0.0033

CHLOROFORM(1)-BENZENE(2)

KIREEV V.A.,SITNIKOV I.P.:ZH.OBSHCH.KHIM.16,979(1946)

ANTOINE VAPOR PRESSURE CONSTANTS
```
          A          B          C
(1) 6.90238    1163.000    227.000
(2) 6.90565    1211.000    220.790
```

T =   25.05

EQUATION	ORDER	A12	A21	D	A122	A211
VAN LAAR	3	-0.2316	-0.0971	-	-	-
MARGULES	3	-0.1943	-0.0582	-	-	-
MARGULES	4	-0.0776	0.0941	0.3883	-	-
ALPHA	2	0.6866	-0.7598	-	-	-
ALPHA	3	-0.9959	-1.6179	-	1.7149	0.9388

			DEVIATION IN VAPOR PHASE COMPOSITION				
X	Y	PRESS.	LAAR 3	MARG 3	MARG 4	ALPHA2	ALPHA3
0.1340	0.2180	105.90	-0.0239	-0.0227	-0.0127	-0.0025	-0.0021
0.2840	0.4240	118.00	0.0010	0.0002	-0.0032	0.0060	0.0048
0.3600	0.5300	125.50	-0.0022	-0.0016	-0.0049	-0.0029	-0.0037
0.6280	0.8040	155.40	-0.0088	-0.0064	0.0006	-0.0019	0.0011
0.7400	0.8860	170.00	-0.0141	-0.0138	-0.0113	-0.0014	-0.0010
0.8660	0.9500	185.90	-0.0089	-0.0106	-0.0157	0.0038	0.0004
		MEAN	0.0098	0.0092	0.0081	0.0031	0.0022

CHLOROFORM(1)-BENZENE(2)

NAGATA I.:J.CHEM.ENG.DATA 7(3),360(1962).

ANTOINE VAPOR PRESSURE CONSTANTS

	A	B	C
(1)	6.93708	1171.200	227.000
(2)	6.90565	1211.033	220.790

P =   760.00

EQUATION	ORDER	A12	A21	D	A122	A211
VAN LAAR	3	-0.0679	-0.1462	-	-	-
MARGULES	3	-0.0566	-0.1221	-	-	-
MARGULES	4	-0.0555	-0.1207	0.0039	-	-
ALPHA	2	0.4667	-0.5621	-	-	-
ALPHA	3	-0.9700	-1.4310	-	1.4631	0.9113

			DEVIATION IN VAPOR PHASE COMPOSITION				
X	Y	TEMP.	LAAR 3	MARG 3	MARG 4	ALPHA2	ALPHA3
0.0600	0.0890	79.20	-0.0010	0.0003	0.0004	-0.0022	-0.0013
0.0680	0.1000	79.00	-0.0005	0.0008	0.0010	-0.0018	-0.0009
0.1160	0.1670	78.40	0.0001	0.0013	0.0014	-0.0015	-0.0008
0.1330	0.1900	77.90	0.0006	0.0017	0.0018	-0.0011	-0.0005
0.1930	0.2700	76.90	0.0012	0.0016	0.0016	-0.0003	-0.0003
0.2290	0.3160	76.20	0.0021	0.0020	0.0020	0.0007	0.0003
0.2660	0.3610	75.70	0.0039	0.0035	0.0034	0.0029	0.0022
0.3180	0.4290	74.70	-0.0003	-0.0010	-0.0011	-0.0009	-0.0018
0.3330	0.4430	74.40	0.0036	0.0029	0.0029	0.0031	0.0022
0.3880	0.5080	73.30	0.0026	0.0020	0.0020	0.0024	0.0017
0.4430	0.5700	72.20	0.0016	0.0014	0.0014	0.0016	0.0014
0.4670	0.6010	71.60	-0.0035	-0.0035	-0.0035	-0.0036	-0.0035
0.5170	0.6520	70.80	-0.0029	-0.0024	-0.0024	-0.0030	-0.0023
0.5700	0.7020	69.70	-0.0007	0.0002	0.0002	-0.0012	-0.0001
0.6370	0.7620	68.30	0.0009	0.0021	0.0022	-0.0003	0.0010
0.7000	0.8140	67.00	0.0023	0.0034	0.0035	0.0003	0.0011
0.7830	0.8750	65.40	0.0044	0.0049	0.0049	0.0015	0.0010
0.8530	0.9220	64.10	0.0037	0.0034	0.0034	0.0006	-0.0010
0.9340	0.9680	62.60	0.0027	0.0020	0.0019	0.0005	-0.0012
		MEAN	0.0020	0.0021	0.0022	0.0016	0.0013

CHLOROFORM(1)-BENZENE(2)

REINDERS W.,DE MINJER C.H.:REC.TRAV.CHIM.PAYS BAS 59,369(1940).

ANTOINE VAPOR PRESSURE CONSTANTS
	A	B	C
(1)	6.93708	1171.200	227.000
(2)	6.90565	1211.033	220.790

P = 760.00

EQUATION	ORDER	A12	A21	D	A122	A211
VAN LAAR	3	-0.0717	-0.2228	-	-	-
MARGULES	3	-0.0521	-0.1507	-	-	-
MARGULES	4	-0.0738	-0.1808	-0.0794	-	-
ALPHA	2	0.4277	-0.5549	-	-	-
ALPHA	3	-0.3015	-0.9970	-	0.7476	0.4606

			DEVIATION IN VAPOR PHASE COMPOSITION				
X	Y	TEMP.	LAAR 3	MARG 3	MARG 4	ALPHA2	ALPHA3
0.0532	0.0755	79.60	0.0019	0.0040	0.0018	-0.0003	0.0003
0.0559	0.0791	79.55	0.0021	0.0043	0.0021	-0.0001	0.0005
0.1313	0.1846	78.25	0.0014	0.0034	0.0018	-0.0019	-0.0013
0.1459	0.2052	78.05	0.0005	0.0022	0.0009	-0.0028	-0.0023
0.2404	0.3224	76.45	0.0060	0.0059	0.0065	0.0037	0.0035
0.2512	0.3386	76.20	0.0053	0.0046	0.0038	0.0012	0.0010
0.2853	0.3850	75.60	-0.0011	-0.0017	-0.0007	-0.0026	-0.0029
0.3542	0.4594	74.35	0.0064	0.0058	0.0066	0.0062	0.0057
0.4216	0.5475	73.00	-0.0055	-0.0053	-0.0053	-0.0048	-0.0051
0.4777	0.6053	71.95	-0.0028	-0.0016	-0.0024	-0.0017	-0.0018
0.6353	0.7526	68.75	0.0047	0.0085	0.0066	0.0041	0.0047
0.7099	0.8222	67.20	0.0000	0.0034	0.0021	-0.0028	-0.0023
0.7994	0.8869	65.40	0.0046	0.0059	0.0060	-0.0008	-0.0009
0.9126	0.9549	63.10	0.0074	0.0058	0.0069	0.0015	0.0008
		MEAN	0.0034	0.0043	0.0038	0.0025	0.0024

CHLOROFORM(1)-CARBON TETRACHLORIDE(2)

KAPLAN S.I.,MONAKHOVA Z.D.:ZH.OBSHCH.KHIM.7,2499(1937).

ANTOINE VAPOR PRESSURE CONSTANTS
	A	B	C
(1)	6.90328	1163.000	227.000
(2)	6.93390	1242.430	230.000

P = 760.00

EQUATION	ORDER	A12	A21	D	A122	A211
VAN LAAR	3	-0.2681	-0.0611	-	-	-
MARGULES	3	-0.1491	-0.0114	-	-	-
MARGULES	4	-0.1215	0.0188	0.0855	-	-
ALPHA	2	0.3958	-0.5702	-	-	-
ALPHA	3	1.2951	-0.1997	-	-1.0261	-0.3276

			DEVIATION IN VAPOR PHASE COMPOSITION				
X	Y	TEMP.	LAAR 3	MARG 3	MARG 4	ALPHA2	ALPHA3
0.1000	0.1350	74.70	-0.0177	-0.0132	-0.0107	0.0028	-0.0032
0.2000	0.2650	72.60	-0.0044	-0.0073	-0.0070	0.0059	0.0020
0.3000	0.3950	70.60	0.0031	-0.0013	-0.0027	0.0027	0.0028
0.4000	0.5200	68.60	0.0009	-0.0004	-0.0015	-0.0034	-0.0004
0.5000	0.6350	66.90	-0.0066	-0.0049	-0.0045	-0.0087	-0.0054
0.6000	0.7250	65.30	-0.0026	-0.0003	0.0011	0.0004	0.0021
0.7000	0.8100	63.90	-0.0051	-0.0046	-0.0035	0.0029	0.0021
0.8000	0.8850	62.60	-0.0074	-0.0095	-0.0097	0.0031	0.0004
0.9000	0.9500	61.50	-0.0078	-0.0110	-0.0126	0.0005	-0.0022
		MEAN	0.0062	0.0058	0.0059	0.0034	0.0023

CHLOROFORM(1)-CARBON TETRACHLORIDE(2)

MC GLASHAN M.L.,PRUE J.E.,SAINSBURY J.E.J.:TRANS.FARAD.SOC.
50,1284(1954).

ANTOINE VAPOR PRESSURE CONSTANTS
	A	B	C
(1)	6.90328	1163.000	227.000
(2)	6.93390	1242.430	230.000

T =   25.00

EQUATION	ORDER	A12	A21	D	A122	A211
VAN LAAR	3	0.0654	0.0815	-	-	-
MARGULES	3	0.0643	0.0802	-	-	-
MARGULES	4	0.0652	0.0811	0.0026	-	-
ALPHA	2	0.9449	-0.2767	-	-	-
ALPHA	3	1.4211	0.0985	-	-0.3830	-0.3982

			DEVIATION IN VAPOR PHASE COMPOSITION				
X	Y	PRESS.	LAAR 3	MARG 3	MARG 4	ALPHA2	ALPHA3
0.0923	0.1636	124.24	-0.0007	-0.0008	-0.0007	-0.0012	0.0028
0.1006	0.1825	125.36	-0.0065	-0.0066	-0.0065	-0.0070	-0.0029
0.2280	0.3545	138.15	-0.0016	-0.0016	-0.0016	-0.0017	0.0007
0.2731	0.4080	142.70	-0.0015	-0.0014	-0.0015	-0.0013	0.0002
0.3185	0.4580	146.70	-0.0012	-0.0012	-0.0012	-0.0007	-0.0002
0.3193	0.4603	146.88	-0.0027	-0.0026	-0.0027	-0.0022	-0.0016
0.3982	0.5377	153.67	-0.0002	-0.0001	-0.0002	0.0007	0.0001
0.4450	0.5817	157.61	-0.0005	-0.0004	-0.0005	0.0006	-0.0004
0.4516	0.5869	158.20	0.0003	0.0003	0.0003	0.0014	0.0003
0.4665	0.5993	159.14	0.0012	0.0012	0.0012	0.0023	0.0012
0.4755	0.6079	160.10	0.0005	0.0006	0.0006	0.0017	0.0005
0.5006	0.6292	162.01	0.0010	0.0010	0.0010	0.0022	0.0010
0.5645	0.6817	166.98	0.0015	0.0015	0.0015	0.0027	0.0017
0.5689	0.6896	167.90	-0.0029	-0.0029	-0.0029	-0.0017	-0.0027
0.5765	0.6929	168.09	-0.0001	-0.0001	-0.0001	0.0011	0.0001
0.6096	0.7192	170.50	-0.0003	-0.0004	-0.0003	0.0008	0.0000
0.6608	0.7603	174.13	-0.0023	-0.0024	-0.0023	-0.0014	-0.0017
0.6669	0.7622	174.73	0.0004	0.0003	0.0003	0.0013	0.0011
0.6841	0.7767	175.92	-0.0013	-0.0014	-0.0013	-0.0005	-0.0005
0.7133	0.7994	178.12	-0.0026	-0.0026	-0.0026	-0.0018	-0.0016
0.7842	0.8485	182.85	-0.0007	-0.0007	-0.0007	-0.0003	0.0006
0.8592	0.9026	188.08	-0.0019	-0.0018	-0.0019	-0.0019	-0.0006
0.8629	0.9024	188.05	0.0009	0.0010	0.0009	0.0009	0.0022
0.8786	0.9153	189.23	-0.0010	-0.0009	-0.0010	-0.0011	0.0002
0.8796	0.9146	189.10	0.0004	0.0005	0.0004	0.0003	0.0016
0.8805	0.9139	188.99	0.0018	0.0018	0.0018	0.0017	0.0030
0.9074	0.9391	190.76	-0.0046	-0.0045	-0.0046	-0.0048	-0.0035
0.9488	0.9658	193.81	-0.0022	-0.0021	-0.0021	-0.0024	-0.0015
0.9494	0.9668	193.71	-0.0027	-0.0027	-0.0027	-0.0029	-0.0021
		MEAN	0.0016	0.0016	0.0016	0.0017	0.0013

CHLOROFORM(1)-CARBON TETRACHLORIDE(2)

MC GLASHAN M.L.,PRUE J.E.,SAINSBURY J.E.J.:TRANS.FARAD.SOC.
50,1284(1954).

ANTOINE VAPOR PRESSURE CONSTANTS
	A	B	C
(1)	6.90328	1163.000	227.000
(2)	6.93390	1242.430	230.000

T =   40.00

EQUATION	ORDER	A12	A21	D	A122	A211
VAN LAAR	3	0.0600	0.0914	-	-	-
MARGULES	3	0.0572	0.0863	-	-	-
MARGULES	4	0.0641	0.0941	0.0238	-	-
ALPHA	2	0.9078	-0.2658	-	-	-
ALPHA	3	1.4366	0.1021	-	-0.4763	-0.3766

DEVIATION IN VAPOR PHASE COMPOSITION

X	Y	PRESS.	LAAR 3	MARG 3	MARG 4	ALPHA2	ALPHA3
0.0319	0.0614	220.25	-0.0033	-0.0035	-0.0029	-0.0026	-0.0013
0.0322	0.0589	220.23	-0.0002	-0.0005	0.0002	0.0004	0.0017
0.0806	0.1448	229.76	-0.0049	-0.0052	-0.0044	-0.0036	-0.0016
0.0827	0.1466	230.07	-0.0033	-0.0037	-0.0029	-0.0021	0.0000
0.1840	0.2936	248.71	-0.0030	-0.0031	-0.0031	-0.0015	0.0001
0.3074	0.4401	269.47	-0.0010	-0.0008	-0.0013	0.0004	0.0004
0.3108	0.4424	269.81	0.0004	0.0006	0.0001	0.0018	0.0018
0.4218	0.5560	286.98	-0.0009	-0.0009	-0.0011	0.0001	-0.0006
0.4241	0.5594	287.35	-0.0022	-0.0021	-0.0024	-0.0011	-0.0018
0.4279	0.5608	287.89	-0.0000	0.0000	-0.0002	0.0010	0.0003
0.5200	0.6428	301.34	-0.0003	-0.0005	-0.0002	0.0007	-0.0000
0.5242	0.6456	301.84	0.0005	0.0003	0.0005	0.0014	0.0007
0.6258	0.7288	315.51	-0.0010	-0.0014	-0.0008	0.0000	-0.0002
0.6268	0.7302	315.72	-0.0016	-0.0020	-0.0015	-0.0006	-0.0008
0.7266	0.8038	328.52	-0.0006	-0.0009	-0.0005	0.0006	0.0009
0.7283	0.8056	328.98	-0.0012	-0.0015	-0.0011	-0.0000	0.0004
0.8438	0.8887	343.00	-0.0017	-0.0016	-0.0018	-0.0004	0.0003
0.8449	0.8895	342.84	-0.0017	-0.0016	-0.0018	-0.0004	0.0003
0.9145	0.9395	351.26	-0.0021	-0.0019	-0.0023	-0.0011	-0.0005
0.9152	0.9401	351.33	-0.0022	-0.0020	-0.0024	-0.0012	-0.0007
0.9273	0.9483	352.81	-0.0017	-0.0014	-0.0019	-0.0008	-0.0003
0.9599	0.9708	356.28	-0.0005	-0.0003	-0.0007	0.0001	0.0004
		MEAN	0.0016	0.0016	0.0015	0.0010	0.0007

CHLOROFORM(1)-CARBON TETRACHLORIDE(2)

MC GLASHAN M.L.,PRUE J.F.,SAINSBURY J.E.J.:TRANS.FARAD.SOC.
50,1284(1954)

ANTOINE VAPOR PRESSURE CONSTANTS

	A	B	C
(1)	6.90328	1163.000	227.000
(2)	6.93390	1242.430	230.000

T =   55.00

EQUATION	ORDER	A12	A21	D	A122	A211
VAN LAAR	3	0.0507	0.0849	-	-	-
MARGULES	3	0.0470	0.0781	-	-	-
MARGULES	4	0.0556	0.0878	0.0299	-	-
ALPHA	2	0.8250	-0.2715	-	-	-
ALPHA	3	0.5988	-0.3839	-	0.2382	0.0979

			DEVIATION IN VAPOR PHASE COMPOSITION				
X	Y	PRESS.	LAAR 3	MARG 3	MARG 4	ALPHA2	ALPHA3
0.0227	0.0424	380.86	-0.0024	-0.0027	-0.0020	-0.0019	-0.0017
0.0281	0.0497	382.46	-0.0004	-0.0007	-0.0000	0.0002	0.0005
0.0394	0.0691	386.02	-0.0007	-0.0011	-0.0002	0.0001	0.0004
0.0502	0.0871	389.74	-0.0008	-0.0012	-0.0003	0.0001	0.0006
0.0597	0.1035	392.78	-0.0017	-0.0022	-0.0012	-0.0007	-0.0002
0.1592	0.2505	422.66	-0.0009	-0.0011	-0.0002	0.0005	0.0012
0.1840	0.2868	430.97	-0.0039	-0.0040	-0.0040	-0.0026	-0.0020
0.3516	0.4786	475.71	-0.0000	0.0002	-0.0005	0.0006	0.0006
0.4118	0.5398	491.04	-0.0009	-0.0008	-0.0012	-0.0005	-0.0006
0.4313	0.5557	496.09	0.0018	0.0019	0.0016	0.0022	0.0020
0.4819	0.6033	508.76	0.0007	0.0006	0.0007	0.0011	0.0008
0.4994	0.6184	512.78	0.0011	0.0010	0.0011	0.0015	0.0012
0.5420	0.6584	523.10	-0.0022	-0.0025	-0.0021	-0.0018	-0.0021
0.5520	0.6677	526.17	-0.0032	-0.0034	-0.0030	-0.0027	-0.0030
0.6700	0.7578	551.34	0.0007	0.0003	0.0009	0.0017	0.0017
0.7209	0.7990	563.01	-0.0022	-0.0026	-0.0021	-0.0010	-0.0008
0.7272	0.8017	564.14	-0.0003	-0.0006	-0.0001	0.0010	0.0013
0.9199	0.9418	603.62	-0.0010	-0.0007	-0.0012	0.0005	0.0011
0.9222	0.9457	604.27	-0.0033	-0.0029	-0.0035	-0.0018	-0.0012
0.9730	0.9822	613.76	-0.0025	-0.0022	-0.0026	-0.0018	-0.0015
		MEAN	0.0015	0.0016	0.0015	0.0012	0.0012

CHLOROFORM(1)-ETHYL ACETATE(2)

NAGATA I.:J.CHEM.ENG.DATA 7(3),367(1962).

ANTOINE VAPOR PRESSURE CONSTANTS
	A	B	C
(1)	6.93708	1171.200	227.000
(2)	7.10232	1245.239	217.911

P =   760.00

EQUATION	ORDER	A12	A21	D	A122	A211
VAN LAAR	3	-0.2827	-0.4750	-	-	-
MARGULES	3	-0.2566	-0.4370	-	-	-
MARGULES	4	-0.2725	-0.4547	-0.0500	-	-
ALPHA	2	-0.2196	-0.7966	-	-	-
ALPHA	3	-0.6720	-0.9900	-	0.5412	0.1810

			DEVIATION IN VAPOR PHASE COMPOSITION				
X	Y	TEMP.	LAAR 3	MARG 3	MARG 4	ALPHA2	ALPHA3
0.0710	0.0640	77.50	-0.0008	0.0013	0.0001	-0.0034	0.0007
0.1100	0.1020	77.60	-0.0006	0.0015	0.0004	-0.0037	0.0011
0.1400	0.1340	77.70	-0.0015	0.0002	-0.0006	-0.0046	0.0000
0.1740	0.1710	77.80	-0.0016	-0.0004	-0.0009	-0.0042	-0.0003
0.2230	0.2270	77.80	-0.0012	-0.0010	-0.0009	-0.0025	-0.0004
0.2590	0.2700	77.50	-0.0002	-0.0007	-0.0003	-0.0006	-0.0001
0.3010	0.3230	77.30	0.0001	-0.0011	-0.0004	0.0011	-0.0003
0.3650	0.4080	76.80	-0.0000	-0.0016	-0.0008	0.0029	-0.0009
0.4480	0.5220	76.00	-0.0005	-0.0014	-0.0012	0.0038	-0.0011
0.5040	0.5960	75.10	0.0019	0.0019	0.0016	0.0060	0.0017
0.5280	0.6280	74.70	0.0020	0.0023	0.0019	0.0058	0.0020
0.5810	0.7000	73.50	-0.0016	-0.0005	-0.0013	0.0009	-0.0011
0.6500	0.7800	71.80	-0.0002	0.0014	0.0006	0.0005	0.0010
0.7040	0.8390	70.40	-0.0032	-0.0019	-0.0025	-0.0039	-0.0019
0.7510	0.8790	68.90	-0.0009	-0.0001	-0.0003	-0.0025	0.0003
0.7900	0.9100	67.70	-0.0017	-0.0014	-0.0014	-0.0036	-0.0006
0.8560	0.9500	65.60	-0.0006	-0.0011	-0.0007	-0.0024	0.0002
0.9220	0.9780	63.70	0.0006	-0.0002	0.0002	-0.0003	0.0010
		MEAN	0.0011	0.0011	0.0009	0.0029	0.0008

CHLOROFORM(1)-ETHYL ALCOHOL(2)

ROECK H.,SCHROEDER W.:Z.PHYS.CHEM.(FRANKFURT)9,277(1956).

ANTOINE VAPOR PRESSURE CONSTANTS

	A	B	C
(1)	6.48083	923.3370	195.233
(2)	8.16290	1623.220	228.980

T =   45.00

EQUATION	ORDER	A12	A21	D	A122	A211
VAN LAAR	3	0.3172	0.9240	-	-	-
MARGULES	3	0.2707	0.6384	-	-	-
MARGULES	4	0.2294	0.5226	-0.2567	-	-
ALPHA	2	4.3222	-0.1397	-	-	-
ALPHA	3	47.3552	5.5286	-	-44.6534	2.5213

			DEVIATION IN VAPOR PHASE COMPOSITION				
X	Y	PRESS.	LAAR 3	MARG 3	MARG 4	ALPHA2	ALPHA3
0.0210	0.0830	184.40	0.0164	0.0088	0.0028	0.0182	-0.0024
0.0650	0.2380	211.00	0.0209	0.0107	0.0032	0.0233	0.0025
0.1580	0.4730	261.00	-0.0024	-0.0039	-0.0037	-0.0020	-0.0013
0.2130	0.5570	302.50	-0.0052	-0.0034	-0.0011	-0.0059	-0.0008
0.2590	0.6070	327.40	-0.0022	0.0004	0.0025	-0.0032	0.0016
0.3060	0.6510	350.60	-0.0021	-0.0003	0.0005	-0.0028	-0.0005
0.3450	0.6780	376.50	0.0015	0.0018	0.0010	0.0015	0.0006
0.3860	0.7040	382.90	0.0032	0.0011	-0.0017	0.0043	-0.0006
		MEAN	0.0067	0.0038	0.0021	0.0076	0.0013

CHLOROFORM(1)-ETHYL ALCOHOL(2)

ROECK H.,SCHROEDER W.:Z.PHYS.CHEM.(FRANKFURT)9,277(1956).

ANTOINE VAPOR PRESSURE CONSTANTS

	A	B	C
(1)	6.48083	923.3370	195.233
(2)	8.16290	1623.220	228.980

T =   55.00

EQUATION	ORDER	A12	A21	D	A122	A211
VAN LAAR	3	0.2951	0.9322	-	-	-
MARGULES	3	0.2465	0.6129	-	-	-
MARGULES	4	0.2056	0.5046	-0.2419	-	-
ALPHA	2	3.4525	-0.1824	-	-	-
ALPHA	3	-4.0645	-2.7591	-	6.9514	2.7688

			DEVIATION IN VAPOR PHASE COMPOSITION				
X	Y	PRESS.	LAAR 3	MARG 3	MARG 4	ALPHA2	ALPHA3
0.0210	0.0750	296.00	0.0097	0.0028	-0.0023	0.0112	0.0023
0.1060	0.3210	368.50	0.0082	0.0014	-0.0026	0.0097	-0.0017
0.1600	0.4280	412.70	0.0052	0.0037	0.0037	0.0055	0.0018
0.2110	0.5130	454.70	-0.0036	-0.0019	0.0001	-0.0042	-0.0013
0.2590	0.5720	490.00	-0.0048	-0.0020	0.0001	-0.0058	0.0005
0.3030	0.6160	519.90	-0.0047	-0.0026	-0.0016	-0.0055	0.0002
0.3430	0.6470	541.20	-0.0015	-0.0010	-0.0016	-0.0016	-0.0003
0.3850	0.6700	562.20	0.0066	0.0045	0.0020	0.0077	0.0001
		MEAN	0.0055	0.0025	0.0018	0.0064	0.0010

CHLOROFORM(1)-ETHYL ALCOHOL(2)

SCATCHARD G.,RAYMOND C.L.:J.AM.CHEM.SOC.60,1278(1938).

ANTOINE VAPOR PRESSURE CONSTANTS

	A	B	C
(1)	6.93708	1171.200	227.000
(2)	8.16290	1623.220	228.980

T =   35.00

EQUATION	ORDER	A12	A21	D	A122	A211
VAN LAAR	3	0.3287	0.8731	-	-	-
MARGULES	3	0.2185	0.6912	-	-	-
MARGULES	4	0.2144	0.6854	-0.0175	-	-
ALPHA	2	6.5714	0.4806	-	-	-
ALPHA	3	7.5907	-0.9183	-	-3.6871	1.9858

			DEVIATION IN VAPOR PHASE COMPOSITION				
X	Y	PRESS.	LAAR 3	MARG 3	MARG 4	ALPHA2	ALPHA3
0.0062	0.0254	104.87	0.0108	0.0032	0.0030	0.0193	0.0044
0.0241	0.0911	111.31	0.0373	0.0152	0.0145	0.0621	0.0188
0.0297	0.1210	113.61	0.0329	0.0082	0.0074	0.0611	0.0122
0.0542	0.2154	123.54	0.0358	0.0057	0.0048	0.0718	0.0109
0.0594	0.2443	125.82	0.0250	-0.0053	-0.0061	0.0617	0.0001
0.1109	0.3885	148.26	0.0244	0.0019	0.0014	0.0591	0.0068
0.1730	0.5304	177.60	-0.0009	-0.0088	-0.0087	0.0237	-0.0055
0.2361	0.6207	205.68	-0.0090	-0.0062	-0.0058	0.0049	-0.0043
0.2873	0.6747	225.06	-0.0132	-0.0055	-0.0052	-0.0066	-0.0045
0.3014	0.6870	229.24	-0.0137	-0.0053	-0.0049	-0.0089	-0.0044
0.3227	0.7009	236.50	-0.0111	-0.0019	-0.0016	-0.0088	-0.0013
0.3845	0.7370	253.39	-0.0068	0.0024	0.0026	-0.0102	0.0026
0.3922	0.7412	255.28	-0.0066	0.0024	0.0026	-0.0106	0.0025
0.4384	0.7646	267.65	-0.0062	0.0011	0.0011	-0.0131	0.0012
0.4827	0.7797	274.04	-0.0020	0.0029	0.0028	-0.0105	0.0032
0.4846	0.7812	274.46	-0.0027	0.0021	0.0020	-0.0113	0.0024
0.6185	0.8181	291.95	0.0038	0.0001	-0.0003	-0.0039	0.0024
0.6783	0.8327	296.93	0.0035	-0.0030	-0.0034	-0.0007	0.0006
0.7746	0.8554	303.05	0.0005	-0.0062	-0.0064	0.0060	-0.0012
0.8265	0.8698	305.39	-0.0031	-0.0064	-0.0063	0.0097	-0.0018
0.8423	0.8752	305.12	-0.0047	-0.0063	-0.0062	0.0104	-0.0022
0.8483	0.8783	306.25	-0.0063	-0.0072	-0.0071	0.0097	-0.0032
0.9315	0.9161	306.05	-0.0101	0.0017	0.0022	0.0155	0.0002
0.9560	0.9363	304.87	-0.0098	0.0042	0.0046	0.0142	0.0007
0.9586	0.9385	304.17	-0.0092	0.0047	0.0052	0.0143	0.0012
0.9600	0.9403	303.69	-0.0095	0.0044	0.0049	0.0137	0.0008
0.9616	0.9464	303.91	-0.0138	0.0001	0.0006	0.0091	-0.0036
		MEAN	0.0116	0.0045	0.0045	0.0204	0.0038

CHLOROFORM(1)-ETHYL ALCOHOL(2)

SCATCHARD G.,RAYMOND C.L.:J.AM.CHEM.SOC.60,1278(1938).

ANTOINE VAPOR PRESSURE CONSTANTS
```
          A          B          C
(1) 6.93708    1171.200    227.000
(2) 8.16290    1623.220    228.980
```

T =   45.00

EQUATION	ORDER	A12	A21	D	A122	A211
VAN LAAR	3	0.3304	0.7847	-	-	-
MARGULES	3	0.2434	0.6622	-	-	-
MARGULES	4	0.2358	0.6535	-0.0290	-	-
ALPHA	2	5.5640	0.6006	-	-2.5221	1.9744
ALPHA	3	5.9081	-0.8711	-	-2.5221	1.9744

			DEVIATION IN VAPOR PHASE COMPOSITION				
X	Y	PRESS.	LAAR 3	MARG 3	MARG 4	ALPHA2	ALPHA3
0.0157	0.0600	180.96	0.0171	0.0053	0.0044	0.0328	0.0060
0.0189	0.0716	182.63	0.0198	0.0064	0.0053	0.0379	0.0072
0.0476	0.1717	199.62	0.0321	0.0108	0.0093	0.0626	0.0121
0.0712	0.2467	214.44	0.0319	0.0105	0.0090	0.0653	0.0118
0.1260	0.3974	249.92	0.0131	0.0001	-0.0005	0.0426	0.0012
0.1997	0.5366	298.08	-0.0059	-0.0059	-0.0055	0.0119	-0.0051
0.2011	0.5395	299.63	-0.0069	-0.0068	-0.0064	0.0106	-0.0060
0.2569	0.6060	329.62	-0.0090	-0.0026	-0.0019	-0.0000	-0.0021
0.3116	0.6557	355.66	-0.0095	-0.0002	0.0006	-0.0076	-0.0000
0.3298	0.6714	365.07	-0.0112	-0.0015	-0.0008	-0.0113	-0.0014
0.4015	0.7143	391.51	-0.0074	0.0015	0.0020	-0.0137	0.0013
0.4439	0.7340	403.91	-0.0047	0.0026	0.0028	-0.0133	0.0023
0.5140	0.7603	420.63	-0.0002	0.0032	0.0030	-0.0107	0.0031
0.5405	0.7703	425.28	-0.0002	0.0016	0.0013	-0.0108	0.0016
0.6283	0.7954	438.89	0.0031	-0.0004	-0.0010	-0.0053	0.0008
0.7148	0.8191	448.17	0.0025	-0.0040	-0.0046	0.0003	-0.0015
0.8206	0.8516	455.56	-0.0020	-0.0060	-0.0061	0.0080	-0.0030
0.8852	0.8783	455.79	-0.0047	-0.0022	-0.0017	0.0136	-0.0011
0.9100	0.8933	454.54	-0.0062	-0.0006	0.0000	0.0142	-0.0009
0.9125	0.8946	454.02	-0.0060	-0.0001	0.0006	0.0146	-0.0005
0.9163	0.8974	453.76	-0.0063	0.0001	0.0008	0.0145	-0.0006
0.9557	0.9319	448.49	-0.0065	0.0031	0.0039	0.0128	0.0003
0.9677	0.9454	445.38	-0.0051	0.0043	0.0050	0.0118	0.0013
0.9758	0.9579	443.07	-0.0057	0.0028	0.0034	0.0086	-0.0001
0.9866	0.9727	439.89	-0.0018	0.0042	0.0046	0.0076	0.0020
		MEAN	0.0087	0.0035	0.0034	0.0177	0.0029

CHLOROFORM(1)-ETHYL ALCOHOL(2)

SCATCHARD G.,RAYMOND C.L.:J.AM.CHEM.SOC.60,1278(1938).

ANTOINE VAPOR PRESSURE CONSTANTS

	A	B	C
(1)	6.93708	1171.200	227.000
(2)	8.16290	1623.220	228.980

T = 55.00

EQUATION	ORDER	A12	A21	D	A122	A211
VAN LAAR	3	0.3287	0.7236	-	-	-
MARGULES	3	0.2589	0.6139	-	-	-
MARGULES	4	0.2524	0.6050	-0.0256	-	-
ALPHA	2	4.4899	0.5901	-	-	-
ALPHA	3	4.5615	-0.7699	-	-1.6514	1.8605

			DEVIATION IN VAPOR PHASE COMPOSITION				
X	Y	PRESS.	LAAR 3	MARG 3	MARG 4	ALPHA2	ALPHA3
0.0331	0.1162	306.38	0.0179	0.0040	0.0029	0.0359	0.0032
0.0712	0.2302	339.89	0.0219	0.0053	0.0041	0.0454	0.0042
0.0802	0.2533	346.89	0.0224	0.0063	0.0052	0.0460	0.0052
0.1029	0.3123	367.01	0.0173	0.0033	0.0024	0.0399	0.0023
0.1479	0.4035	407.90	0.0138	0.0057	0.0053	0.0317	0.0050
0.1869	0.4795	441.04	-0.0017	-0.0048	-0.0047	0.0110	-0.0051
0.2201	0.5271	469.41	-0.0061	-0.0057	-0.0054	0.0022	-0.0058
0.2806	0.5942	508.78	-0.0093	-0.0048	-0.0043	-0.0082	-0.0046
0.3412	0.6417	543.53	-0.0068	-0.0007	-0.0003	-0.0112	-0.0007
0.3445	0.6419	545.72	-0.0045	0.0015	0.0019	-0.0092	0.0015
0.3767	0.6641	560.25	-0.0046	0.0012	0.0015	-0.0115	0.0011
0.3904	0.6720	566.74	-0.0039	0.0018	0.0020	-0.0115	0.0016
0.3965	0.6760	569.02	-0.0041	0.0014	0.0016	-0.0121	0.0011
0.4794	0.7161	599.03	-0.0001	0.0026	0.0024	-0.0106	0.0019
0.5730	0.7527	623.67	0.0019	0.0003	-0.0003	-0.0074	-0.0004
0.6211	0.7639	632.14	0.0073	0.0038	0.0031	0.0003	0.0033
0.6851	0.7857	641.49	0.0056	0.0004	-0.0002	0.0033	0.0004
0.7269	0.8010	646.79	0.0029	-0.0025	-0.0030	0.0047	-0.0022
0.7764	0.8181	650.96	0.0011	-0.0032	-0.0034	0.0086	-0.0028
0.8390	0.8417	653.11	-0.0001	-0.0004	-0.0002	0.0155	-0.0005
0.9037	0.8798	650.38	-0.0044	0.0019	0.0026	0.0178	0.0001
0.9430	0.9150	644.24	-0.0066	0.0029	0.0038	0.0153	-0.0000
0.9652	0.9408	626.79	-0.0054	0.0039	0.0046	0.0125	0.0008
		MEAN	0.0074	0.0030	0.0028	0.0162	0.0023

CHLOROFORM(1)-METHYL ALCOHOL(2)

KIREEV V.A.,SITNIKOV I.P.:ZH.FIZ.KHIM.15,492(1941).

ANTOINE VAPOR PRESSURE CONSTANTS

	A	B	C
(1)	6.90328	1163.000	227.000
(2)	8.07246	1574.990	238.860

T =   35.00

EQUATION	ORDER	A12	A21	D	A122	A211
VAN LAAR	3	0.4145	1.0281	-	-	-
MARGULES	3	0.2941	0.8235	-	-	-
MARGULES	4	0.3352	0.8680	0.1360	-	-
ALPHA	2	4.7052	1.9532	-	-	-
ALPHA	3	14.9763	1.0149	-	-13.4593	5.0582

			DEVIATION IN VAPOR PHASE COMPOSITION				
X	Y	PRESS.	LAAR 3	MARG 3	MARG 4	ALPHA2	ALPHA3
0.0500	0.1430	230.30	0.0113	-0.0121	-0.0052	0.0648	-0.0107
0.1000	0.2630	253.30	0.0063	-0.0168	-0.0108	0.0643	-0.0065
0.1500	0.3570	274.80	0.0010	-0.0136	-0.0108	0.0486	0.0013
0.2000	0.4310	295.60	-0.0028	-0.0078	-0.0080	0.0303	0.0062
0.2500	0.4900	313.20	-0.0052	-0.0025	-0.0047	0.0136	0.0074
0.3000	0.5390	328.90	-0.0079	-0.0003	-0.0034	-0.0019	0.0045
0.3500	0.5800	340.40	-0.0107	-0.0010	-0.0039	-0.0152	-0.0008
0.4000	0.6100	350.60	-0.0088	0.0005	-0.0016	-0.0214	-0.0026
0.4500	0.6350	357.60	-0.0073	-0.0002	-0.0011	-0.0253	-0.0049
0.5000	0.6540	362.50	-0.0041	-0.0006	-0.0002	-0.0249	-0.0051
0.5500	0.6680	364.90	0.0002	-0.0005	0.0011	-0.0205	-0.0031
0.6000	0.6840	366.70	-0.0007	-0.0058	-0.0032	-0.0184	-0.0050
0.6500	0.6920	367.90	0.0036	-0.0053	-0.0022	-0.0079	-0.0000
0.7000	0.7040	368.90	0.0015	-0.0097	-0.0067	-0.0001	0.0006
0.7500	0.7140	368.30	-0.0002	-0.0112	-0.0091	0.0119	0.0037
0.8000	0.7240	366.50	-0.0022	-0.0091	-0.0090	0.0279	0.0085
0.8500	0.7440	364.00	-0.0116	-0.0090	-0.0118	0.0402	0.0075
0.9000	0.7760	355.20	-0.0228	-0.0045	-0.0108	0.0516	0.0037
0.9500	0.8450	340.00	-0.0375	-0.0021	-0.0109	0.0465	-0.0119
		MEAN	0.0077	0.0059	0.0060	0.0282	0.0049

CHLOROFORM(1)-METHYL ALCOHOL(2)

KIREEV V.A.,SITNIKOV I.P.:ZH.FIZ.KHIM.15,492(1941).

ANTOINE VAPOR PRESSURE CONSTANTS

	A	B	C
(1)	6.90328	1163.000	227.000
(2)	8.07246	1574.990	238.860

T =   49.30

EQUATION	ORDER	A12	A21	D	A122	A211
VAN LAAR	3	0.4505	1.0088	-	-	-
MARGULES	3	0.3514	0.8420	-	-	-
MARGULES	4	0.3798	0.8725	0.0934	-	-
ALPHA	2	4.8098	2.5841	-	-	-
ALPHA	3	12.5906	0.9064	-	-10.8707	5.5939

			DEVIATION IN VAPOR PHASE COMPOSITION				
X	Y	PRESS.	LAAR 3	MARG 3	MARG 4	ALPHA2	ALPHA3
0.0500	0.1380	441.60	0.0095	-0.0091	-0.0044	0.0681	-0.0027
0.1000	0.2540	482.70	0.0034	-0.0148	-0.0107	0.0660	0.0007
0.1500	0.3500	524.80	-0.0080	-0.0192	-0.0173	0.0429	-0.0001
0.2000	0.4240	559.60	-0.0153	-0.0187	-0.0189	0.0201	-0.0013
0.2500	0.4760	586.10	-0.0136	-0.0109	-0.0125	0.0067	0.0019
0.3000	0.5200	606.10	-0.0138	-0.0075	-0.0095	-0.0068	-0.0001
0.3500	0.5500	621.30	-0.0079	-0.0001	-0.0021	-0.0115	0.0025
0.4000	0.5770	633.30	-0.0052	0.0021	0.0006	-0.0168	0.0011
0.4500	0.5980	642.90	-0.0016	0.0035	0.0029	-0.0183	0.0009
0.5000	0.6190	646.80	-0.0023	-0.0004	-0.0002	-0.0213	-0.0028
0.5500	0.6340	650.20	-0.0007	-0.0026	-0.0014	-0.0190	-0.0030
0.6000	0.6470	652.60	-0.0003	-0.0059	-0.0040	-0.0147	-0.0027
0.6500	0.6590	654.90	-0.0014	-0.0101	-0.0079	-0.0087	-0.0021
0.7000	0.6700	656.90	-0.0035	-0.0140	-0.0118	-0.0001	-0.0005
0.7500	0.6800	656.80	-0.0057	-0.0155	-0.0140	0.0122	0.0031
0.8000	0.6940	654.20	-0.0113	-0.0169	-0.0168	0.0250	0.0052
0.8500	0.7160	647.50	-0.0206	-0.0173	-0.0193	0.0372	0.0045
0.9000	0.7540	632.60	-0.0329	-0.0153	-0.0200	0.0463	-0.0010
0.9500	0.8240	601.30	-0.0385	-0.0057	-0.0124	0.0482	-0.0090
		MEAN	0.0103	0.0100	0.0098	0.0258	0.0024

CHLOROFORM(1)-METHYL ALCOHOL(2)

NAGATA I.:J.CHEM.ENG.DATA 7(3),367(1962).

ANTOINE VAPOR PRESSURE CONSTANTS

	A	B	C
(1)	6.93708	1171.200	227.000
(2)	8.07246	1574.990	238.860

P = 760.00

EQUATION	ORDER	A12	A21	D	A122	A211
VAN LAAR	3	0.4122	0.8500	-	-	-
MARGULES	3	0.3418	0.7369	-	-	-
MARGULES	4	0.3812	0.7784	0.1321	-	-
ALPHA	2	3.5023	2.0056.	-	-	-
ALPHA	3	10.6348	1.5322	-	-9.1939	3.7869

			DEVIATION IN VAPOR PHASE COMPOSITION				
X	Y	TEMP.	LAAR 3	MARG 3	MARG 4	ALPHA2	ALPHA3
0.0400	0.1020	63.00	0.0008	-0.0096	-0.0043	0.0420	-0.0055
0.0650	0.1540	62.00	0.0042	-0.0080	-0.0020	0.0542	-0.0007
0.0950	0.2150	60.90	0.0021	-0.0094	-0.0041	0.0538	0.0009
0.1460	0.3040	59.30	-0.0023	-0.0092	-0.0068	0.0414	0.0028
0.1960	0.3780	57.80	-0.0082	-0.0098	-0.0102	0.0224	0.0002
0.2300	0.4200	57.00	-0.0109	-0.0094	-0.0113	0.0104	-0.0017
0.2870	0.4720	55.90	-0.0073	-0.0022	-0.0054	-0.0001	0.0012
0.3320	0.5070	55.30	-0.0059	0.0005	-0.0029	-0.0079	0.0010
0.3830	0.5400	54.70	-0.0038	0.0024	-0.0003	-0.0140	0.0006
0.4250	0.5640	54.30	-0.0030	0.0020	0.0002	-0.0180	-0.0007
0.4590	0.5800	54.00	-0.0013	0.0023	0.0014	-0.0188	-0.0005
0.5200	0.6070	53.80	-0.0013	-0.0011	-0.0002	-0.0199	-0.0021
0.5570	0.6190	53.70	0.0008	-0.0012	0.0006	-0.0166	-0.0004
0.6280	0.6430	53.50	0.0003	-0.0056	-0.0024	-0.0106	-0.0002
0.6360	0.6460	53.50	-0.0003	-0.0065	-0.0033	-0.0101	-0.0006
0.6670	0.6550	53.50	-0.0003	-0.0077	-0.0044	-0.0051	0.0005
0.7530	0.6840	53.70	-0.0055	-0.0129	-0.0106	0.0097	0.0009
0.7970	0.7010	53.90	-0.0086	-0.0133	-0.0127	0.0201	0.0013
0.8550	0.7300	54.40	-0.0125	-0.0100	-0.0126	0.0361	0.0017
0.9040	0.7680	55.20	-0.0144	-0.0025	-0.0086	0.0493	0.0009
0.9370	0.8120	56.30	-0.0167	0.0016	-0.0064	0.0511	-0.0037
0.9700	0.8750	57.90	-0.0062	0.0133	0.0059	0.0488	0.0001
		MEAN	0.0053	0.0064	0.0053	0.0255	0.0013

CHLOROFORM(1)-METHYL ALCOHOL(2)

TYRER D.:J.CHEM.SOC.101,1104(1912).

ANTOINE VAPOR PRESSURE CONSTANTS
```
          A          B          C
(1) 6.90328    1163.000    227.000
(2) 8.07246    1574.990    238.860
```

P =   757.00

EQUATION	ORDER	A12	A21	D	A122	A211
VAN LAAR	3	0.4516	0.9178	-	-	-
MARGULES	3	0.4059	0.7763	-	-	-
MARGULES	4	0.4624	0.8661	0.2366	-	-
ALPHA	2	2.8949	1.4303	-	-	-
ALPHA	3	-3.5739	-2.2023	-	5.5164	2.0936

			DEVIATION IN VAPOR PHASE COMPOSITION				
X	Y	TEMP.	LAAR 3	MARG 3	MARG 4	ALPHA2	ALPHA3
0.0289	0.0829	64.05	-0.0017	-0.0076	-0.0004	0.0153	-0.0032
0.0629	0.1610	62.38	0.0020	-0.0058	0.0031	0.0251	0.0005
0.1032	0.2403	60.68	0.0038	-0.0026	0.0041	0.0247	0.0038
0.1517	0.3230	59.07	0.0006	-0.0024	-0.0000	0.0138	0.0037
0.2116	0.4119	57.52	-0.0107	-0.0101	-0.0119	-0.0085	-0.0040
0.2870	0.4876	55.94	-0.0121	-0.0096	-0.0132	-0.0211	-0.0042
0.3850	0.5417	54.52	0.0029	0.0034	0.0023	-0.0129	0.0043
0.5177	0.5885	53.66	0.0168	0.0100	0.0154	0.0079	-0.0006
0.7072	0.6781	53.65	-0.0234	-0.0358	-0.0283	0.0112	-0.0003
		MEAN	0.0082	0.0097	0.0088	0.0156	0.0027

DICHLOROMETHANE(1)-CARBON TETRACHLORIDE(2)

KAPLAN S.I.,MONAKHOVA Z.D.:ZH.OBSHCH.KHIM.7,2499(1937).

ANTOINE VAPOR PRESSURE CONSTANTS

	A	B	C
(1)	7.07138	1134.600	231.000
(2)	6.93390	1242.430	230.000

P = 760.00

EQUATION	ORDER	A12	A21	D	A122	A211
VAN LAAR	3	0.0008	-0.0017	-	-	-
MARGULES	3	-0.0502	0.1169	-	-	-
MARGULES	4	-0.2498	-0.1342	-0.6925	-	-
ALPHA	2	2.6203	-0.6780	-	-	-
ALPHA	3	-.6014/ 05	-.1410/ 05	-	.6274/ 05	.1264/ 0

			DEVIATION IN VAPOR PHASE COMPOSITION				
X	Y	TEMP.	LAAR 3	MARG 3	MARG 4	ALPHA2	ALPHA3
---	---	---	---	---	---	---	---
0.1000	0.2200	69.60	0.0449	0.0392	0.0100	0.0659	0.0028
0.2000	0.4600	61.40	-0.0060	0.0029	0.0022	0.0124	0.0024
0.3000	0.6200	55.80	-0.0278	-0.0092	0.0014	-0.0161	-0.0085
0.4000	0.7150	52.80	-0.0195	0.0004	0.0051	-0.0133	-0.0017
0.5000	0.7850	50.80	-0.0090	0.0058	0.0009	-0.0075	0.0022
0.6000	0.8420	49.10	0.0011	0.0054	-0.0046	-0.0038	0.0013
0.7000	0.8800	47.85	0.0932	0.0119	0.0044	0.0080	0.0073
0.8000	0.9200	46.60	0.0144	0.0096	0.0101	0.0102	0.0027
0.9000	0.9670	44.60	0.0025	-0.0025	0.0041	-0.0002	-0.0151
		MEAN	0.0243	0.0097	0.0048	0.0153	0.0049

DICHLOROMETHANE(1)-CHLOROFORM(2)

KAPLAN S.I.,MONAKHOVA Z.D.:ZH.OBSHCH.KHIM.,7,2499(1937)

ANTOINE VAPOR PRESSURE CONSTANTS

	A	B	C
(1)	7.07138	1134.600	231.000
(2)	6.90328	1163.000	227.000

P = 760.00

EQUATION	ORDER	A12	A21	D	A122	A211
VAN LAAR	3	0.0003	-0.0001	-	-	-
MARGULES	3	-0.1211	0.0345	-	-	-
MARGULES	4	0.0261	0.2091	0.4825	-	-
ALPHA	2	1.0012	-0.7263	-	-	-
ALPHA	3	-1.0259	-1.5646	-	2.1179	0.8674

			DEVIATION IN VAPOR PHASE COMPOSITION				
X	Y	TEMP.	LAAR 3	MARG 3	MARG 4	ALPHA2	ALPHA3
0.1000	0.1970	58.45	-0.0074	-0.0313	-0.0138	-0.0115	-0.0076
0.2000	0.3450	55.80	0.0037	-0.0145	-0.0128	0.0000	0.0021
0.3000	0.4770	53.40	0.0027	0.0007	-0.0068	0.0054	0.0046
0.4000	0.6000	51.10	-0.0130	0.0008	-0.0037	0.0007	-0.0015
0.5000	0.7020	48.80	-0.0206	-0.0016	0.0017	0.0001	-0.0015
0.6000	0.7900	47.70	-0.0273	-0.0100	-0.0021	-0.0017	-0.0015
0.7000	0.8600	44.75	-0.0259	-0.0143	-0.0084	0.0006	0.0020
0.8000	0.9200	42.95	-0.0237	-0.0189	-0.0198	-0.0003	0.0006
0.9000	0.9670	41.10	-0.0157	-0.0159	-0.0234	-0.0008	-0.0014
		MEAN	0.0155	0.0120	0.0103	0.0023	0.0025

NITROMETHANE(1)-BENZENE(2)

BROWN I.,SMITH F.:AUSTR.J.CHEM.8,501(1955).

ANTOINE VAPOR PRESSURE CONSTANTS
	A	B	C
(1)	7.15540	1366.483	218.611
(2)	6.90565	1211.033	220.790

T =   45.00

EQUATION	ORDER	A12	A21	D	A122	A211
VAN LAAR	3	0.4462	0.5179	-	-	-
MARGULES	3	0.4442	0.5137	-	-	-
MARGULES	4	0.4849	0.5515	0.1312	-	-
ALPHA	2	1.0402	5.2758	-	-	-
ALPHA	3	2.4714	4.4307	-	-2.0649	2.8894

			DEVIATION IN VAPOR PHASE COMPOSITION				
X	Y	PRESS.	LAAR 3	MARG 3	MARG 4	ALPHA2	ALPHA3
0.0445	0.0548	226.43	-0.0068	-0.0070	-0.0041	0.0151	-0.0007
0.0966	0.1026	227.83	-0.0074	-0.0075	-0.0048	0.0182	-0.0005
0.1979	0.1694	226.92	-0.0024	-0.0023	-0.0028	0.0119	0.0008
0.2927	0.2173	223.50	0.0010	0.0011	-0.0015	0.0029	0.0008
0.3921	0.2602	218.27	0.0018	0.0018	-0.0007	-0.0047	0.0005
0.4737	0.2942	213.14	-0.0003	-0.0004	-0.0014	-0.0095	-0.0008
0.5259	0.3166	208.62	-0.0027	-0.0030	-0.0027	-0.0116	-0.0021
0.6180	0.3543	200.11	-0.0030	-0.0035	-0.0010	-0.0076	0.0001
0.7118	0.4023	188.14	-0.0039	-0.0043	-0.0011	0.0007	0.0015
0.8102	0.4765	170.13	-0.0036	-0.0036	-0.0032	0.0157	0.0021
0.9051	0.6144	142.18	-0.0037	-0.0029	-0.0100	0.0303	-0.0032
0.9601	0.7708	117.19	0.0027	0.0039	-0.0067	0.0343	-0.0018
		MEAN	0.0033	0.0034	0.0033	0.0135	0.0012

NITROMETHANE(1)-CARBON TETRACHLORIDE(2)

BROWN I.,SMITH F.:AUSTR.J.CHEM.8,501(1955).

ANTOINE VAPOR PRESSURE CONSTANTS

	A	B	C
(1)	7.15540	1366.483	218.611
(2)	6.93390	1242.430	230.000

T =   45.00

EQUATION	ORDER	A12	A21	D	A122	A211
VAN LAAR	3	0.9093	0.8108	-	-	-
MARGULES	3	0.9026	0.8113	-	-	-
MARGULES	4	1.0211	0.9223	0.4091	-	-
ALPHA	2	7.0145	24.3760	-	-	-
ALPHA	3	4.8114	10.8501	-	-1.0543	11.0182

			DEVIATION IN VAPOR PHASE COMPOSITION				
X	Y	PRESS.	LAAR 3	MARG 3	MARG 4	ALPHA2	ALPHA3
0.0459	0.1296	287.42	-0.0278	-0.0287	-0.0121	0.0191	-0.0019
0.0918	0.1780	297.19	-0.0162	-0.0171	-0.0043	0.0091	0.0008
0.1954	0.2217	302.91	0.0044	0.0044	0.0011	-0.0030	0.0015
0.2829	0.2368	302.56	0.0088	0.0094	-0.0004	-0.0053	0.0006
0.3656	0.2458	301.42	0.0059	0.0068	-0.0025	-0.0052	-0.0007
0.4659	0.2532	298.86	0.0001	0.0009	-0.0025	-0.0022	-0.0008
0.5366	0.2598	296.88	-0.0055	-0.0049	-0.0028	-0.0008	-0.0017
0.6065	0.2657	293.27	-0.0081	-0.0076	-0.0007	0.0029	-0.0001
0.6835	0.2773	287.10	-0.0102	-0.0100	-0.0001	0.0052	0.0006
0.8043	0.3138	264.65	-0.0046	-0.0048	-0.0004	0.0074	0.0030
0.9039	0.4078	214.63	0.0120	0.0116	-0.0071	-0.0018	0.0004
0.9488	0.5283	170.95	0.0219	0.0215	-0.0156	-0.0176	-0.0076
		MEAN	0.0105	0.0106	0.0041	0.0066	0.0016

METHYL ALCOHOL(1)-BENZENE(2)

FRITZWEILER R.,DIETRICH K.R.:ANGEW.CHEM.A.CHEM.FABRIK.,
NO.4,BERLIN,W 35(1933)

ANTOINE VAPOR PRESSURE CONSTANTS

	A	B	C
(1)	8.07246	1574.990	238.860
(2)	6.90565	1211.033	220.790

P = 760.00

EQUATION	ORDER	A12	A21	D	A122	A211
VAN LAAR	3	0.9964	0.8526	-	-	-
MARGULES	3	0.9868	0.8528	-	-	-
MARGULES	4	1.0718	0.9310	0.3709	-	-
ALPHA	2	18.8376	10.9780	-	-	-
ALPHA	3	2.1418	13.4177	-	22.6171	-10.3244

DEVIATION IN VAPOR PHASE COMPOSITION

X	Y	TEMP.	LAAR 3	MARG 3	MARG 4	ALPHA2	ALPHA3
0.0280	0.3100	69.40	-0.0149	-0.0183	0.0114	-0.0115	0.0299
0.0500	0.3950	66.80	0.0071	0.0040	0.0286	-0.0040	0.0336
0.0570	0.4200	65.70	0.0054	0.0025	0.0247	-0.0091	0.0268
0.0900	0.4850	61.40	0.0148	0.0130	0.0238	-0.0106	0.0173
0.1180	0.5650	59.00	-0.0285	-0.0295	-0.0268	-0.0584	-0.0360
0.1200	0.5600	59.20	-0.0210	-0.0219	-0.0198	-0.0516	-0.0295
0.2700	0.5750	58.00	0.0279	0.0294	0.0126	0.0042	0.0078
0.4400	0.5850	57.80	0.0146	0.0162	0.0086	0.0238	0.0131
0.5860	0.6100	57.70	-0.0158	-0.0152	-0.0067	0.0162	-0.0046
0.6950	0.6250	57.60	-0.0197	-0.0201	-0.0045	0.0154	-0.0091
0.8170	0.6550	58.10	0.0011	0.0003	0.0101	0.0107	-0.0007
0.8830	0.7000	58.90	0.0188	0.0180	0.0169	-0.0066	0.0105
0.9020	0.7300	59.60	0.0153	0.0146	0.0097	-0.0239	0.0064
0.9340	0.8010	60.40	0.0007	0.0002	-0.0107	-0.0627	-0.0060
0.9450	0.8220	61.20	0.0039	0.0035	-0.0087	-0.0674	-0.0015
0.9680	0.9000	62.40	-0.0140	-0.0142	-0.0269	-0.0934	-0.0155
0.9880	0.9420	63.40	0.0101	0.0100	0.0023	-0.0470	0.0110
		MEAN	0.0137	0.0136	0.0149	0.0304	0.0153

METHYL ALCOHOL(1)-BENZENE(2)

LEE S.C.: J.PHYS.CHEM. 35,3558(1931).

ANTOINE VAPOR PRESSURE CONSTANTS

	A	B	C
(1)	8.07246	1574.990	238.860
(2)	6.90565	1211.033	220.790

T = 40.00

EQUATION	ORDER	A12	A21	D	A122	A211
VAN LAAR	3	1.1147	0.7699	-	-	-
MARGULES	3	1.0539	0.7523	-	-	-
MARGULES	4	1.2349	0.8870	0.5163	-	-
ALPHA	2	13.6178	10.3338	-	-	-
ALPHA	3	2.1616	41.9390	-	48.6761	-40.5163

			DEVIATION IN VAPOR PHASE COMPOSITION				
X	Y	PRESS.	LAAR 3	MARG 3	MARG 4	ALPHA2	ALPHA3
0.1410	0.5070	349.00	0.0373	0.0317	0.0488	-0.0480	0.0033
0.2270	0.5270	356.60	0.0364	0.0391	0.0312	-0.0241	-0.0019
0.3040	0.5310	362.50	0.0295	0.0355	0.0184	-0.0060	0.0014
0.4020	0.5400	364.20	0.0111	0.0175	0.0016	0.0039	-0.0007
0.4680	0.5430	365.60	0.0034	0.0087	-0.0014	0.0111	0.0008
0.5520	0.5480	366.00	-0.0016	0.0015	0.0009	0.0182	0.0026
0.6430	0.5660	366.20	-0.0078	-0.0070	0.0009	0.0140	-0.0044
0.7020	0.5800	362.50	-0.0048	-0.0050	0.0049	0.0108	-0.0070
0.7500	0.5780	357.50	0.0186	0.0182	0.0265	0.0236	0.0092
0.8340	0.6410	345.20	0.0173	0.0179	0.0148	-0.0109	-0.0059
0.8780	0.6700	334.00	0.0393	0.0409	0.0278	-0.0147	0.0147
0.8960	0.7230	325.20	0.0125	0.0147	-0.0029	-0.0531	-0.0092
0.9150	0.7530	322.50	0.0145	0.0173	-0.0046	-0.0634	-0.0011
		MEAN	0.0180	0.0196	0.0142	0.0232	0.0048

69

METHYL ALCOHOL(1)-BENZENE(2)

SCATCHARD G.,WOOD S.E.,MOCHEL J.M.:J.AM.CHEM.SOC.68,1960(1946).

ANTOINE VAPOR PRESSURE CONSTANTS
	A	B	C
(1)	8.07246	1574.990	238.860
(2)	6.90565	1211.033	220.790

T =   35.00

EQUATION	ORDER	A12	A21	D	A122	A211
VAN LAAR	3	0.9744	0.7302	-	-	-
MARGULES	3	0.9463	0.7218	-	-	-
MARGULES	4	1.0952	0.8612	0.5157	-	-
ALPHA	2	13.1416	9.5751	-	-	-
ALPHA	3	5.2724	20.4934	-	20.2503	-17.1916

			DEVIATION IN VAPOR PHASE COMPOSITION				
X	Y	PRESS.	LAAR 3	MARG 3	MARG 4	ALPHA2	ALPHA3
0.0242	0.2733	203.29	-0.0507	-0.0594	-0.0102	-0.0556	0.0246
0.0254	0.3128	211.10	-0.0827	-0.0916	-0.0419	-0.0883	-0.0082
0.1302	0.4858	274.25	0.0039	0.0017	0.0142	-0.0328	-0.0048
0.3107	0.5304	288.47	0.0161	0.0200	0.0027	0.0024	-0.0008
0.4989	0.5546	292.50	-0.0028	-0.0007	-0.0022	0.0120	0.0012
0.5191	0.5571	292.70	-0.0040	-0.0022	-0.0011	0.0125	0.0017
0.6305	0.5790	292.49	-0.0101	-0.0104	0.0027	0.0078	0.0000
0.7965	0.6421	283.58	-0.0001	-0.0013	0.0059	-0.0170	-0.0023
0.9197	0.7688	255.82	0.0189	0.0197	-0.0019	-0.0629	0.0015
		MEAN	0.0210	0.0230	0.0092	0.0324	0.0050

70

METHYL ALCOHOL(1)-BENZENE(2)

SCATCHARD G.,WOOD S.E.,MOCHEL J.M.:J.AM.CHEM.SOC.68,1960(1946).

ANTOINE VAPOR PRESSURE CONSTANTS
	A	B	C
(1)	8.07246	1574.990	238.860
(2)	6.90565	1211.033	220.790

T =   55.00

EQUATION	ORDER	A12	A21	D	A122	A211
VAN LAAR	3	0.9532	0.7163	-	-	-
MARGULES	3	0.9298	0.7083	-	-	-
MARGULES	4	1.0566	0.8528	0.5264	-	-
ALPHA	2	13.5071	8.4506	-	-	-
ALPHA	3	6.4434	17.5444	-	17.7464	-14.6006

			DEVIATION IN VAPOR PHASE COMPOSITION				
X	Y	PRESS.	LAAR 3	MARG 3	MARG 4	ALPHA2	ALPHA3
0.0304	0.3019	465.84	-0.0302	-0.0378	0.0049	-0.0417	0.0286
0.0493	0.4051	527.12	-0.0482	-0.0552	-0.0169	-0.0688	-0.0087
0.1031	0.4841	597.48	-0.0052	-0.0081	0.0071	-0.0379	-0.0071
0.3297	0.5540	664.24	0.0182	0.0220	0.0049	0.0124	0.0031
0.4874	0.5845	675.62	-0.0053	-0.0036	-0.0029	0.0109	-0.0009
0.4984	0.5858	675.99	-0.0059	-0.0044	-0.0021	0.0114	-0.0003
0.6076	0.6078	678.44	-0.0138	-0.0144	0.0014	0.0061	-0.0005
0.7896	0.6716	664.91	-0.0045	-0.0062	0.0054	-0.0189	0.0006
0.9014	0.7697	622.29	0.0117	0.0118	-0.0035	-0.0576	-0.0002
		MEAN	0.0159	0.0181	0.0055	0.0295	0.0056

METHYL ALCOHOL(1)-BENZENE(2)

SHIRAI H.,NAKANISHI K.:KAGAKU KOGAKU 29,180(1965).

ANTOINE VAPOR PRESSURE CONSTANTS

	A	B	C
(1)	8.07246	1574.990	238.860
(2)	6.90565	1211.033	220.790

P = 730.00

EQUATION	ORDER	A12	A21	D	A122	A211
VAN LAAR	3	0.9068	0.7157	-	-	-
MARGULES	3	0.8887	0.7134	-	-	-
MARGULES	4	0.9743	0.8075	0.3533	-	-
ALPHA	2	12.6256	7.2203	-	-	-
ALPHA	3	3.8528	12.2891	-	15.5765	-9.9709

			DEVIATION IN VAPOR PHASE COMPOSITION				
X	Y	TEMP.	LAAR 3	MARG 3	MARG 4	ALPHA2	ALPHA3
0.0200	0.2090	71.33	-0.0060	-0.0115	0.0150	-0.0164	0.0359
0.0200	0.3200	67.34	-0.1203	-0.1257	-0.0995	-0.1274	-0.0751
0.0400	0.3240	65.87	-0.0107	-0.0166	0.0107	-0.0262	0.0258
0.0630	0.4410	60.97	-0.0513	-0.0562	-0.0345	-0.0688	-0.0262
0.0670	0.4085	63.37	-0.0049	-0.0095	0.0110	-0.0263	0.0145
0.0875	0.4555	60.86	-0.0094	-0.0128	0.0017	-0.0316	0.0007
0.1240	0.4945	59.72	0.0029	0.0014	0.0063	-0.0206	-0.0008
0.1900	0.5315	58.55	0.0133	0.0145	0.0076	-0.0054	-0.0008
0.2990	0.5550	56.98	0.0175	0.0205	0.0079	0.0158	0.0065
0.4554	0.5870	56.51	-0.0024	0.0000	-0.0031	0.0187	0.0007
0.5290	0.6070	56.39	-0.0162	-0.0148	-0.0108	0.0112	-0.0078
0.6654	0.6265	56.65	-0.0095	-0.0100	0.0029	0.0151	0.0009
0.7328	0.6490	56.51	-0.0070	-0.0081	0.0042	0.0070	0.0004
0.7895	0.6735	57.27	0.0011	-0.0002	0.0078	-0.0016	0.0029
0.8395	0.7110	57.34	0.0031	0.0020	0.0033	-0.0195	0.0002
0.9155	0.8015	57.50	0.0057	0.0053	-0.0063	-0.0549	-0.0014
0.9275	0.8135	59.68	0.0100	0.0098	-0.0033	-0.0575	0.0016
0.9665	0.9065	61.22	0.0017	0.0018	-0.0116	-0.0692	-0.0037
		MEAN	0.0163	0.0178	0.0137	0.0330	0.0114

METHYL ALCOHOL(1)-BENZENE(2)

SODAY F.,BENNETT G.W.:J.CHEM.EDUC. 7,1336(1930).

ANTOINE VAPOR PRESSURE CONSTANTS

	A	B	C
(1)	8.07246	1574.990	238.860
(2)	6.90565	1211.033	220.790

P = 725.00

EQUATION	ORDER	A12	A21	D	A122	A211
VAN LAAR	3	1.1401	0.7356	-	-	-
MARGULES	3	1.0959	0.7102	-	-	-
MARGULES	4	1.1737	0.8194	0.4295	-	-
ALPHA	2	23.4102	12.3435	-	-	-
ALPHA	3	0.4601	16.1639	-	29.6263	-14.6429

			DEVIATION IN VAPOR PHASE COMPOSITION				
X	Y	TEMP.	LAAR 3	MARG 3	MARG 4	ALPHA2	ALPHA3
0.0240	0.1750	68.20	0.1466	0.1310	0.1592	0.1365	0.1709
0.0360	0.3010	63.90	0.0899	0.0758	0.1014	0.0776	0.1081
0.0470	0.4350	60.30	-0.0016	-0.0134	0.0083	-0.0139	0.0123
0.0470	0.4870	58.80	-0.0555	-0.0673	-0.0457	-0.0659	-0.0397
0.0590	0.5110	57.70	-0.0437	-0.0530	-0.0358	-0.0558	-0.0341
0.0630	0.5110	57.50	-0.0339	-0.0423	-0.0267	-0.0465	-0.0261
0.0630	0.5340	57.25	-0.0572	-0.0657	-0.0500	-0.0695	-0.0491
0.0920	0.5460	56.75	-0.0195	-0.0223	-0.0166	-0.0324	-0.0204
0.2490	0.5990	56.40	-0.0220	-0.0124	-0.0303	0.0030	-0.0065
0.6190	0.5990	56.60	-0.0258	-0.0316	-0.0134	0.0520	0.0235
0.7850	0.6660	56.90	-0.0209	-0.0270	-0.0115	0.0073	-0.0071
0.8470	0.7130	57.60	-0.0121	-0.0151	-0.0108	-0.0241	-0.0130
0.9020	0.7710	58.30	0.0021	0.0029	-0.0059	-0.0568	-0.0015
0.9410	0.8440	59.55	0.0001	0.0031	-0.0127	-0.0933	0.0035
0.9830	0.9360	61.90	0.0115	0.0138	0.0029	-0.0758	0.0195
		MEAN	0.0362	0.0385	0.0354	0.0540	0.0357

F

METHYL ALCOHOL(1)-BUTYL ALCOHOL(2)

HILL W.D.,VAN WINKLE M.:IND.ENG.CHEM.44,205(1952).

ANTOINE VAPOR PRESSURE CONSTANTS
	A	B	C
(1)	8.07246	1574.990	238.860
(2)	7.54472	1405.873	183.908

P = 760.00

EQUATION	ORDER	A12	A21	D	A122	A211
VAN LAAR	3	0.2101	1.9354	-	-	-
MARGULES	3	0.1917	0.5208	-	-	-
MARGULES	4	0.3968	0.8552	0.9179	-	-
ALPHA	2	10.8791	-0.7748	-	-	-
ALPHA	3	24.3197	0.8545	-	-11.9997	-1.7018

DEVIATION IN VAPOR PHASE COMPOSITION

X	Y	TEMP.	LAAR 3	MARG 3	MARG 4	ALPHA2	ALPHA3
0.1080	0.5960	102.50	-0.0331	-0.0299	-0.0044	-0.0102	0.0014
0.2050	0.7500	95.30	-0.0081	-0.0041	-0.0124	-0.0027	-0.0013
0.3660	0.8600	89.00	0.0075	0.0023	-0.0023	0.0042	0.0005
0.4980	0.9090	84.20	0.0084	-0.0045	0.0016	0.0036	0.0007
0.6070	0.9380	81.20	0.0043	-0.0126	-0.0031	0.0010	-0.0002
0.6980	0.9570	81.00	-0.0016	-0.0194	-0.0113	-0.0013	-0.0008
0.7760	0.9700	79.40	-0.0063	-0.0212	-0.0179	-0.0024	-0.0006
0.8440	0.9790	74.80	-0.0102	-0.0178	-0.0207	-0.0022	0.0005
0.9020	0.9870	71.40	-0.0183	-0.0139	-0.0219	-0.0026	0.0003
		MEAN	0.0109	0.0140	0.0106	0.0033	0.0007

METHYL ALCOHOL(1)-CARBON TETRACHLORIDE(2)

HIPKIN H.,MYERS H.S.:IND.ENG.CHEM.46,2524(1954).

ANTOINE VAPOR PRESSURE CONSTANTS
```
           A          B          C
(1) 8.07246    1574.990    238.860
(2) 6.93390    1242.430    230.000
```

P =   760.00

EQUATION	ORDER	A12	A21	D	A122	A211
VAN LAAR	3	1.0302	0.7875	-	-	-
MARGULES	3	1.0020	0.7824	-	-	-
MARGULES	4	1.1526	0.8918	0.4858	-	-
ALPHA	2	12.6804	9.5793	-	-	-
ALPHA	3	5.6054	36.3904	-	37.1621	-33.3755

DEVIATION IN VAPOR PHASE COMPOSITION

X	Y	TEMP.	LAAR 3	MARG 3	MARG 4	ALPHA2	ALPHA3
0.0020	0.0200	76.10	0.0123	0.0104	0.0222	0.0061	0.0553
0.0020	0.0270	75.85	0.0053	0.0033	0.0151	-0.0009	0.0483
0.0040	0.1200	72.35	-0.0593	-0.0628	-0.0419	-0.0699	0.0122
0.0130	0.2415	67.60	-0.0786	-0.0863	-0.0416	-0.1048	0.0357
0.0170	0.2640	66.85	-0.0652	-0.0739	-0.0242	-0.0972	0.0492
0.0300	0.3830	62.00	-0.0986	-0.1084	-0.0538	-0.1409	0.0001
0.0505	0.4450	59.40	-0.0722	-0.0812	-0.0323	-0.1265	-0.0101
0.1070	0.4900	57.20	-0.0022	-0.0065	0.0154	-0.0683	-0.0037
0.1240	0.5000	56.95	0.0053	0.0024	0.0171	-0.0607	-0.0064
0.2480	0.5220	56.25	0.0305	0.0335	0.0174	-0.0148	-0.0016
0.4010	0.5365	55.80	0.0137	0.0177	0.0009	0.0066	0.0002
0.4525	0.5410	55.75	0.0067	0.0101	-0.0016	0.0107	0.0007
0.4980	0.5450	55.75	0.0015	0.0042	-0.0020	0.0137	0.0016
0.5050	0.5485	55.70	-0.0022	0.0004	-0.0048	0.0113	-0.0011
0.5500	0.5520	55.65	-0.0050	-0.0033	-0.0027	0.0146	0.0009
0.5655	0.5520	55.70	-0.0042	-0.0027	-0.0002	0.0170	0.0030
0.5965	0.5580	55.70	-0.0077	-0.0069	-0.0007	0.0159	0.0017
0.0030	0.5610	55.70	-0.5191	-0.5216	-0.5066	-0.5226	-0.4554
0.6250	0.5630	55.75	-0.0091	-0.0087	0.0004	0.0156	0.0015
0.6760	0.5760	55.75	-0.0118	-0.0122	0.0005	0.0118	-0.0007
0.7250	0.5910	56.00	-0.0105	-0.0116	0.0022	0.0072	-0.0019
0.7270	0.5950	56.00	-0.0137	-0.0148	-0.0010	0.0037	-0.0052
0.7640	0.6030	56.35	-0.0038	-0.0051	0.0071	0.0054	0.0011
0.8130	0.6300	56.75	0.0021	0.0009	0.0078	-0.0048	0.0006
0.8380	0.6490	57.10	0.0055	0.0044	0.0071	-0.0126	0.0003
0.8680	0.6770	57.70	0.0109	0.0102	0.0066	-0.0233	0.0014
0.8830	0.6955	58.20	0.0126	0.0121	0.0049	-0.0309	0.0012
0.8970	0.7160	58.60	0.0131	0.0128	0.0023	-0.0392	0.0007
0.9175	0.7530	59.50	0.0119	0.0119	-0.0032	-0.0538	-0.0010
0.9380	0.8030	60.40	0.0044	0.0048	-0.0140	-0.0728	-0.0067
0.9480	0.8230	60.85	0.0082	0.0087	-0.0112	-0.0729	-0.0012
0.9620	0.8640	61.80	0.0046	0.0052	-0.0146	-0.0780	-0.0017
0.9790	0.9100	62.80	0.0111	0.0117	-0.0041	-0.0595	0.0094
0.9860	0.9390	63.50	0.0066	0.0071	-0.0052	-0.0509	0.0062
0.9930	0.9665	64.10	0.0053	0.0056	-0.0015	-0.0302	0.0057
0.9970	0.9880	64.50	-0.0003	-0.0002	-0.0035	-0.0177	-0.0000
0.9990	0.9950	64.60	0.0008	0.0009	-0.0003	-0.0053	0.0010
		MEAN	0.0307	0.0320	0.0243	0.0513	0.0199

METHYL ALCOHOL(1)-CARBON TETRACHLORIDE(2)

SCATCHARD G.,TICKNOR L.B.:J.AM.CHEM.SOC.74,3724(1952).

ANTOINE VAPOR PRESSURE CONSTANTS
	A	B	C
(1)	8.07246	1574.990	238.860
(2)	6.93390	1242.430	230.000

T =   55.00

EQUATION	ORDER	A12	A21	D	A122	A211
VAN LAAR	3	1.0952	0.7877	-	-	-
MARGULES	3	1.0550	0.7915	-	-	-
MARGULES	4	1.2973	1.0408	1.1316	-	-
ALPHA	2	23.7267	17.6897	-	-	-
ALPHA	3	7.6977	36.4749	-	34.9952	-32.0850

			DEVIATION IN VAPOR PHASE COMPOSITION				
X	Y	PRESS.	LAAR 3	MARG 3	MARG 4	ALPHA2	ALPHA3
0.0234	0.3692	568.90	-0.1105	-0.1241	-0.0387	-0.0786	-0.0157
0.0343	0.3926	592.60	-0.0681	-0.0818	0.0000	-0.0468	0.0046
0.0525	0.4326	644.10	-0.0337	-0.0454	0.0200	-0.0298	0.0051
0.1734	0.5084	721.60	0.0354	0.0384	0.0087	0.0069	-0.0006
0.5450	0.5535	746.30	-0.0171	-0.0137	0.0035	0.0169	0.0000
0.8699	0.6724	687.30	0.0098	0.0053	0.0086	-0.0473	-0.0000
		MEAN	0.0458	0.0514	0.0132	0.0377	0.0044

METHYL ALCOHOL(1)-CARBON TETRACHLORIDE(2)

SCATCHARD G.,WOOD S.E.,MOCHEL J.M.:J.AM.CHEM.SOC.68,1960(1946).

ANTOINE VAPOR PRESSURE CONSTANTS
```
        A          B          C
(1) 8.07246    1574.990    238.860
(2) 6.93390    1242.430    230.000
```

T =   35.00

EQUATION	ORDER	A12	A21	D	A122	A211
VAN LAAR	3	1.0733	0.7620	-	-	-
MARGULES	3	1.0304	0.7465	-	-	-
MARGULES	4	1.2187	0.9347	0.6502	-	-
ALPHA	2	15.2691	14.1510	-	-	-
ALPHA	3	8.6735	59.6431	-	48.2036	-55.3299

			DEVIATION IN VAPOR PHASE COMPOSITION				
X	Y	PRESS.	LAAR 3	MARG 3	MARG 4	ALPHA2	ALPHA3
0.0169	0.3297	259.13	-0.1523	-0.1642	-0.1065	-0.1480	-0.0035
0.0189	0.3374	262.31	-0.1448	-0.1573	-0.0973	-0.1420	0.0008
0.1349	0.4630	315.12	0.0172	0.0141	0.0294	-0.0306	0.0007
0.3560	0.4915	324.64	0.0152	0.0206	0.0024	0.0065	-0.0003
0.4776	0.5030	325.71	-0.0027	0.0006	-0.0030	0.0110	0.0002
0.4939	0.5056	325.71	-0.0054	-0.0026	-0.0036	0.0104	-0.0005
0.6557	0.5302	323.81	-0.0127	-0.0137	0.0051	0.0067	0.0008
0.7912	0.5792	312.61	0.0011	-0.0003	0.0084	-0.0138	-0.0005
0.9120	0.7024	277.37	0.0260	0.0280	-0.0044	-0.0665	0.0002
		MEAN	0.0419	0.0446	0.0289	0.0484	0.0008

METHYL ALCOHOL(1)-CARBON TETRACHLORIDE(2)

SCATCHARD G.,WOOD S.E.,MOCHEL J.M.:J.AM.CHEM.SOC.68,1960(1946).

ANTOINE VAPOR PRESSURE CONSTANTS
	A	B	C
(1)	8.07246	1574.990	238.860
(2)	6.93390	1242.430	230.000

T = 55.00

EQUATION	ORDER	A12	A21	D	A122	A211
VAN LAAR	3	1.0147	0.7500	-	-	-
MARGULES	3	0.9867	0.7371	-	-	-
MARGULES	4	1.0964	0.8521	0.4132	-	-
ALPHA	2	14.6800	11.3592	-	-	-
ALPHA	3	4.5480	20.6092	-	20.7948	-16.9666

			DEVIATION IN VAPOR PHASE COMPOSITION				
X	Y	PRESS.	LAAR 3	MARG 3	MARG 4	ALPHA2	ALPHA3
0.0254	0.3619	580.66	-0.1215	-0.1305	-0.0939	-0.1255	-0.0593
0.0579	0.3639	591.16	0.0161	0.0083	0.0398	-0.0091	0.0419
0.1493	0.4981	716.95	0.0121	0.0116	0.0142	-0.0306	-0.0122
0.3647	0.5284	741.36	0.0149	0.0185	0.0069	0.0071	0.0013
0.4893	0.5431	745.60	-0.0011	0.0004	0.0001	0.0107	0.0010
0.4946	0.5438	745.72	-0.0017	-0.0003	-0.0001	0.0108	0.0009
0.6448	0.5686	744.54	-0.0093	-0.0107	0.0024	0.0068	-0.0006
0.7903	0.6187	724.28	0.0040	0.0025	0.0102	-0.0123	-0.0010
0.9087	0.7337	658.37	0.0223	0.0237	0.0076	-0.0608	0.0007
		MEAN	0.0226	0.0229	0.0195	0.0304	0.0132

METHYL ALCOHOL(1)-1,2-DICHLOROETHANE(2)

FORDYCE C.T.,SIMONSON D.R.:IND.ENG.CHEM.41,104(1949).

ANTOINE VAPOR PRESSURE CONSTANTS

	A	B	C
(1)	8.07246	1574.990	238.860
(2)	7.18431	1358.500	232.000

P = 760.00

EQUATION	ORDER	A12	A21	D	A122	A211
VAN LAAR	3	0.7731	0.7547	-	-	-
MARGULES	3	0.7721	0.7554	-	-	-
MARGULES	4	0.9113	0.8831	0.5012	-	-
ALPHA	2	12.0354	4.9440	-	-	-
ALPHA	3	7.9968	11.8466	-	14.5570	-9.4774

			DEVIATION IN VAPOR PHASE COMPOSITION				
X	Y	TEMP.	LAAR 3	MARG 3	MARG 4	ALPHA2	ALPHA3
0.0180	0.2560	73.40	-0.0895	-0.0898	-0.0492	-0.0785	0.0015
0.0650	0.4740	65.20	-0.0916	-0.0919	-0.0551	-0.0820	-0.0139
0.1950	0.5730	61.40	-0.0023	-0.0023	-0.0104	-0.0044	0.0078
0.3400	0.6220	60.60	0.0045	0.0047	-0.0123	0.0101	0.0011
0.5720	0.6560	60.20	0.0011	0.0013	0.0083	0.0263	0.0136
0.5820	0.6930	60.10	-0.0346	-0.0344	-0.0264	-0.0089	-0.0212
0.8200	0.7470	60.40	-0.0125	-0.0125	-0.0081	-0.0065	0.0064
0.9030	0.8140	61.20	-0.0038	-0.0040	-0.0163	-0.0271	0.0037
0.9560	0.9020	62.30	-0.0090	-0.0091	-0.0277	-0.0490	-0.0109
		MEAN	0.0277	0.0278	0.0238	0.0325	0.0089

METHYL ALCOHOL(1)-1,2-DICHLOROETHANE(2)

UDOVENKO V.V.,FRID TS.B.:ZH.FIZ.KHIM.22,1263(1948).

ANTOINE VAPOR PRESSURE CONSTANTS

	A	B	C
(1)	8.07246	1574.990	238.860
(2)	7.18431	1358.500	232.000

T =   40.00

EQUATION	ORDER	A12	A21	D	A122	A211
VAN LAAR	3	0.9191	0.7107	-	-	-
MARGULES	3	0.9037	0.6926	-	-	-
MARGULES	4	0.8563	0.6454	-0.1521	-	-
ALPHA	2	13.5323	7.4290	-	-	-
ALPHA	3	-1.4675	7.9610	-	18.5292	-7.4353

			DEVIATION IN VAPOR PHASE COMPOSITION				
X	Y	PRESS.	LAAR 3	MARG 3	MARG 4	ALPHA2	ALPHA3
0.1000	0.4790	265.40	0.0048	0.0023	-0.0069	-0.0224	0.0076
0.2000	0.5690	303.50	-0.0031	-0.0028	-0.0023	-0.0258	-0.0090
0.3000	0.5920	319.90	-0.0027	-0.0018	0.0019	-0.0104	-0.0041
0.4000	0.6030	325.00	-0.0057	-0.0054	-0.0025	0.0019	-0.0012
0.5000	0.6050	326.80	-0.0010	-0.0018	-0.0018	0.0173	0.0058
0.6000	0.6140	327.90	0.0033	0.0017	-0.0011	0.0240	0.0064
0.7000	0.6400	327.40	0.0046	0.0032	-0.0003	0.0157	-0.0005
0.8000	0.6970	320.40	-0.0000	0.0001	-0.0003	-0.0159	-0.0077
0.9000	0.8100	301.70	-0.0137	-0.0111	-0.0053	-0.0763	0.0035
		MEAN	0.0043	0.0034	0.0025	0.0233	0.0051

80

METHYL ALCOHOL(1)-1,2-DICHLOROETHANE(2)

UDOVENKO V.V.,FRID TS.B.:ZH.FIZ.KHIM.22,1263(1948).

ANTOINE VAPOR PRESSURE CONSTANTS

	A	B	C
(1)	8.07246	1574.990	238.860
(2)	7.18431	1358.500	232.000

T =   50.00

EQUATION	ORDER	A12	A21	D	A122	A211
VAN LAAR	3	0.9058	0.6738	-	-	-
MARGULES	3	0.8832	0.6539	-	-	-
MARGULES	4	0.8811	0.6518	-0.0065	-	-
ALPHA	2	11.9702	6.2503	-	-	-
ALPHA	3	0.2271	8.8565	-	17.3602	-8.0972

			DEVIATION IN VAPOR PHASE COMPOSITION				
X	Y	PRESS.	LAAR 3	MARG 3	MARG 4	ALPHA2	ALPHA3
0.1000	0.4780	404.20	0.0065	0.0027	0.0023	-0.0320	0.0065
0.2000	0.5620	457.60	0.0043	0.0048	0.0048	-0.0218	-0.0053
0.3000	0.5910	483.80	-0.0001	0.0016	0.0017	-0.0080	-0.0051
0.4000	0.6020	493.20	-0.0007	0.0003	0.0004	0.0071	0.0006
0.5000	0.6120	499.90	-0.0009	-0.0012	-0.0012	0.0167	0.0037
0.6000	0.6250	503.30	0.0031	0.0017	0.0016	0.0214	0.0056
0.7000	0.6570	501.40	0.0022	0.0008	0.0007	0.0089	-0.0019
0.8000	0.7110	492.80	0.0035	0.0036	0.0036	-0.0175	-0.0054
0.9000	0.8140	469.70	-0.0012	0.0014	0.0017	-0.0650	0.0030
		MEAN	0.0025	0.0020	0.0020	0.0220	0.0041

METHYL ALCOHOL(1)-1,2-DICHLOROETHANE(2)

UDOVENKO V.V.,FRID TS.B.:ZH.FIZ.KHIM.22,1263(1948).

ANTOINE VAPOR PRESSURE CONSTANTS
	A	B	C
(1)	8.07246	1574.990	238.860
(2)	7.18431	1358.500	232.000

T = 60.00

EQUATION	ORDER	A12	A21	D	A122	A211
VAN LAAR	3	0.9178	0.6301	-	-	-
MARGULES	3	0.8803	0.6004	-	-	-
MARGULES	4	0.8826	0.6027	0.0072	-	-
ALPHA	2	10.4822	5.3250	-	-	-
ALPHA	3	0.7121	8.0180	-	15.1328	-7.2451

			DEVIATION IN VAPOR PHASE COMPOSITION				
X	Y	PRESS.	LAAR 3	MARG 3	MARG 4	ALPHA2	ALPHA3
0.1000	0.4640	586.20	0.0268	0.0204	0.0209	-0.0333	0.0077
0.2000	0.5560	667.30	0.0108	0.0117	0.0117	-0.0241	-0.0073
0.3000	0.5840	695.50	0.0056	0.0084	0.0082	-0.0049	-0.0029
0.4000	0.5990	712.40	0.0019	0.0037	0.0036	0.0092	0.0016
0.5000	0.6130	719.70	0.0007	0.0005	0.0005	0.0172	0.0034
0.6000	0.6320	726.40	0.0031	0.0011	0.0013	0.0180	0.0025
0.7000	0.6640	724.30	0.0071	0.0052	0.0054	0.0077	-0.0016
0.8000	0.7190	710.10	0.0116	0.0119	0.0119	-0.0171	-0.0033
0.9000	0.8220	680.00	0.0070	0.0107	0.0104	-0.0611	0.0021
		MEAN	0.0083	0.0082	0.0082	0.0214	0.0036

METHYL ALCOHOL(1)-ETHYL ACETATE(2)

AKITA K.,YOSHIDA F.:J.CHEM.ENG.DATA 8(4),484(1963).

ANTOINE VAPOR PRESSURE CONSTANTS

	A	B	C
(1)	8.07246	1574.990	238.860
(2)	7.10232	1245.239	217.911

P = 760.00

EQUATION	ORDER	A12	A21	D	A122	A211
VAN LAAR	3	0.4201	0.4292	-	-	-
MARGULES	3	0.4201	0.4290	-	-	-
MARGULES	4	0.4546	0.4609	0.0980	-	-
ALPHA	2	2.8237	1.1085	-	-	-
ALPHA	3	2.9256	1.5681	-	0.2587	-0.5456

			DEVIATION IN VAPOR PHASE COMPOSITION				
X	Y	TEMP.	LAAR 3	MARG 3	MARG 4	ALPHA2	ALPHA3
0.0190	0.0790	74.80	-0.0063	-0.0063	-0.0015	-0.0123	-0.0071
0.0240	0.0930	74.00	-0.0034	-0.0034	0.0020	-0.0105	-0.0044
0.0560	0.1810	72.30	0.0007	0.0007	0.0081	-0.0111	-0.0020
0.1810	0.3840	67.10	0.0063	0.0063	0.0078	-0.0052	-0.0001
0.3110	0.4920	64.70	0.0105	0.0104	0.0083	0.0052	0.0048
0.3500	0.5200	64.20	0.0072	0.0072	0.0050	0.0038	0.0024
0.4030	0.5570	63.60	0.0003	0.0003	-0.0016	-0.0008	-0.0031
0.5660	0.6400	62.60	-0.0019	-0.0020	-0.0012	0.0007	-0.0016
0.6160	0.6750	62.40	-0.0124	-0.0124	-0.0111	-0.0098	-0.0115
0.6460	0.6780	62.40	0.0000	0.0000	0.0015	0.0021	0.0009
0.7080	0.7110	62.30	0.0012	0.0012	0.0025	0.0015	0.0014
0.7200	0.7160	62.10	0.0032	0.0031	0.0043	0.0032	0.0033
0.7340	0.7170	62.30	0.0108	0.0108	0.0118	0.0101	0.0105
0.7430	0.7320	62.50	0.0016	0.0016	0.0025	0.0003	0.0009
0.7440	0.7350	62.60	0.0013	0.0013	0.0022	-0.0001	0.0005
0.8100	0.7790	62.50	0.0014	0.0014	0.0007	-0.0036	-0.0017
0.8150	0.7840	62.40	0.0003	0.0003	-0.0006	-0.0051	-0.0031
0.8890	0.8460	62.80	0.0049	0.0049	0.0019	-0.0048	-0.0018
0.9390	0.9030	63.30	0.0056	0.0056	0.0020	-0.0046	-0.0018
		MEAN	0.0042	0.0042	0.0040	0.0050	0.0033

METHYL ALCOHOL(1)-ETHYL ACETATE(2)

MURTI P.S.,VAN WINKLE M.:CHEM.ENG.DATA SERIES 3,72(1958).

ANTOINE VAPOR PRESSURE CONSTANTS
	A	B	C
(1)	8.07246	1574.990	238.860
(2)	7.10232	1245.239	217.911

T =   40.00

EQUATION	ORDER	A12	A21	D	A122	A211
VAN LAAR	3	0.5109	0.4710	-	-	-
MARGULES	3	0.5096	0.4703	-	-	-
MARGULES	4	0.5383	0.5088	0.1033	-	-
ALPHA	2	3.4429	1.7932	-	-	-
ALPHA	3	1.7942	1.9980	-	2.1628	-0.9835

			DEVIATION IN VAPOR PHASE COMPOSITION				
X	Y	PRESS.	LAAR 3	MARG 3	MARG 4	ALPHA2	ALPHA3
0.0500	0.2115	231.50	-0.0362	-0.0365	-0.0305	-0.0405	-0.0290
0.0670	0.2130	231.50	0.0044	0.0042	0.0100	-0.0004	0.0116
0.0970	0.2620	240.00	0.0163	0.0161	0.0208	0.0112	0.0225
0.1540	0.3718	259.00	-0.0098	-0.0098	-0.0079	-0.0136	-0.0062
0.2175	0.4242	268.50	0.0019	0.0019	0.0015	0.0006	0.0033
0.2620	0.4695	284.00	-0.0094	-0.0093	-0.0106	-0.0085	-0.0089
0.3000	0.4912	289.50	-0.0067	-0.0066	-0.0081	-0.0040	-0.0065
0.3820	0.5356	296.50	-0.0076	-0.0075	-0.0084	-0.0013	-0.0073
0.4500	0.5360	298.00	0.0227	0.0228	0.0232	0.0311	0.0237
0.5680	0.6150	304.00	-0.0056	-0.0055	-0.0030	0.0033	-0.0032
0.6560	0.6600	305.00	-0.0089	-0.0089	-0.0059	-0.0032	-0.0057
0.7190	0.6940	302.00	-0.0079	-0.0079	-0.0056	-0.0066	-0.0043
0.7800	0.7300	303.50	-0.0031	-0.0031	-0.0026	-0.0080	0.0006
0.8100	0.7495	301.50	0.0008	0.0008	0.0002	-0.0075	0.0045
		MEAN	0.0101	0.0101	0.0099	0.0100	0.0098

METHYL ALCOHOL(1)-ETHYL ACETATE(2)

MURTI P.S.,VAN WINKLE M.:CHEM.ENG.DATA SERIES 3,72(1958).

ANTOINE VAPOR PRESSURE CONSTANTS

	A	B	C
(1)	8.07246	1574.990	238.860
(2)	7.10232	1245.239	217.911

T =   50.00

EQUATION	ORDER	A12	A21	D	A122	A211
VAN LAAR	3	0.4582	0.4961	-	-	-
MARGULES	3	0.4585	0.4938	-	-	-
MARGULES	4	0.5572	0.5828	0.3078	-	-
ALPHA	2	3.2537	1.5095	-	-	-
ALPHA	3	6.5130	3.8767	-	-2.3619	-1.8625

			DEVIATION IN VAPOR PHASE COMPOSITION				
X	Y	PRESS.	LAAR 3	MARG 3	MARG 4	ALPHA2	ALPHA3
0.0525	0.1700	330.50	0.0037	0.0038	0.0246	0.0033	0.0193
0.1260	0.3375	373.50	-0.0166	-0.0165	-0.0058	-0.0198	-0.0118
0.2315	0.4360	405.50	0.0075	0.0075	0.0026	0.0026	-0.0002
0.3435	0.5125	435.00	0.0105	0.0103	0.0012	0.0069	0.0009
0.4500	0.5685	455.50	0.0087	0.0084	0.0038	0.0076	0.0035
0.5425	0.6170	459.50	0.0008	0.0004	0.0018	0.0017	0.0010
0.5680	0.6325	460.50	-0.0038	-0.0041	-0.0012	-0.0025	-0.0022
0.6350	0.6640	461.50	-0.0055	-0.0059	-0.0000	-0.0038	-0.0012
0.7060	0.6975	463.00	-0.0040	-0.0042	0.0021	-0.0029	0.0011
0.7580	0.7290	461.50	-0.0055	-0.0056	-0.0013	-0.0059	-0.0021
0.8215	0.7655	457.50	0.0026	0.0027	0.0020	-0.0009	0.0009
0.8755	0.8100	452.50	0.0065	0.0067	0.0003	-0.0003	-0.0021
0.9250	0.8550	444.50	0.0188	0.0191	0.0085	0.0099	0.0038
		MEAN	0.0073	0.0073	0.0042	0.0052	0.0038

METHYL ALCOHOL(1)-ETHYL ACETATE(2)

MURTI P.S.,VAN WINKLE M.:CHEM.ENG.DATA SERIES 3,72(1958).

ANTOINE VAPOR PRESSURE CONSTANTS
	A	B	C
(1)	8.07246	1574.990	238.860
(2)	7.10232	1245.239	217.911

T =   60.00

EQUATION	ORDER	A12	A21	D	A122	A211
VAN LAAR	3	0.4700	0.4476	-	-	-
MARGULES	3	0.4689	0.4482	-	-	-
MARGULES	4	0.5420	0.5271	0.2522	-	-
ALPHA	2	3.2336	1.3881	-	-	-
ALPHA	3	5.2987	3.7490	-	-0.8606	-2.3532

			DEVIATION IN VAPOR PHASE COMPOSITION				
X	Y	PRESS.	LAAR 3	MARG 3	MARG 4	ALPHA2	ALPHA3
0.0190	0.0950	456.00	-0.0183	-0.0185	-0.0079	-0.0220	-0.0064
0.0495	0.1785	490.50	-0.0061	-0.0063	0.0090	-0.0129	0.0099
0.1090	0.3100	545.00	-0.0102	-0.0103	-0.0006	-0.0180	-0.0014
0.1360	0.3450	558.00	-0.0034	-0.0035	0.0023	-0.0106	0.0013
0.1900	0.4160	591.50	-0.0092	-0.0092	-0.0099	-0.0142	-0.0107
0.2375	0.4350	611.00	0.0156	0.0157	0.0113	0.0131	0.0108
0.3590	0.5320	644.00	-0.0009	-0.0007	-0.0065	0.0026	-0.0063
0.4020	0.5500	660.00	0.0035	0.0037	-0.0006	0.0087	-0.0004
0.4950	0.5940	673.00	0.0033	0.0035	0.0040	0.0107	0.0038
0.5900	0.6430	684.50	-0.0022	-0.0020	0.0030	0.0049	0.0025
0.6990	0.7020	690.00	-0.0052	-0.0051	0.0010	-0.0027	0.0009
0.7350	0.7280	687.50	-0.0097	-0.0096	-0.0046	-0.0097	-0.0045
0.7480	0.7320	688.00	-0.0054	-0.0054	-0.0009	-0.0065	-0.0007
0.8980	0.8470	677.00	0.0065	0.0064	-0.0009	-0.0080	-0.0018
0.9100	0.8535	674.50	0.0138	0.0137	0.0056	-0.0013	0.0043
		MEAN	0.0075	0.0076	0.0045	0.0097	0.0044

METHYL ALCOHOL(1)-ETHYL ACETATE(2)

MURTI P.S.,VAN WINKLE M.:CHEM.ENG.DATA SERIES 3,72(1958).

ANTOINE VAPOR PRESSURE CONSTANTS
	A	B	C
(1)	8.07246	1574.990	238.860
(2)	7.10232	1245.239	217.911

P = 760.00

EQUATION	ORDER	A12	A21	D	A122	A211
VAN LAAR	3	0.4396	0.4480	-	-	-
MARGULES	3	0.4397	0.4477	-	-	-
MARGULES	4	0.5006	0.5048	0.1944	-	-
ALPHA	2	3.0588	1.2509	-	-	-
ALPHA	3	4.2917	2.6600	-	-0.4572	-1.3994

			DEVIATION IN VAPOR PHASE COMPOSITION				
X	Y	TEMP.	LAAR 3	MARG 3	MARG 4	ALPHA2	ALPHA3
0.0125	0.0475	76.10	0.0041	0.0042	0.0107	0.0002	0.0080
0.0320	0.1330	74.15	-0.0134	-0.0133	-0.0019	-0.0212	-0.0070
0.0800	0.2475	71.24	-0.0041	-0.0041	0.0074	-0.0158	0.0003
0.1550	0.3650	67.75	0.0009	0.0009	0.0041	-0.0098	-0.0011
0.2510	0.4550	65.60	0.0082	0.0082	0.0040	0.0013	0.0014
0.3465	0.5205	64.10	0.0078	0.0078	0.0022	0.0054	0.0013
0.4020	0.5560	64.00	0.0031	0.0031	-0.0012	0.0026	-0.0022
0.4975	0.5970	63.25	0.0083	0.0083	0.0076	0.0106	0.0066
0.5610	0.6380	62.97	-0.0034	-0.0035	-0.0016	-0.0004	-0.0028
0.5890	0.6560	62.50	-0.0086	-0.0087	-0.0059	-0.0053	-0.0069
0.6220	0.6670	62.65	-0.0036	-0.0037	-0.0000	-0.0008	-0.0013
0.6960	0.7000	62.50	0.0017	0.0017	0.0058	0.0026	0.0044
0.7650	0.7420	62.35	0.0015	0.0015	0.0039	-0.0011	0.0025
0.8250	0.7890	62.60	-0.0014	-0.0014	-0.0020	-0.0082	-0.0039
0.8550	0.8070	62.80	0.0065	0.0065	0.0040	-0.0026	0.0016
0.9160	0.8600	63.21	0.0167	0.0167	0.0108	0.0046	0.0074
0.9550	0.9290	63.90	-0.0016	-0.0015	-0.0075	-0.0123	-0.0111
		MEAN	0.0056	0.0056	0.0047	0.0062	0.0041

METHYL ALCOHOL(1)-ETHYL ACETATE(2)

NAGATA I.:J.CHEM.ENG.DATA 7(3),367(1963).

ANTOINE VAPOR PRESSURE CONSTANTS
```
           A           B           C
(1) 8.07246    1574.990     238.860
(2) 7.10232    1245.239     217.911
```

P =   760.00

EQUATION	ORDER	A12	A21	D	A122	A211
VAN LAAR	3	0.4239	0.4180	-	-	-
MARGULES	3	0.4238	0.4180	-	-	-
MARGULES	4	0.4401	0.4363	0.0536	-	-
ALPHA	2	2.9112	1.1469	-	-	-
ALPHA	3	2.4444	1.5405	-	0.9027	-0.6736

			DEVIATION IN VAPOR PHASE COMPOSITION				
X	Y	TEMP.	LAAR 3	MARG 3	MARG 4	ALPHA2	ALPHA3
0.0280	0.1200	74.40	-0.0167	-0.0167	-0.0139	-0.0234	-0.0154
0.0370	0.1330	74.00	-0.0019	-0.0020	0.0011	-0.0100	-0.0007
0.0730	0.2200	71.50	0.0021	0.0020	0.0052	-0.0082	0.0030
0.1230	0.3100	69.30	0.0038	0.0037	0.0057	-0.0060	0.0034
0.2110	0.4200	66.40	0.0006	0.0006	0.0005	-0.0048	-0.0011
0.2360	0.4420	66.00	0.0016	0.0016	0.0011	-0.0026	-0.0004
0.2390	0.4400	65.80	0.0061	0.0061	0.0055	0.0021	0.0041
0.2650	0.4660	65.30	-0.0007	-0.0007	-0.0016	-0.0033	-0.0027
0.3520	0.5260	64.00	0.0001	0.0001	-0.0009	0.0016	-0.0014
0.4080	0.5580	63.70	-0.0002	-0.0002	-0.0008	0.0031	-0.0013
0.4400	0.5730	63.60	0.0016	0.0016	0.0013	0.0057	0.0008
0.5330	0.6200	63.10	0.0006	0.0006	0.0012	0.0057	0.0009
0.5850	0.6470	62.90	-0.0009	-0.0009	0.0001	0.0037	-0.0002
0.6640	0.6870	62.40	-0.0001	-0.0001	0.0012	0.0024	0.0007
0.7080	0.7110	62.40	0.0011	0.0011	0.0022	0.0013	0.0014
0.7480	0.7370	62.40	0.0001	0.0001	0.0008	-0.0023	-0.0003
0.7930	0.7680	62.40	0.0005	0.0005	0.0006	-0.0052	-0.0010
0.8220	0.7900	62.50	0.0011	0.0011	0.0007	-0.0070	-0.0013
0.8830	0.8420	62.80	0.0045	0.0045	0.0030	-0.0079	0.0003
0.9610	0.9340	64.00	0.0057	0.0057	0.0040	-0.0052	0.0017
		MEAN	0.0025	0.0025	0.0026	0.0056	0.0021

METHYL ALCOHOL(1)-ETHYL ALCOHOL(2)

AMER H.H.,PAXTON R.R.,VAN WINKLE M.:IND.ENG.CHEM.48,142(1956).

ANTOINE VAPOR PRESSURE CONSTANTS

	A	B	C
(1)	8.07246	1574.990	238.860
(2)	8.16290	1623.220	228.980

P = 760.00

EQUATION	ORDER	A12	A21	D	A122	A211
VAN LAAR	3	-0.0201	1.2664	-	-	-
MARGULES	3	0.0019	-0.0242	-	-	-
MARGULES	4	0.0448	0.0224	0.1326	-	-
ALPHA	2	0.5932	-0.4172	-	-	-
ALPHA	3	-5.5924	-4.1890	-	6.2213	3.7983

			DEVIATION IN VAPOR PHASE COMPOSITION				
X	Y	TEMP.	LAAR 3	MARG 3	MARG 4	ALPHA2	ALPHA3
0.1340	0.1830	76.60	0.0171	0.0231	0.0267	0.0158	0.0177
0.2420	0.3260	75.00	0.0157	0.0226	0.0216	0.0138	0.0132
0.3200	0.4280	73.60	0.0066	0.0137	0.0113	0.0045	-0.0101
0.4010	0.5290	72.30	-0.0056	0.0017	0.0000	-0.0076	0.0038
0.4350	0.5660	71.70	-0.0074	0.0000	-0.0010	-0.0093	-0.0023
0.5420	0.6760	70.00	-0.0144	-0.0063	-0.0049	-0.0156	-0.0137
0.6520	0.7590	68.60	-0.0027	0.0062	0.0084	-0.0026	-0.0031
0.7280	0.8130	67.70	0.0028	0.0120	0.0133	0.0040	0.0036
0.7900	0.8580	66.90	0.0030	0.0122	0.0121	0.0052	0.0021
0.8140	0.8750	66.60	0.0027	0.0119	0.0112	0.0054	0.0026
0.8730	0.9190	65.80	-0.0027	0.0068	0.0049	0.0018	-0.0004
0.9100	0.9370	65.60	0.0006	0.0117	0.0095	0.0080	0.0062
		MEAN	0.0068	0.0107	0.0104	0.0078	0.0066

89

G

METHYL ALCOHOL(1)-HEPTANE(2)

BENEDICT M.,JOHNSON C.A.,SOLOMON E.,RUBIN L.C.:TRANS.AM.INST.
        CHEM.ENGRS. 41,371(1945).

ANTOINE VAPOR PRESSURE CONSTANTS
         A   B   C
(1) 8.07246   1574.990   238.860
(2) 6.90240   1268.115   216.900

P =   760.00

EQUATION	ORDER	A12	A21	D	A122	A211
VAN LAAR	3	0.9775	1.0702	-	-	-
MARGULES	3	0.9819	1.0624	-	-	-
MARGULES	4	1.2818	1.2753	0.9362	-	-
ALPHA	2	109.1542	36.0505	-	-	-
ALPHA	3	19.3278	40.5350	-	107.1288	-33.0575

			DEVIATION IN VAPOR PHASE COMPOSITION				
X	Y	TEMP.	LAAR 3	MARG 3	MARG 4	ALPHA2	ALPHA3
0.1380	0.7200	60.60	0.0006	0.0014	0.0182	-0.0019	0.0030
0.1780	0.7330	59.47	0.0125	0.0131	0.0105	-0.0071	-0.0035
0.3900	0.7390	58.93	0.0325	0.0317	0.0063	0.0025	0.0018
0.6680	0.7460	58.82	-0.0058	-0.0070	0.0099	0.0032	-0.0022
0.8100	0.7480	58.81	-0.0004	-0.0005	0.0094	0.0062	0.0019
0.8850	0.7650	59.01	0.0190	0.0199	0.0065	-0.0048	-0.0002
0.9460	0.8090	59.90	0.0483	0.0496	0.0149	-0.0342	-0.0003
		MEAN	0.0170	0.0176	0.0108	0.0086	0.0018

METHYL ALCOHOL(1)-HEXANE(2)

FERGUSON J.B.:J.PHYS.CHEM. 36,1125(1932).

ANTOINE VAPOR PRESSURE CONSTANTS
```
          A          B          C
(1) 8.07246    1574.990    238.860
(2) 6.87776    1171.530    224.366
```

T =   45.00

EQUATION	ORDER	A12	A21	D	A122	A211
VAN LAAR	3	1.0260	1.0215	-	-	-
MARGULES	3	1.0258	1.0218	-	-	-
MARGULES	4	1.3359	1.3309	1.0125	-	-
ALPHA	2	179.5596	182.3873	-	-	-
ALPHA	3	-.4139/ 10	-.4064/ 10	-	.2721/ 09	-.7783/ 06

			DEVIATION IN VAPOR PHASE COMPOSITION				
X	Y	PRESS.	LAAR 3	MARG 3	MARG 4	ALPHA2	ALPHA3
0.1000	0.4810	615.00	-0.0602	-0.0603	-0.0021	0.0033	0.0083
0.2000	0.4920	626.00	0.0124	0.0124	0.0075	-0.0009	-0.0010
0.3000	0.4940	627.00	0.0269	0.0270	0.0003	-0.0005	-0.0012
0.4000	0.4960	627.00	0.0173	0.0174	-0.0031	-0.0010	-0.0015
0.5000	0.4960	627.00	0.0007	0.0008	0.0007	0.0001	0.0002
0.6000	0.4980	627.00	-0.0177	-0.0176	0.0026	-0.0007	-0.0001
0.7000	0.4980	627.00	-0.0248	-0.0248	0.0017	0.0007	0.0015
0.8000	0.4980	627.00	-0.0076	-0.0076	-0.0026	0.0033	0.0032
0.9000	0.5110	616.00	0.0642	0.0642	0.0062	-0.0029	-0.0081
		MEAN	0.0258	0.0258	0.0030	0.0015	0.0028

METHYL ALCOHOL(1)-HEXANE(2)

VILIM O.:COLLECTION CZECH.CHEM.COMMUN.26,2124(1961).

ANTOINE VAPOR PRESSURE CONSTANTS

	A	B	C
(1)	8.07246	1574.990	238.860
(2)	6.87776	1171.530	224.366

P =   745.00

EQUATION	ORDER	A12	A21	D	A122	A211
VAN LAAR	3	0.9412	1.0015	-	-	-
MARGULES	3	0.9432	0.9973	-	-	-
MARGULES	4	1.0616	1.1440	0.6455	-	-
ALPHA	2	16.4013	15.5911	-	-	-
ALPHA	3	15.7699	10.3615	-	-2.9642	6.4179

DEVIATION IN VAPOR PHASE COMPOSITION

X	Y	TEMP.	LAAR 3	MARG 3	MARG 4	ALPHA2	ALPHA3
0.0330	0.2560	62.00	-0.0295	-0.0289	0.0024	0.0193	-0.0017
0.0560	0.3320	58.00	-0.0217	-0.0210	0.0029	0.0110	-0.0027
0.0900	0.3880	54.20	-0.0001	0.0005	0.0087	0.0079	0.0032
0.2500	0.4830	49.50	0.0298	0.0296	-0.0009	-0.0079	-0.0006
0.5120	0.5120	48.70	-0.0002	-0.0011	0.0056	0.0006	0.0001
0.9420	0.6520	52.00	0.0397	0.0410	0.0086	0.0169	0.0008
0.9800	0.7900	56.60	0.0634	0.0644	0.0282	0.0099	-0.0019
		MEAN	0.0263	0.0266	0.0082	0.0105	0.0016

92

METHYL ALCOHOL(1)-ISOPRENE(2)

VILIM O.:COLLECTION CZECH.CHEM.COMMUN.26,2124(1961).

ANTOINE VAPOR PRESSURE CONSTANTS
```
          A         B         C
(1) 8.07246   1574.990   238.860
(2) 6.90334   1080.996   234.670
```

P =   745.00

EQUATION	ORDER	A12	A21	D	A122	A211
VAN LAAR	3	0.9704	0.9019	-	-	-
MARGULES	3	0.9664	0.9030	-	-	-
MARGULES	4	1.1725	0.9682	0.5139	-	-
ALPHA	2	5.5223	32.4236	-	-	-
ALPHA	3	4.1819	60.7063	-	6.9817	-34.3272

			DEVIATION IN VAPOR PHASE COMPOSITION				
X	Y	TEMP.	LAAR 3	MARG 3	MARG 4	ALPHA2	ALPHA3
0.0140	0.0840	32.60	-0.0530	-0.0532	-0.0379	-0.0248	0.0000
0.1540	0.1540	30.10	0.0157	0.0155	0.0210	-0.0070	0.0000
0.6380	0.1900	32.80	0.0064	0.0067	0.0070	0.0060	0.0000
0.8100	0.2400	37.50	-0.0020	-0.0021	0.0061	0.0027	-0.0001
0.9100	0.3420	44.00	0.0102	0.0098	0.0065	-0.0104	0.0005
0.9670	0.5500	51.50	0.0315	0.0309	0.0089	-0.0329	-0.0012
		MEAN	0.0198	0.0197	0.0146	0.0140	0.0003

METHYL ALCOHOL(1)-METHYLCYCLOPENTANE(2)

VILIM O.:COLLECTION CZECH.CHEM.COMMUN.,26,2124(1961)

ANTOINE VAPOR PRESSURE CONSTANTS

	A	B	C
(1)	8.07246	1574.990	238.860
(2)	6.86283	1186.059	226.042

P =   745.00

EQUATION	ORDER	A12	A21	D	A122	A211
VAN LAAR	3	0.9058	0.9054	-	-	-
MARGULES	3	0.9058	0.9054	-	-	-
MARGULES	4	0.8585	0.8301	-0.2129	-	-
ALPHA	2	19.6342	17.2622	-	12.5798	-1.3463
ALPHA	3	-3.7103	2.8196	-	12.5798	-1.3463

			DEVIATION IN VAPOR PHASE COMPOSITION				
X	Y	TEMP.	LAAR 3	MARG 3	MARG 4	ALPHA2	ALPHA3
0.0790	0.3890	60.00	-0.0119	-0.0119	-0.0214	0.0202	-0.0207
0.1420	0.4410	54.80	0.0227	0.0227	0.0210	0.0201	0.0193
0.3120	0.5380	50.20	-0.0084	-0.0084	-0.0042	-0.0305	-0.0049
0.5220	0.5220	49.80	0.0076	0.0076	0.0025	0.0091	-0.0005
0.8010	0.5600	50.60	0.0061	0.0061	0.0042	0.0112	0.0009
		MEAN	0.0113	0.0113	0.0106	0.0182	0.0093

METHYL ALCOHOL(1)-2-METHYL PENTANE(2)

VILIM O.:COLLECTION CZECH.CHEM.COMMUN.26,2124(1961).

ANTOINE VAPOR PRESSURE CONSTANTS

	A	B	C
(1)	8.07246	1574.990	238.860
(2)	6.83910	1135.410	226.572

P = 745.00

EQUATION	ORDER	A12	A21	D	A122	A211
VAN LAAR	3	0.9650	0.9039	-	-	-
MARGULES	3	0.9627	0.9047	-	-	-
MARGULES	4	1.1203	1.0451	0.6336	-	-
ALPHA	2	13.6661	20.8786	-	-	-
ALPHA	3	18.4755	20.7485	-	-5.5471	5.8661

			DEVIATION IN VAPOR PHASE COMPOSITION				
X	Y	TEMP.	LAAR 3	MARG 3	MARG 4	ALPHA2	ALPHA3
0.0400	0.2340	54.00	-0.0314	-0.0320	0.0072	0.0088	0.0018
0.1270	0.3360	47.80	0.0222	0.0220	0.0269	0.0040	-0.0012
0.3950	0.3950	44.70	0.0262	0.0267	0.0066	0.0005	0.0009
0.5660	0.4170	44.90	-0.0063	-0.0060	0.0037	-0.0034	-0.0014
0.5680	0.4150	44.90	-0.0044	-0.0041	0.0059	-0.0012	0.0008
0.8810	0.4920	48.40	0.0362	0.0359	0.0306	0.0086	0.0002
		MEAN	0.0211	0.0211	0.0135	0.0044	0.0011

METHYL ALCOHOL(1)-3-METHYL PENTANE(2)

VILIM O.:COLLECTION CZECH.CHEM.COMMUN.26,2124(1961).

ANTOINE VAPOR PRESSURE CONSTANTS
```
          A          B          C
(1) 8.07246    1574.990    238.860
(2) 6.84887    1152.368    227.129
```

P =   745.00

EQUATION	ORDER	A12	A21	D	A122	A211
VAN LAAR	3	0.9308	0.9019	-	-	-
MARGULES	3	0.9280	0.9039	-	-	-
MARGULES	4	1.1610	1.1850	0.9962	-	-
ALPHA	2	14.8781	20.5066	-	-	-
ALPHA	3	18.6866	19.3754	-	-4.9991	5.1310

			DEVIATION IN VAPOR PHASE COMPOSITION				
X	Y	TEMP.	LAAR 3	MARG 3	MARG 4	ALPHA2	ALPHA3
0.0790	0.3210	52.70	-0.0109	-0.0117	0.0294	0.0039	-0.0024
0.1900	0.3800	47.80	0.0402	0.0400	0.0219	0.0047	0.0026
0.4250	0.4250	46.20	0.0256	0.0262	0.0130	-0.0041	-0.0017
0.7820	0.4720	47.40	0.0033	0.0033	0.0305	-0.0001	0.0015
0.9320	0.5690	56.40	0.1012	0.1006	0.0325	0.0092	-0.0019
		MEAN	0.0362	0.0364	0.0255	0.0044	0.0020

METHYL ALCOHOL(1)-NITROMETHANE(2)

DESSEIGNE G.,BELLIOT CH.:J.CHIM.PHYS.49,46(1952).

ANTOINE VAPOR PRESSURE CONSTANTS
```
            A           B          C
  (1) 8.07246     1574.990    238.860
  (2) 7.15541     1366.483    218.611
```

P =   760.00

EQUATION	ORDER	A12	A21	D	A122	A211
VAN LAAR	3	0.7453	1.0138	-	-	-
MARGULES	3	0.7512	0.9591	-	-	-
MARGULES	4	0.9062	1.2189	0.7501	-	-
ALPHA	2	28.0791	3.3134	-	-	-
ALPHA	3	59.5169	18.3037	-	11.9925	-14.5115

DEVIATION IN VAPOR PHASE COMPOSITION

X	Y	TEMP.	LAAR 3	MARG 3	MARG 4	ALPHA2	ALPHA3
0.0630	0.7160	83.90	-0.1835	-0.1802	-0.1404	-0.1130	-0.0325
0.1190	0.7410	78.00	-0.0787	-0.0764	-0.0668	-0.0273	-0.0007
0.2060	0.7600	71.50	-0.0147	-0.0142	-0.0245	0.0222	0.0171
0.3090	0.7950	68.90	-0.0099	-0.0115	-0.0228	0.0234	0.0074
0.3970	0.8350	67.50	-0.0353	-0.0387	-0.0418	0.0010	-0.0154
0.5780	0.8580	66.20	-0.0520	-0.0575	-0.0401	-0.0001	-0.0062
0.7270	0.8780	65.30	-0.0721	-0.0755	-0.0575	-0.0067	0.0012
0.8420	0.8900	64.85	-0.0675	-0.0652	-0.0706	-0.0057	0.0130
0.9300	0.9300	64.55	-0.0544	-0.0473	-0.0809	-0.0239	-0.0031
0.9640	0.9620	64.63	-0.0416	-0.0350	-0.0697	-0.0343	-0.0195
		MEAN	0.0610	0.0601	0.0615	0.0258	0.0116

METHYL ALCOHOL(1)-PROPYL ALCOHOL(2)

HILL W.D.,VAN WINKLE M.:IND.ENG.CHEM.44,205,208(1952).

ANTOINE VAPOR PRESSURE CONSTANTS
	A	B	C
(1)	8.07246	1574.990	238.860
(2)	7.61924	1375.140	193.000

P = 760.00

EQUATION	ORDER	A12	A21	D	A122	A211
VAN LAAR	3	0.0009	-0.0031	-	-	-
MARGULES	3	-0.0649	0.2347	-	-	-
MARGULES	4	-0.0034	0.3170	0.2326	-	-
ALPHA	2	3.0769	-0.7980	-	-	-
ALPHA	3	-15.1514	-5.1731	-	18.2228	4.4317

			DEVIATION IN VAPOR PHASE COMPOSITION				
X	Y	TEMP.	LAAR 3	MARG 3	MARG 4	ALPHA2	ALPHA3
0.0900	0.2600	89.00	-0.0104	-0.0136	-0.0040	0.0282	0.0254
0.1720	0.4550	83.20	-0.0348	-0.0163	-0.0150	0.0057	-0.0057
0.3190	0.6680	79.70	-0.0420	-0.0026	-0.0064	-0.0075	0.0029
0.4450	0.7770	78.30	-0.0337	-0.0002	-0.0001	-0.0060	-0.0046
0.5550	0.8420	76.70	-0.0216	-0.0022	0.0008	-0.0006	-0.0018
0.6520	0.8890	73.90	-0.0127	-0.0077	-0.0044	0.0010	-0.0011
0.7380	0.9210	71.30	0.0049	-0.0097	-0.0078	0.0042	0.0020
0.8140	0.9510	69.40	0.0037	-0.0159	-0.0162	0.0005	-0.0011
0.8830	0.9710	66.70	-0.0028	-0.0138	-0.0159	0.0010	0.0013
		MEAN	0.0185	0.0091	0.0078	0.0061	0.0051

METHYL ALCOHOL(1)-ISOPROPYL ALCOHOL(2)

BALLARD L.H.,VAN WINKLE M.:IND.ENG.CHEM.44,2450(1952).

ANTOINE VAPOR PRESSURE CONSTANTS
	A	B	C
(1)	8.07246	1574.990	238.860
(2)	7.75634	1366.142	197.970

P = 760.00

EQUATION	ORDER	A12	A21	D	A122	A211
VAN LAAR	3	-0.0492	-0.1203	-	-	-
MARGULES	3	-0.0426	-0.0920	-	-	-
MARGULES	4	-0.0561	-0.1082	-0.0453	-	-
ALPHA	2	0.7081	-0.6074	-	-	-
ALPHA	3	0.6922	-0.7423	-	-0.0692	0.1835

			DEVIATION IN VAPOR PHASE COMPOSITION				
X	Y	TEMP.	LAAR 3	MARG 3	MARG 4	ALPHA2	ALPHA3
0.0810	0.1320	81.00	0.0018	0.0026	0.0009	0.0007	-0.0033
0.1950	0.2960	78.90	0.0070	0.0071	0.0070	0.0054	0.0022
0.2930	0.4285	77.10	0.0039	0.0037	0.0044	0.0022	0.0018
0.4080	0.5700	74.80	-0.0028	-0.0025	-0.0021	-0.0047	-0.0025
0.5220	0.6850	72.70	-0.0015	-0.0001	-0.0006	-0.0034	-0.0007
0.6605	0.8000	70.20	0.0036	0.0053	0.0045	0.0012	0.0019
0.7900	0.8910	67.90	0.0043	0.0049	0.0049	0.0016	-0.0004
0.9010	0.9535	66.20	0.0043	0.0037	0.0042	0.0021	-0.0006
		MEAN	0.0037	0.0037	0.0036	0.0027	0.0017

METHANOL(1)-ISOPROPANOL(2)

DUNLOP J.G.:M.S.THESIS,BROOKLYN POLYTECHN.INST.,1948.

ANTOINE VAPOR PRESSURE CONSTANTS

	A	B	C
(1)	8.07246	1574.990	238.860
(2)	7.75634	1366.142	197.970

P = 760.00

EQUATION	ORDER	A12	A21	D	A122	A211
VAN LAAR	3	0.0891	0.1027	-	-	-
MARGULES	3	0.0879	0.1030	-	-	-
MARGULES	4	0.0616	0.0693	-0.1054	-	-
ALPHA	2	1.3511	-0.3611	-	-	-
ALPHA	3	-0.9456	-1.4981	-	2.2244	1.1051

			DEVIATION IN VAPOR PHASE COMPOSITION				
X	Y	TEMP.	LAAR 3	MARG 3	MARG 4	ALPHA2	ALPHA3
0.0680	0.1460	79.20	0.0005	0.0003	-0.0031	-0.0014	-0.0039
0.1060	0.2060	78.50	0.0103	0.0102	0.0077	0.0080	0.0056
0.1410	0.2680	78.50	0.0059	0.0058	0.0048	0.0040	0.0022
0.1840	0.3360	76.50	0.0040	0.0040	0.0047	0.0008	0.0000
0.3100	0.5010	74.10	-0.0036	-0.0034	-0.0007	-0.0067	-0.0049
0.4250	0.6120	72.20	-0.0013	-0.0011	-0.0002	-0.0039	-0.0019
0.5240	0.6840	70.50	0.0093	0.0094	0.0080	0.0065	0.0072
0.6520	0.7850	68.80	0.0007	0.0007	-0.0019	-0.0023	-0.0037
0.8110	0.8860	67.10	0.0010	0.0010	0.0007	-0.0020	-0.0029
0.8240	0.8890	66.50	0.0062	0.0062	0.0062	0.0029	0.0022
0.8780	0.9240	65.90	0.0037	0.0036	0.0047	0.0007	0.0009
0.9470	0.9670	65.30	0.0016	0.0016	0.0029	-0.0002	0.0006
		MEAN	0.0040	0.0039	0.0038	0.0033	0.0030

METHYL ALCOHOL(1)-TOLUENE(2)

BENEDICT M.,JOHNSON C.A.,SOLOMON E.,RUBIN L.C.:TRANS.AM.INST.
        CHEM.ENGRS.41,371(1945).

ANTOINE VAPOR PRESSURE CONSTANTS
         A   B   C
(1) 8.07246   1574.990   238.860
(2) 6.95334   1343.943   219.377

P =   760.00

EQUATION	ORDER	A12	A21	D	A122	A211
VAN LAAR	3	0.8740	0.7949	-	-	-
MARGULES	3	0.8674	0.7965	-	-	-
MARGULES	4	1.0240	0.9123	0.4724	-	-
ALPHA	2	28.8824	4.6598	-	-	-
ALPHA	3	10.8896	12.4602	-	47.8002	-11.6283

			DEVIATION IN VAPOR PHASE COMPOSITION				
X	Y	TEMP.	LAAR 3	MARG 3	MARG 4	ALPHA2	ALPHA3
0.1300	0.7420	70.25	0.0038	0.0030	0.0152	-0.0334	0.0001
0.2660	0.7820	66.44	0.0187	0.0190	0.0093	0.0002	0.0009
0.4070	0.8030	65.58	0.0090	0.0097	0.0004	0.0082	-0.0023
0.5930	0.8190	64.47	-0.0010	-0.0004	0.0023	0.0126	-0.0000
0.6920	0.8290	64.10	-0.0009	-0.0007	0.0056	0.0113	0.0023
0.7790	0.8450	63.79	0.0019	0.0019	0.0061	0.0040	0.0022
0.8430	0.8690	63.67	0.0017	0.0016	0.0007	-0.0112	-0.0034
0.8820	0.8830	63.58	0.0082	0.0080	0.0031	-0.0173	-0.0008
0.9270	0.9110	63.62	0.0117	0.0115	0.0029	-0.0298	-0.0005
0.9690	0.9500	93.94	0.0135	0.0133	0.0058	-0.0352	0.0025
		MEAN	0.0070	0.0069	0.0051	0.0163	0.0015

METHYL ALCOHOL(1)-TOLUENE(2)

BURKE D.E.,WILLIAMS G.C.,PLANK C.A.:J.CHEM.ENG.DATA 9(2),212(1964).

ANTOINE VAPOR PRESSURE CONSTANTS
	A	B	C
(1)	8.07246	1574.990	238.860
(2)	6.95334	1343.943	219.377

P = 760.00

EQUATION	ORDER	A12	A21	D	A122	A211
VAN LAAR	3	0.9542	0.8998	-	-	-
MARGULES	3	0.9531	0.8996	-	-	-
MARGULES	4	0.9798	0.9494	0.1617	-	-
ALPHA	2	51.4927	7.8020	-	-	-
ALPHA	3	3.7445	7.3332	-	50.4981	-6.7815

			DEVIATION IN VAPOR PHASE COMPOSITION				
X	Y	TEMP.	LAAR 3	MARG 3	MARG 4	ALPHA2	ALPHA3
0.0460	0.5190	89.90	0.1042	0.1039	0.1111	0.1211	0.1291
0.0580	0.6270	84.80	0.0381	0.0379	0.0432	0.0503	0.0571
0.0700	0.7040	80.40	-0.0083	-0.0085	-0.0048	0.0001	0.0061
0.0940	0.7770	74.75	-0.0399	-0.0400	-0.0386	-0.0366	-0.0322
0.1140	0.7930	71.30	-0.0333	-0.0333	-0.0334	-0.0325	-0.0291
0.1320	0.8010	69.70	-0.0264	-0.0264	-0.0275	-0.0271	-0.0244
0.2340	0.8130	66.75	-0.0001	0.0000	-0.0034	0.0009	0.0006
0.3300	0.8220	65.75	-0.0014	-0.0012	-0.0038	0.0083	0.0058
0.4390	0.8280	65.10	-0.0091	-0.0090	-0.0089	0.0129	0.0082
0.6750	0.8420	64.15	-0.0262	-0.0263	-0.0211	0.0126	0.0038
0.8300	0.8660	63.70	-0.0191	-0.0192	-0.0183	-0.0017	-0.0037
0.8700	0.8780	63.60	-0.0121	-0.0122	-0.0135	-0.0094	-0.0035
0.9300	0.9120	63.70	-0.0021	-0.0021	-0.0063	-0.0309	-0.0008
0.9740	0.9570	64.10	0.0029	0.0029	-0.0006	-0.0464	0.0042
		MEAN	0.0231	0.0231	0.0239	0.0279	0.0220

METHYL ALCOHOL(1)-TRICHLORETHYLENE(2)

FRITZWEILER R.,DIETRICH K.R.:ANGEW.CHEM.A.CHEM.FABRIK.,
         NO.4,BERLIN,W 35(1933)

ANTOINE VAPOR PRESSURE CONSTANTS
         A   B   C
  (1) 8.07246   1574.990   238.860
  (2) 7.02808   1315.000   230.000

P =   760.00

EQUATION	ORDER	A12	A21	D	A122	A211
VAN LAAR	3	0.9546	0.8153	-	-	-
MARGULES	3	0.9454	0.8146	-	-	-
MARGULES	4	0.9993	0.8714	0.2412	-	-
ALPHA	2	18.3491	7.7086	-	-	-
ALPHA	3	2.3944	8.5332	-	20.2220	-6.3972

			DEVIATION IN VAPOR PHASE COMPOSITION				
X	Y	TEMP.	LAAR 3	MARG 3	MARG 4	ALPHA2	ALPHA3
0.0010	0.0450	83.50	-0.0252	-0.0256	-0.0231	-0.0262	-0.0221
0.0280	0.1880	78.20	0.1436	0.1400	0.1599	0.1206	0.1557
0.0320	0.2540	75.30	0.1013	0.0977	0.1172	0.0782	0.1136
0.0320	0.3280	72.60	0.0247	0.0211	0.0405	0.0042	0.0396
0.0400	0.3600	70.10	0.0348	0.0314	0.0495	0.0122	0.0471
0.0450	0.4430	67.30	-0.0276	-0.0309	-0.0138	-0.0498	-0.0155
0.0510	0.4720	66.10	-0.0331	-0.0362	-0.0205	-0.0567	-0.0233
0.0510	0.5230	64.10	-0.0863	-0.0894	-0.0737	-0.1077	-0.0743
0.0940	0.6000	60.90	-0.0563	-0.0580	-0.0517	-0.0854	-0.0598
0.2370	0.6440	60.40	-0.0001	0.0009	-0.0079	-0.0216	-0.0140
0.3000	0.6420	60.10	0.0103	0.0118	0.0023	-0.0003	0.0018
0.3480	0.6400	60.10	0.0144	0.0160	0.0077	0.0126	0.0108
0.5760	0.6600	59.60	-0.0063	-0.0058	-0.0005	0.0268	0.0093
0.6460	0.6700	59.40	-0.0102	-0.0102	-0.0017	0.0258	0.0054
0.7240	0.6890	59.40	-0.0126	-0.0131	-0.0037	0.0184	-0.0019
0.8200	0.7290	59.50	-0.0081	-0.0088	-0.0042	-0.0009	-0.0090
0.8880	0.7780	60.10	0.0033	0.0028	0.0003	-0.0226	-0.0045
0.9020	0.8000	60.40	-0.0016	-0.0019	-0.0060	-0.0359	-0.0098
0.9240	0.8200	60.70	0.0096	0.0094	0.0031	-0.0381	0.0019
0.9440	0.8460	61.00	0.0175	0.0174	0.0098	-0.0410	0.0115
0.9510	0.8700	61.40	0.0071	0.0070	-0.0009	-0.0544	0.0017
0.9640	0.9020	61.80	0.0023	0.0023	-0.0053	-0.0614	-0.0012
0.9800	0.9330	62.50	0.0098	0.0098	0.0039	-0.0462	0.0084
0.9920	0.9630	63.30	0.0128	0.0128	0.0098	-0.0203	0.0126
		MEAN	0.0275	0.0275	0.0257	0.0403	0.0273

METHYL ALCOHOL(1)-WATER(2)

BENNETT G.W.: J.CHEM.EDUC. 6,1544(1929).

ANTOINE VAPOR PRESSURE CONSTANTS

	A	B	C
(1)	8.07246	1574.990	238.860
(2)	7.96681	1668.210	228.000

P = 735.00

EQUATION	ORDER	A12	A21	D	A122	A211
VAN LAAR	3	0.4476	0.4088	-	-	-
MARGULES	3	0.4459	0.4095	-	-	-
MARGULES	4	0.5051	0.4875	0.2452	-	-
ALPHA	2	8.3561	0.4141	-	-	-
ALPHA	3	.1055/ 10	.1233/ 09	-	-.1045/ 10	-.1075/ 04

			DEVIATION IN VAPOR PHASE COMPOSITION				
X	Y	TEMP.	LAAR 3	MARG 3	MARG 4	ALPHA2	ALPHA3
0.0084	0.1030	96.50	-0.0266	-0.0269	-0.0176	-0.0303	0.0281
0.0258	0.2270	92.30	-0.0250	-0.0255	-0.0082	-0.0338	0.0029
0.0680	0.3910	87.50	0.0024	0.0019	0.0166	-0.0069	0.0053
0.1370	0.5680	80.10	-0.0113	-0.0114	-0.0086	-0.0157	-0.0141
0.2400	0.6800	75.90	-0.0084	-0.0081	-0.0132	-0.0014	-0.0010
0.4800	0.7900	70.60	0.0001	0.0003	0.0013	0.0145	0.0159
0.5720	0.8200	68.70	0.0014	0.0016	0.0055	0.0118	0.0114
0.7410	0.9060	66.40	-0.0272	-0.0272	-0.0239	-0.0321	-0.0418
		MEAN	0.0128	0.0129	0.0119	0.0183	0.0151

METHYL ALCOHOL(1)-WATER(2)

BREDIG G.,BAYER R.:Z.PHYS.CHEM.130,1(1927).

ANTOINE VAPOR PRESSURE CONSTANTS
```
        A          B          C
(1) 8.07246    1574.990    238.860
(2) 7.96681    1668.210    228.000
```

T =   39.76

EQUATION	ORDER	A12	A21	D	A122	A211
VAN LAAR	3	0.3212	0.4939	-	-	-
MARGULES	3	0.3040	0.4681	-	-	-
MARGULES	4	0.2600	0.3915	-0.1951	-	-
ALPHA	2	8.8793	-0.0160	-	-	-
ALPHA	3	-3.2701	-1.3329	-	12.1730	0.4823

DEVIATION IN VAPOR PHASE COMPOSITION

X	Y	PRESS.	LAAR 3	MARG 3	MARG 4	ALPHA2	ALPHA3
0.0478	0.2559	68.10	0.0700	0.0644	0.0512	0.0661	0.0674
0.0925	0.4562	85.60	0.0276	0.0238	0.0156	0.0242	0.0263
0.0925	0.4628	86.30	0.0210	0.0172	0.0090	0.0176	0.0197
0.1335	0.6214	97.60	-0.0460	-0.0479	-0.0513	-0.0483	-0.0460
0.1523	0.6164	103.40	-0.0090	-0.0101	-0.0119	-0.0108	-0.0085
0.1809	0.6486	109.80	-0.0006	-0.0009	-0.0008	-0.0017	0.0005
0.2032	0.6734	118.40	0.0009	0.0011	0.0022	0.0004	0.0024
0.2027	0.6796	119.10	-0.0059	-0.0057	-0.0046	-0.0063	-0.0044
0.2228	0.6954	122.40	-0.0010	-0.0004	0.0013	-0.0009	0.0008
0.2557	0.7263	132.00	-0.0031	-0.0021	-0.0000	-0.0022	-0.0010
0.2866	0.7383	138.20	0.0076	0.0088	0.0108	0.0092	0.0098
0.3065	0.7612	142.70	-0.0025	-0.0013	0.0005	-0.0005	-0.0002
0.3716	0.8053	155.30	-0.0119	-0.0109	-0.0104	-0.0088	-0.0101
0.4172	0.8048	161.50	0.0080	0.0086	0.0080	0.0117	0.0093
0.4362	0.8238	167.40	-0.0038	-0.0033	-0.0044	0.0001	-0.0028
0.5033	0.8457	175.40	-0.0034	-0.0035	-0.0061	0.0011	-0.0034
0.5933	0.8619	188.20	0.0052	0.0045	0.0008	0.0098	0.0044
0.6917	0.8835	202.50	0.0077	0.0067	0.0037	0.0110	0.0095
0.6949	0.8974	206.40	-0.0054	-0.0064	-0.0094	-0.0022	-0.0034
0.8002	0.9536	223.10	-0.0350	-0.0353	-0.0354	-0.0352	-0.0213
0.9270	0.9761	244.30	-0.0146	-0.0137	-0.0106	-0.0211	0.0071
		MEAN	0.0138	0.0132	0.0118	0.0138	0.0123

H

METHYL ALCOHOL(1)-WATER(2)

BREDIG G.,BAYER R.:Z.PHYS.CHEM.130,1(1927).

ANTOINE VAPOR PRESSURE CONSTANTS
	A	B	C
(1)	8.07246	1574.990	238.860
(2)	7.96681	1668.210	228.000

T = 49.76

EQUATION	ORDER	A12	A21	D	A122	A211
VAN LAAR	3	0.2226	0.3419	-	-	-
MARGULES	3	0.1999	0.3370	-	-	-
MARGULES	4	0.1042	0.2004	-0.3825	-	-
ALPHA	2	6.2810	-0.3319	-	-	-
ALPHA	3	-1.4110	-1.5153	-	7.6446	0.8436

DEVIATION IN VAPOR PHASE COMPOSITION

X	Y	PRESS.	LAAR 3	MARG 3	MARG 4	ALPHA2	ALPHA3
0.0486	0.2741	119.50	-0.0027	-0.0096	-0.0349	-0.0082	-0.0084
0.1218	0.4741	157.00	0.0153	0.0116	0.0009	0.0109	0.0124
0.1478	0.5220	169.70	0.0182	0.0160	0.0109	0.0147	0.0167
0.2131	0.6294	196.00	0.0056	0.0063	0.0099	0.0045	0.0070
0.2693	0.7106	217.70	-0.0178	-0.0158	-0.0098	-0.0171	-0.0149
0.3252	0.7580	236.60	-0.0212	-0.0185	-0.0134	-0.0191	-0.0177
0.5143	0.8203	283.00	0.0135	0.0153	0.0113	0.0177	0.0145
0.6279	0.8654	306.40	0.0080	0.0085	0.0022	0.0114	0.0062
0.7083	0.9007	324.10	-0.0026	-0.0027	-0.0076	-0.0008	-0.0049
0.8037	0.9406	348.40	-0.0138	-0.0142	-0.0144	-0.0149	-0.0134
0.9007	0.9627	373.50	-0.0039	-0.0040	0.0001	-0.0081	0.0010
0.9461	0.9736	391.10	0.0025	0.0025	0.0066	-0.0020	0.0074
		MEAN	0.0104	0.0104	0.0102	0.0108	0.0104

METHYL ALCOHOL(1)-WATER(2)

BREDIG G.,BAYER R.:Z.PHYS.CHEM. 130,1(1927).

ANTOINE VAPOR PRESSURE CONSTANTS
	A	B	C
(1)	8.07246	1574.990	238.860
(2)	7.96681	1668.210	228.000

P =   760.00

EQUATION	ORDER	A12	A21	D	A122	A211
VAN LAAR	3	0.4219	0.3210	-	-	-
MARGULES	3	0.4157	0.3100	-	-	-
MARGULES	4	0.4004	0.2847	-0.0666	-	-
ALPHA	2	6.9067	0.3221	-	-	-
ALPHA	3	5.9424	0.9916	-	2.1412	-1.1247

			DEVIATION IN VAPOR PHASE COMPOSITION				
X	Y	TEMP.	LAAR 3	MARG 3	MARG 4	ALPHA2	ALPHA3
0.0531	0.2834	92.90	0.0353	0.0337	0.0294	0.0102	0.0301
0.0767	0.4001	90.30	-0.0017	-0.0029	-0.0064	-0.0259	-0.0078
0.0926	0.4353	88.90	0.0056	0.0046	0.0018	-0.0166	-0.0006
0.1257	0.4831	86.60	0.0267	0.0263	0.0248	0.0099	0.0212
0.1315	0.5455	85.00	-0.0242	-0.0246	-0.0258	-0.0414	-0.0310
0.1674	0.5585	83.20	0.0153	0.0153	0.0151	0.0044	0.0099
0.1818	0.5775	82.30	0.0140	0.0141	0.0143	0.0052	0.0090
0.2083	0.6273	81.60	-0.0082	-0.0079	-0.0073	-0.0127	-0.0117
0.2319	0.6485	80.20	-0.0072	-0.0069	-0.0061	-0.0093	-0.0105
0.2818	0.6775	78.00	0.0023	0.0025	0.0035	0.0044	0.0000
0.2909	0.6801	77.80	0.0056	0.0059	0.0068	0.0086	0.0037
0.3333	0.6918	76.70	0.0195	0.0196	0.0202	0.0249	0.0185
0.3513	0.7347	76.20	-0.0135	-0.0135	-0.0130	-0.0075	-0.0143
0.4620	0.7756	73.80	-0.0024	-0.0030	-0.0036	0.0034	-0.0031
0.5292	0.7971	72.70	0.0035	0.0028	0.0016	0.0062	0.0017
0.5937	0.8183	71.30	0.0078	0.0070	0.0056	0.0053	0.0037

METHANOL(1)-WATER(2)

DUNLOP J.G.:M.S.THESIS,BROOKLYN POLYTECHN.INST.,1948.

ANTOINE VAPOR PRESSURE CONSTANTS

	A	B	C
(1)	8.07246	1574.990	238.860
(2)	7.96681	1668.210	228.000

P = 760.00

EQUATION	ORDER	A12	A21	D	A122	A211
VAN LAAR	3	0.3785	0.2116	-	-	-
MARGULES	3	0.3574	0.1743	-	-	-
MARGULES	4	0.3349	0.1384	-0.1051	-	-
ALPHA	2	5.3481	-0.0768	-	-	-
ALPHA	3	4.0263	0.8471	-	2.8643	-1.4861

			DEVIATION IN VAPOR PHASE COMPOSITION				
X	Y	TEMP.	LAAR 3	MARG 3	MARG 4	ALPHA2	ALPHA3
0.0200	0.1340	96.40	0.0061	0.0017	-0.0031	-0.0209	-0.0001
0.0400	0.2300	93.50	0.0112	0.0063	0.0004	-0.0259	0.0026
0.0600	0.3040	91.20	0.0135	0.0095	0.0042	-0.0252	0.0047
0.0800	0.3650	89.30	0.0124	0.0096	0.0054	-0.0238	0.0044
0.1000	0.4180	87.70	0.0078	0.0062	0.0034	-0.0237	0.0010
0.1500	0.5170	84.40	-0.0017	-0.0011	-0.0011	-0.0195	-0.0054
0.2000	0.5790	81.70	-0.0006	0.0009	0.0026	-0.0063	-0.0021
0.3000	0.6650	78.00	0.0012	0.0018	0.0039	0.0104	0.0014
0.4000	0.7290	75.30	0.0013	-0.0002	0.0004	0.0142	0.0002
0.5000	0.7790	73.10	0.0046	0.0018	0.0004	0.0136	0.0010
0.6000	0.8250	71.20	0.0065	0.0036	0.0012	0.0065	0.0005
0.7000	0.8700	69.30	0.0063	0.0047	0.0027	-0.0047	-0.0007
0.8000	0.9150	67.50	0.0040	0.0042	0.0038	-0.0168	-0.0014
0.9000	0.9580	66.00	0.0021	0.0035	0.0044	-0.0211	0.0004
0.9500	0.9790	65.00	0.0012	0.0023	0.0032	-0.0161	0.0012
		MEAN	0.0054	0.0038	0.0027	0.0166	0.0018

METHYL ALCOHOL(1)-WATER(2)

HUGES H.E.,MALONEY J.O.:CHEM.ENG.PROG. 48,192(1952).

ANTOINE VAPOR PRESSURE CONSTANTS

	A	B	C
(1)	8.07246	1574.990	238.860
(2)	7.96681	1668.210	228.000

P = 760.00

EQUATION	ORDER	A12	A21	D	A122	A211
VAN LAAR	3	0.3554	0.1876	-	-	-
MARGULES	3	0.3292	0.1542	-	-	-
MARGULES	4	0.3840	0.3916	0.4978	-	-
ALPHA	2	5.1719	-0.1490	-	-	-
ALPHA	3	16.6605	3.9462	-	-9.2108	-4.2185

DEVIATION IN VAPOR PHASE COMPOSITION

X	Y	TEMP.	LAAR 3	MARG 3	MARG 4	ALPHA2	ALPHA3
0.0321	0.1900	95.30	0.0060	0.0000	0.0122	-0.0232	0.0107
0.0372	0.2220	94.00	-0.0023	-0.0083	0.0037	-0.0335	0.0016
0.0523	0.2940	92.50	-0.0144	-0.0198	-0.0094	-0.0469	-0.0117
0.0595	0.3080	91.50	-0.0032	-0.0082	0.0011	-0.0356	-0.0015
0.0750	0.3520	89.90	0.0001	-0.0036	0.0027	-0.0303	0.0002
0.0876	0.3900	88.10	-0.0040	-0.0068	-0.0028	-0.0325	-0.0058
0.1540	0.4900	85.10	0.0194	0.0207	0.0157	0.0104	0.0166
0.1580	0.5160	83.90	0.0003	0.0017	-0.0035	-0.0088	-0.0037
0.1820	0.5520	82.90	-0.0043	-0.0022	-0.0085	-0.0073	-0.0080
0.2250	0.5930	82.10	0.0010	0.0036	-0.0025	0.0077	-0.0007
0.2900	0.6430	78.70	0.0095	0.0114	0.0097	0.0231	0.0084
0.3490	0.7030	76.70	-0.0076	-0.0069	-0.0025	0.0088	-0.0074
0.8130	0.9180	67.40	0.0090	0.0092	0.0084	-0.0112	0.0027
0.9180	0.9630	65.60	0.0059	0.0070	-0.0013	-0.0144	-0.0027
		MEAN	0.0062	0.0078	0.0060	0.0210	0.0058

109

METHYL ALCOHOL(1)-WATER(2)

RAMALHO R.S.,TILLER F.M.,JAMES W.J.,BUNCH D.W.:IND.ENG.CHEM.:53,895(1961).

ANTOINE VAPOR PRESSURE CONSTANTS

	A	B	C
(1)	8.07246	1574.990	238.860
(2)	7.96681	1668.210	228.000

P = 760.00

EQUATION	ORDER	A12	A21	D	A122	A211
VAN LAAR	3	0.3625	0.2418	-	-	-
MARGULES	3	0.3567	0.2236	-	-	-
MARGULES	4	0.3362	0.1574	-0.1734	-	-
ALPHA	2	5.5439	-0.0186	-	-	-
ALPHA	3	-15.4867	-4.0508	-	20.2397	2.4846

DEVIATION IN VAPOR PHASE COMPOSITION

X	Y	TEMP.	LAAR 3	MARG 3	MARG 4	ALPHA2	ALPHA3
0.0293	0.1831	95.20	0.0046	0.0035	-0.0009	-0.0215	-0.0362
0.0346	0.2107	94.50	0.0029	0.0019	-0.0024	-0.0252	-0.0412
0.0406	0.2363	93.70	0.0046	0.0037	-0.0004	-0.0252	-0.0424
0.0422	0.2652	92.80	-0.0168	-0.0177	-0.0218	-0.0475	-0.0650
0.0557	0.2978	91.80	0.0032	0.0027	-0.0005	-0.0287	-0.0476
0.0644	0.3265	90.90	0.0044	0.0041	0.0016	-0.0276	-0.0468
0.0737	0.3608	90.00	-0.0010	-0.0010	-0.0027	-0.0325	-0.0517
0.0838	0.3861	89.10	0.0020	0.0023	0.0014	-0.0284	-0.0472
0.0948	0.4142	89.20	0.0005	0.0011	0.0011	-0.0272	-0.0451
0.2801	0.6621	78.80	-0.0027	-0.0018	0.0018	-0.0008	0.0251
0.3004	0.6882	77.60	-0.0132	-0.0125	-0.0095	-0.0101	0.1191
0.3212	0.6882	77.60	0.0006	0.0010	0.0032	0.0058	-0.0248
0.3435	0.7002	76.90	0.0031	0.0032	0.0046	0.0094	0.0024
0.3664	0.7178	76.20	-0.0004	-0.0006	-0.0002	0.0065	0.0073
0.3909	0.7274	75.70	0.0040	0.0034	0.0029	0.0115	0.0168
0.4141	0.7428	75.10	0.0014	0.0005	-0.0010	0.0088	0.0171
0.4391	0.7597	74.60	-0.0025	-0.0036	-0.0061	0.0047	0.0154
0.4637	0.7668	74.00	0.0028	0.0015	-0.0019	0.0092	0.0217
0.8457	0.9360	67.20	-0.0005	-0.0002	0.0003	-0.0242	-0.0514
0.8867	0.9632	66.60	-0.0107	-0.0101	-0.0085	-0.0349	-0.1047
0.9293	0.9771	65.70	-0.0068	-0.0061	-0.0042	-0.0281	-0..130
		MEAN	0.0042	0.0039	0.0037	0.0199	0.0933

METHYL ALCOHOL(1)-WATER(2)

SCHROEDER W.,:CHEM.ING.TECHN.30,523(1959).

ANTOINE VAPOR PRESSURE CONSTANTS

	A	B	C
(1)	8.07246	1574.990	238.860
(2)	7.96681	1668.210	228.000

T = 140.00

EQUATION	ORDER	A12	A21	D	A122	A211
VAN LAAR	3	0.3577	0.1987	-	-	-
MARGULES	3	0.3241	0.1774	-	-	-
MARGULES	4	0.3587	0.2194	0.1255	-	-
ALPHA	2	3.6182	-0.0412	-	-	-
ALPHA	3	8.2634	2.4275	-	-3.0880	-2.5486

			DEVIATION IN VAPOR PHASE COMPOSITION				
X	Y	PRESS.	LAAR 3	MARG 3	MARG 4	ALPHA2	ALPHA3
0.0960	0.3440	3822.00	0.0185	0.0135	0.0197	-0.0311	-0.0010
0.2750	0.5790	5100.00	0.0089	0.0116	0.0090	0.0026	0.0009
0.4440	0.7020	5910.00	0.0014	0.0022	0.0018	0.0081	-0.0003
0.5950	0.7860	6620.00	0.0023	0.0013	0.0036	0.0018	-0.0007
0.7910	0.8800	7450.00	0.0114	0.0114	0.0118	-0.0071	0.0010
0.9460	0.9620	7940.00	0.0099	0.0107	0.0091	-0.0059	-0.0008
		MEAN	0.0087	0.0085	0.0092	0.0094	0.0008

111

PHOSGENE(1)-TOLUENE(2)

KIREEV V.A.,KAPLAN S.I.,VASNEVA K.I.:ZH.OBSHCH.KHIM.6,799(1936).

ANTOINE VAPOR PRESSURE CONSTANTS

	A	B	C
(1)	6.84297	941.2500	230.000
(2)	6.95334	1343.943	219.377

T =   20.00

EQUATION	ORDER	A12	A21	D	A122	A211
VAN LAAR	3	0.2196	0.7894	-	-	-
MARGULES	3	0.1962	0.4842	-	-	-
MARGULES	4	0.2438	0.6402	0.3914	-	-
ALPHA	2	90.3274	-0.6624	-	-	-
ALPHA	3	.1940/ 06	.2195/ 04	-	-.1933/ 06	-.1598/ 04

			DEVIATION IN VAPOR PHASE COMPOSITION				
X	Y	PRESS.	LAAR 3	MARG 3	MARG 4	ALPHA2	ALPHA3
0.0891	0.8966	184.80	-0.0002	-0.0006	0.0001	-0.0002	0.0001
0.2298	0.9615	421.00	-0.0001	0.0008	-0.0006	-0.0002	-0.0005
0.3532	0.9760	567.00	0.0010	0.0011	0.0009	0.0009	0.0009
0.4191	0.9812	651.00	0.0004	0.0000	0.0005	0.0004	0.0004
0.4530	0.9836	702.00	-0.0001	-0.0007	-0.0000	-0.0001	-0.0000
0.4838	0.9855	747.00	-0.0006	-0.0013	-0.0005	-0.0005	-0.0004
0.5104	0.9868	780.00	-0.0008	-0.0016	-0.0007	-0.0006	-0.0005
		MEAN	0.0005	0.0009	0.0005	0.0004	0.0004

TRICHLOROETHYLENE(1)-NITROMETHANE(2)

YU A.H.,HICKMAN J.B.:J.CHEM.EDUC.26,207(1949).

ANTOINE VAPOR PRESSURE CONSTANTS
```
          A         B         C
(1) 7.02808   1315.000   230.000
(2) 7.15541   1366.483   218.611
```

P = 760.00

EQUATION	ORDER	A12	A21	D	A122	A211
VAN LAAR	3	0.6366	0.7534	-	-	-
MARGULES	3	0.6343	0.7439	-	-	-
MARGULES	4	0.6846	0.8197	0.2114	-	-
ALPHA	2	7.4113	3.4827	-	-	-
ALPHA	3	5.2293	1.9465	-	1.1036	1.1246

			DEVIATION IN VAPOR PHASE COMPOSITION				
X	Y	TEMP.	LAAR 3	MARG 3	MARG 4	ALPHA2	ALPHA3
0.0250	0.1550	96.30	-0.0181	-0.0185	-0.0076	0.0075	-0.0063
0.0500	0.2300	94.00	0.0055	0.0050	0.0175	0.0349	0.0199
0.0700	0.3350	90.50	-0.0374	-0.0378	-0.0267	-0.0118	-0.0249
0.1250	0.4200	88.50	-0.0075	-0.0076	-0.0030	0.0069	0.0011
0.2900	0.5500	83.00	0.0160	0.0161	0.0121	0.0100	0.0147
0.3400	0.5850	82.10	0.0033	0.0032	0.0000	-0.0035	0.0014
0.3950	0.6150	81.50	-0.0082	-0.0086	-0.0099	-0.0139	-0.0097
0.4850	0.6400	80.90	-0.0113	-0.0121	-0.0093	-0.0122	-0.0104
0.6400	0.6550	80.30	0.0041	0.0030	0.0105	0.0138	0.0103
0.7100	0.6850	80.40	-0.0086	-0.0094	-0.0029	0.0045	-0.0008
0.8650	0.7600	81.10	-0.0043	-0.0035	-0.0094	0.0016	-0.0017
		MEAN	0.0113	0.0113	0.0099	0.0110	0.0092

1,1,1-TRICHLOROETHANE(1)-1,2-DICHLOROETHANE(2)

BIGG D.C.,BANERJEE S.C.,DORAISWAMY L.K.:J.CHEM.ENG.DATA 9(1),17(1964).

ANTOINE VAPOR PRESSURE CONSTANTS

	A	B	C
(1)	6.94983	1217.000	225.000
(2)	6.95222	1247.800	223.000

P = 760.00

EQUATION	ORDER	A12	A21	D	A122	A211
VAN LAAR	3	0.1023	1.0247	-	-	-
MARGULES	3	0.0353	0.2640	-	-	-
MARGULES	4	0.3004	0.5796	0.8710	-	-
ALPHA	2	0.8522	-0.1790	-	-	-
ALPHA	3	5.5745	3.4790	-	-4.2531	-3.8363

			DEVIATION IN VAPOR PHASE COMPOSITION				
X	Y	TEMP.	LAAR 3	MARG 3	MARG 4	ALPHA2	ALPHA3
0.0450	0.0800	80.00	-0.0064	-0.0139	0.0163	-0.0007	0.0111
0.0750	0.1400	79.30	-0.0198	-0.0293	0.0037	-0.0119	0.0014
0.1200	0.2150	78.40	-0.0284	-0.0379	-0.0125	-0.0189	-0.0077
0.2050	0.3150	77.00	-0.0135	-0.0187	-0.0172	-0.0051	-0.0019
0.3150	0.4450	75.60	-0.0125	-0.0139	-0.0275	-0.0094	-0.0142
0.3500	0.4520	75.20	0.0185	0.0169	0.0040	0.0196	0.0135
0.4700	0.5650	74.10	0.0239	0.0160	0.0182	0.0193	0.0129
0.5750	0.6550	73.20	0.0240	0.0067	0.0230	0.0176	0.0152
0.6100	0.7300	73.00	-0.0236	-0.0440	-0.0254	-0.0295	-0.0301
0.6750	0.7600	72.65	-0.0058	-0.0306	-0.0128	-0.0090	-0.0059
0.7750	0.8250	72.40	-0.0058	-0.0296	-0.0264	0.0016	0.0096
		MEAN	0.0166	0.0234	0.0170	0.0130	0.0112

ACETONITRILE(1)-BENZENE(2)

BROWN I.,SMITH F.:AUSTR.J.CHEM.8,62(1955).

ANTOINE VAPOR PRESSURE CONSTANTS

	A	B	C
(1)	7.24299	1397.929	238.894
(2)	6.90565	1211.033	220.790

T = 45.00

EQUATION	ORDER	A12	A21	D	A122	A211
VAN LAAR	3	0.4140	0.4254	-	-	-
MARGULES	3	0.4142	0.4251	-	-	-
MARGULES	4	0.4633	0.4721	0.1632	-	-
ALPHA	2	1.6445	1.8922	-	-	-
ALPHA	3	2.7579	2.9605	-	-0.9524	-0.7936

DEVIATION IN VAPOR PHASE COMPOSITION

X	Y	PRESS.	LAAR 3	MARG 3	MARG 4	ALPHA2	ALPHA3
0.0455	0.1056	239.70	-0.0098	-0.0098	-0.0033	-0.0042	-0.0014
0.0940	0.1818	251.67	-0.0078	-0.0078	-0.0021	-0.0019	-0.0003
0.1829	0.2783	264.66	-0.0005	-0.0005	-0.0007	0.0021	0.0006
0.2909	0.3607	273.45	0.0043	0.0042	-0.0002	0.0036	0.0009
0.3980	0.4274	277.49	0.0028	0.0027	-0.0010	0.0013	-0.0002
0.5069	0.4885	278.03	-0.0008	-0.0009	-0.0009	-0.0013	-0.0006
0.5458	0.5098	277.36	-0.0019	-0.0020	-0.0005	-0.0018	-0.0004
0.5946	0.5375	275.86	-0.0035	-0.0036	-0.0006	-0.0026	-0.0004
0.7206	0.6157	268.46	-0.0041	-0.0042	-0.0000	-0.0014	0.0008
0.8145	0.6913	257.81	-0.0012	-0.0012	-0.0008	0.0015	0.0011
0.8972	0.7869	242.50	0.0028	0.0028	-0.0025	0.0040	-0.0004
0.9573	0.8916	225.30	0.0037	0.0038	-0.0031	0.0035	-0.0021
		MEAN	0.0036	0.0036	0.0013	0.0024	0.0008

ACETONITRILE(1)-NITROMETHANE(2)

BROWN I.,SMITH F.:AUSTR.J.CHEM.8,62(1955).

ANTOINE VAPOR PRESSURE CONSTANTS
	A	B	C
(1)	7.24299	1397.929	238.894
(2)	7.15540	1366.483	218.611

T =   60.00

EQUATION	ORDER	A12	A21	D	A122	A211
VAN LAAR	3	-0.0014	0.0033	-	-	-
MARGULES	3	0.0114	-0.0339	-	-	-
MARGULES	4	0.0138	-0.0310	0.0086	-	-
ALPHA	2	1.0057	-0.5020	-	6.1977	
ALPHA	3	-5.1924	-3.5923	-	6.1977	3.0912

			DEVIATION IN VAPOR PHASE COMPOSITION				
X	Y	PRESS.	LAAR 3	MARG 3	MARG 4	ALPHA2	ALPHA3
0.0951	0.1741	195.50	0.0038	0.0049	0.0052	0.0000	-0.0000
0.1940	0.3255	213.70	0.0058	0.0039	0.0039	0.0001	0.0001
0.2930	0.4539	232.24	0.0062	0.0016	0.0014	0.0001	-0.0000
0.3939	0.5658	251.20	0.0057	0.0005	0.0004	0.0002	0.0001
0.4614	0.6325	264.24	0.0043	0.0001	0.0001	-0.0003	-0.0012
0.5001	0.6674	271.45	0.0038	0.0008	0.0009	0.0001	0.0001
0.6051	0.7548	291.64	-0.0028	0.0020	0.0022	-0.0002	-0.0001
0.8055	0.8925	329.97	-0.0022	0.0046	0.0046	0.0001	-0.0001
0.9025	0.9488	349.17	0.0008	0.0036	0.0035	0.0001	0.0000
0.9486	0.9737	358.27	0.0005	0.0022	0.0021	0.0000	-0.0000
		MEAN	0.0036	0.0024	0.0024	0.0001	0.0002

116

ACETONITRILE(1)-ALPHA PICOLINE(2)

OGAWA S.,KISHIDA H.,KUYAMA H.:KAGAKU KOGAKU 22,157(1958).

ANTOINE VAPOR PRESSURE CONSTANTS

	A	B	C
(1)	7.24299	1397.929	238.894
(2)	7.04278	1422.874	212.467

P = 760.00

EQUATION	ORDER	A12	A21	D	A122	A211
VAN LAAR	3	0.1913	0.2347	-	-	-
MARGULES	3	0.1909	0.2297	-	-	-
MARGULES	4	0.2401	0.2957	0.1829	-	-
ALPHA	2	4.8017	-0.5230	-	-	-
ALPHA	3	8.6625	0.4299	-	-3.4413	-0.9302

			DEVIATION IN VAPOR PHASE COMPOSITION				
X	Y	TEMP.	LAAR 3	MARG 3	MARG 4	ALPHA2	ALPHA3
0.0500	0.2420	121.90	-0.0110	-0.0110	0.0016	-0.0109	-0.0017
0.1000	0.3900	116.20	-0.0044	-0.0044	0.0043	-0.0058	0.0024
0.1500	0.5010	111.20	-0.0053	-0.0053	-0.0020	-0.0078	-0.0026
0.2000	0.5780	107.20	-0.0007	-0.0006	-0.0011	-0.0032	-0.0008
0.2500	0.6380	103.80	0.0023	0.0023	-0.0001	0.0003	0.0005
0.3000	0.6870	100.80	0.0034	0.0033	0.0006	0.0021	0.0010
0.3500	0.7280	98.40	0.0032	0.0030	0.0009	0.0029	0.0009
0.4000	0.7630	96.30	0.0022	0.0019	0.0010	0.0029	0.0007
0.4500	0.7940	94.60	0.0001	-0.0002	0.0001	0.0017	-0.0004
0.5000	0.8200	93.10	-0.0006	-0.0010	0.0005	0.0016	-0.0001
0.5500	0.8450	92.60	-0.0038	-0.0042	-0.0019	-0.0007	-0.0018
0.6000	0.8650	90.20	-0.0025	-0.0029	-0.0002	-0.0004	-0.0009
0.6500	0.8830	89.00	-0.0017	-0.0020	0.0006	-0.0001	0.0002
0.7000	0.9000	87.90	-0.0010	-0.0012	0.0009	-0.0002	0.0008
0.7500	0.9170	86.80	-0.0010	-0.0011	0.0000	-0.0013	0.0002
0.8000	0.9320	85.70	0.0006	0.0006	0.0006	-0.0011	0.0007
0.8500	0.9470	84.60	0.0020	0.0021	0.0010	-0.0009	0.0008
0.9000	0.9640	83.50	0.0015	0.0016	-0.0002	-0.0022	-0.0010
0.9500	0.9800	82.50	0.0024	0.0025	0.0008	-0.0009	-0.0006
		MEAN	0.0026	0.0027	0.0010	0.0025	0.0010

ACETONITRILE(1)-WATER(2)

BLACKFORD D.S.,YORK R.:J.CHEM.ENG.DATA 10(4),313(1965).

ANTOINE VAPOR PRESSURE CONSTANTS
	A	B	C
(1)	7.24299	1397.929	238.894
(2)	7.96681	1668.210	228.000

P = 760.00

EQUATION	ORDER	A12	A21	D	A122	A211
VAN LAAR	3	0.8187	0.7376	-	-	-
MARGULES	3	0.8145	0.7375	-	-	-
MARGULES	4	0.9344	0.8793	0.4697	-	-
ALPHA	2	13.5373	5.3282	-	-	-
ALPHA	3	9.9341	8.6222	-	8.6752	-5.3311

DEVIATION IN VAPOR PHASE COMPOSITION

X	Y	TEMP.	LAAR 3	MARG 3	MARG 4	ALPHA2	ALPHA3
0.0290	0.2630	86.50	-0.0117	-0.0131	0.0273	0.0049	0.0480
0.0930	0.5050	81.10	-0.0236	-0.0245	-0.0034	-0.0285	-0.0047
0.1420	0.5590	80.00	-0.0050	-0.0054	-0.0004	-0.0159	-0.0047
0.2540	0.6170	78.60	0.0081	0.0085	-0.0033	-0.0008	-0.0040
0.4020	0.6550	77.40	0.0009	0.0016	-0.0056	0.0056	-0.0031
0.5070	0.6640	76.70	0.0035	0.0040	0.0075	0.0167	0.0090
0.5270	0.6730	76.60	-0.0032	-0.0028	0.0028	0.0112	0.0039
0.7180	0.7280	76.00	-0.0204	-0.0205	-0.0072	-0.0103	-0.0078
0.8390	0.7800	76.60	-0.0117	-0.0119	-0.0139	-0.0282	-0.0134
0.8560	0.7610	76.80	0.0204	0.0202	0.0149	-0.0018	0.0152
0.9860	0.9450	80.40	0.0197	0.0197	0.0085	-0.0145	0.0040
		MEAN	0.0117	0.0120	0.0086	0.0126	0.0107

118

ACETONITRILE(1)-WATER(2)

OTHMER D.F.,JOSEFOWITZ S.:IND.ENG.CHEM.39,1175(1947)

ANTOINE VAPOR PRESSURE CONSTANTS

	A	B	C
(1)	7.07354	1279.200	224.000
(2)	7.96681	1668.210	228.000

P =   150.00

EQUATION	ORDER	A12	A21	D	A122	A211
VAN LAAR	3	0.7526	0.9830	-	-	-
MARGULES	3	0.7490	0.9579	-	-	-
MARGULES	4	0.8781	1.0702	0.5371	-	-
ALPHA	2	25.7378	4.4822	-	-	-
ALPHA	3	9.7827	2.3124	-	13.2249	-0.1323

			DEVIATION IN VAPOR PHASE COMPOSITION				
X	Y	TEMP.	LAAR 3	MARG 3	MARG 4	ALPHA2	ALPHA3
0.0030	0.0640	58.70	-0.0233	-0.0236	-0.0108	0.0094	0.0027
0.0520	0.5070	44.80	-0.0679	-0.0681	-0.0300	0.0235	0.0135
0.1680	0.7320	36.70	-0.0386	-0.0372	-0.0462	-0.0112	-0.0047
0.5130	0.8100	34.50	-0.0219	-0.0237	-0.0238	0.0021	0.0034
0.7720	0.8350	34.10	-0.0390	-0.0408	-0.0273	0.0041	-0.0044
0.9000	0.8600	34.60	-0.0171	-0.0151	-0.0198	0.0047	0.0046
0.9550	0.9100	36.00	-0.0083	-0.0051	-0.0181	-0.0134	-0.0002
0.9800	0.9550	36.60	-0.0070	-0.0047	-0.0152	-0.0227	-0.0056
		MEAN	0.0279	0.0273	0.0239	0.0114	0.0049

ACETONITRILE(1)-WATER(2)

OTHMER D.F.,JOSEFOWITZ S.:IND.ENG.CHEM.39,1175(1947)

ANTOINE VAPOR PRESSURE CONSTANTS
```
          A          B          C
(1) 7.07354    1279.200    224.000
(2) 7.96681    1668.210    228.000
```


P =   300.00

EQUATION	ORDER	A12	A21	D	A122	A211
VAN LAAR	3	0.8795	0.9364	-	-	-
MARGULES	3	0.8822	0.9320	-	-	-
MARGULES	4	1.0431	1.1004	0.6531	-	-
ALPHA	2	31.4426	8.6848	-	-	-
ALPHA	3	15.2012	11.4865	-	24.8390	-7.3919

			DEVIATION IN VAPOR PHASE COMPOSITION				
X	Y	TEMP.	LAAR 3	MARG 3	MARG 4	ALPHA2	ALPHA3
0.0080	0.1070	73.50	0.0078	0.0084	0.0481	0.0883	0.1234
0.0300	0.4200	64.70	-0.0885	-0.0873	-0.0256	0.0159	0.0548
0.1180	0.6860	54.00	-0.0567	-0.0561	-0.0451	-0.0310	-0.0190
0.3110	0.7320	51.70	0.0001	-0.0003	-0.0199	0.0023	-0.0006
0.5200	0.7460	51.40	-0.0100	-0.0108	-0.0071	0.0136	0.0073
0.7000	0.7720	51.10	-0.0354	-0.0359	-0.0163	0.0027	-0.0015
0.8600	0.8080	51.20	-0.0236	-0.0233	-0.0257	-0.0113	-0.0039
0.9140	0.8350	51.60	-0.0040	-0.0033	-0.0212	-0.0200	-0.0024
0.9800	0.9140	52.30	0.0294	0.0299	0.0107	-0.0205	0.0118
0.9900	0.9530	53.20	0.0166	0.0169	0.0048	-0.0217	0.0041
		MEAN	0.0272	0.0272	0.0225	0.0227	0.0229

ACETONITRILE(1)-WATER(2)

VIERK A.L.:Z.ANORG.CHEM. 261,283(1950)

ANTOINE VAPOR PRESSURE CONSTANTS
```
         A          B          C
(1) 7.70735    1279.200    224.000
(2) 7.96681    1668.200    228.000
```

T =   20.00

EQUATION	ORDER	A12	A21	D	A122	A211
VAN LAAR	3	7.5793	0.8639	-	-	-
MARGULES	3	1.0018	0.6293	-	-	-
MARGULES	4	0.8910	0.5495	-0.3010	-	-
ALPHA	2	119.0763	17.3311	-	-	-
ALPHA	3	.2476/ 10	.3323/ 09	-	-.9433/ 09	-.2553/ 07

DEVIATION IN VAPOR PHASE COMPOSITION

X	Y	PRESS.	LAAR 3	MARG 3	MARG 4	ALPHA2	ALPHA3
0.0700	0.7940	63.00	0.2041	0.0978	0.0843	-0.0023	0.0340
0.1500	0.8290	71.50	0.1253	0.0952	0.0924	0.0046	0.0055
0.1900	0.8470	73.20	0.0647	0.0817	0.0813	-0.0051	-0.0094
0.3500	0.8580	74.20	-0.0593	0.0733	0.0760	-0.0014	-0.0090
0.6500	0.8590	74.70	-0.0094	0.0791	0.0782	0.0070	0.0075
0.7600	0.8700	74.90	0.0247	0.0805	0.0799	-0.0014	0.0019
0.9200	0.8890	74.80	0.0759	0.0916	0.0931	-0.0118	-0.0099
		MEAN	0.0805	0.0856	0.0836	0.0048	0.0110

121

I

ACETONITRILE(1)-WATER(2)

VIERK A.L.:Z.ANORG.CHEM. 261,283(1950)

ANTOINE VAPOR PRESSURE CONSTANTS

	A	B	C
(1)	7.07354	1279.200	224.000
(2)	7.96681	1668.200	228.000

T = 30.00

EQUATION	ORDER	A12	A21	D	A122	A211
VAN LAAR	3	1.0016	0.9832	-	-	-
MARGULES	3	1.0007	0.9842	-	-	-
MARGULES	4	1.1708	1.1835	0.8123	-	-
ALPHA	2	72.5338	16.1575	-	-	-
ALPHA	3	0.5850	18.7242	-	85.8584	-18.2857

DEVIATION IN VAPOR PHASE COMPOSITION

X	Y	PRESS.	LAAR 3	MARG 3	MARG 4	ALPHA2	ALPHA3
0.0300	0.6300	80.00	-0.1489	-0.1493	-0.0820	-0.0322	-0.0076
0.0700	0.7200	101.00	-0.0683	-0.0686	-0.0384	-0.0126	-0.0002
0.1700	0.7700	115.00	-0.0085	-0.0086	-0.0220	-0.0001	0.0026
0.4400	0.7970	119.00	-0.0183	-0.0181	-0.0257	0.0043	-0.0013
0.6500	0.8000	120.00	-0.0385	-0.0384	-0.0139	0.0099	-0.0001
0.8700	0.8300	119.40	-0.0197	-0.0198	-0.0217	-0.0074	0.0006
0.9200	0.8700	119.00	-0.0156	-0.0157	-0.0347	-0.0386	-0.0003
		MEAN	0.0454	0.0455	0.0341	0.0150	0.0018

122

1,2-ETHYLENE DIBROMIDE(1)-CHLOROBENZENE(2)

LACHER J.R.,HUNT R.E.:J.AM.CHEM.SOC.,63,1752(1941).

ANTOINE VAPOR PRESSURE CONSTANTS

	A	B	C
(1)	7.06245	1489.700	220.000
(2)	6.94504	1413.120	216.000

T =   75.00

EQUATION	ORDER	A12	A21	D	A122	A211
VAN LAAR	3	0.0015	-0.0037	-	-	-
MARGULES	3	-0.0623	0.2235	-	-	-
MARGULES	4	-0.1402	0.1531	-0.2237	-	-
ALPHA	2	0.2529	0.2773	-	-	-
ALPHA	3	0.4457	0.3829	-	-0.2230	-0.0669

			DEVIATION IN VAPOR PHASE COMPOSITION				
X	Y	PRESS.	LAAR 3	MARG 3	MARG 4	ALPHA2	ALPHA3
0.1684	0.1882	125.40	-0.0425	-0.0340	-0.0365	0.0015	-0.0000
0.3576	0.3707	128.00	-0.0509	-0.0127	-0.0077	-0.0001	0.0000
0.5132	0.5101	128.40	-0.0368	-0.0010	-0.0005	-0.0011	-0.0001
0.6443	0.6268	127.70	0.0067	-0.0071	-0.0108	-0.0006	0.0000
0.7895	0.7633	125.60	0.0098	-0.0249	-0.0249	0.0009	0.0000
0.9214	0.9034	122.30	0.0054	-0.0287	-0.0206	0.0016	-0.0000
		MEAN	0.0254	0.0181	0.0168	0.0010	0.0000

1,2-ETHYLENE DIBROMIDE(1)-CHLOROBENZENE(2)

LACHER J.R.,HUNT R.E.:J.AM.CHEM.SOC.,63,1752(1941).

ANTOINE VAPOR PRESSURE CONSTANTS

	A	B	C
(1)	7.06245	1489.700	220.000
(2)	6.94504	1413.120	216.000

T = 100.00

EQUATION	ORDER	A12	A21	D	A122	A211
VAN LAAR	3	0.0015	-0.0036	-	-	-
MARGULES	3	-0.0633	0.2031	-	-	-
MARGULES	4	-0.1351	0.1382	-0.2064	-	-
ALPHA	2	0.2296	0.2266	-	-	-
ALPHA	3	0.6896	0.6043	-	-0.4786	-0.3333

			DEVIATION IN VAPOR PHASE COMPOSITION				
X	Y	PRESS.	LAAR 3	MARG 3	MARG 4	ALPHA2	ALPHA3
0.1693	0.1881	304.40	-0.0386	-0.0315	-0.0338	0.0013	0.0000
0.3568	0.3707	309.50	-0.0465	-0.0117	-0.0070	-0.0000	-0.0001
0.5129	0.5130	311.20	-0.0342	-0.0009	-0.0004	-0.0011	-0.0001
0.6434	0.6306	310.00	0.0061	-0.0064	-0.0098	-0.0005	-0.0003
0.7901	0.7695	306.40	0.0080	-0.0233	-0.0233	0.0004	-0.0003
0.9207	0.9053	300.30	0.0047	-0.0253	-0.0182	0.0019	0.0002
		MEAN	0.0230	0.0165	0.0154	0.0009	0.0002

124

1,1-DICHLOROETHANE(1)-CARBON TETRACHLORIDE(2)

KAPLAN S.I.,MONAKHOVA Z.D.:ZH.OBSHCH.KHIM.7,2499(1937)

ANTOINE VAPOR PRESSURE CONSTANTS

	A	B	C
(1)	6.98530	1171.420	228.120
(2)	6.93390	1242.430	230.000

P = 760.00

EQUATION	ORDER	A12	A21	D	A122	A211
VAN LAAR	3	0.0646	0.3266	-	-	-
MARGULES	3	0.0137	0.1678	-	-	-
MARGULES	4	-0.0199	0.1298	-0.1085	-	-
ALPHA	2	1.2594	-0.3783	-	-	-
ALPHA	3	-5.5951	-3.7810	-	6.6225	3.2778

			DEVIATION IN VAPOR PHASE COMPOSITION				
X	Y	TEMP.	LAAR 3	MARG 3	MARG 4	ALPHA2	ALPHA3
0.0500	0.1120	74.80	-0.0122	-0.0195	-0.0235	-0.0066	-0.0155
0.1000	0.1850	73.10	0.0045	-0.0040	-0.0080	0.0127	-0.0010
0.1500	0.2620	71.20	0.0086	0.0020	-0.0003	0.0172	0.0019
0.2000	0.3470	69.15	-0.0029	-0.0063	-0.0066	0.0049	-0.0100
0.2500	0.4200	67.50	-0.0091	-0.0094	-0.0082	-0.0028	-0.0156
0.3000	0.4820	66.20	-0.0103	-0.0083	-0.0064	-0.0057	-0.0150
0.3500	0.5370	65.10	-0.0097	-0.0065	-0.0046	-0.0068	-0.0108
0.4000	0.5870	64.10	-0.0087	-0.0053	-0.0040	-0.0073	0.0126
0.4500	0.6300	63.30	-0.0049	-0.0022	-0.0018	-0.0047	-0.0138
0.5000	0.6670	62.60	0.0013	0.0026	0.0020	0.0008	-0.0006
0.5500	0.7040	61.90	0.0043	0.0037	0.0024	0.0034	0.0065
0.6000	0.7450	61.30	0.0004	-0.0021	-0.0039	-0.0002	0.0063
0.6500	0.7750	60.70	0.0048	0.0008	-0.0011	0.0051	0.0145
0.7000	0.8100	60.15	0.0020	-0.0029	-0.0045	0.0039	0.0159
0.7500	0.8450	59.60	-0.0028	-0.0077	-0.0085	0.0013	0.0049
0.8000	0.8770	59.10	-0.0061	-0.0098	-0.0096	0.0008	0.0118
0.8500	0.9100	58.60	-0.0114	-0.0125	-0.0111	-0.0015	0.0092
0.9000	0.9420	58.10	-0.0156	-0.0129	-0.0108	-0.0031	0.0056
0.9500	0.9720	57.65	-0.0148	-0.0092	-0.0072	-0.0027	0.0025
		MEAN	0.0071	0.0067	0.0066	0.0048	0.0092

1,1-DICHLORETHANE(1)-CHLOROFORM(2)

KAPLAN S.I.,MONAKHOVA Z.D.:ZH.OBSHCH.KHIM.,7,2499(1937)

ANTOINE VAPOR PRESSURE CONSTANTS
```
            A          B          C
(1) 6.98530    1171.420    228.120
(2) 6.90328    1163.000    227.000
```

P = 760.00

EQUATION	ORDER	A12	A21	D	A122	A211
VAN LAAR	3	0.0034	-0.0046	-	-	-
MARGULES	3	-0.0700	0.0854	-	-	-
MARGULES	4	0.0354	0.1980	0.3235	-	-
ALPHA	2	0.2893	-0.2954	-	-	-
ALPHA	3	-2.2338	-1.7366	-	2.7700	1.2855

			DEVIATION IN VAPOR PHASE COMPOSITION				
X	Y	TEMP.	LAAR 3	MARG 3	MARG 4	ALPHA2	ALPHA3
0.1000	0.1520	60.90	-0.0359	-0.0436	-0.0346	-0.0259	-0.0116
0.2000	0.2520	60.45	-0.0234	-0.0249	-0.0237	-0.0055	0.0073
0.3000	0.3600	60.05	-0.0217	-0.0138	-0.0189	0.0012	0.0038
0.4000	0.4670	59.65	-0.0192	-0.0082	-0.0126	0.0031	-0.0055
0.5000	0.5600	59.30	0.0225	0.0015	0.0025	0.0132	-0.0004
0.6000	0.6570	58.95	0.3074	-0.0026	0.0030	0.0134	0.0051
0.7000	0.7700	58.53	-0.0225	-0.0303	-0.0251	-0.0083	-0.0046
0.8000	0.8600	58.12	-0.0320	-0.0385	-0.0397	-0.0129	-0.0003
0.9000	0.9350	57.65	-0.0205	-0.0296	-0.0383	-0.0085	0.0030
		MEAN	0.0561	0.0215	0.0221	0.0102	0.0046

1,2-DICHLOROETHANE(1)-PROPYL ALCOHOL(2)

UDOVENKO V.V.,FRID TS.B.:ZH.FIZ.KHIM.22,1263(1948).

ANTOINE VAPOR PRESSURE CONSTANTS
	A	B	C
(1)	7.18431	1358.500	232.000
(2)	7.85418	1497.910	204.112

T = 50.00

EQUATION	ORDER	A12	A21	D	A122	A211
VAN LAAR	3	0.3860	1.1027	-	-	-
MARGULES	3	0.2659	0.8047	-	-	-
MARGULES	4	0.3132	0.8606	0.1631	-	-
ALPHA	2	7.2993	0.7485	-	-	-
ALPHA	3	5.6606	-0.6908	-	-0.3591	1.6046

			DEVIATION IN VAPOR PHASE COMPOSITION				
X	Y	PRESS.	LAAR 3	MARG 3	MARG 4	ALPHA2	ALPHA3
0.1000	0.4090	132.80	-0.0171	-0.0451	-0.0366	0.0300	-0.0039
0.2000	0.5880	170.70	-0.0168	-0.0217	-0.0220	0.0099	0.0006
0.3000	0.6830	198.30	-0.0105	-0.0049	-0.0077	-0.0017	0.0034
0.4000	0.7450	218.00	-0.0088	-0.0038	-0.0053	-0.0110	-0.0013
0.5000	0.7820	230.90	-0.0038	-0.0050	-0.0041	-0.0101	-0.0021
0.6000	0.8050	238.00	0.0011	-0.0076	-0.0049	-0.0027	0.0003
0.7000	0.8270	243.20	-0.0033	-0.0164	-0.0137	0.0030	-0.0002
0.8000	0.8500	245.00	-0.0159	-0.0236	-0.0236	0.0102	0.0018
0.9000	0.8940	246.00	-0.0453	-0.0282	-0.0336	0.0089	-0.0006
		MEAN	0.0136	0.0174	0.0168	0.0097	0.0016

1,2-DICHLOROETHANE(1)-PROPYL ALCOHOL(2)

UDOVENKO V.V.,FRID TS.B.:ZH.FIZ.KHIM.22,1263(1948).

ANTOINE VAPOR PRESSURE CONSTANTS
	A	B	C
(1)	7.18431	1358.500	232.000
(2)	7.85418	1497.910	204.112

T =   60.00

EQUATION	ORDER	A12	A21	D	A122	A211
VAN LAAR	3	0.3816	1.1326	-	-	-
MARGULES	3	0.2478	0.8111	-	-	-
MARGULES	4	0.2641	0.8300	0.0551	-	-
ALPHA	2	6.3943	0.8235	-	-	-
ALPHA	3	4.4556	-0.9499	-	-0.2006	1.9475

			DEVIATION IN VAPOR PHASE COMPOSITION				
X	Y	PRESS.	LAAR 3	MARG 3	MARG 4	ALPHA2	ALPHA3
0.1000	0.3730	210.00	-0.0141	-0.0448	-0.0419	0.0365	-0.0049
0.2000	0.5500	269.20	-0.0127	-0.0192	-0.0193	0.0176	0.0054
0.3000	0.6620	298.00	-0.0195	-0.0134	-0.0143	-0.0090	-0.0027
0.4000	0.7210	326.10	-0.0107	-0.0044	-0.0049	-0.0129	-0.0006
0.5000	0.7580	344.00	-0.0021	-0.0025	-0.0022	-0.0097	0.0005
0.6000	0.7830	354.10	0.0034	-0.0055	-0.0046	-0.0017	0.0018
0.7000	0.8100	359.70	-0.0044	-0.0188	-0.0178	0.0020	-0.0025
0.8000	0.8340	362.50	-0.0179	-0.0266	-0.0266	0.0120	0.0012
0.9000	0.8830	361.70	-0.0537	-0.0339	-0.0359	0.0114	-0.0001
		MEAN	0.0154	0.0188	0.0186	0.0125	0.0022

1,2-DICHLOROETHANE(1)-PROPYL ALCOHOL(2)

UDOVENKO V.V.,FRID TS.B.:ZH.FIZ.KHIM.22,1263(1948).

ANTOINE VAPOR PRESSURE CONSTANTS
```
           A           B          C
(1) 7.18431      1358.500    232.000
(2) 7.85418      1497.910    204.112
```

T =   70.00

EQUATION	ORDER	A12	A21	D	A122	A211
VAN LAAR	3	0.3671	1.0412	-	-	-
MARGULES	3	0.2440	0.7660	-	-	-
MARGULES	4	0.2619	0.7863	0.0597	-	-
ALPHA	2	5.3298	0.8607	-	-	-
ALPHA	3	4.2949	-0.7955	-	-0.8166	1.9486

			DEVIATION IN VAPOR PHASE COMPOSITION				
X	Y	PRESS.	LAAR 3	MARG 3	MARG 4	ALPHA2	ALPHA3
0.1000	0.3380	320.40	-0.0140	-0.0409	-0.0379	0.0343	-0.0060
0.2000	0.5120	386.50	-0.0134	-0.0196	-0.0196	0.0169	0.0037
0.3000	0.6200	435.40	-0.0137	-0.0078	-0.0090	-0.0029	0.0026
0.4000	0.6920	470.00	-0.0141	-0.0076	-0.0083	-0.0165	-0.0042
0.5000	0.7300	493.50	-0.0026	-0.0023	-0.0020	-0.0107	-0.0001
0.6000	0.7580	507.60	0.0038	-0.0041	-0.0030	-0.0020	0.0022
0.7000	0.7890	515.30	-0.0039	-0.0168	-0.0157	0.0020	-0.0019
0.8000	0.8180	518.50	-0.0168	-0.0244	-0.0244	0.0123	0.0014
0.9000	0.8730	515.50	-0.0492	-0.0310	-0.0333	0.0130	-0.0001
		MEAN	0.0146	0.0172	0.0170	0.0123	0.0025

1,2-DICHLOROETHANE(1)-PROPYL ALCOHOL(2)

UDOVENKO V.V.,FRID TS.B.:ZH.FIZ.KHIM.22,1263(1948).

ANTOINE VAPOR PRESSURE CONSTANTS
	A	B	C
(1)	7.18431	1358.500	232.000
(2)	7.85418	1497.910	204.112

T =   80.00

EQUATION	ORDER	A12	A21	D	A122	A211
VAN LAAR	3	0.3553	0.9882	-	-	-
MARGULES	3	0.2359	0.7372	-	-	-
MARGULES	4	0.2555	0.7591	0.0647	-	-
ALPHA	2	4.5589	0.9268	-	-	-
ALPHA	3	5.0602	-0.5743	-	-2.3124	2.1164

			DEVIATION IN VAPOR PHASE COMPOSITION				
X	Y	PRESS.	LAAR 3	MARG 3	MARG 4	ALPHA2	ALPHA3
0.1000	0.3060	480.00	-0.0113	-0.0361	-0.0330	0.0356	-0.0080
0.2000	0.4730	562.80	-0.0083	-0.0143	-0.0143	0.0220	0.0061
0.3000	0.5870	628.00	-0.0131	-0.0071	-0.0084	-0.0027	0.0023
0.4000	0.6640	667.00	-0.0153	-0.0083	-0.0091	-0.0190	-0.0058
0.5000	0.7030	695.00	-0.0014	-0.0003	-0.0000	-0.0116	0.0009
0.6000	0.7340	712.00	0.0054	-0.0016	-0.0004	-0.0027	0.0035
0.7000	0.7730	719.00	-0.0068	-0.0189	-0.0176	-0.0028	-0.0057
0.8000	0.7970	722.00	-0.0105	-0.0175	-0.0176	0.0175	0.0050
0.9000	0.8600	717.20	-0.0444	-0.0270	-0.0295	0.0172	-0.0014
		MEAN	0.0129	0.0146	0.0144	0.0146	0.0043

1,2-DICHLOROETHANE(1)-TOLUENE(2)

ALPERT N.,ELVING P.J.:IND.ENG.CHEM.,41,2864(1949).

ANTOINE VAPOR PRESSURE CONSTANTS
	A	B	C
(1)	7.18431	1358.500	232.000
(2)	6.95334	1343.943	219.377

P = 760.00

EQUATION	ORDER	A12	A21	D	A122	A211
VAN LAAR	3	-0.0005	0.0001	-	-	-
MARGULES	3	0.0483	-0.0101	-	-	-
MARGULES	4	0.1404	0.1012	0.3158	-	-
ALPHA	2	1.2624	-0.4968	-	-	-
ALPHA	3	.3307/ 06	.1530/ 06	-	-.3272/ 06	-.1512/ 06

			DEVIATION IN VAPOR PHASE COMPOSITION				
X	Y	TEMP.	LAAR 3	MARG 3	MARG 4	ALPHA2	ALPHA3
0.0180	0.0660	109.10	-0.0275	-0.0235	-0.0155	-0.0262	-0.0070
0.0700	0.1690	108.10	-0.0277	-0.0176	-0.0034	-0.0240	-0.0124
0.1400	0.2750	105.70	-0.0123	-0.0014	0.0059	-0.0077	-0.0027
0.2000	0.3480	102.70	0.0005	0.0148	0.0148	0.0101	0.0117
0.2620	0.4240	101.50	-0.0002	0.0190	0.0143	0.0168	0.0164
0.3640	0.5500	97.80	0.0104	0.0076	0.0028	0.0075	0.0059
0.3980	0.6030	97.10	-0.0069	-0.0114	-0.0147	-0.0111	-0.0127
0.4820	0.6630	94.60	0.0134	0.0066	0.0080	0.0063	0.0050
0.5400	0.7250	92.90	0.0008	-0.0064	-0.0022	-0.0078	-0.0087
0.6050	0.7820	91.30	-0.0059	-0.0126	-0.0067	-0.0155	-0.0160
0.6980	0.8210	89.10	0.0191	0.0144	0.0191	0.0091	0.0091
0.8010	0.8920	87.50	0.0098	0.0077	0.0075	0.0012	0.0008
0.8970	0.9420	85.50	0.0103	0.0101	0.0059	0.0047	0.0027
		MEAN	0.0111	0.0118	0.0093	0.0114	0.0085

1,2-DICHLOROETHANE(1)-TOLUENE(2)

JONES C.A.,SCHOENBORN E.M.,COLBURN A.P.:IND.ENG.CHEM.35,666(1943)

ANTOINE VAPOR PRESSURE CONSTANTS
	A	B	C
(1)	7.18431	1358.500	232.000
(2)	6.95334	1343.943	219.377

P = 760.00

EQUATION	ORDER	A12	A21	D	A122	A211
VAN LAAR	3	0.0000	-0.0000	-	-	-
MARGULES	3	-0.0158	0.0036	-	-	-
MARGULES	4	0.0022	0.0258	0.0585	-	-
ALPHA	2	1.1903	-0.5910	-	-	-
ALPHA	3	-1.2083	-1.7054	-	2.3673	1.1466

			DEVIATION IN VAPOR PHASE COMPOSITION				
X	Y	TEMP.	LAAR 3	MARG 3	MARG 4	ALPHA2	ALPHA3
0.0450	0.0950	108.40	-0.0015	-0.0040	-0.0016	-0.0013	-0.0024
0.1000	0.1970	107.00	-0.0009	-0.0043	-0.0019	-0.0005	-0.0023
0.2350	0.4010	102.00	0.0041	0.0026	0.0024	0.0043	0.0028
0.2520	0.4240	100.80	0.0041	0.0029	0.0025	0.0040	0.0027
0.3650	0.5650	99.30	0.0226	-0.0024	-0.0030	-0.0023	-0.0023
0.3750	0.5780	97.80	0.0005	-0.0040	-0.0045	-0.0046	-0.0044
0.4150	0.6140	96.80	0.0002	0.0014	0.0012	0.0006	0.0011
0.4460	0.6450	96.00	-0.0013	0.0006	0.0006	-0.0004	0.0005
0.4790	0.6740	95.90	-0.0005	0.0017	0.0020	0.0011	0.0021
0.5680	0.7500	92.60	-0.0020	0.0003	0.0013	-0.0002	0.0009
0.5850	0.7650	92.20	-0.0039	-0.0017	-0.0007	-0.0020	-0.0010
0.7000	0.8440	90.20	-0.0028	-0.0013	-0.0006	-0.0002	-0.0004
0.7840	0.8930	87.70	-0.0007	0.0001	0.0000	0.0018	0.0004
0.8120	0.9090	87.10	-0.0010	-0.0004	-0.0007	0.0014	-0.0002
		MEAN	0.0033	0.0020	0.0016	0.0018	0.0017

ETHYLENE OXIDE(1)-WATER(2)

COLES K.F.,POPPER F.:IND.ENG.CHEM.42,1434(1950).

ANTOINE VAPOR PRESSURE CONSTANTS
	A	B	C
(1)	7.12886	1054.738	237.783
(2)	7.96681	1668.210	228.000

P = 760.00

EQUATION	ORDER	A12	A21	D	A122	A211
VAN LAAR	3	0.8026	1.0385	-	-	-
MARGULES	3	0.7945	1.0315	-	-	-
MARGULES	4	0.7565	0.9332	-0.3594	-	-
ALPHA	2	246.8795	1.1790	-	-	-
ALPHA	3	.6348/ 10	.2977/ 08	-	-.6356/ 10	-.6027/ 04

			DEVIATION IN VAPOR PHASE COMPOSITION				
X	Y	TEMP.	LAAR 3	MARG 3	MARG 4	ALPHA2	ALPHA3
0.0400	0.8600	50.00	0.0211	0.0203	0.0168	0.0445	0.0320
0.0650	0.9370	37.60	0.0004	0.0002	-0.0001	0.0004	-0.0055
0.0820	0.9595	31.50	-0.0049	-0.0049	-0.0044	-0.0107	-0.0144
0.0950	0.9648	31.00	-0.0047	-0.0046	-0.0038	-0.0099	-0.0124
0.2100	0.9816	16.40	0.0023	0.0025	0.0035	-0.0050	-0.0035
0.2320	0.9841	15.10	0.0011	0.0014	0.0023	-0.0057	-0.0040
0.2740	0.9845	15.00	0.0019	0.0021	0.0029	-0.0036	-0.0014
0.4320	0.9853	14.30	0.0028	0.0029	0.0029	0.0008	0.0039
0.5600	0.9845	13.70	0.0037	0.0037	0.0029	0.0037	0.0072
0.6150	0.9853	13.20	0.0029	0.0028	0.0018	0.0036	0.0071
0.8750	0.9888	12.00	0.0016	0.0015	0.0016	0.0022	0.0059
0.8900	0.9905	11.90	0.0005	0.0004	0.0006	0.0006	0.0043
0.9100	0.9900	11.80	0.0020	0.0019	0.0023	0.0012	0.0049
0.9330	0.9934	11.70	-0.0001	-0.0001	0.0005	-0.0019	0.0016
0.9510	0.9927	11.50	0.0020	0.0020	0.0026	-0.0010	0.0024
		MEAN	0.0035	0.0034	0.0033	0.0063	0.0074

ETHYL BROMIDE(1)-BENZENE(2)

TYRER D.:J.CHEM.SOC.,101,81(1912).

ANTOINE VAPOR PRESSURE CONSTANTS

	A	B	C
(1)	6.91995	1090.810	231.710
(2)	6.90565	1211.033	220.790

P = 760.00

EQUATION	ORDER	A12	A21	D	A122	A211
VAN LAAR	3	-0.0011	0.0003	-	-	-
MARGULES	3	-0.0399	0.0137	-	-	-
MARGULES	4	-0.0755	-0.0380	-0.1380	-	-
ALPHA	2	2.6212	-0.8474	-	-	-
ALPHA	3	-13.1990	-4.9079	-	16.0311	4.2104

DEVIATION IN VAPOR PHASE COMPOSITION

X	Y	TEMP.	LAAR 3	MARG 3	MARG 4	ALPHA2	ALPHA3
0.0740	0.2060	74.86	0.0150	0.0053	-0.0009	0.0201	0.0281
0.1520	0.3870	70.00	0.0042	0.0008	-0.0012	0.0117	0.0182
0.2350	0.5330	65.39	-0.1184	-0.0002	0.0013	0.0025	-0.0083
0.3230	0.6480	60.84	-0.0026	0.0016	0.0034	-0.0023	0.0098
0.4170	0.7460	56.52	-0.0079	-0.0028	-0.0027	-0.0094	-0.0068
0.5180	0.8200	52.59	-0.0071	-0.0021	-0.0038	-0.0074	-0.0099
0.6260	0.8700	48.78	0.0040	0.0076	0.0055	0.0059	0.0004
0.7410	0.9260	45.10	-0.0021	-0.0003	-0.0013	0.0021	-0.0040
0.8660	0.9670	41.49	-0.0013	-0.0012	-0.0004	0.0034	0.0041
		MEAN	0.0181	0.0024	0.0023	0.0072	0.0099

ETHYL BROMIDE(1)-ETHYL ALCOHOL(2)

SMITH C.P.,ENGEL E.W.:J.AM.CHEM.SOC.51,2660(1929).

ANTOINE VAPOR PRESSURE CONSTANTS

	A	B	C
(1)	6.91995	1090.810	231.710
(2)	8.16290	1623.220	228.980

T =   30.00

EQUATION	ORDER	A12	A21	D	A122	A211
VAN LAAR	3	0.5706	0.8850	-	-	-
MARGULES	3	0.5533	0.8224	-	-	-
MARGULES	4	0.6400	0.9603	0.3761	-	-
ALPHA	2	29.3098	1.7067	-	-	-
ALPHA	3	31.7099	1.5039	-	-3.5444	0.5040

DEVIATION IN VAPOR PHASE COMPOSITION

X	Y	PRESS.	LAAR 3	MARG 3	MARG 4	ALPHA2	ALPHA3
0.1044	0.7256	253.30	-0.0045	-0.0064	0.0037	0.0038	0.0014
0.1337	0.7680	291.40	-0.0041	-0.0049	-0.0005	0.0003	-0.0010
0.1447	0.7783	308.60	-0.0020	-0.0024	0.0004	0.0013	0.0003
0.1914	0.8164	358.40	-0.0011	-0.0007	-0.0022	-0.0013	-0.0013
0.2438	0.8419	403.60	0.0011	0.0017	-0.0016	-0.0013	-0.0007
0.2453	0.8409	402.20	0.0028	0.0033	0.0000	0.0003	0.0009
0.3216	0.8648	450.50	0.0034	0.0036	0.0008	-0.0001	0.0009
0.3790	0.8755	474.90	0.0046	0.0042	0.0030	0.0013	0.0023
0.4569	0.8933	501.40	-0.0024	-0.0036	-0.0022	-0.0044	-0.0036
0.5834	0.9018	525.80	-0.0003	-0.0027	0.0021	0.0008	0.0010
0.7003	0.9142	549.30	-0.0058	-0.0081	-0.0033	-0.0018	-0.0023
0.7190	0.9109	552.20	-0.0012	-0.0033	0.0010	0.0030	0.0023
0.8825	0.9283	567.80	0.0043	0.0063	0.0011	0.0019	-0.0002
		MEAN	0.0029	0.0039	0.0017	0.0017	0.0014

ETHYL BROMIDE(1)-HEPTANE(2)

SMITH C.P,ENGEL E.W.:J.AM.CHEM.SOC.,51,2646(1929).

ANTOINE VAPOR PRESSURE CONSTANTS

	A	B	C
(1)	6.91995	1090.810	231.710
(2)	6.90240	1268.115	216.900

T =   30.00

EQUATION	ORDER	A12	A21	D	A122	A211
VAN LAAR	3	0.1082	0.1416	-	-	-
MARGULES	3	0.1072	0.1373	-	-	-
MARGULES	4	0.1220	0.1641	0.0673	-	-
ALPHA	2	11.0851	-0.6448	-	-	-
ALPHA	3	17.9350	-0.1018	-	-6.9952	-0.4135

			DEVIATION IN VAPOR PHASE COMPOSITION				
X	Y	PRESS.	LAAR 3	MARG 3	MARG 4	ALPHA2	ALPHA3
0.0923	0.5434	122.00	0.0056	0.0055	0.0082	0.0012	-0.0008
0.2176	0.7566	199.70	0.0028	0.0029	0.0024	0.0012	0.0007
0.2843	0.8161	239.80	-0.0022	-0.0022	-0.0027	-0.0032	0.0041
0.3246	0.8337	261.00	0.0047	0.0047	0.0044	-0.0014	-0.0009
0.4621	0.8964	337.10	-0.0001	-0.0002	0.0003	-0.0014	-0.0009
0.4723	0.9001	340.30	-0.0005	-0.0006	-0.0001	-0.0020	-0.0015
0.6219	0.9339	404.00	0.0040	0.0039	0.0046	0.0005	0.0010
0.7985	0.9633	474.20	0.0067	0.0067	0.0068	0.0002	-0.0001
0.9540	0.9862	543.90	0.0072	0.0072	0.0070	0.0017	-0.0000
		MEAN	0.0038	0.0038	0.0040	0.0017	0.0014

ETHYL ALCOHOL(1)-ACETONITRILE(2)

VIERK A.L.:Z.ANORG.CHEM. 261,283(1950).

ANTOINE VAPOR PRESSURE CONSTANTS

	A	B	C
(1)	8.16290	1623.220	228.980
(2)	7.07354	1279.200	224.000

T =   20.00

EQUATION	ORDER	A12	A21	D	A122	A211
VAN LAAR	3	0.6602	0.6781	-	-	-
MARGULES	3	0.6600	0.6782	-	-	-
MARGULES	4	0.6303	0.6458	-0.0994	-	-
ALPHA	2	3.3768	6.4532	-	-	-
ALPHA	3	1.3205	0.7500	-	0.3151	5.1547

			DEVIATION IN VAPOR PHASE COMPOSITION				
X	Y	PRESS.	LAAR 3	MARG 3	MARG 4	ALPHA2	ALPHA3
0.0500	0.1110	75.00	0.0078	0.0077	0.0033	0.0324	0.0019
0.0900	0.1800	77.80	0.0035	0.0035	-0.0004	0.0231	0.0000
0.1900	0.2890	81.00	-0.0057	-0.0057	-0.0058	-0.0066	-0.0023
0.3200	0.3500	81.90	-0.0013	-0.0012	0.0009	-0.0140	0.0023
0.5000	0.4000	81.50	-0.0038	-0.0038	-0.0041	-0.0112	-0.0050
0.5900	0.4050	80.50	0.0118	0.0119	0.0098	0.0115	0.0098
0.6800	0.4500	78.00	-0.0063	-0.0063	-0.0090	0.0007	-0.0074
0.8100	0.5170	72.50	-0.0016	-0.0016	-0.0014	0.0122	0.0019
		MEAN	0.0052	0.0052	0.0043	0.0140	0.0038

137

J

ETHYL ALCOHOL(1)-BENZENE(2)

BROWN I.,SMITH F.:AUSTR.J.CHEM.7,264(1954).

ANTOINE VAPOR PRESSURE CONSTANTS

	A	B	C
(1)	8.16290	1623.220	228.980
(2)	6.90565	1211.033	220.790

T = 45.00

EQUATION	ORDER	A12	A21	D	A122	A211
VAN LAAR	3	0.8911	0.6060	-	-	-
MARGULES	3	0.8475	0.5880	-	-	-
MARGULES	4	0.9672	0.6976	0.3891	-	-
ALPHA	2	4.7128	7.5827	-	6.6773	-13.3299
ALPHA	3	3.6303	18.0065	-	6.6773	-13.3299

			DEVIATION IN VAPOR PHASE COMPOSITION				
X	Y	PRESS.	LAAR 3	MARG 3	MARG 4	ALPHA2	ALPHA3
0.0374	0.1965	271.01	-0.0383	-0.0472	-0.0209	-0.0530	0.0040
0.0972	0.2895	296.53	-0.0168	-0.0228	-0.0045	-0.0438	-0.0044
0.2183	0.3370	306.55	0.0132	0.0157	0.0103	-0.0070	0.0028
0.3141	0.3625	309.33	0.0081	0.0128	0.0014	0.0018	-0.0000
0.4150	0.3842	309.59	0.0002	0.0041	-0.0040	0.0071	-0.0010
0.5199	0.4065	307.46	-0.0045	-0.0029	-0.0025	0.0104	0.0004
0.5284	0.4101	306.99	-0.0063	-0.0049	-0.0037	0.0089	-0.0010
0.6155	0.4343	302.05	-0.0063	-0.0069	0.0010	0.0084	0.0007
0.7087	0.4751	291.81	-0.0053	-0.0070	0.0034	0.0002	-0.0005
0.8102	0.5456	271.08	0.0033	0.0026	0.0046	-0.0146	0.0008
0.9193	0.7078	227.72	0.0116	0.0147	-0.0065	-0.0442	-0.0001
0.9591	0.8201	203.28	0.0095	0.0133	-0.0115	-0.0484	-0.0011
		MEAN	0.0103	0.0129	0.0062	0.0206	0.0014

ETHYL ALCOHOL(1)-BENZENE(2)

HO J.C.K.,LU B.C.Y.:J.CHEM.ENG.DATA 8(4),549(1963).

ANTOINE VAPOR PRESSURE CONSTANTS
```
          A           B          C
(1) 8.16290     1623.220    228.980
(2) 6.90565     1211.033    220.790
```

T =   55.00

EQUATION	ORDER	A12	A21	D	A122	A211
VAN LAAR	3	0.8563	0.5726	-	-	-
MARGULES	3	0.8114	0.5505	-	-	-
MARGULES	4	0.8959	0.6296	0.2691	-	-
ALPHA	2	4.3618	6.0539	-	-	-
ALPHA	3	2.4272	13.0191	-	6.2202	-9.7636

			DEVIATION IN VAPOR PHASE COMPOSITION				
X	Y	PRESS.	LAAR 3	MARG 3	MARG 4	ALPHA2	ALPHA3
0.0570	0.2460	422.80	-0.0339	-0.0432	-0.0245	-0.0592	-0.0084
0.1590	0.3260	457.90	0.0087	0.0075	0.0112	-0.0158	0.0066
0.2660	0.3690	463.90	0.0075	0.0113	0.0052	-0.0006	0.0017
0.3670	0.3980	470.00	-0.0014	0.0029	-0.0041	0.0056	-0.0029
0.5260	0.4340	471.00	-0.0050	-0.0040	-0.0033	0.0146	-0.0006
0.6320	0.4720	460.00	-0.0077	-0.0088	-0.0029	0.0091	-0.0034
0.7430	0.5230	437.40	0.0035	0.0020	0.0077	0.0043	0.0038
0.8300	0.6020	411.60	0.0060	0.0066	0.0047	-0.0167	0.0028
0.9160	0.7470	358.80	-0.0024	0.0015	-0.0127	-0.0522	-0.0051
		MEAN	0.0085	0.0098	0.0084	0.0198	0.0039

ETHYL ALCOHOL(1)-BENZENE(2)

LEHFELDT N.:PHIL.MAG.46,42(1898).

ANTOINE VAPOR PRESSURE CONSTANTS

	A	B	C
(1)	8.16290	1623.220	228.980
(2)	6.90565	1211.033	220.790

T = 50.00

EQUATION	ORDER	A12	A21	D	A122	A211
VAN LAAR	3	0.8045	0.6256	-	-	-
MARGULES	3	0.7857	0.6214	-	-	-
MARGULES	4	0.9377	0.7941	0.5690	-	-
ALPHA	2	5.0918	6.8914	-	-	-
ALPHA	3	7.3436	21.7044	-	4.9638	-16.4694

			DEVIATION IN VAPOR PHASE COMPOSITION				
X	Y	PRESS.	LAAR 3	MARG 3	MARG 4	ALPHA2	ALPHA3
0.0880	0.2810	350.40	-0.0303	-0.0331	-0.0097	-0.0278	0.0098
0.1210	0.3220	369.00	-0.0295	-0.0312	-0.0179	-0.0307	-0.0073
0.2150	0.3570	396.90	0.0009	0.0022	-0.0067	-0.0015	-0.0028
0.3550	0.3910	405.90	0.0069	0.0096	-0.0039	0.0152	0.0021
0.4440	0.4170	404.40	-0.0033	-0.0010	-0.0057	0.0127	-0.0001
0.5610	0.4510	397.60	-0.0135	-0.0126	-0.0016	0.0084	0.0020
0.6970	0.5130	378.40	-0.0275	-0.0281	-0.0089	-0.0110	-0.0028
0.8860	0.6680	315.00	-0.0064	-0.0066	-0.0279	-0.0345	0.0013
		MEAN	0.0148	0.0156	0.0103	0.0177	0.0035

ETHYL ALCOHOL(1)-BENZENE(2)

TYRER D.:J.CHEM.SOC.101,1104(1912).

ANTOINE VAPOR PRESSURE CONSTANTS
	A	B	C
(1)	8.16290	1623.220	228.980
(2)	6.90565	1211.033	220.790

P =   750.00

EQUATION	ORDER	A12	A21	D	A122	A211
VAN LAAR	3	0.8123	0.5863	-	-	-
MARGULES	3	0.7781	0.5753	-	-	-
MARGULES	4	0.9371	0.7004	0.4514	-	-
ALPHA	2	4.1238	4.9068	-	-	-
ALPHA	3	7.5244	21.7020	-	6.1963	-17.5281

			DEVIATION IN VAPOR PHASE COMPOSITION				
X	Y	TEMP.	LAAR 3	MARG 3	MARG 4	ALPHA2	ALPHA3
0.1585	0.3531	69.54	0.0077	0.0059	0.0157	-0.0317	0.0025
0.2977	0.4045	68.20	0.0129	0.0162	0.0032	-0.0029	-0.0032
0.4208	0.4358	67.76	0.0066	0.0101	-0.0014	0.0096	-0.0012
0.5367	0.4662	67.97	0.0027	0.0046	0.0033	0.0151	0.0038
0.6290	0.5053	68.41	-0.0054	-0.0049	0.0014	0.0068	-0.0002
0.7178	0.5491	69.00	-0.0042	-0.0045	0.0035	-0.0001	0.0003
0.7982	0.6063	70.26	0.0019	0.0017	0.0031	-0.0105	-0.0013
0.8715	0.6833	71.86	0.0127	0.0136	0.0014	-0.0204	-0.0032
0.9385	0.7874	74.40	0.0331	0.0350	0.0111	-0.0140	0.0054
		MEAN	0.0097	0.0107	0.0049	0.0124	0.0023

ETHYL ALCOHOL(1)-BENZENE(2)

UDOVENKO V.V.,FATKULINA L.G.:ZH.FIZ.KHIM.26,719(1952).

ANTOINE VAPOR PRESSURE CONSTANTS

	A	B	C
(1)	8.16290	1623.220	228.980
(2)	6.90565	1211.033	220.790

T = 40.00

EQUATION	ORDER	A12	A21	D	A122	A211
VAN LAAR	3	0.9548	0.6414	-	-	-
MARGULES	3	0.9069	0.6212	-	-	-
MARGULES	4	0.9993	0.6992	0.3006	-	-
ALPHA	2	5.1405	8.9110	-	-	-
ALPHA	3	2.5897	17.9179	-	7.7136	-12.7573

			DEVIATION IN VAPOR PHASE COMPOSITION				
X	Y	PRESS.	LAAR 3	MARG 3	MARG 4	ALPHA2	ALPHA3
0.0200	0.1450	208.40	-0.0388	-0.0471	-0.0303	-0.0503	-0.0033
0.0950	0.2800	239.80	-0.0000	-0.0067	0.0076	-0.0368	0.0016
0.2040	0.3320	249.10	0.0155	0.0179	0.0141	-0.0155	-0.0014
0.3780	0.3620	252.30	0.0078	0.0128	0.0038	0.0067	0.0016
0.4900	0.3840	248.80	-0.0035	-0.0011	-0.0037	0.0094	-0.0017
0.5920	0.4050	245.70	-0.0050	-0.0054	-0.0011	0.0123	-0.0005
0.7020	0.4400	237.30	0.0010	-0.0012	0.0071	0.0112	0.0025
0.8020	0.5070	219.40	0.0052	0.0039	0.0077	-0.0062	-0.0005
0.8800	0.6050	196.30	0.0081	0.0099	0.0023	-0.0322	-0.0028
0.9430	0.7470	169.50	0.0102	0.0147	-0.0033	-0.0523	0.0005
0.9870	0.9120	145.60	0.0170	0.0196	0.0088	-0.0199	0.0142
		MEAN	0.0102	0.0127	0.0082	0.0230	0.0028

ETHYL ALCOHOL(1)-BENZENE(2)

UDOVENKO V.V.,FATKULINA L.G.:ZH.FIZ.KHIM.26,719(1952).

ANTOINE VAPOR PRESSURE CONSTANTS
```
        A          B          C
(1) 8.16290    1623.220    228.980
(2) 6.90565    1211.033    220.790
```

T =   50.00

EQUATION	ORDER	A12	A21	D	A122	A211
VAN LAAR	3	0.9302	0.6471	-	-	-
MARGULES	3	0.8905	0.6310	-	-	-
MARGULES	4	0.9926	0.7210	0.3387	-	-
ALPHA	2	5.6193	8.2626	-	-	-
ALPHA	3	2.9533	16.7986	-	8.1028	-12.1438

			DEVIATION IN VAPOR PHASE COMPOSITION				
X	Y	PRESS.	LAAR 3	MARG 3	MARG 4	ALPHA2	ALPHA3
0.0250	0.1650	314.70	-0.0324	-0.0404	-0.0186	-0.0440	0.0087
0.0890	0.3000	358.70	-0.0123	-0.0186	-0.0011	-0.0438	-0.0022
0.2060	0.3600	378.30	0.0111	0.0132	0.0086	-0.0160	-0.0027
0.3850	0.3920	384.60	0.0059	0.0103	0.0007	0.0082	0.0018
0.4860	0.4110	383.20	-0.0026	-0.0002	-0.0030	0.0120	0.0006
0.5860	0.4340	378.10	-0.0071	-0.0072	-0.0021	0.0123	-0.0003
0.6940	0.4700	366.90	-0.0051	-0.0068	0.0029	0.0080	-0.0000
0.7900	0.5260	344.40	0.0017	0.0004	0.0058	-0.0046	0.0004
0.8660	0.6100	316.80	0.0047	0.0053	-0.0007	-0.0286	-0.0027
0.9360	0.7450	276.80	0.0099	0.0130	-0.0063	-0.0502	0.0011
0.9840	0.9090	239.60	0.0121	0.0144	0.0008	-0.0289	0.0089
		MEAN	0.0095	0.0118	0.0046	0.0233	0.0027

ETHYL ALCOHOL(1)-BENZENE(2)

UDOVENKO V.V.,FATKULINA L.G.:ZH.FIZ.KHIM.26,719(1952).

ANTOINE VAPOR PRESSURE CONSTANTS
```
          A         B          C
(1)  8.16290   1623.220   228.980
(2)  6.90565   1211.033   220.790
```

T =   60.00

EQUATION	ORDER	A12	A21	D	A122	A211
VAN LAAR	3	0.9178	0.6442	-	-	-
MARGULES	3	0.8814	0.6258	-	-	-
MARGULES	4	0.9729	0.7161	0.3116	-	-
ALPHA	2	5.9613	7.6695	-	-	-
ALPHA	3	2.6752	14.1126	-	8.0792	-10.0629

			DEVIATION IN VAPOR PHASE COMPOSITION				
X	Y	PRESS.	LAAR 3	MARG 3	MARG 4	ALPHA2	ALPHA3
0.0260	0.1610	452.70	-0.0157	-0.0235	-0.0028	-0.0294	0.0201
0.0800	0.3100	518.20	-0.0200	-0.0267	-0.0086	-0.0510	-0.0075
0.1850	0.3750	553.70	0.0090	0.0097	0.0087	-0.0203	-0.0024
0.3220	0.4080	565.00	0.0072	0.0112	0.0021	0.0003	-0.0001
0.3980	0.4200	568.00	0.0030	0.0064	-0.0004	0.0081	0.0016
0.4840	0.4360	566.60	-0.0031	-0.0014	-0.0027	0.0119	0.0011
0.5680	0.4550	562.60	-0.0070	-0.0071	-0.0025	0.0124	0.0000
0.6800	0.4900	548.70	-0.0055	-0.0072	0.0020	0.0085	-0.0004
0.7710	0.5370	524.20	0.0012	-0.0001	0.0059	-0.0017	0.0005
0.8590	0.6250	485.00	0.0042	0.0053	-0.0006	-0.0286	-0.0032
0.9290	0.7470	431.60	0.0094	0.0130	-0.0053	-0.0508	0.0010
0.9810	0.9040	377.40	0.0123	0.0150	0.0010	-0.0331	0.0094
		MEAN	0.0081	0.0106	0.0035	0.0214	0.0039

ETHYL ALCOHOL(1)-BUTYL ALCOHOL(2)

BRUNJES A.S.,BOGART M.J.P.:IND.ENG.CHEM.35,255(1943).

ANTOINE VAPOR PRESSURE CONSTANTS
```
            A          B          C
(1)  8.16290    1623.220    228.980
(2)  7.54472    1405.873    183.908
```

P =   760.00

EQUATION	ORDER	A12	A21	D	A122	A211
VAN LAAR	3	0.0011	-0.0013	-	-	-
MARGULES	3	-0.1427	0.1393	-	-	-
MARGULES	4	-0.0575	0.2547	0.3219	-	-
ALPHA	2	3.1242	-1.0222	-	-	-
ALPHA	3	-.5085/ 05	-.1011/ 05	-	.5230/ 05	.1080/ 05

			DEVIATION IN VAPOR PHASE COMPOSITION			
X	Y	TEMP.	LAAR 3	MARG 3	MARG 4	ALPHA2	ALPHA3
0.0620	0.1810	114.30	0.0293	-0.0046	0.0111	0.0361	-0.0212
0.1130	0.3160	111.10	0.0268	-0.0031	0.0077	0.0360	0.0182
0.1780	0.4960	105.90	-0.0194	-0.0309	-0.0297	-0.0103	-0.0062
0.2520	0.5970	101.30	-0.0044	0.0040	-0.0007	0.0053	0.0165
0.3230	0.7080	96.40	-0.0279	-0.0085	-0.0132	-0.0187	-0.0081
0.4270	0.7840	93.50	-0.0083	0.0117	0.0112	0.0028	0.0075
0.4880	0.8280	90.60	0.0109	0.0088	0.0107	0.0038	0.0046
0.5550	0.8810	88.10	0.0379	-0.0095	-0.0060	-0.0077	-0.0109
0.5740	0.8850	88.30	-0.0007	-0.0057	-0.0020	-0.0013	-0.0054
0.6010	0.9000	86.40	-0.0159	-0.0090	-0.0051	-0.0023	-0.0077
0.6660	0.9270	85.70	-0.0212	-0.0138	-0.0104	0.0003	-0.0074
0.7430	0.9460	83.90	-0.0129	-0.0104	-0.0086	0.0099	0.0012
0.7380	0.9760	83.30	-0.0442	-0.0414	-0.0395	-0.0218	-0.0305
0.8450	0.9820	81.00	-0.0180	-0.0206	-0.0218	0.0014	-0.0045
0.8870	0.9870	80.30	-0.0122	-0.0156	-0.0178	0.0043	0.0014
0.9380	0.9970	79.10	-0.0101	-0.0132	-0.0157	0.0007	0.0035
		MEAN	0.0188	0.0132	0.0132	0.0102	0.0097

ETHYL ALCOHOL(1)-BUTYL ALCOHOL(2)

HELLWIG L.R.,VAN WINKLE M.:IND.ENG.CHEM.45,624(1953).

ANTOINE VAPOR PRESSURE CONSTANTS
	A	B	C
(1)	8.16290	1623.220	228.980
(2)	7.54472	1405.873	183.908

P = 760.00

EQUATION	ORDER	A12	A21	D	A122	A211
VAN LAAR	3	-0.0046	0.0073	-	-	-
MARGULES	3	0.0560	-0.0900	-	-	-
MARGULES	4	0.0783	-0.0608	0.0763	-	-
ALPHA	2	3.0627	-0.7879	-	-	-
ALPHA	3	.3270/ 05	.7976/ 04	-	-.3267/ 05	-.8078/ 04

			DEVIATION IN VAPOR PHASE COMPOSITION				
X	Y	TEMP.	LAAR 3	MARG 3	MARG 4	ALPHA2	ALPHA3
0.1290	0.3840	107.60	-0.0063	0.0029	0.0057	-0.0072	-0.0048
0.2330	0.5630	102.10	-0.0022	-0.0062	-0.0065	-0.0082	-0.0069
0.2820	0.6020	99.60	0.0232	0.0155	0.0147	0.0156	0.0164
0.3570	0.6960	95.80	0.0089	-0.0006	-0.0012	0.0004	0.0005
0.4530	0.7790	92.90	-0.0011	-0.0048	-0.0046	-0.0040	-0.0045
0.4680	0.7860	92.40	0.0006	-0.0009	-0.0005	-0.0004	-0.0011
0.6100	0.8710	87.70	-0.8710	0.0015	0.0025	-0.0024	-0.0035
0.7090	0.9120	84.80	-0.0059	0.0068	0.0073	0.0006	-0.0004
0.8800	0.9680	80.80	0.0041	0.0072	0.0067	0.0023	0.0027
		MEAN	0.1026	0.0051	0.0055	0.0046	0.0045

146

ETHYL ALCOHOL(1)-DECANE(2)

ELLIS S.R.E.,SPURR M.J.:BRIT.CHEM.ENG.6,95(1961).

ANTOINE VAPOR PRESSURE CONSTANTS
```
          A          B          C
(1) 8.16290    1623.220    228.980
(2) 6.95367    1501.268    194.480
```

P =   760.00

EQUATION	ORDER	A12	A21	D	A122	A211
VAN LAAR	3	0.7386	0.9487	-	-	-
MARGULES	3	0.7265	0.9573	-	-	-
MARGULES	4	0.6631	0.6989	-0.6716	-	-
ALPHA	2	132.5746	2.8453	-	-	-
ALPHA	3	.6363/ 12	.9210/ 10	-	-.6161/ 12	-.1598/ 05

DEVIATION IN VAPOR PHASE COMPOSITION

X	Y	TEMP.	LAAR 3	MARG 3	MARG 4	ALPHA2	ALPHA3
0.0120	0.3250	160.00	0.2212	0.2153	0.1861	0.2829	0.4253
0.0180	0.4600	150.00	0.1880	0.1831	0.1590	0.2358	0.3130
0.0220	0.5800	140.00	0.1179	0.1136	0.0930	0.1545	0.2060
0.0370	0.7760	120.00	0.0276	0.0251	0.0147	0.0413	0.0479
0.0500	0.8460	110.00	0.0038	0.0023	-0.0030	0.0080	0.0013
0.0700	0.9000	100.00	-0.0105	-0.0110	-0.0119	-0.0136	-0.0268
0.0860	0.9200	95.00	-0.0118	-0.0119	-0.0108	-0.0177	-0.0317
0.1180	0.9350	90.00	-0.0053	-0.0049	-0.0019	-0.0131	-0.0251
0.2000	0.9480	85.00	0.0044	0.0053	0.0088	-0.0034	-0.0083
0.3700	0.9580	82.00	0.0072	0.0080	0.0080	0.0023	0.0062
0.6200	0.9620	80.80	0.0061	0.0060	0.0000	0.0061	0.0156
0.7900	0.9650	80.00	0.0058	0.0051	0.0027	0.0059	0.0172
0.9100	0.9770	79.10	0.0032	0.0027	0.0065	-0.0037	0.0074
		MEAN	0.0471	0.0457	0.0390	0.0606	0.0871

ETHYL ALCOHOL(1)-1,2-DICHLOROETHANE(2)

UDOVENKO V.V.,FATKULINA L.G.:ZH.FIZ.KHIM.26,719(1952).

ANTOINE VAPOR PRESSURE CONSTANTS
	A	B	C
(1)	8.16290	1623.220	228.980
(2)	7.18431	1358.500	232.000

T = 40.00

EQUATION	ORDER	A12	A21	D	A122	A211
VAN LAAR	3	0.9430	0.6139	-	-	-
MARGULES	3	0.8904	0.5849	-	-	-
MARGULES	4	0.9761	0.6736	0.2980	-	-
ALPHA	2	5.2257	7.3513	-	-	-
ALPHA	3	2.2740	13.6164	-	7.3123	-9.6117

			DEVIATION IN VAPOR PHASE COMPOSITION				
X	Y	PRESS.	LAAR 3	MARG 3	MARG 4	ALPHA2	ALPHA3
0.0330	0.1920	184.00	-0.0187	-0.0307	-0.0103	-0.0495	0.0003
0.1240	0.3250	208.50	0.0119	0.0074	0.0158	-0.0327	-0.0010
0.2320	0.3680	217.00	0.0171	0.0207	0.0152	-0.0091	0.0013
0.3460	0.3950	220.00	0.0059	0.0108	0.0030	0.0024	-0.0006
0.4000	0.4020	220.00	0.0043	0.0084	0.0025	0.0096	0.0023
0.4920	0.4250	219.40	-0.0064	-0.0049	-0.0051	0.0090	-0.0035
0.6420	0.4590	214.80	-0.0001	-0.0023	0.0061	0.0147	0.0015
0.7800	0.5330	201.30	0.0086	0.0074	0.0123	-0.0021	-0.0006
0.8870	0.6560	179.00	0.0179	0.0219	0.0108	-0.0319	0.0017
0.9480	0.8000	158.70	0.0094	0.0156	-0.0034	-0.0558	-0.0026
		MEAN	0.0100	0.0130	0.0085	0.0217	0.0015

ETHYL ALCOHOL(1)-1,2-DICHLOROETHANE(2)

UDOVENKO V.V.,FATKULINA L.G.:ZH.FIZ.KHIM.26,719(1952).

ANTOINE VAPOR PRESSURE CONSTANTS

	A	B	C
(1)	8.16290	1623.220	228.980
(2)	7.18431	1358.500	232.000

T =   50.00

EQUATION	ORDER	A12	A21	D	A122	A211
VAN LAAR	3	0.8756	0.6082	-	-	-
MARGULES	3	0.8411	0.5870	-	-	-
MARGULES	4	0.9060	0.6544	0.2283	-	-
ALPHA	2	5.3174	6.3624	-	-	-
ALPHA	3	2.1196	10.3839	-	6.4861	-7.0333

			DEVIATION IN VAPOR PHASE COMPOSITION				
X	Y	PRESS.	LAAR 3	MARG 3	MARG 4	ALPHA2	ALPHA3
0.0390	0.1940	274.60	-0.0056	-0.0136	0.0020	-0.0283	0.0141
0.1320	0.3510	317.80	-0.0020	-0.0043	0.0008	-0.0340	-0.0077
0.1940	0.3780	328.70	0.0098	0.0110	0.0092	-0.0152	-0.0007
0.3400	0.4200	336.70	0.0045	0.0080	0.0017	0.0034	-0.0001
0.4020	0.4310	338.00	0.0027	0.0054	0.0008	0.0103	0.0020
0.4880	0.4500	337.90	-0.0023	-0.0014	-0.0017	0.0136	0.0010
0.6360	0.4900	332.50	-0.0019	-0.0037	0.0028	0.0141	0.0013
0.7660	0.5600	316.80	0.0004	-0.0009	0.0039	-0.0043	-0.0034
0.8760	0.6720	287.10	0.0091	0.0113	0.0048	-0.0313	-0.0000
0.9430	0.8030	258.50	0.0097	0.0139	0.0006	-0.0481	0.0033
		MEAN	0.0048	0.0073	0.0028	0.0202	0.0034

ETHYL ALCOHOL(1)-1,2-DICHLOROETHANE(2)

UDOVENKO V.V.,FATKULINA L.G.:ZH.FIZ.KHIM.26,719(1952).

ANTOINE VAPOR PRESSURE CONSTANTS
	A	B	C
(1)	8.16290	1623.220	228.980
(2)	7.18431	1358.500	232.000

T =   60.00

EQUATION	ORDER	A12	A21	D	A122	A211
VAN LAAR	3	0.8725	0.6831	-	-	-
MARGULES	3	0.8508	0.6777	-	-	-
MARGULES	4	0.9979	0.8214	0.5026	-	-
ALPHA	2	7.1390	7.7256	-	-	-
ALPHA	3	4.6795	8.2808	-	3.4022	-2.4083

			DEVIATION IN VAPOR PHASE COMPOSITION				
X	Y	PRESS.	LAAR 3	MARG 3	MARG 4	ALPHA2	ALPHA3
0.0430	0.1940	400.70	0.0238	0.0182	0.0576	0.0150	0.0282
0.1470	0.3810	470.80	0.0136	0.0121	0.0198	-0.0171	-0.0082
0.2030	0.4200	491.50	0.0088	0.0094	0.0036	-0.0212	-0.0153
0.3550	0.4380	500.50	0.0270	0.0297	0.0153	0.0138	0.0133
0.4210	0.4600	503.50	0.0122	0.0148	0.0053	0.0074	0.0049
0.4830	0.4740	504.10	0.0050	0.0071	0.0046	0.0068	0.0028
0.6260	0.5130	499.50	-0.0077	-0.0072	0.0054	-0.0002	-0.0055
0.7420	0.5700	486.00	-0.0180	-0.0184	-0.0056	-0.0217	-0.0239
0.8680	0.5770	444.00	0.0872	0.0873	0.0761	0.0467	0.0567
0.9390	0.8040	403.50	-0.0105	-0.0096	-0.0396	-0.0760	-0.0538
		MEAN	0.0214	0.0214	0.0233	0.0226	0.0213

ETHYL ALCOHOL(1)-ETHYL ACETATE(2)

MURTI P.S.,VAN WINKLE M.:CHEM.ENG.DATA SERIES 3,72(1958).

ANTOINE VAPOR PRESSURE CONSTANTS

	A	B	C
(1)	8.16290	1623.220	228.980
(2)	7.10232	1245.239	217.911

T =   40.00

EQUATION	ORDER	A12	A21	D	A122	A211
VAN LAAR	3	0.4373	0.4237	-	-	-
MARGULES	3	0.4373	0.4235	-	-	-
MARGULES	4	0.4216	0.4093	-0.0487	-	-
ALPHA	2	1.3008	2.6791	-	-	-
ALPHA	3	0.5816	1.2721	-	0.4031	1.2929

			DEVIATION IN VAPOR PHASE COMPOSITION				
X	Y	PRESS.	LAAR 3	MARG 3	MARG 4	ALPHA2	ALPHA3
0.0757	0.1226	206.00	-0.0022	-0.0022	-0.0039	0.0078	-0.0001
0.1285	0.1874	212.00	-0.0063	-0.0063	-0.0074	0.0022	-0.0033
0.2256	0.2745	213.50	-0.0107	-0.0107	-0.0102	-0.0074	-0.0074
0.2316	0.2614	216.00	0.0066	0.0067	0.0073	0.0097	0.0099
0.3441	0.3414	216.00	-0.0046	-0.0047	-0.0034	-0.0055	-0.0024
0.3554	0.3307	215.50	0.0121	0.0121	0.0134	0.0111	0.0142
0.4389	0.4022	216.00	-0.0171	-0.0172	-0.0164	-0.0188	-0.0161
0.5410	0.4317	211.00	0.0035	0.0035	0.0033	0.0028	0.0037
0.6359	0.4910	204.50	-0.0045	-0.0045	-0.0055	-0.0033	-0.0044
0.7166	0.5315	200.50	0.0081	0.0081	0.0070	0.0109	0.0088
0.7576	0.5760	191.50	-0.0037	-0.0037	-0.0045	-0.0003	-0.0024
0.7715	0.5830	191.50	0.0016	0.0016	0.0009	0.0052	0.0031
0.8260	0.6442	184.00	-0.0037	-0.0037	-0.0035	0.0002	-0.0011
0.8638	0.6921	176.00	-0.0033	-0.0033	-0.0021	0.0005	0.0002
0.9009	0.7544	165.00	-0.0072	-0.0072	-0.0052	-0.0040	-0.0031
0.9312	0.8164	158.00	-0.0104	-0.0104	-0.0079	-0.0080	-0.0063
0.9636	0.8868	151.50	-0.0026	-0.0025	-0.0002	-0.0012	0.0008
0.9776	0.9192	146.00	0.0054	0.0054	0.0073	0.0062	0.0079
		MEAN	0.0063	0.0063	0.0061	0.0058	0.0053

ETHYL ALCOHOL(1)-ETHYL ACETATE(2)

MURTI P.S.,VAN WINKLE M.:CHEM.ENG.DATA SERIES 3,72(1958).

ANTOINE VAPOR PRESSURE CONSTANTS
	A	B	C
(1)	8.16290	1623.220	228.980
(2)	7.10232	1245.239	217.911

T =   60.00

EQUATION	ORDER	A12	A21	D	A122	A211
VAN LAAR	3	0.3962	0.3447	-	-	-
MARGULES	3	0.3919	0.3450	-	-	-
MARGULES	4	0.4777	0.4177	0.2694	-	-
ALPHA	2	1.1142	1.7783	-	-	-
ALPHA	3	3.5469	5.3769	-	-1.7630	-3.3042

			DEVIATION IN VAPOR PHASE COMPOSITION				
X	Y	PRESS.	LAAR 3	MARG 3	MARG 4	ALPHA2	ALPHA3
0.0505	0.1107	444.00	-0.0196	-0.0202	-0.0097	-0.0194	-0.0060
0.0595	0.1100	444.50	-0.0053	-0.0058	0.0049	-0.0051	0.0084
0.1319	0.2023	464.00	-0.0069	-0.0072	-0.0017	-0.0074	-0.0004
0.2286	0.2801	478.00	0.0019	0.0021	-0.0020	0.0014	-0.0010
0.2286	0.2889	478.50	-0.0069	-0.0067	-0.0108	-0.0074	-0.0098
0.3279	0.3257	484.50	0.0240	0.0246	0.0164	0.0245	0.0183
0.4437	0.4244	485.00	-0.0071	-0.0064	-0.0116	-0.0051	-0.0099
0.5011	0.4578	481.00	-0.0080	-0.0074	-0.0092	-0.0053	-0.0081
0.5229	0.4625	479.50	-0.0001	0.0004	0.0000	0.0027	0.0008
0.5860	0.4865	474.00	0.0137	0.0141	0.0175	0.0167	0.0176
0.6200	0.5294	473.00	-0.0075	-0.0071	-0.0021	-0.0046	-0.0022
0.6870	0.5880	466.00	-0.0189	-0.0188	-0.0121	-0.0167	-0.0123
0.7541	0.6285	454.00	-0.0039	-0.0039	0.0013	-0.0032	0.0018
0.8064	0.6800	444.50	-0.0038	-0.0039	-0.0022	-0.0046	-0.0007
0.8559	0.7260	421.50	0.0086	0.0085	0.0051	0.0061	0.0075
0.8940	0.7730	411.00	0.0151	0.0149	0.0074	0.0115	0.0102
0.9247	0.8491	389.00	-0.0108	-0.0110	-0.0211	-0.0148	-0.0184
0.9760	0.9393	361.50	0.0018	0.0017	-0.0061	-0.0008	-0.0050
0.9565	0.8849	375.00	0.0138	0.0137	0.0034	0.0102.	0.0053
		MEAN	0.0094	0.0094	0.0076	0.0088	0.0076

ETHYL ALCOHOL(1)-ETHYL ACETATE(2)

MURTI P.S.,VAN WINKLE M.:CHEM.ENG.DATA SERIES 3,72(1958).

ANTOINE VAPOR PRESSURE CONSTANTS

	A	B	C
(1)	8.16290	1623.220	228.980
(2)	7.10232	1245.239	217.911

P = 760.00

EQUATION	ORDER	A12	A21	D	A122	A211
VAN LAAR	3	0.3700	0.3393	-	-	-
MARGULES	3	0.3683	0.3596	-	-	-
MARGULES	4	0.4270	0.3961	0.1862	-	-
ALPHA	2	1.1946	1.4633	-	-	-
ALPHA	3	2.4898	3.6207	-	-0.7679	-2.1767

			DEVIATION IN VAPOR PHASE COMPOSITION				
X	Y	TEMP.	LAAR 3	MARG 3	MARG 4	ALPHA2	ALPHA3
0.0505	0.1036	75.55	-0.0070	-0.0073	0.0003	-0.0080	0.0041
0.1260	0.2146	73.82	-0.0149	-0.0150	-0.0103	-0.0154	-0.0066
0.1343	0.2146	73.78	-0.0054	-0.0055	-0.0014	-0.0059	0.0021
0.2271	0.2960	73.04	0.0012	0.0013	-0.0008	0.0018	0.0016
0.3128	0.3634	72.50	-0.0033	-0.0031	-0.0077	-0.0009	-0.0057
0.3358	0.3643	72.28	0.0107	0.0110	0.0063	0.0138	0.0083
0.5052	0.4803	72.18	-0.0028	-0.0025	-0.0026	0.0026	-0.0017
0.5441	0.5074	72.35	-0.0063	-0.0060	-0.0046	-0.0009	-0.0040
0.6442	0.5618	72.70	0.0040	0.0041	0.0084	0.0087	0.0096
0.6828	0.6092	72.90	-0.0156	-0.0155	-0.0110	-0.0117	-0.0092
0.7860	0.6819	74.14	0.0003	0.0003	0.0017	0.0001	0.0055
0.8774	0.7908	75.50	-0.0043	-0.0044	-0.0094	-0.0085	-0.0035
0.9482	0.8924	76.70	0.0031	0.0030	-0.0047	-0.0018	0.0004
		MEAN	0.0061	0.0061	0.0053	0.0062	0.0048

K

ETHYL ALCOHOL(1)-ETHYLBENZENE(2)

ELLIS S.R.M.,SPURR J.M.:BRIT.CHEM.ENG.6,93(1961).

ANTOINE VAPOR PRESSURE CONSTANTS
```
          A          B          C
(1) 8.16290    1623.220    228.980
(2) 6.95719    1424.255    213.206
```

P = 760.00

EQUATION	ORDER	A12	A21	D	A122	A211
VAN LAAR	3	0.6967	0.5913	-	-	-
MARGULES	3	0.7014	0.5662	-	-	-
MARGULES	4	0.5463	0.3380	-0.7157	-	-
ALPHA	2	28.8410	3.1452	-	-	-
ALPHA	3	-33.4984	-0.5077	-	58.6067	-3.8103

			DEVIATION IN VAPOR PHASE COMPOSITION				
X	Y	TEMP.	LAAR 3	MARG 3	MARG 4	ALPHA2	ALPHA3
0.0240	0.3400	120.00	0.0897	0.0920	0.0263	0.0599	0.0358
0.0480	0.5250	110.00	0.0579	0.0596	0.0100	0.0299	0.0118
0.0820	0.6600	100.00	0.0238	0.0247	-0.0007	0.0011	-0.0071
0.1080	0.7130	95.00	0.0134	0.0139	0.0015	-0.0058	-0.0083
0.1500	0.7650	90.00	0.0034	0.0032	0.0041	-0.0113	-0.0071
0.2600	0.8150	85.00	0.0040	0.0026	0.0142	-0.0030	0.0120
0.5100	0.8490	82.00	0.0112	0.0091	0.0029	0.0091	-0.0007
0.6160	0.8630	80.80	0.0122	0.0107	-0.0024	0.0053	0.0090
0.7100	0.8750	80.00	0.0174	0.0169	0.0048	0.0014	0.0036
0.8200	0.9030	79.10	0.0177	0.0186	0.0172	-0.0156	-0.0248
		MEAN	0.0251	0.0251	0.0084	0.0143	0.0120

ETHYL ALCOHOL(1)-HEPTANE(2)

FERGUSON J.B.,FREED M.,MORRIS A.C.:J.PHYS.CHEM.37,87(1933).

ANTOINE VAPOR PRESSURE CONSTANTS
	A	B	C
(1)	8.16290	1623.220	228.980
(2)	6.90240	1268.115	216.900

T = 30.00

EQUATION	ORDER	A12	A21	D	A122	A211
VAN LAAR	3	1.0933	0.8661	-	-	-
MARGULES	3	1.0611	0.8727	-	-	-
MARGULES	4	1.2325	1.0782	0.7257	-	-
ALPHA	2	29.0415	21.4143	-	-	-
ALPHA	3	7.0653	43.2640	-	48.7230	-37.0847

			DEVIATION IN VAPOR PHASE COMPOSITION				
X	Y	PRESS.	LAAR 3	MARG 3	MARG 4	ALPHA2	ALPHA3
0.0140	0.3950	95.00	-0.2168	-0.2263	-0.1730	-0.1504	-0.0654
0.0470	0.4720	109.00	-0.0918	-0.1031	-0.0485	-0.0586	-0.0072
0.1050	0.5120	117.10	-0.0044	-0.0102	0.0092	-0.0184	0.0024
0.1170	0.5020	115.20	0.0186	0.0139	0.0271	-0.0002	0.0170
0.2550	0.5420	120.50	0.0253	0.0288	0.0059	0.0033	0.0003
0.3460	0.5520	122.00	0.0090	0.0147	-0.0064	0.0051	-0.0036
0.6630	0.5710	122.00	-0.0293	-0.0276	0.0008	0.0117	-0.0038
0.8220	0.5990	119.40	0.0063	0.0044	0.0117	0.0060	0.0051
0.9080	0.6770	110.00	0.0322	0.0293	-0.0018	-0.0377	0.0001
0.9650	0.8180	94.30	0.0316	0.0296	-0.0180	-0.0984	-0.0037
		MEAN	0.0465	0.0488	0.0302	0.0390	0.0109

ETHYL ALCOHOL(1)-HEPTANE(2)

SMITH C.P.,ENGEL E.W.:J.AM.CHEM.SOC.51,2660(1929).

ANTOINE VAPOR PRESSURE CONSTANTS
```
          A          B          C
(1) 8.16290    1623.220    228.980
(2) 6.90240    1268.115    216.900
```

T =   30.00

EQUATION	ORDER	A12	A21	D	A122	A211
VAN LAAR	3	1.0074	0.7550	-	-	-
MARGULES	3	0.9856	0.7384	-	-	-
MARGULES	4	0.9956	0.7479	0.0362	-	-
ALPHA	2	12.4426	8.9910	-	-	-
ALPHA	3	-2.0354	13.9730	-	23.5265	-12.9754

			DEVIATION IN VAPOR PHASE COMPOSITION				
X	Y	PRESS.	LAAR 3	MARG 3	MARG 4	ALPHA2	ALPHA3
0.0400	0.2820	78.00	0.0271	0.0203	0.0236	0.0020	0.0776
0.0684	0.4556	103.60	-0.0553	-0.0604	-0.0579	-0.0916	-0.0250
0.1236	0.4986	112.50	-0.0142	-0.0156	-0.0148	-0.0556	-0.0078
0.2803	0.5344	117.70	0.0043	0.0070	0.0058	-0.0103	0.0051
0.3342	0.5381	118.00	0.0018	0.0044	0.0031	-0.0003	0.0070
0.5151	0.5496	119.90	-0.0106	-0.0108	-0.0108	0.0206	0.0037
0.5934	0.5679	119.90	-0.0223	-0.0238	-0.0231	0.0147	-0.0113
0.7174	0.5828	119.60	-0.0053	-0.0074	-0.0063	0.0233	-0.0091
0.7687	0.5971	118.90	0.0061	0.0045	0.0054	0.0225	-0.0038
0.8154	0.6282	117.00	0.0083	0.0078	0.0081	0.0079	-0.0024
0.8550	0.6454	114.50	0.0299	0.0305	0.0302	0.0102	0.0260
0.8902	0.7116	111.30	0.0094	0.0112	0.0103	-0.0310	0.0198
0.9173	0.8269	106.30	-0.0607	-0.0581	-0.0595	-0.1181	-0.0354
0.9545	0.8958	97.10	-0.0474	-0.0443	-0.0460	-0.1216	-0.0072
0.9913	0.9430	82.50	0.0220	0.0232	0.0225	-0.0158	0.0386
		MEAN	0.0216	0.0220	0.0218	0.0364	0.0187

ETHYL ALCOHOL(1)-HEPTANE(2)

SMITH C.P.,ENGEL E.W.:J.AM.CHEM.SOC.51,2660(1929).

ANTOINE VAPOR PRESSURE CONSTANTS
          A   B   C
(1) 8.16290   1623.220   228.980
(2) 6.90240   1268.115   216.900

T =   50.00

EQUATION	ORDER	A12	A21	D	A122	A211
VAN LAAR	3	1.0279	0.7664	-	-	-
MARGULES	3	1.0002	0.7514	-	-	-
MARGULES	4	1.0454	0.7903	0.1457	-	-
ALPHA	2	16.0903	10.0551	-	-	-
ALPHA	3	-0.8925	17.6127	-	30.0303	-16.7146

			DEVIATION IN VAPOR PHASE COMPOSITION				
X	Y	PRESS.	LAAR 3	MARG 3	MARG 4	ALPHA2	ALPHA3
0.0514	0.4582	244.20	-0.0620	-0.0708	-0.0558	-0.0907	-0.0165
0.1180	0.5383	289.40	-0.0147	-0.0178	-0.0120	-0.0566	-0.0126
0.3022	0.5755	307.40	0.0032	0.0063	0.0015	-0.0082	-0.0002
0.4382	0.5824	311.30	-0.0078	-0.0057	-0.0086	0.0092	0.0031
0.5862	0.5933	311.30	-0.0146	-0.0151	-0.0131	0.0181	0.0011
0.6646	0.6074	309.50	-0.0149	-0.0163	-0.0127	0.0152	-0.0045
0.7327	0.6145	311.30	0.0012	-0.0003	0.0032	0.0201	0.0032
0.7720	0.6279	308.80	0.0078	0.0066	0.0092	0.0155	0.0047
0.8230	0.6516	304.80	0.0203	0.0200	0.0203	0.0070	0.0135
0.8788	0.7554	293.60	-0.0251	-0.0240	-0.0272	-0.0704	-0.0251
0.9274	0.8324	281.70	-0.0264	-0.0240	-0.0302	-0.1040	-0.0072
0.9769	0.9127	247.40	0.0105	0.0124	0.0073	-0.0699	0.0365
		MEAN	0.0174	0.0183	0.0168	0.0404	0.0107

ETHYL ALCOHOL(1)-HEPTANE(2)

SMITH C.P.,ENGEL E.W.:J.AM.CHEM.SOC.51,2660(1929).

ANTOINE VAPOR PRESSURE CONSTANTS

	A	B	C
(1)	8.16290	1623.220	228.980
(2)	6.90240	1268.115	216.900

T = 70.00

EQUATION	ORDER	A12	A21	D	A122	A211
VAN LAAR	3	0.9910	0.8830	-	-	-
MARGULES	3	0.9838	0.8830	-	-	-
MARGULES	4	1.0948	0.9765	0.3721	-	-
ALPHA	2	22.4938	12.4905	-	-	-
ALPHA	3	1.2321	12.7292	-	25.7717	-9.9132

			DEVIATION IN VAPOR PHASE COMPOSITION				
X	Y	PRESS.	LAAR 3	MARG 3	MARG 4	ALPHA2	ALPHA3
0.0567	0.4685	500.30	-0.0278	-0.0303	0.0047	-0.0297	0.0031
0.1180	0.5531	614.30	0.0110	0.0098	0.0223	-0.0232	0.0030
0.1573	0.5790	648.00	0.0195	0.0190	0.0214	-0.0222	0.0003
0.2575	0.6000	689.60	0.0324	0.0330	0.0218	-0.0072	0.0070
0.3633	0.6114	705.90	0.0241	0.0251	0.0130	0.0008	0.0065
0.4290	0.6213	712.10	0.0103	0.0113	0.0030	-0.0007	-0.0004
0.5069	0.6241	715.70	0.0019	0.0028	0.0010	0.0049	-0.0014
0.5968	0.6280	717.70	-0.0053	-0.0048	0.0012	0.0101	-0.0037
0.6648	0.6383	717.70	-0.0124	-0.0121	-0.0024	0.0072	-0.0112
0.7174	0.6483	717.70	-0.0141	-0.0140	-0.0038	0.0039	-0.0161
0.8089	0.6483	713.80	0.0206	0.0204	0.0247	0.0206	0.0087
0.8200	0.6516	704.90	0.0240	0.0238	0.0268	0.0201	0.0109
0.8640	0.6713	693.40	0.0393	0.0391	0.0352	0.0149	0.0231
0.8940	0.7293	676.50	0.0150	0.0148	0.0054	-0.0282	-0.0005
0.9250	0.8013	651.80	-0.0100	-0.0101	-0.0249	-0.0760	-0.0208
0.9564	0.8639	610.00	-0.0066	-0.0067	-0.0240	-0.0935	-0.0090
0.9827	0.9211	569.40	0.0128	0.0128	0.0007	-0.0651	0.0158
		MEAN	0.0169	0.0171	0.0139	0.0252	0.0083

ETHYL ALCOHOL(1)-HEXANE(2)

SINOR J.E.,WEBER J.H.:J.CHEM.ENG.DATA 5,244(1960).

ANTOINE VAPOR PRESSURE CONSTANTS

	A	B	C
(1)	8.16290	1623.220	228.980
(2)	6.87776	1171.530	224.366

P =   760.00

EQUATION	ORDER	A12	A21	D	A122	A211
VAN LAAR	3	1.0237	0.7852	-	-	-
MARGULES	3	0.9889	0.7817	-	-	-
MARGULES	4	1.1487	0.9361	0.5483	-	-
ALPHA	2	8.8090	17.1742	-	-	-
ALPHA	3	3.7045	31.3435	-	13.0333	-21.5236

			DEVIATION IN VAPOR PHASE COMPOSITION				
X	Y	TEMP.	LAAR 3	MARG 3	MARG 4	ALPHA2	ALPHA3
0.0060	0.0650	66.70	-0.0260	-0.0287	-0.0144	-0.0143	0.0170
0.0100	0.1600	63.50	-0.0998	-0.1039	-0.0832	-0.0827	-0.0416
0.0450	0.2550	60.20	-0.0658	-0.0739	-0.0365	-0.0549	-0.0086
0.1020	0.2900	59.15	-0.0041	-0.0098	0.0122	-0.0211	0.0078
0.2350	0.3250	58.45	0.0219	0.0237	0.0126	-0.0043	0.0041
0.2750	0.3300	58.25	0.0197	0.0226	0.0080	-0.0013	0.0034
0.3300	0.3400	58.00	0.0093	0.0133	-0.0022	-0.0023	-0.0019
0.4120	0.3500	58.10	-0.0034	0.0008	-0.0097	-0.0010	-0.0058
0.5480	0.3600	58.35	-0.0145	-0.0117	-0.0066	0.0071	-0.0041
0.6670	0.3700	58.70	-0.0095	-0.0085	0.0065	0.0175	0.0041
0.7550	0.3950	59.40	-0.0022	-0.0022	0.0105	0.0157	0.0057
0.8480	0.4680	61.80	0.0040	0.0036	-0.0023	-0.0119	-0.0058
0.9200	0.5800	65.90	0.0283	0.0284	-0.0064	-0.0415	-0.0027
0.9400	0.6350	67.40	0.0348	0.0353	-0.0077	-0.0522	-0.0003
0.9800	0.8070	73.20	0.0506	0.0513	0.0116	-0.0433	0.0202
0.9900	0.9050	76.00	0.0191	0.0196	-0.0063	-0.0482	-0.0011
		MEAN	0.0258	0.0273	0.0148	0.0262	0.0084

159

ETHYL ALCOHOL(1)-CYCLOHEXANE(2)

DESHPANDE A.K.,LU B.C.Y.:J.ENG.CHEM.DATA 8(4),549(1963).

ANTOINE VAPOR PRESSURE CONSTANTS

	A	B	C
(1)	8.16290	1623.220	228.980
(2)	6.84498	1203.526	222.863

P = 760.00

EQUATION	ORDER	A12	A21	D	A122	A211
VAN LAAR	3	1.0766	0.7262	-	-	-
MARGULES	3	1.0219	0.7039	-	-	-
MARGULES	4	1.1390	0.8173	0.4013	-	-
ALPHA	2	9.4925	11.6195	-	-	-
ALPHA	3	2.6997	29.3634	-	20.2510	-25.3651

			DEVIATION IN VAPOR PHASE COMPOSITION				
X	Y	TEMP.	LAAR 3	MARG 3	MARG 4	ALPHA2	ALPHA3
0.0200	0.1750	73.99	0.0038	-0.0109	0.0221	-0.0293	0.0551
0.0300	0.3020	69.08	-0.0751	-0.0910	-0.0554	-0.1123	-0.0280
0.0650	0.3580	66.94	-0.0210	-0.0346	-0.0042	-0.0768	-0.0130
0.0810	0.3630	66.08	0.0020	-0.0093	0.0161	-0.0567	-0.0020
0.0860	0.3650	66.37	0.0082	-0.0023	0.0215	-0.0521	-0.0001
0.1250	0.3880	65.59	0.0241	0.0190	0.0308	-0.0365	-0.0013
0.1510	0.3960	65.23	0.0302	0.0282	0.0333	-0.0269	-0.0001
0.2060	0.4080	65.12	0.0332	0.0359	0.0306	-0.0131	0.0007
0.2580	0.4150	64.93	0.0294	0.0346	0.0239	-0.0034	0.0017
0.2830	0.4180	64.87	0.0261	0.0320	0.0201	0.0001	0.0019
0.3150	0.4260	64.84	0.0170	0.0233	0.0108	-0.0006	-0.0025
0.3660	0.4300	64.78	0.0104	0.0165	0.0052	0.0054	-0.0013
0.4030	0.4310	64.77	0.0077	0.0131	0.0041	0.0110	0.0013
0.4310	0.4310	64.77	0.0068	0.0116	0.0047	0.0156	0.0041
0.4440	0.4380	64.78	-0.0004	0.0040	-0.0018	0.0107	-0.0016
0.5000	0.4430	64.81	-0.0047	-0.0020	-0.0026	0.0146	-0.0005
0.5570	0.4550	64.88	-0.0124	-0.0115	-0.0066	0.0119	-0.0051
0.6130	0.4600	65.01	-0.0083	-0.0089	0.0003	0.0168	-0.0007
0.6210	0.4580	64.99	-0.0046	-0.0054	0.0043	0.0203	0.0029
0.6780	0.4750	65.25	-0.0046	-0.0065	0.0055	0.0153	-0.0003
0.7380	0.5050	65.56	-0.0068	-0.0089	0.0019	0.0014	-0.0087
0.7630	0.4960	66.03	0.0186	0.0167	0.0256	0.0187	0.0126
0.7760	0.5150	65.93	0.0084	0.0067	0.0143	0.0045	0.0011
0.7810	0.4980	66.40	0.0302	0.0286	0.0357	0.0235	0.0212
0.8090	0.5450	66.90	0.0079	0.0070	0.0101	-0.0111	-0.0056
0.8330	0.5780	67.26	-0.0000	-0.0001	-0.0013	-0.0310	-0.0167
0.8530	0.5950	67.98	0.0085	0.0092	0.0037	-0.0348	-0.0113
0.8810	0.6230	68.86	0.0226	0.0247	0.0125	-0.0391	0.0007
0.8980	0.6530	69.44	0.0231	0.0260	0.0095	-0.0505	0.0013
0.9090	0.6780	70.11	0.0207	0.0242	0.0050	-0.0611	-0.0009
0.9290	0.7250	71.42	0.0205	0.0248	0.0015	-0.0750	0.0010
0.9510	0.7780	72.48	0.0283	0.0330	0.0075	-0.0756	0.0150
		MEAN	0.0164	0.0191	0.0135	0.0299	0.0069

ETHYL ALCOHOL(1)-METHYLCYCLOHEXANE(2)

ISII N.:J.SOC.CHEM.IND.JAPAN 38,659(1935)

ANTOINE VAPOR PRESSURE CONSTANTS

	A	B	C
(1)	8.16290	1623.220	228.980
(2)	6.82689	1272.864	221.630

T =   0.00

EQUATION	ORDER	A12	A21	D	A122	A211
VAN LAAR	3	1.0139	0.8556	-	-	-
MARGULES	3	0.9997	0.8541	-	-	-
MARGULES	4	1.1824	1.0357	0.5963	-	-
ALPHA	2	21.8622	23.2408	-	-	-
ALPHA	3	5.5582	44.6373	-	36.4516	-37.2257

DEVIATION IN VAPOR PHASE COMPOSITION

X	Y	PRESS.	LAAR 3	MARG 3	MARG 4	ALPHA2	ALPHA3
0.0500	0.3950	19.67	-0.1070	-0.1113	-0.0554	-0.0486	0.0015
0.1000	0.4355	21.09	-0.0398	-0.0426	-0.0091	-0.0268	0.0012
0.1500	0.4544	21.70	-0.0089	-0.0101	0.0020	-0.0193	-0.0023
0.2000	0.4647	22.10	0.0048	0.0050	0.0020	-0.0146	-0.0045
0.2500	0.4697	22.31	0.0105	0.0116	-0.0003	-0.0098	-0.0044
0.3000	0.4719	22.40	0.0115	0.0132	-0.0026	-0.0048	-0.0029
0.3500	0.4719	22.40	0.0105	0.0125	-0.0030	0.0009	-0.0001
0.4000	0.4725	22.37	0.0068	0.0088	-0.0033	0.0051	0.0018
0.4500	0.4737	22.31	0.0017	0.0036	-0.0031	0.0082	0.0030
0.5000	0.4750	22.25	-0.0032	-0.0015	-0.0017	0.0110	0.0042
0.5500	0.4881	22.13	-0.0185	-0.0172	-0.0108	0.0020	-0.0060
0.6000	0.4820	21.97	-0.0123	-0.0114	0.0005	0.0125	0.0037
0.6500	0.4869	21.77	-0.0136	-0.0131	0.0022	0.0125	0.0038
0.7000	0.4935	21.58	-0.0117	-0.0116	0.0041	0.0118	0.0042
0.7500	0.5080	21.16	-0.0108	-0.0110	0.0009	0.0048	0.0004
0.8000	0.5295	20.64	-0.0067	-0.0070	-0.0040	-0.0064	-0.0039
0.8500	0.5633	19.83	0.0011	0.0008	-0.0111	-0.0245	-0.0081
0.9000	0.6156	18.60	0.0174	0.0173	-0.0152	-0.0488	-0.0050
0.9500	0.7111	16.65	0.0429	0.0431	-0.0080	-0.0783	0.0151
		MEAN	0.0179	0.0186	0.0073	0.0184	0.0040

ETHYL ALCOHOL(1)-METHYLCYCLOHEXANE(2)

ISII N.:J.SOC.CHEM.IND.JAPAN 38,659(1935).

ANTOINE VAPOR PRESSURE CONSTANTS
	A	B	C
(1)	8.16290	1623.220	228.980
(2)	6.82689	1272.864	221.630

T =   10.00

EQUATION	ORDER	A12	A21	D	A122	A211
VAN LAAR	3	1.0912	0.8297	-	-	-
MARGULES	3	1.0603	0.8177	-	-	-
MARGULES	4	1.1931	0.9437	0.4220	-	-
ALPHA	2	22.9527	23.3901	-		
ALPHA	3	2.0833	26.1915	-	26.2402	-20.5551

DEVIATION IN VAPOR PHASE COMPOSITION

X	Y	PRESS.	LAAR 3	MARG 3	MARG 4	ALPHA2	ALPHA3
0.0500	0.3975	36.10	-0.0579	-0.0677	-0.0243	-0.0413	-0.0143
0.1000	0.4360	38.30	0.0078	0.0020	0.0273	-0.0172	0.0039
0.1500	0.4608	39.50	0.0248	0.0228	0.0320	-0.0155	0.0012
0.2000	0.4739	40.30	0.0285	0.0291	0.0272	-0.0137	-0.0007
0.2500	0.4815	40.70	0.0255	0.0278	0.0193	-0.0115	-0.0018
0.3000	0.4860	41.05	0.0195	0.0227	0.0112	-0.0089	-0.0023
0.3500	0.4878	41.20	0.0131	0.0165	0.0050	-0.0051	-0.0016
0.4000	0.4891	41.30	0.0061	0.0093	0.0001	-0.0017	-0.0013
0.4500	0.4891	41.30	0.0006	0.0034	-0.0022	0.0025	-0.0002
0.5000	0.4891	41.30	-0.0036	-0.0016	-0.0026	0.0066	0.0008
0.5500	0.4891	41.30	-0.0056	-0.0044	-0.0008	0.0106	0.0019
0.6000	0.4903	41.20	-0.0057	-0.0053	0.0022	0.0136	0.0024
0.6500	0.4920	41.05	-0.0022	-0.0024	0.0076	0.0167	0.0037
0.7000	0.4981	40.65	0.0024	0.0018	0.0123	0.0163	0.0030
0.7500	0.5100	40.00	0.0088	0.0081	0.0161	0.0116	0.0008
0.8000	0.5334	38.90	0.0143	0.0139	0.0159	-0.0019	-0.0048
0.8500	0.5695	37.40	0.0230	0.0235	0.0154	-0.0229	-0.0077
0.9000	0.6254	35.10	0.0378	0.0396	0.0182	-0.0518	0.0002
0.9500	0.7431	30.75	0.0379	0.0407	0.0087	-0.1058	0.0089
		MEAN	0.0171	0.0180	0.0131	0.0197	0.0032

ETHYL ALCOHOL(1)-METHYLCYCLOHEXANE(2)

ISII N.:J.SOC.CHEM.JAPAN 38,659(1935).

ANTOINE VAPOR PRESSURE CONSTANTS
```
            A          B          C
(1)  8.16290    1623.220    228.980
(2)  6.82689    1272.864    221.630
```

T = 20.00

EQUATION	ORDER	A12	A21	D	A122	A211
VAN LAAR	3	1.0723	0.8119	-	-	-
MARGULES	3	1.0414	0.7996	-	-	-
MARGULES	4	1.1784	0.9315	0.4386	-	-
ALPHA	2	21.9646	19.5526	-	-	-
ALPHA	3	2.3006	33.0238	-	35.5802	-29.0418

			DEVIATION IN VAPOR PHASE COMPOSITION				
X	Y	PRESS.	LAAR 3	MARG 3	MARG 4	ALPHA2	ALPHA3
0.0500	0.4180	62.20	-0.0636	-0.0736	-0.0279	-0.0501	0.0008
0.1000	0.4692	66.50	-0.0082	-0.0139	0.0122	-0.0307	-0.0002
0.1500	0.4927	68.80	0.0115	0.0095	0.0189	-0.0239	-0.0044
0.2000	0.5036	70.10	0.0184	0.0191	0.0171	-0.0176	-0.0052
0.2500	0.5092	70.70	0.0186	0.0209	0.0121	-0.0119	-0.0046
0.3000	0.5098	71.20	0.0175	0.0208	0.0089	-0.0043	-0.0010
0.3500	0.5133	71.30	0.0106	0.0140	0.0023	-0.0014	-0.0014
0.4000	0.5154	71.40	0.0039	0.0071	-0.0023	0.0019	-0.0011
0.4500	0.5154	71.40	-0.0005	0.0022	-0.0033	0.0066	0.0011
0.5000	0.5154	71.40	-0.0036	-0.0016	-0.0024	0.0111	0.0032
0.5500	0.5175	71.30	-0.0066	-0.0055	-0.0015	0.0134	0.0036
0.6000	0.5210	71.20	-0.0080	-0.0077	0.0004	0.0145	0.0033
0.6500	0.5246	71.10	-0.0054	-0.0057	0.0050	0.0160	0.0043
0.7000	0.5340	70.60	-0.0030	-0.0037	0.0073	0.0126	0.0019
0.7500	0.5481	69.70	0.0020	0.0013	0.0097	0.0061	-0.0009
0.8000	0.5697	68.10	0.0100	0.0096	0.0117	-0.0052	-0.0034
0.8500	0.6086	65.40	0.0158	0.0163	0.0081	-0.0286	-0.0081
0.9000	0.6694	61.40	0.0239	0.0257	0.0042	-0.0620	-0.0046
0.9500	0.7745	55.00	0.0298	0.0324	0.0014	-0.1038	0.0133
		MEAN	0.0137	0.0153	0.0082	0.0222	0.0035

ETHYL ALCOHOL(1)-METHYLCYCLOHEXANE(2)

ISII N.:J.SOC.CHEM.JAPAN 38,659(1935).

ANTOINE VAPOR PRESSURE CONSTANTS
	A	B	C
(1)	8.16290	1623.220	228.980
(2)	6.82689	1272.864	221.630

T =   30.00

EQUATION	ORDER	A12	A21	D	A122	A211
VAN LAAR	3	1.0791	0.7921	-	-	-
MARGULES	3	1.0372	0.7814	-	-	-
MARGULES	4	1.2587	0.9901	0.6988	-	-
ALPHA	2	21.9927	17.9965	-	-	-
ALPHA	3	12.4841	60.9206	-	54.8972	-53.3836

			DEVIATION IN VAPOR PHASE COMPOSITION				
X	Y	PRESS.	LAAR 3	MARG 3	MARG 4	ALPHA2	ALPHA3
0.0500	0.4433	104.90	-0.0651	-0.0792	-0.0027	-0.0658	0.0131
0.1000	0.4857	112.20	-0.0013	-0.0094	0.0334	-0.0336	0.0043
0.1500	0.5060	115.80	0.0199	0.0170	0.0327	-0.0216	-0.0029
0.2000	0.5157	117.30	0.0268	0.0276	0.0248	-0.0130	-0.0050
0.2500	0.5207	118.30	0.0268	0.0299	0.0161	-0.0060	-0.0047
0.3000	0.5235	118.80	0.0231	0.0277	0.0088	-0.0001	-0.0031
0.3500	0.5260	119.40	0.0171	0.0222	0.0032	0.0042	-0.0017
0.4000	0.5275	119.80	0.0113	0.0163	0.0009	0.0083	0.0007
0.4500	0.5300	120.20	0.0050	0.0094	0.0001	0.0108	0.0021
0.5000	0.5345	120.30	-0.0019	0.0016	-0.0002	0.0109	0.0020
0.5500	0.5383	120.20	-0.0058	-0.0033	0.0024	0.0117	0.0031
0.6000	0.5437	120.10	-0.0079	-0.0065	0.0056	0.0110	0.0036
0.6500	0.5527	119.60	-0.0095	-0.0090	0.0072	0.0072	0.0018
0.7000	0.5626	118.90	-0.0063	-0.0066	0.0102	0.0033	0.0012
0.7500	0.5769	117.70	-0.0002	-0.0008	0.0119	-0.0034	-0.0007
0.8000	0.5981	115.70	0.0092	0.0086	0.0116	-0.0143	-0.0042
0.8500	0.6273	112.70	0.0251	0.0252	0.0124	-0.0281	-0.0064
0.9000	0.6682	108.20	0.0517	0.0528	0.0200	-0.0421	-0.0015
0.9500	0.7470	100.00	0.0778	0.0797	0.0328	-0.0591	0.0108
		MEAN	0.0206	0.0228	0.0125	0.0187	0.0038

ETHYL ALCOHOL(1)-METHYLCYCLOHEXANE(2)

KRETSCHMER C.B.,WIEBE R.:J.AM.CHEM.SOC.71,3176(1949).

ANTOINE VAPOR PRESSURE CONSTANTS
	A	B	C
(1)	8.16290	1623.220	228.980
(2)	6.82689	1272.864	221.630

T =   35.00

EQUATION	ORDER	A12	A21	D	A122	A211
VAN LAAR	3	1.0553	0.7886	-	-	-
MARGULES	3	1.0195	0.7815	-	-	-
MARGULES	4	1.2164	0.9933	0.6979	-	-
ALPHA	2	21.3778	16.0089	-	-	-
ALPHA	3	10.5441	46.4409	-	45.1093	-40.4893

			DEVIATION IN VAPOR PHASE COMPOSITION				
X	Y	PRESS.	LAAR 3	MARG 3	MARG 4	ALPHA2	ALPHA3
0.0526	0.4645	135.40	-0.0766	-0.0882	-0.0241	-0.0740	-0.0027
0.1446	0.5118	146.97	0.0169	0.0144	0.0263	-0.0160	0.0025
0.2878	0.5362	151.27	0.0216	0.0258	0.0070	0.0028	-0.0009
0.4052	0.5471	152.36	0.0043	0.0090	-0.0034	0.0081	-0.0013
0.5403	0.5575	152.93	-0.0106	-0.0081	0.0005	0.0112	0.0013
0.6914	0.5817	152.22	-0.0132	-0.0135	0.0077	0.0034	0.0004
0.8450	0.6423	145.73	0.0181	0.0175	0.0089	-0.0242	-0.0009
0.9676	0.8369	120.04	0.0447	0.0458	0.0031	-0.0829	0.0009
		MEAN	0.0257	0.0278	0.0101	0.0278	0.0014

ETHYL ALCOHOL(1)-METHYLCYCLOHEXANE(2)

KRETSCHMER C.B.,WIEBE R.:J.AM.CHEM.SOC.71,3176(1949).

ANTOINE VAPOR PRESSURE CONSTANTS
	A	B	C
(1)	8.16290	1623.220	228.980
(2)	6.82689	1272.864	221.630

T =   55.00

EQUATION	ORDER	A12	A21	D	A122	A211
VAN LAAR	3	1.0171	0.7693	-	-	-
MARGULES	3	0.9840	0.7655	-	-	-
MARGULES	4	1.1574	0.9490	0.6060	-	-
ALPHA	2	22.0139	13.3912	-	-	-
ALPHA	3	9.4741	31.3093	-	36.1683	-26.8156

			DEVIATION IN VAPOR PHASE COMPOSITION				
X	Y	PRESS.	LAAR 3	MARG 3	MARG 4	ALPHA2	ALPHA3
0.0528	0.4835	319.83	-0.0711	-0.0821	-0.0247	-0.0671	-0.0051
0.1251	0.5375	352.80	0.0065	0.0028	0.0213	-0.0176	0.0029
0.2205	0.5645	368.00	0.0236	0.0258	0.0165	0.0006	0.0010
0.3621	0.5846	376.34	0.0098	0.0145	0.0000	0.0089	-0.0012
0.5071	0.5988	379.83	-0.0073	-0.0040	-0.0016	0.0113	-0.0004
0.6832	0.6244	380.06	-0.0101	-0.0102	0.0073	0.0045	0.0006
0.7792	0.6528	375.78	0.0025	0.0015	0.0103	-0.0088	-0.0001
0.9347	0.7879	337.52	0.0411	0.0412	0.0083	-0.0671	-0.0001
		MEAN	0.0215	0.0228	0.0113	0.0232	0.0014

ETHYL ALCOHOL(1)-METHYLCYCLOPENTANE(2)

SINOR J.E.,WEBER J.H.:J.CHEM.ENG.DATA 5,244(1960).

ANTOINE VAPOR PRESSURE CONSTANTS

	A	B	C
(1)	8.16290	1623.220	228.980
(2)	6.86283	1186.059	226.042

P = 760.00

EQUATION	ORDER	A12	A21	D	A122	A211
VAN LAAR	3	1.0542	0.7228	-	-	-
MARGULES	3	1.0018	0.7074	-	-	-
MARGULES	4	1.1397	0.8360	0.4825	-	-
ALPHA	2	6.8185	12.4437	-	-	-
ALPHA	3	2.4289	32.0315	-	15.1803	-26.3753

			DEVIATION IN VAPOR PHASE COMPOSITION				
X	Y	TEMP.	LAAR 3	MARG 3	MARG 4	ALPHA2	ALPHA3
0.0150	0.1500	66.30	-0.0489	-0.0580	-0.0322	-0.0599	0.0076
0.0300	0.2220	63.70	-0.0551	-0.0672	-0.0337	-0.0756	-0.0023
0.0850	0.2950	61.25	-0.0051	-0.0143	0.0100	-0.0487	0.0003
0.2160	0.3320	60.30	0.0269	0.0301	0.0210	-0.0104	0.0031
0.3480	0.3500	60.05	0.0109	0.0172	0.0025	0.0029	0.0001
0.4670	0.3610	60.10	-0.0020	0.0023	-0.0027	0.0125	0.0009
0.5800	0.3820	60.30	-0.0144	-0.0133	-0.0057	0.0117	-0.0047
0.7130	0.4130	61.20	-0.0063	-0.0080	0.0067	0.0136	-0.0005
0.8000	0.4600	62.80	0.0061	0.0046	0.0119	0.0032	0.0021
0.8570	0.5190	64.60	0.0147	0.0145	0.0091	-0.0153	0.0035
0.8980	0.5930	67.00	0.0154	0.0168	-0.0017	-0.0422	-0.0000
0.9650	0.8150	73.65	0.0010	0.0045	-0.0282	-0.0908	-0.0095
0.9850	0.9080	76.10	0.0038	0.0062	-0.0161	-0.0628	-0.0005
		MEAN	0.0162	0.0198	0.0140	0.0346	0.0027

167

ETHYL ALCOHOL(1)-ISOOCTANE(2)

KRETSCHMER C.B.,NOWAKOWSKA J.,WIEBE R.:J.AM.CHEM.SOC.70,
        1785(1948).

ANTOINE VAPOR PRESSURE CONSTANTS
            A   B   C
    (1) 8.16290   1623.220   228.980
    (2) 6.81189   1257.840   220.735

T =   25.00

EQUATION	ORDER	A12	A21	D	A122	A211
VAN LAAR	3	1.0388	0.8508	-	-	-
MARGULES	3	1.0150	0.8549	-	-	-
MARGULES	4	1.2364	1.0977	0.8329	-	-
ALPHA	2	23.4856	20.3106	-	-	-
ALPHA	3	14.5876	46.0371	-	34.7897	-34.4336

			DEVIATION IN VAPOR PHASE COMPOSITION				
X	Y	PRESS.	LAAR 3	MARG 3	MARG 4	ALPHA2	ALPHA3
0.0565	0.4441	86.56	-0.0821	-0.0897	-0.0243	-0.0517	-0.0045
0.1182	0.4762	91.81	-0.0014	-0.0051	0.0183	-0.0151	0.0030
0.1700	0.4910	93.57	0.0209	0.0201	0.0180	-0.0058	0.0013
0.2748	0.5073	95.22	0.0258	0.0286	0.0040	0.0021	-0.0013
0.3773	0.5153	95.85	0.0136	0.0178	-0.0021	0.0069	-0.0006
0.5416	0.5285	96.14	-0.0122	-0.0089	0.0027	0.0082	-0.0004
0.7225	0.5501	95.25	-0.0149	-0.0146	0.0125	0.0052	0.0025
0.8511	0.5994	91.49	0.0143	0.0128	0.0033	-0.0151	-0.0028
0.9603	0.7471	75.71	0.0759	0.0745	0.0125	-0.0520	-0.0011
0.9757	0.8023	70.41	0.0778	0.0768	0.0212	-0.0505	0.0055
		MEAN	0.0339	0.0349	0.0119	0.0212	0.0023

ETHYL ALCOHOL(1)-ISOOCTANE(2)

KRETSCHMER C.B.,NOWAKOWSKA J.,WIEBE R.:J.AM.CHEM.SOC.70,
    1785(1948).

ANTOINE VAPOR PRESSURE CONSTANTS

	A	B	C
(1)	8.16290	1623.220	228.980
(2)	6.81189	1257.840	220.735

T =   50.00

EQUATION	ORDER	A12	A21	D	A122	A211
VAN LAAR	3	1.0073	0.8857	-	-	-
MARGULES	3	0.9971	0.8897	-	-	-
MARGULES	4	1.1965	1.0437	0.7347	-	-
ALPHA	2	21.3106	13.9188	-	-	-
ALPHA	3	10.5114	31.0615	-	32.9420	-24.8152

			DEVIATION IN VAPOR PHASE COMPOSITION				
X	Y	PRESS.	LAAR 3	MARG 3	MARG 4	ALPHA2	ALPHA3
0.0113	0.2938	207.31	-0.1520	-0.1546	-0.0979	-0.1148	-0.0239
0.0340	0.4238	250.15	-0.1126	-0.1163	-0.0449	-0.0836	-0.0027
0.0579	0.4752	271.87	-0.0644	-0.0677	-0.0108	-0.0575	0.0008
0.1240	0.5254	296.29	0.0090	0.0077	0.0196	-0.0201	0.0037
0.3428	0.5701	315.21	0.0265	0.0286	-0.0031	0.0055	-0.0007
0.5176	0.5863	318.26	-0.0030	-0.0012	-0.0038	0.0097	-0.0007
0.5943	0.5941	318.75	-0.0140	-0.0128	0.0002	0.0097	0.0000
0.6144	0.5969	318.82	-0.0167	-0.0156	0.0009	0.0090	-0.0002
0.7713	0.6279	315.10	-0.0189	-0.0194	0.0042	0.0001	0.0012
0.8799	0.6881	301.38	0.0027	0.0013	-0.0026	-0.0250	-0.0006
0.9319	0.7526	282.86	0.0222	0.0208	-0.0034	-0.0466	0.0002
0.9516	0.7942	271.27	0.0272	0.0260	-0.0036	-0.0571	-0.0001
0.9829	0.9008	242.85	0.0223	0.0216	-0.0019	-0.0585	-0.0003
		MEAN	0.0378	0.0380	0.0151	0.0382	0.0027

L

ETHYL ALCOHOL(1)-PROPYL ALCOHOL(2)

UDOVENKO V.V.,FRID TS.B.:ZH.FIZ.KHIM.22,1135(1948).

ANTOINE VAPOR PRESSURE CONSTANTS

	A	B	C
(1)	8.16290	1623.220	228.980
(2)	7.85418	1497.910	204.112

T =   50.00

EQUATION	ORDER	A12	A21	D	A122	A211
VAN LAAR	3	0.0623	0.1553	-	-	-
MARGULES	3	0.0507	0.1189	-	-	-
MARGULES	4	0.0706	0.1423	0.0661	-	-
ALPHA	2	1.8576	-0.4856	-	-	-
ALPHA	3	1.2227	-0.6320	-	0.7194	0.1000

			DEVIATION IN VAPOR PHASE COMPOSITION				
X	Y	PRESS.	LAAR 3	MARG 3	MARG 4	ALPHA2	ALPHA3
0.1000	0.2420	108.00	-0.0069	-0.0089	-0.0060	-0.0042	-0.0008
0.2000	0.4090	123.60	-0.0039	-0.0043	-0.0041	-0.0013	0.0009
0.3000	0.5360	139.10	-0.0021	-0.0016	-0.0028	-0.0003	-0.0000
0.4000	0.6360	153.70	-0.0009	-0.0008	-0.0015	0.0003	-0.0009
0.5000	0.7160	167.40	0.0011	0.0002	0.0006	0.0021	0.0004
0.6000	0.7850	179.00	0.0004	-0.0013	-0.0002	0.0018	0.0006
0.7000	0.8470	190.20	-0.0030	-0.0046	-0.0037	-0.0007	-0.0007
0.8000	0.9010	200.60	-0.0046	-0.0051	-0.0052	-0.0013	0.0002
0.9000	0.9520	211.80	-0.0056	-0.0046	-0.0056	-0.0021	0.0000
		MEAN	0.0032	0.0035	0.0033	0.0016	0.0005

ETHYL ALCOHOL(1)-PROPYL ALCOHOL(2)

UDOVENKO V.V.,FRID TS.B.:ZH.FIZ.KHIM.22,1135(1948).

ANTOINE VAPOR PRESSURE CONSTANTS

	A	B	C
(1)	8.16290	1623.220	228.980
(2)	7.85418	1497.910	204.112

T = 60.00

EQUATION	ORDER	A12	A21	D	A122	A211
VAN LAAR	3	0.0618	0.1603	-	-	-
MARGULES	3	0.0492	0.1203	-	-	-
MARGULES	4	0.0684	0.1426	0.0632	-	-
ALPHA	2	1.7256	-0.4628	-	-	-
ALPHA	3	1.3584	-0.5223	-	0.4440	0.0203

			DEVIATION IN VAPOR PHASE COMPOSITION				
X	Y	PRESS.	LAAR 3	MARG 3	MARG 4	ALPHA2	ALPHA3
0.1000	0.2320	178.10	-0.0060	-0.0080	-0.0053	-0.0027	0.0003
0.2000	0.3980	202.90	-0.0050	-0.0054	-0.0052	-0.0019	0.0000
0.3000	0.5240	226.50	-0.0025	-0.0020	-0.0031	-0.0005	-0.0003
0.4000	0.6240	248.50	-0.0003	-0.0002	-0.0009	0.0008	-0.0003
0.5000	0.7060	269.10	0.0012	0.0002	0.0006	0.0019	0.0004
0.6000	0.7770	287.50	0.0002	-0.0017	-0.0006	0.0013	0.0003
0.7000	0.8400	304.80	-0.0024	-0.0043	-0.0034	-0.0003	-0.0002
0.8000	0.8970	321.40	-0.0053	-0.0059	-0.0060	-0.0018	-0.0005
0.9000	0.9490	338.20	-0.0054	-0.0042	-0.0052	-0.0014	0.0005
		MEAN	0.0031	0.0035	0.0034	0.0014	0.0003

ETHYL ALCOHOL(1)-PROPYL ALCOHOL(2)

UDOVENKO V.V.,FRID TS.B.:ZH.FIZ.KHIM.22,1135(1948).

ANTOINE VAPOR PRESSURE CONSTANTS

	A	B	C
(1)	8.16290	1623.220	228.980
(2)	7.85418	1497.910	204.112

T =   70.00

EQUATION	ORDER	A12	A21	D	A122	A211
VAN LAAR	3	0.0617	0.1414	-	-	-
MARGULES	3	0.0515	0.1134	-	-	-
MARGULES	4	0.0663	0.1304	0.0484	-	-
ALPHA	2	1.5986	-0.4395	-	-	-
ALPHA	3	1.0496	-0.5883	-	0.6145	0.1080

			DEVIATION IN VAPOR PHASE COMPOSITION				
X	Y	PRESS.	LAAR 3	MARG 3	MARG 4	ALPHA2	ALPHA3
0.1000	0.2230	286.50	-0.0048	-0.0064	-0.0044	-0.0022	0.0006
0.2000	0.3870	324.10	-0.0051	-0.0054	-0.0053	-0.0025	-0.0006
0.3000	0.5110	360.10	-0.0015	-0.0011	-0.0020	0.0003	0.0005
0.4000	0.6130	392.50	-0.0010	-0.0009	-0.0014	0.0001	-0.0009
0.5000	0.6950	424.20	0.0016	0.0009	0.0012	0.0025	0.0011
0.6000	0.7690	451.60	-0.0007	-0.0020	-0.0012	0.0006	-0.0005
0.7000	0.8330	476.50	-0.0022	-0.0035	-0.0028	-0.0000	-0.0000
0.8000	0.8920	501.00	-0.0045	-0.0049	-0.0049	-0.0014	-0.0001
0.9000	0.9470	526.00	-0.0049	-0.0041	-0.0048	-0.0017	0.0002
		MEAN	0.0029	0.0032	0.0031	0.0013	0.0005

ETHYL ALCOHOL(1)-PROPYL ALCOHOL(2)

UDOVENKO V.V.,FRID TS.B.:ZH.FIZ.KHIM.22,1135(1948).

ANTOINE VAPOR PRESSURE CONSTANTS

	A	B	C
(1)	8.16290	1623.220	228.980
(2)	7.85418	1497.910	204.112

T = 80.00

EQUATION	ORDER	A12	A21	D	A122	A211
VAN LAAR	3	0.0614	0.1524	-	-	-
MARGULES	3	0.0480	0.1181	-	-	-
MARGULES	4	0.0612	0.1332	0.0429	-	-
ALPHA	2	1.5130	-0.4206	-	-	-
ALPHA	3	0.2247	-0.9399	-	1.3054	0.4781

			DEVIATION IN VAPOR PHASE COMPOSITION				
X	Y	PRESS.	LAAR 3	MARG 3	MARG 4	ALPHA2	ALPHA3
0.1000	0.2180	437.00	-0.0062	-0.0084	-0.0066	-0.0030	-0.0019
0.2000	0.3760	491.30	-0.0028	-0.0033	-0.0032	0.0003	0.0014
0.3000	0.5020	544.10	-0.0015	-0.0009	-0.0017	0.0005	0.0010
0.4000	0.6060	592.70	-0.0023	-0.0019	-0.0024	-0.0014	-0.0017
0.5000	0.6890	637.00	0.0003	-0.0002	0.0000	0.0009	-0.0000
0.6000	0.7610	674.00	0.0012	-0.0002	0.0005	0.0021	0.0011
0.7000	0.8280	709.60	-0.0021	-0.0037	-0.0031	-0.0002	-0.0005
0.8000	0.8880	744.80	-0.0043	-0.0049	-0.0049	-0.0010	-0.0001
0.9000	0.9450	780.10	-0.0055	-0.0045	-0.0052	-0.0016	0.0001
		MEAN	0.0029	0.0031	0.0031	0.0012	0.0009

ETHYL ALCOHOL(1)-ISOPROPYL ALCOHOL(2)

BALLARD L.H.,VAN WINKLE M.:IND.ENG.CHEM.44,2450(1952).

ANTOINE VAPOR PRESSURE CONSTANTS
```
          A          B          C
(1) 8.16290    1623.220    228.980
(2) 7.75634    1366.142    197.970
```

P = 760.00

EQUATION	ORDER	A12	A21	D	A122	A211
VAN LAAR	3	-0.0061	-0.0397	-	-	-
MARGULES	3	0.0044	-0.0157	-	-	-
MARGULES	4	-0.0185	-0.0379	-0.0672	-	-
ALPHA	2	0.1435	-0.1558	-	-	-
ALPHA	3	7.6492	6.5245	-	-7.4265	-6.7779

			DEVIATION IN VAPOR PHASE COMPOSITION				
X	Y	TEMP.	LAAR 3	MARG 3	MARG 4	ALPHA2	ALPHA3
0.1240	0.1410	81.90	-0.0010	0.0004	-0.0013	-0.0012	0.0015
0.2400	0.2665	81.40	0.0002	0.0009	0.0014	0.0003	0.0016
0.2425	0.2730	81.30	-0.0036	-0.0029	-0.0024	-0.0035	-0.0022
0.3480	0.3845	81.00	-0.0034	-0.0036	-0.0022	-0.0029	-0.0031
0.4555	0.4850	80.50	0.0068	0.0061	0.0067	0.0076	0.0064
0.4570	0.4955	80.50	-0.0022	-0.0029	-0.0023	-0.0014	-0.0026
0.5520	0.5895	80.20	-0.0012	-0.0017	-0.0023	-0.0002	-0.0016
0.5590	0.5970	80.10	-0.0018	-0.0023	-0.0029	-0.0008	-0.0022
0.6135	0.6460	79.90	0.0026	0.0023	0.0012	0.0034	0.0023
0.7425	0.7700	79.40	0.0017	0.0019	0.0013	0.0018	0.0022
0.8295	0.8545	79.10	-0.0023	-0.0023	-0.0017	-0.0032	-0.0013
0.8335	0.8580	79.10	-0.0022	-0.0022	-0.0015	-0.0031	-0.0012
0.9150	0.9295	78.80	-0.0007	-0.0014	0.0001	-0.0024	0.0004
		MEAN	0.0023	0.0024	0.0021	0.0025	0.0022

ETHYL ALCOHOL(1)-TOLUENE(2)

LEHFELDT N.:PHIL.MAG.46,59(1898).

ANTOINE VAPOR PRESSURE CONSTANTS

	A	B	C
(1)	8.16290	1623.220	228.980
(2)	6.95334	1343.943	219.377

T = 50.00

EQUATION	ORDER	A12	A21	D	A122	A211
VAN LAAR	3	0.8309	0.7031	-	-	-
MARGULES	3	0.8225	0.6989	-	-	-
MARGULES	4	0.8732	0.7440	0.1599	-	-
ALPHA	2	12.4421	4.1234	-	-	-
ALPHA	3	2.9714	7.8977	-	18.7039	-7.0175

			DEVIATION IN VAPOR PHASE COMPOSITION				
X	Y	PRESS.	LAAR 3	MARG 3	MARG 4	ALPHA2	ALPHA3
0.1380	0.5910	199.50	-0.0031	-0.0038	0.0006	-0.0463	-0.0027
0.3340	0.6590	235.00	0.0126	0.0134	0.0090	0.0031	0.0055
0.4370	0.6810	241.00	0.0030	0.0036	0.0010	0.0082	0.0011
0.5330	0.7050	245.00	-0.0106	-0.0104	-0.0100	0.0035	-0.0085
0.6340	0.7130	247.00	-0.0020	-0.0021	0.0007	0.0139	0.0010
0.7360	0.7390	249.00	0.0021	0.0019	0.0046	0.0085	0.0019
0.8060	0.7660	246.50	0.0091	0.0090	0.0096	0.0004	0.0055
0.8850	0.8390	241.50	-0.0041	-0.0038	-0.0072	-0.0381	-0.0084
0.9460	0.9020	233.50	0.0043	0.0048	-0.0004	-0.0453	0.0047
		MEAN	0.0056	0.0059	0.0048	0.0186	0.0044

ETHYL ALCOHOL(1)-TOLUENE(2)

KRETSCHMER C.B.,WIEBE R.:J.AM.CHEM.SOC.71,1793(1949).

ANTOINE VAPOR PRESSURE CONSTANTS
```
          A           B          C
(1) 8.16290    1623.220   228.980
(2) 6.95334    1343.943   219.377
```

T =   35.00

EQUATION	ORDER	A12	A21	D	A122	A211
VAN LAAR	3	0.9796	0.6435	-	-	-
MARGULES	3	0.9278	0.6188	-	-	-
MARGULES	4	1.0527	0.7744	0.4839	-	-
ALPHA	2	16.6393	7.0665	-	-	-
ALPHA	3	9.7263	21.8067	-	28.6018	-20.2834

			DEVIATION IN VAPOR PHASE COMPOSITION				
X	Y	PRESS.	LAAR 3	MARG 3	MARG 4	ALPHA2	ALPHA3
0.0330	0.4216	79.38	-0.0560	-0.0753	-0.0272	-0.1005	0.0059
0.0468	0.4749	86.34	-0.0436	-0.0610	-0.0176	-0.0914	-0.0010
0.1214	0.5662	102.09	0.0101	0.0055	0.0181	-0.0288	-0.0000
0.2079	0.6014	108.93	0.0175	0.0207	0.0145	-0.0002	-0.0011
0.3620	0.6346	114.26	0.0010	0.0063	-0.0028	0.0148	-0.0029
0.4160	0.6384	115.34	0.0005	0.0047	-0.0000	0.0212	0.0024
0.5930	0.6730	117.90	-0.0088	-0.0098	0.0025	0.0127	0.0021
0.7263	0.7164	118.57	-0.0046	-0.0069	0.0056	-0.0102	-0.0014
0.8519	0.7848	116.56	0.0128	0.0133	0.0071	-0.0453	-0.0006
0.9701	0.9318	107.64	0.0134	0.0156	0.0003	-0.0710	0.0012
		MEAN	0.0168	0.0219	0.0096	0.0396	0.0019

ETHYL ALCOHOL(1)-TOLUENE(2)

KRETSCHMER C.B.,WIEBE R.:J.AM.CHEM.SOC.71,1793(1949).

ANTOINE VAPOR PRESSURE CONSTANTS
```
         A          B          C
(1) 8.16290    1623.220    228.980
(2) 6.95334    1343.943    219.377
```

T =   55.00

EQUATION	ORDER	A12	A21	D	A122	A211
VAN LAAR	3	0.8896	0.6353	-	-	-
MARGULES	3	0.8548	0.6224	-	-	-
MARGULES	4	0.9642	0.7388	0.3974	-	-
ALPHA	2	12.3397	4.0593	-	-	-
ALPHA	3	8.2364	13.2637	-	19.6968	-12.0568

			DEVIATION IN VAPOR PHASE COMPOSITION				
X	Y	PRESS.	LAAR 3	MARG 3	MARG 4	ALPHA2	ALPHA3
0.0439	0.4369	196.64	-0.0298	-0.0422	-0.0031	-0.1041	0.0050
0.1157	0.5679	247.70	0.0056	0.0013	0.0142	-0.0532	-0.0016
0.2497	0.6319	279.24	0.0154	0.0183	0.0091	-0.0029	-0.0012
0.4034	0.6649	294.17	0.0044	0.0078	0.0009	0.0170	0.0008
0.4142	0.6673	294.75	0.0032	0.0065	0.0003	0.0172	0.0005
0.6282	0.7150	305.48	-0.0072	-0.0075	0.0019	0.0117	-0.0002
0.7186	0.7431	307.81	-0.0035	-0.0044	0.0051	0.0015	0.0000
0.8423	0.8049	306.23	0.0078	0.0078	0.0058	-0.0242	-0.0003
0.9163	0.8685	299.53	0.0137	0.0147	0.0036	-0.0435	-0.0001
0.9635	0.9307	290.47	0.0114	0.0125	0.0016	-0.0444	0.0005
		MEAN	0.0102	0.0123	0.0046	0.0320	0.0010

ETHYL ALCOHOL(1)-TOLUENE(2)

WRIGHT W.A.:J.PHYS.CHEM.37,233(1933).

ANTOINE VAPOR PRESSURE CONSTANTS
```
        A          B          C
(1) 8.16290    1623.220   228.980
(2) 6.95334    1343.943   219.377
```

T =   60.00

EQUATION	ORDER	A12	A21	D	A122	A211
VAN LAAR	3	0.8891	0.7321	-	-	-
MARGULES	3	0.8760	0.7273	-	-	-
MARGULES	4	0.9689	0.8112	0.2950	-	-
ALPHA	2	17.4874	5.6422	-	-	-
ALPHA	3	7.4840	22.1789	-	44.5503	-21.4750

			DEVIATION IN VAPOR PHASE COMPOSITION				
X	Y	PRESS.	LAAR 3	MARG 3	MARG 4	ALPHA2	ALPHA3
0.1070	0.6180	240.00	-0.0340	-0.0362	-0.0211	-0.0641	0.0054
0.2310	0.6750	367.00	-0.0059	-0.0052	-0.0099	-0.0217	-0.0086
0.3520	0.6830	373.00	0.0058	0.0072	-0.0004	0.0087	0.0016
0.4430	0.6900	382.00	0.0046	0.0057	0.0015	0.0194	0.0056
0.5430	0.7110	387.00	-0.0093	-0.0088	-0.0074	0.0134	-0.0026
0.6250	0.7230	390.00	-0.0105	-0.0105	-0.0055	0.0126	-0.0013
0.7260	0.7440	395.00	-0.0059	-0.0061	-0.0007	0.0067	0.0016
0.7670	0.7580	397.00	-0.0035	-0.0038	-0.0000	0.0002	0.0016
0.8450	0.8020	397.00	-0.0019	-0.0019	-0.0044	-0.0239	-0.0022
0.9040	0.8530	388.00	-0.0000	0.0003	-0.0079	-0.0477	-0.0016
0.9570	0.9220	375.00	-0.0003	0.0002	-0.0094	-0.0630	0.0024
		MEAN	0.0074	0.0078	0.0062	0.0256	0.0031

ETHYL ALCOHOL(1)-TOLUENE(2)

WRIGHT W.A.:J.PHYS.CHEM.37,233(1933).

ANTOINE VAPOR PRESSURE CONSTANTS
```
          A          B          C
(1) 8.16290    1623.220    228.980
(2) 6.95334    1343.943    219.377
```

T =   65.00

EQUATION	ORDER	A12	A21	D	A122	A211
VAN LAAR	3	0.8638	0.7391	-	-	-
MARGULES	3	0.8553	0.7362	-	-	-
MARGULES	4	0.9380	0.8114	0.2640	-	-
ALPHA	2	17.6777	5.4143	-	-	-
ALPHA	3	6.5516	17.8393	-	38.6286	-17.0931

			DEVIATION IN VAPOR PHASE COMPOSITION				
X	Y	PRESS.	LAAR 3	MARG 3	MARG 4	ALPHA2	ALPHA3
0.1070	0.6190	301.00	-0.0358	-0.0372	-0.0238	-0.0588	0.0055
0.2310	0.6810	455.00	-0.0073	-0.0068	-0.0110	-0.0202	-0.0077
0.3520	0.6930	466.00	0.0034	0.0043	-0.0023	0.0066	0.0002
0.4430	0.7000	472.00	0.0034	0.0042	0.0005	0.0174	0.0045
0.5430	0.7180	477.00	-0.0073	-0.0068	-0.0056	0.0145	-0.0007
0.6250	0.7310	481.00	-0.0099	-0.0098	-0.0054	0.0126	-0.0008
0.7260	0.7520	486.00	-0.0067	-0.0069	-0.0021	0.0065	0.0013
0.7670	0.7660	487.00	-0.0052	-0.0053	-0.0020	-0.0002	0.0009
0.8450	0.8080	488.00	-0.0036	-0.0036	-0.0058	-0.0227	-0.0023
0.9040	0.8570	480.00	-0.0014	-0.0012	-0.0084	-0.0454	-0.0015
0.9570	0.9230	466.00	-0.0002	0.0000	-0.0085	-0.0593	0.0027
		MEAN	0.0076	0.0078	0.0069	0.0240	0.0025

ETHYL ALCOHOL(1)-TOLUENE(2)

WRIGHT W.A.:J.PHYS.CHEM.37,233(1933).

ANTOINE VAPOR PRESSURE CONSTANTS
           A   B   C
(1) 8.16290   1623.220   228.980
(2) 6.95334   1343.943   219.377

T =   70.00

EQUATION	ORDER	A12	A21	D	A122	A211
VAN LAAR	3	0.8390	0.7457	-	-	-
MARGULES	3	0.8339	0.7443	-	-	-
MARGULES	4	0.9069	0.8110	0.2337	-	-
ALPHA	2	17.7934	5.1696	-	-	-
ALPHA	3	5.9725	14.7161	-	34.2324	-13.9458

			DEVIATION IN VAPOR PHASE COMPOSITION				
X	Y	PRESS.	LAAR 3	MARG 3	MARG 4	ALPHA2	ALPHA3
0.1070	0.6200	367.00	-0.0378	-0.0387	-0.0269	-0.0542	0.0053
0.2310	0.6860	557.00	-0.0080	-0.0077	-0.0114	-0.0182	-0.0062
0.3520	0.7030	569.00	0.0006	0.0012	-0.0046	0.0041	-0.0017
0.4430	0.7100	572.00	0.0017	0.0022	-0.0009	0.0151	0.0031
0.5430	0.7240	584.00	-0.0045	-0.0042	-0.0031	0.0163	0.0020
0.6250	0.7390	590.00	-0.0095	-0.0094	-0.0055	0.0124	-0.0004
0.7260	0.7600	592.00	-0.0076	-0.0077	-0.0035	0.0061	0.0010
0.7670	0.7740	598.00	-0.0069	-0.0070	-0.0041	-0.0007	0.0001
0.8450	0.8140	598.00	-0.0052	-0.0052	-0.0071	-0.0216	-0.0024
0.9040	0.8610	591.00	-0.0027	-0.0027	-0.0090	-0.0430	-0.0015
0.9570	0.9240	575.00	-0.0002	-0.0000	-0.0075	-0.0555	0.0031
		MEAN	0.0077	0.0078	0.0076	0.0225	0.0024

ETHYL ALCOHOL(1)-TOLUENE(2)

WRIGHT W.A.:J.PHYS.CHEM.37,233(1933).

ANTOINE VAPOR PRESSURE CONSTANTS
	A	B	C
(1)	8.16290	1623.220	228.980
(2)	6.95334	1343.943	219.377

T =   75.00

EQUATION	ORDER	A12	A21	D	A122	A211
VAN LAAR	3	0.8113	0.7563	-	-	-
MARGULES	3	0.8093	0.7561	-	-	-
MARGULES	4	0.8685	0.8106	0.1904	-	-
ALPHA	2	18.0040	4.9405	-	-	-
ALPHA	3	5.3683	11.9764	-	30.4812	-11.2036

			DEVIATION IN VAPOR PHASE COMPOSITION				
X	Y	PRESS.	LAAR 3	MARG 3	MARG 4	ALPHA2	ALPHA3
0.1070	0.6200	444.00	-0.0397	-0.0400	-0.0305	-0.0476	0.0062
0.2310	0.6930	677.00	-0.0107	-0.0106	-0.0136	-0.0174	-0.0060
0.3520	0.7140	688.00	-0.0028	-0.0026	-0.0072	0.0013	-0.0037
0.4430	0.7200	698.00	0.0006	0.0009	-0.0016	0.0134	0.0025
0.5430	0.7310	707.00	-0.0022	-0.0021	-0.0011	0.0176	0.0043
0.6250	0.7470	715.00	-0.0088	-0.0087	-0.0055	0.0126	0.0005
0.7260	0.7680	722.00	-0.0085	-0.0085	-0.0052	0.0061	0.0009
0.7670	0.7830	724.00	-0.0098	-0.0098	-0.0075	-0.0018	-0.0015
0.8450	0.8200	724.00	-0.0073	-0.0074	-0.0089	-0.0203	-0.0025
0.9040	0.8650	716.00	-0.0047	-0.0047	-0.0098	-0.0404	-0.0014
0.9570	0.9250	699.00	-0.0006	-0.0006	-0.0066	-0.0517	0.0037
		MEAN	0.0087	0.0087	0.0089	0.0209	0.0030

ETHYL ALCOHOL(1)-TOLUENE(2)

WRIGHT W.A.:J.PHYS.CHEM.37,233(1933).

ANTOINE VAPOR PRESSURE CONSTANTS
	A	B	C
(1)	8.16290	1623.220	228.980
(2)	6.95334	1343.943	219.377

T =   80.00

EQUATION	ORDER	A12	A21	D	A122	A211
VAN LAAR	3	0.7825	0.7733	-	-	-
MARGULES	3	0.7824	0.7733	-	-	-
MARGULES	4	0.8209	0.8091	0.1245	-	-
ALPHA	2	18.3939	4.7546	-	-	-
ALPHA	3	4.4675	9.4355	-	27.3625	-8.7087

			DEVIATION IN VAPOR PHASE COMPOSITION				
X	Y	PRESS.	LAAR 3	MARG 3	MARG 4	ALPHA2	ALPHA3
0.1070	0.6210	537.00	-0.0427	-0.0427	-0.0365	-0.0405	0.0061
0.2310	0.6990	818.00	-0.0122	-0.0122	-0.0141	-0.0146	-0.0038
0.3520	0.7290	832.00	-0.0097	-0.0097	-0.0126	-0.0049	-0.0085
0.4430	0.7300	844.00	0.0001	0.0001	-0.0014	0.0122	0.0030
0.5430	0.7380	856.00	0.0005	0.0005	0.0012	0.0193	0.0073
0.6250	0.7550	864.00	-0.0078	-0.0078	-0.0057	0.0132	0.0017
0.7260	0.7770	874.00	-0.0106	-0.0106	-0.0084	0.0054	-0.0003
0.7670	0.7910	877.00	-0.0121	-0.0121	-0.0106	-0.0018	-0.0025
0.8450	0.8250	880.00	-0.0093	-0.0093	-0.0103	-0.0179	-0.0020
0.9040	0.8700	868.00	-0.0088	-0.0088	-0.0121	-0.0390	-0.0023
0.9570	0.9260	848.00	-0.0020	-0.0020	-0.0059	-0.0482	0.0049
		MEAN	0.0105	0.0105	0.0108	0.0197	0.0039

ETHYL ALCOHOL(1)-TOLUENE(2)

WRIGHT W.A.:J.PHYS.CHEM.37,233(1933).

ANTOINE VAPOR PRESSURE CONSTANTS

	A	B	C
(1)	8.16290	1623.220	228.980
(2)	6.95334	1343.943	219.377

T = 85.00

EQUATION	ORDER	A12	A21	D	A122	A211
VAN LAAR	3	0.7585	0.7790	-	-	-
MARGULES	3	0.7586	0.7787	-	-	-
MARGULES	4	0.7960	0.8137	0.1214	-	-
ALPHA	2	18.2918	4.4600	-	-	-
ALPHA	3	4.8618	8.5093	-	25.5868	-7.7536

			DEVIATION IN VAPOR PHASE COMPOSITION				
X	Y	PRESS.	LAAR 3	MARG 3	MARG 4	ALPHA2	ALPHA3
0.1070	0.6210	642.00	-0.0448	-0.0447	-0.0387	-0.0366	0.0077
0.2310	0.7060	990.00	-0.0162	-0.0162	-0.0181	-0.0157	-0.0055
0.3520	0.7340	1005.00	-0.0088	-0.0089	-0.0117	-0.0030	-0.0068
0.4430	0.7400	1016.00	-0.0027	-0.0027	-0.0042	0.0094	0.0004
0.5430	0.7430	1027.00	0.0034	0.0034	0.0040	0.0217	0.0102
0.6250	0.7640	1037.00	-0.0089	-0.0089	-0.0070	0.0117	0.0009
0.7260	0.7850	1047.00	-0.0117	-0.0117	-0.0096	0.0049	-0.0002
0.7670	0.8000	1052.00	-0.0148	-0.0148	-0.0134	-0.0033	-0.0037
0.8450	0.8310	1052.00	-0.0108	-0.0108	-0.0117	-0.0167	-0.0015
0.9040	0.8740	1047.00	-0.0100	-0.0099	-0.0131	-0.0364	-0.0021
0.9570	0.9270	1026.00	-0.0018	-0.0018	-0.0055	-0.0440	0.0049
		MEAN	0.0122	0.0122	0.0125	0.0185	0.0040

ETHYL ALCOHOL(1)-TRICHLORETHYLENE(2)

FRITZWEILER R.,DIETRICH K.R.:ANGEW.CHEM.A.CHEM.FABRIK.,
NO.4,BERLIN,W 35(1933)

ANTOINE VAPOR PRESSURE CONSTANTS

	A	B	C
(1)	8.16290	1623.220	228.980
(2)	7.02808	1315.000	230.000

P = 760.00

EQUATION	ORDER	A12	A21	D	A122	A211
VAN LAAR	3	0.9558	0.6752	-	-	-
MARGULES	3	0.9241	0.6597	-	-	-
MARGULES	4	0.9594	0.6880	0.1269	-	-
ALPHA	2	7.3708	6.0465	-	-	-
ALPHA	3	0.6087	12.3180	-	14.3747	-10.5674

			DEVIATION IN VAPOR PHASE COMPOSITION				
X	Y	TEMP.	LAAR 3	MARG 3	MARG 4	ALPHA2	ALPHA3
0.0144	0.1000	83.40	0.0413	0.0339	0.0423	-0.0000	0.0621
0.0225	0.1480	81.60	0.0478	0.0390	0.0490	-0.0055	0.0692
0.0282	0.2400	78.70	-0.0152	-0.0243	-0.0139	-0.0717	0.0074
0.0260	0.3180	76.60	-0.1079	-0.1168	-0.1067	-0.1593	-0.0816
0.0550	0.3570	74.20	-0.0330	-0.0411	-0.0316	-0.0989	-0.0194
0.0680	0.3720	73.00	-0.0164	-0.0233	-0.0150	-0.0828	-0.0076
0.1130	0.4200	71.60	0.0057	0.0031	0.0067	-0.0567	0.0004
0.2810	0.4810	71.20	0.0165	0.0213	0.0159	-0.0040	0.0077
0.5190	0.5230	70.80	-0.0070	-0.0064	-0.0067	0.0194	-0.0001
0.5200	0.5250	70.80	-0.0088	-0.0083	-0.0085	0.0176	-0.0019
0.5420	0.5280	70.80	-0.0086	-0.0086	-0.0081	0.0197	-0.0016
0.6280	0.5500	70.90	-0.0114	-0.0134	-0.0102	0.0184	-0.0070
0.7160	0.5640	71.00	0.0091	0.0063	0.0107	0.0300	0.0079
0.7720	0.6080	71.30	-0.0006	-0.0031	0.0007	0.0076	-0.0054
0.8080	0.6260	71.40	0.0102	0.0084	0.0110	0.0073	0.0049
0.8220	0.6460	71.70	0.0037	0.0023	0.0043	-0.0046	-0.0016
0.8500	0.6570	72.20	0.0233	0.0227	0.0233	0.0033	0.0194
0.8680	0.7000	72.70	0.0033	0.0033	0.0029	-0.0251	0.0011
0.8860	0.7520	73.00	-0.0232	-0.0225	-0.0241	-0.0597	-0.0224
0.9090	0.7730	73.40	-0.0068	-0.0053	-0.0082	-0.0532	-0.0010
0.9270	0.8220	74.20	-0.0212	-0.0192	-0.0230	-0.0747	-0.0119
0.9440	0.8480	75.00	-0.0103	-0.0080	-0.0123	-0.0678	0.0020
0.9640	0.8890	75.60	-0.0018	0.0006	-0.0037	-0.0567	0.0122
0.9770	0.9210	76.20	0.0031	0.0050	0.0015	-0.0431	0.0154
		MEAN	0.0182	0.0186	0.0183	0.0411	0.0155

ETHYL ALCOHOL(1)-WATER(2)

BEEBE A.H.,COULTER K.E.,LINDSAY R.A.,BAKER E.M.:IND.ENG.CHEM.
     34,1501(1942).

ANTOINE VAPOR PRESSURE CONSTANTS
         A   B   C
 (1) 8.16290   1623.220   228.980
 (2) 7.96681   1668.210   228.000

P =   95.00

EQUATION	ORDER	A12	A21	D	A122	A211
VAN LAAR	3	0.7592	0.3866	-	-	-
MARGULES	3	0.6792	0.3273	-	-	-
MARGULES	4	0.6755	0.3240	-0.0125	-	-
ALPHA	2	5.1697	1.1789	-	-	-
ALPHA	3	3.5272	5.2474	-	6.9889	-5.4181

			DEVIATION IN VAPOR PHASE COMPOSITION				
X	Y	TEMP.	LAAR 3	MARG 3	MARG 4	ALPHA2	ALPHA3
0.0035	0.0205	50.60	0.0244	0.0174	0.0171	0.0006	0.0175
0.0045	0.0275	50.30	0.0292	0.0206	0.0202	-0.0006	0.0206
0.0175	0.1315	48.10	0.0476	0.0267	0.0257	-0.0356	0.0233
0.0585	0.3050	45.10	0.0677	0.0484	0.0473	-0.0507	0.0315
0.0680	0.3615	43.10	0.0372	0.0202	0.0193	-0.0794	0.0005
0.0935	0.4110	41.70	0.0386	0.0282	0.0275	-0.0653	0.0043
0.1650	0.5200	40.50	0.0060	0.0083	0.0082	-0.0521	-0.0165
0.2125	0.5455	40.00	0.0090	0.0151	0.0153	-0.0230	-0.0063
0.2410	0.5675	39.30	0.0008	0.0078	0.0082	-0.0185	-0.0111
0.3615	0.6060	37.50	0.0093	0.0140	0.0144	0.0246	0.0061
0.4740	0.6505	35.90	0.0080	0.0069	0.0070	0.0320	0.0048
0.4985	0.6555	36.20	0.0130	0.0106	0.0107	0.0367	0.0092
0.5815	0.6970	37.20	0.0082	0.0031	0.0029	0.0258	0.0015
0.6460	0.7290	36.50	0.0091	0.0036	0.0033	0.0167	-0.0009
0.6540	0.7310	36.00	0.0115	0.0061	0.0059	0.0175	0.0011
0.7230	0.7760	34.60	0.0067	0.0030	0.0027	-0.0019	-0.0062
0.7900	0.8200	37.30	0.0054	0.0047	0.0046	-0.0175	-0.0062
0.8370	0.8520	36.20	0.0073	0.0091	0.0091	-0.0255	-0.0021
0.8731	0.8817	36.70	0.0053	0.0087	0.0088	-0.0329	-0.0008
0.8830	0.8885	36.30	0.0065	0.0102	0.0104	-0.0327	0.0014
0.8880	0.8930	36.00	0.0061	0.0100	0.0101	-0.0335	0.0014
0.8973	0.9012	37.00	0.0054	0.0095	0.0097	-0.0345	0.0019
0.9489	0.9502	36.30	0.0012	0.0052	0.0054	-0.0330	0.0027
0.9707	0.9715	37.40	0.0001	0.0029	0.0030	-0.0247	0.0023
0.9825	0.9835	35.40	-0.0006	0.0013	0.0014	-0.0176	0.0012
		MEAN	0.0146	0.0121	0.0119	0.0293	0.0073

M

ETHYL ALCOHOL(1)-WATER(2)

BEEBE A.H.,COULTER K.E.,LINDSAY R.A.,BAKER E.M.:IND.ENG.CHEM.
34,1501(1942).

ANTOINE VAPOR PRESSURE CONSTANTS

	A	B	C
(1)	8.16290	1623.220	228.980
(2)	7.96681	1668.210	228.000

P = 190.00

EQUATION	ORDER	A12	A21	D	A122	A211
VAN LAAR	3	0.7807	0.3954	-	-	-
MARGULES	3	0.6927	0.3352	-	-	-
MARGULES	4	0.7135	0.3541	0.0686	-	-
ALPHA	2	5.5023	1.3902	-	-	-
ALPHA	3	5.4688	7.2993	-	6.9744	-7.3784

			DEVIATION IN VAPOR PHASE COMPOSITION				
X	Y	TEMP.	LAAR 3	MARG 3	MARG 4	ALPHA2	ALPHA3
0.0160	0.1460	62.00	0.0248	0.0022	0.0075	-0.0534	0.0148
0.0370	0.2755	60.00	0.0228	-0.0035	0.0032	-0.0884	0.0042
0.0650	0.3650	57.20	0.0291	0.0089	0.0146	-0.0835	0.0053
0.0900	0.4125	55.30	0.0335	0.0201	0.0244	-0.0669	0.0087
0.1580	0.5015	52.20	0.0198	0.0202	0.0210	-0.0372	-0.0002
0.2090	0.5455	53.00	0.0055	0.0111	0.0102	-0.0226	-0.0080
0.2385	0.5650	52.40	-0.0004	0.0066	0.0051	-0.0153	-0.0109
0.3535	0.6045	50.10	0.0033	0.0089	0.0068	0.0211	0.0006
0.4705	0.6445	49.80	0.0060	0.0057	0.0050	0.0332	0.0054
0.4970	0.6540	48.90	0.0075	0.0058	0.0056	0.0337	0.0062
0.5805	0.6925	50.50	0.0052	0.0007	0.0015	0.0245	0.0021
0.6525	0.7260	48.50	0.0090	0.0040	0.0054	0.0154	0.0023
0.7000	0.7550	49.80	0.0066	0.0024	0.0038	0.0031	-0.0014
0.7200	0.7685	48.70	0.0055	0.0020	0.0033	-0.0030	-0.0034
0.7895	0.8152	50.20	0.0041	0.0037	0.0043	-0.0210	-0.0052
0.8416	0.8502	49.10	0.0077	0.0100	0.0098	-0.0293	-0.0004
0.8735	0.8790	49.00	0.0042	0.0080	0.0072	-0.0380	-0.0020
0.8970	0.8990	49.80	0.0038	0.0082	0.0072	-0.0404	-0.0005
0.9485	0.9466	49.50	0.0024	0.0066	0.0055	-0.0359	0.0025
0.9600	0.9580	47.60	0.0021	0.0058	0.0048	-0.0318	0.0027
0.9719	0.9700	50.30	0.0015	0.0045	0.0036	-0.0256	0.0027
0.9812	0.9798	48.60	0.0010	0.0032	0.0026	-0.0192	0.0022
		MEAN	0.0094	0.0069	0.0074	0.0338	0.0042

ETHYL ALCOHOL(1)-WATER(2)

BEEBE A.H.,COULTER K.E.,LINDSAY R.A.,BAKER E.M.:IND.ENG.CHEM.
           34,1501(1942).

ANTOINE VAPOR PRESSURE CONSTANTS
         A   B   C
(1) 8.16290   1623.220   228.980
(2) 7.96681   1668.210   228.000

P =   380.00

EQUATION	ORDER	A12	A21	D	A122	A211
VAN LAAR	3	0.8098	0.3993	-	-	-
MARGULES	3	0.7089	0.3380	-	-	-
MARGULES	4	0.7723	0.3986	0.2123	-	-
ALPHA	2	5.7505	1.5556	-	-	-
ALPHA	3	8.4029	10.7049	-	7.1759	-10.7545

			DEVIATION IN VAPOR PHASE COMPOSITION				
X	Y	TEMP.	LAAR 3	MARG 3	MARG 4	ALPHA2	ALPHA3
0.0160	0.1470	78.10	0.0290	0.0027	0.0195	-0.0515	0.0371
0.0315	0.2505	76.00	0.0271	-0.0037	0.0170	-0.0813	0.0296
0.0600	0.3765	72.40	0.0093	-0.0156	0.0029	-0.1043	0.0011
0.0855	0.4300	69.30	0.0117	-0.0051	0.0088	-0.0894	-0.0022
0.1465	0.5005	67.70	0.0110	0.0095	0.0133	-0.0482	-0.0045
0.2060	0.5415	67.50	0.0048	0.0111	0.0087	-0.0191	-0.0064
0.2360	0.5600	67.10	-0.0007	0.0074	0.0030	-0.0107	-0.0095
0.3495	0.5945	65.30	0.0049	0.0121	0.0061	0.0281	0.0038
0.4675	0.6410	64.70	0.0005	0.0013	-0.0006	0.0327	0.0024
0.4875	0.6425	64.30	0.0071	0.0067	0.0058	0.0386	0.0088
0.5800	0.6890	64.40	0.0012	-0.0030	-0.0001	0.0236	0.0013
0.6525	0.7250	64.20	0.0027	-0.0024	0.0022	0.0114	0.0002
0.7000	0.7495	63.80	0.0058	0.0014	0.0059	0.0032	0.0015
0.7175	0.7680	63.20	-0.0018	-0.0056	-0.0014	-0.0090	-0.0068
0.7890	0.8111	63.80	0.0029	0.0023	0.0040	-0.0231	-0.0033
0.8420	0.8488	62.70	0.0053	0.0076	0.0066	-0.0339	-0.0005
0.8749	0.8768	62.50	0.0042	0.0081	0.0054	-0.0409	-0.0004
0.8967	0.8973	63.60	0.0023	0.0068	0.0033	-0.0448	-0.0009
0.9485	0.9440	63.50	0.0035	0.0078	0.0039	-0.0380	0.0036
0.9727	0.9692	63.00	0.0023	0.0053	0.0025	-0.0267	0.0034
		MEAN	0.0069	0.0063	0.0060	0.0379	0.0064

ETHYL ALCOHOL(1)-WATER(2)

CAREY J.S.,LEWIS W.K.:IND.ENG.CHEM.24,882(1932).

ANTOINE VAPOR PRESSURE CONSTANTS

	A	B	C
(1)	8.16290	1623.220	228.980
(2)	7.96681	1668.210	228.000

P = 760.00

EQUATION	ORDER	A12	A21	D	A122	A211
VAN LAAR	3	0.7715	0.3848	-	-	-
MARGULES	3	0.6787	0.3160	-	-	-
MARGULES	4	0.7352	0.3824	0.2013	-	-
ALPHA	2	6.5780	2.2001	-	-	-
ALPHA	3	7.7105	9.9817	-	6.7599	-10.0836

			DEVIATION IN VAPOR PHASE COMPOSITION				
X	Y	TEMP.	LAAR 3	MARG 3	MARG 4	ALPHA2	ALPHA3
0.0190	0.1700	95.50	0.0135	-0.0109	0.0043	-0.0483	0.0275
0.0721	0.3891	89.00	0.0059	-0.0140	0.0005	-0.0664	0.0053
0.0966	0.4375	86.70	0.0025	-0.0108	-0.0001	-0.0578	-0.0022
0.1238	0.4704	85.30	0.0041	-0.0029	0.0038	-0.0415	-0.0026
0.1661	0.5089	84.10	0.0017	0.0018	0.0035	-0.0228	-0.0051
0.2337	0.5445	82.70	0.0033	0.0088	0.0058	0.0044	-0.0006
0.2608	0.5580	82.30	0.0014	0.0076	0.0037	0.0099	-0.0011
0.3273	0.5826	81.50	0.0023	0.0077	0.0034	0.0228	0.0027
0.3965	0.6122	80.70	-0.0016	0.0009	-0.0018	0.0234	0.0003
0.5079	0.6564	79.80	-0.0007	-0.0036	-0.0020	0.0174	0.0004
0.5198	0.6599	79.70	0.0010	-0.0023	-0.0002	0.0175	0.0019
0.5732	0.6841	79.30	0.0019	-0.0029	0.0009	0.0093	0.0010
0.6763	0.7385	78.74	0.0029	-0.0016	0.0033	-0.0141	-0.0018
0.7472	0.7815	78.41	0.0044	0.0027	0.0058	-0.0329	-0.0021
0.8943	0.8943	78.15	0.0041	0.0097	0.0055	-0.0657	0.0014
		MEAN	0.0034	0.0059	0.0030	0.0303	0.0037

ETHYL ALCOHOL(1)-WATER(2)

DULITSKAYA K.A.:ZH.OBSHCH.KHIM.15,9(1945).

ANTOINE VAPOR PRESSURE CONSTANTS

	A	B	C
(1)	8.16290	1623.220	228.980
(2)	7.96681	1668.210	228.000

T =   50.00

EQUATION	ORDER	A12	A21	D	A122	A211
VAN LAAR	3	0.8719	0.4032	-	-	-
MARGULES	3	0.7528	0.2971	-	-	-
MARGULES	4	0.8439	0.4386	0.3638	-	-
ALPHA	2	8.4788	3.0164	-	-	-
ALPHA	3	7.9103	17.4329	-	17.2002	-19.2619

			DEVIATION IN VAPOR PHASE COMPOSITION				
X	Y	PRESS.	LAAR 3	MARG 3	MARG 4	ALPHA2	ALPHA3
0.0956	0.4796	154.50	0.0010	-0.0160	0.0019	-0.0640	0.0058
0.1600	0.5384	173.30	-0.0031	-0.0043	0.0001	-0.0278	-0.0050
0.2500	0.5749	187.00	-0.0034	0.0024	-0.0015	0.0082	-0.0036
0.3366	0.5934	193.30	0.0042	0.0076	0.0040	0.0318	0.0060
0.4870	0.6509	202.80	-0.0018	-0.0078	-0.0020	0.0220	-0.0019
0.7455	0.7859	216.30	-0.0016	-0.0034	0.0021	-0.0458	0.0001
		MEAN	0.0025	0.0069	0.0019	0.0332	0.0037

ETHYL ALCOHOL(1)-WATER(2)

JONES C.A.,SCHOENBORN E.M.,COLBURN A.P.:IND.ENG.CHEM.
35,666(1943).

ANTOINE VAPOR PRESSURE CONSTANTS

	A	B	C
(1)	8.16290	1623.220	228.980
(2)	7.96681	1668.210	228.000

P = 760.00

EQUATION	ORDER	A12	A21	D	A122	A211
VAN LAAR	3	0.7531	0.3887	-	-	-
MARGULES	3	0.6736	0.3208	-	-	-
MARGULES	4	0.7159	0.3783	0.1610	-	-
ALPHA	2	6.3707	2.1006	-	-	-
ALPHA	3	5.2111	7.9970	-	7.7536	-8.2734

			DEVIATION IN VAPOR PHASE COMPOSITION				
X	Y	TEMP.	LAAR 3	MARG 3	MARG 4	ALPHA2	ALPHA3
0.0180	0.1790	95.50	-0.0077	-0.0279	-0.0170	-0.0654	0.0002
0.0540	0.3375	90.60	0.0047	-0.0162	-0.0037	-0.0727	0.0054
0.1240	0.4700	85.40	0.0027	-0.0033	0.0016	-0.0450	-0.0035
0.1760	0.5140	83.70	0.0031	0.0041	0.0047	-0.0204	-0.0022
0.2300	0.5420	82.75	0.0053	0.0096	0.0077	0.0012	0.0025
0.2880	0.5700	82.00	0.0022	0.0070	0.0041	0.0125	0.0018
0.3850	0.6120	81.00	-0.0033	-0.0016	-0.0031	0.0179	-0.0017
0.4400	0.6330	80.50	-0.0033	-0.0041	-0.0040	0.0180	-0.0020
0.5140	0.6570	79.80	0.0034	-0.0004	0.0020	0.0186	0.0028
0.6730	0.7350	78.90	0.0052	0.0006	0.0049	-0.0106	-0.0010
0.8400	0.8500	78.26	0.0031	0.0066	0.0049	-0.0568	0.0002
		MEAN	0.0040	0.0074	0.0053	0.0308	0.0021

ETHYL ALCOHOL(1)-WATER(2)

KIRSCHBAUM E.,GERTSNER H.:VERFAHRENSTECHNIK 1,10(1910).

ANTOINE VAPOR PRESSURE CONSTANTS
```
          A          B          C
(1)  8.16290   1623.220   228.980
(2)  7.96681   1668.210   228.000
```

P =   50.00

EQUATION	ORDER	A12	A21	D	A122	A211
VAN LAAR	3	0.6845	0.4032	-	-	-
MARGULES	3	0.6481	0.3450	-	-	-
MARGULES	4	0.5930	0.2691	-0.2184	-	-
ALPHA	2	7.1530	1.9817	-	-	-
ALPHA	3	0.5408	3.2071	-	9.3044	-3.6709

			DEVIATION IN VAPOR PHASE COMPOSITION				
X	Y	TEMP.	LAAR 3	MARG 3	MARG 4	ALPHA2	ALPHA3
0.0197	0.1100	38.15	0.0662	0.0569	0.0428	0.0242	0.0554
0.0442	0.2290	32.80	0.0765	0.0666	0.0500	0.0209	0.0609
0.0640	0.3320	30.90	0.0405	0.0325	0.0179	-0.0137	0.0248
0.0860	0.4080	29.40	0.0175	0.0119	0.0005	-0.0307	0.0032
0.1110	0.4680	28.15	0.0009	-0.0022	-0.0098	-0.0384	-0.0110
0.1420	0.5120	27.10	-0.0043	-0.0051	-0.0086	-0.0323	-0.0130
0.1670	0.5430	26.25	-0.0116	-0.0112	-0.0120	-0.0312	-0.0178
0.2020	0.5680	25.60	-0.0107	-0.0093	-0.0075	-0.0200	-0.0138
0.2400	0.5810	25.05	-0.0017	-0.0000	0.0036	-0.0016	-0.0020
0.2780	0.6000	24.60	-0.0023	-0.0011	0.0031	0.0048	-0.0009
0.3220	0.6180	24.30	-0.0019	-0.0017	0.0023	0.0111	0.0007
0.3790	0.6320	23.90	0.0059	0.0043	0.0068	0.0228	0.0083
0.4250	0.6480	23.60	0.0070	0.0039	0.0046	0.0243	0.0081
0.4780	0.6690	23.25	0.0061	0.0015	-0.0002	0.0210	0.0046
0.5440	0.6970	22.90	0.0046	-0.0011	-0.0054	0.0129	-0.0011
0.6120	0.7120	22.60	0.0197	0.0140	0.0080	0.0171	0.0097
0.6870	0.7680	22.25	0.0011	-0.0029	-0.0086	-0.0173	-0.0116
0.7800	0.8220	21.95	0.0014	0.0014	-0.0005	-0.0401	-0.0076
0.8850	0.8980	21.70	-0.0003	0.0041	0.0082	-0.0621	0.0065
		MEAN	0.0147	0.0122	0.0105	0.0235	0.0137

ETHYL ALCOHOL(1)-WATER(2)

RIEDER R.M.,THOMPSON A.R.:IND.ENG.CHEM.41,2905(1949).

ANTOINE VAPOR PRESSURE CONSTANTS

	A	B	C
(1)	8.16290	1623.220	228.980
(2)	7.96681	1668.210	228.000

P = 760.00

EQUATION	ORDER	A12	A21	D	A122	A211
VAN LAAR	3	0.7236	0.3818	-	-	-
MARGULES	3	0.6643	0.3170	-	-	-
MARGULES	4	0.7045	0.3871	0.1904	-	-
ALPHA	2	6.8149	2.3805	-	-	-
ALPHA	3	6.5784	7.4661	-	5.1949	-7.2566

DEVIATION IN VAPOR PHASE COMPOSITION

X	Y	TEMP.	LAAR 3	MARG 3	MARG 4	ALPHA2	ALPHA3
0.0028	0.0320	99.30	-0.0005	-0.0043	-0.0017	-0.0107	0.0018
0.0118	0.1130	96.90	0.0031	-0.0085	-0.0005	-0.0306	0.0084
0.0137	0.1570	96.00	-0.0257	-0.0383	-0.0296	-0.0629	-0.0204
0.0144	0.1350	96.00	0.0017	-0.0112	-0.0023	-0.0367	0.0070
0.0176	0.1560	95.60	0.0040	-0.0101	-0.0002	-0.0391	0.0092
0.0222	0.1860	94.80	0.0044	-0.0110	-0.0001	-0.0442	0.0090
0.0246	0.2120	93.80	-0.0070	-0.0228	-0.0116	-0.0579	-0.0029
0.0302	0.2310	93.50	0.0045	-0.0116	0.0001	-0.0500	0.0076
0.0331	0.2480	92.90	0.0020	-0.0142	-0.0024	-0.0539	0.0043
0.0519	0.3180	90.50	0.0066	-0.0076	0.0034	-0.0514	0.0047
0.0530	0.3140	90.50	0.0142	0.0001	0.0110	-0.0437	0.0120
0.0625	0.3390	89.40	0.0173	0.0049	0.0148	-0.0389	0.0133
0.0673	0.3700	88.40	-0.0011	-0.0126	-0.0032	-0.0561	-0.0060
0.0715	0.3620	88.60	0.0169	0.0062	0.0151	-0.0367	0.0115
0.0871	0.4060	87.20	0.0054	-0.0024	0.0046	-0.0427	-0.0020
0.1260	0.4680	85.40	-0.0005	-0.0022	0.0006	-0.0325	-0.0096
0.1430	0.4870	84.50	-0.0018	-0.0014	-0.0002	-0.0271	-0.0109
0.1720	0.5050	84.00	0.0043	0.0072	0.0063	-0.0103	-0.0038
0.2060	0.5300	83.40	0.0014	0.0060	0.0035	-0.0026	-0.0051
0.2100	0.5270	83.00	0.0068	0.0116	0.0089	0.0038	0.0005
0.2550	0.5520	82.30	0.0048	0.0099	0.0063	0.0121	0.0008
0.2840	0.5670	82.00	0.0026	0.0072	0.0036	0.0148	0.0001
0.3210	0.5860	81.40	-0.0012	0.0022	-0.0008	0.0153	-0.0021
0.3240	0.5860	81.50	-0.0001	0.0033	0.0003	0.0167	-0.0008
0.3450	0.5910	81.20	0.0032	0.0057	0.0033	0.0214	0.0031
0.4050	0.6140	80.90	0.0036	0.0030	0.0028	0.0226	0.0047
0.4300	0.6260	80.50	0.0015	-0.0003	0.0006	0.0195	0.0027
0.4490	0.6330	80.20	0.0023	-0.0005	0.0012	0.0189	0.0032
0.5060	0.6610	80.00	-0.0017	-0.0069	-0.0028	0.0088	-0.0017
0.5450	0.6730	79.50	0.0039	-0.0024	0.0031	0.0083	0.0027
0.6630	0.7330	78.80	0.0044	-0.0019	0.0048	-0.0175	-0.0023
0.7350	0.7760	78.50	0.0048	0.0013	0.0059	-0.0371	-0.0055
0.8040	0.8150	78.40	0.0130	0.0135	0.0142	-0.0482	0.0004
0.9170	0.9060	78.30	0.0133	0.0186	0.0136	-0.0614	0.0032
		MEAN	0.0056	0.0080	0.0054	0.0310	0.0054

192

ETHYL ALCOHOL(1)-WATER(2)

UDOVENKO V.V.,FATKULINA L.G.:ZH.FIZ.KHIM.26,1438(1952).

ANTOINE VAPOR PRESSURE CONSTANTS

	A	B	C
(1)	8.16290	1623.220	228.980
(2)	7.96681	1668.210	228.000

T = 40.00

EQUATION	ORDER	A12	A21	D	A122	A211
VAN LAAR	3	0.8463	0.3144	-	-	-
MARGULES	3	0.6798	0.1950	-	-	-
MARGULES	4	0.7070	0.2306	0.1067	-	-
ALPHA	2	5.1966	1.4477	-	-	-
ALPHA	3	14.1690	13.5292	-	0.3374	-13.3479

			DEVIATION IN VAPOR PHASE COMPOSITION				
X	Y	PRESS.	LAAR 3	MARG 3	MARG 4	ALPHA2	ALPHA3
0.0250	0.1800	66.30	0.0749	0.0273	0.0355	-0.0495	0.0459
0.0580	0.3160	79.60	0.0664	0.0308	0.0385	-0.0651	0.0254
0.0990	0.4240	91.90	0.0270	0.0133	0.0178	-0.0708	-0.0128
0.1300	0.4730	99.60	0.0069	0.0052	0.0075	-0.0632	-0.0282
0.2930	0.5360	115.20	0.0242	0.0359	0.0334	0.0403	0.0124
0.3980	0.5950	121.00	0.0110	0.0124	0.0112	0.0389	0.0041
0.5600	0.6860	127.40	0.0031	-0.0085	-0.0062	0.0119	-0.0069
0.6760	0.7440	130.50	0.0150	0.0052	0.0081	-0.0051	-0.0003
0.7790	0.8080	132.90	0.0203	0.0195	0.0206	-0.0273	0.0016
0.8600	0.8690	134.00	0.0186	0.0252	0.0240	-0.0436	0.0002
		MEAN	0.0268	0.0183	0.0203	0.0416	0.0138

ETHYL ALCOHOL(1)-WATER(2)

UDOVENKO V.V.,FATKULINA L.G.:ZH.FIZ.KHIM.26,1438(1952).

ANTOINE VAPOR PRESSURE CONSTANTS
	A	B	C
(1)	8.16290	1623.220	228.980
(2)	7.96681	1668.210	228.000

T =   50.00

EQUATION	ORDER	A12	A21	D	A122	A211
VAN LAAR	3	0.8882	0.3412	-	-	-
MARGULES	3	0.6906	0.2714	-	-	-
MARGULES	4	0.7927	0.3510	0.3335	-	-
ALPHA	2	4.0607	0.9365	-	-	-
ALPHA	3	9.6487	10.7328	-	3.3439	-10.7823

			DEVIATION IN VAPOR PHASE COMPOSITION				
X	Y	PRESS.	LAAR 3	MARG 3	MARG 4	ALPHA2	ALPHA3
0.0290	0.2080	115.70	0.0845	0.0243	0.0575	-0.0823	0.0287
0.1100	0.4390	161.00	0.0372	0.0220	0.0361	-0.0982	-0.0145
0.2460	0.5210	187.90	0.0248	0.0435	0.0333	-0.0024	0.0053
0.4510	0.6230	204.80	0.0036	0.0109	0.0040	0.0280	-0.0010
0.5810	0.6850	213.40	0.0077	0.0042	0.0074	0.0231	-0.0004
0.6820	0.7400	218.10	0.0133	0.0076	0.0146	0.0099	0.0009
0.8620	0.8700	222.80	0.0121	0.0149	0.0137	-0.0264	-0.0002
0.8850	0.8910	222.90	0.0096	0.0134	0.0107	-0.0304	-0.0012
0.9260	0.9290	223.00	0.0056	0.0103	0.0060	-0.0319	-0.0014
0.9470	0.9450	223.10	0.0077	0.0120	0.0076	-0.0249	0.0029
		MEAN	0.0206	0.0163	0.0191	0.0358	0.0057

ETHYL ALCOHOL(1)-WATER(2)

UDOVENKO V.V.,FATKULINA L.G.:ZH.FIZ.KHIM.26,1438(1952).

ANTOINE VAPOR PRESSURE CONSTANTS

	A	B	C
(1)	8.16290	1623.220	228.980
(2)	7.96681	1668.210	228.000

T =   60.00

EQUATION	ORDER	A12	A21	D	A122	A211
VAN LAAR	3	1.2626	0.3250	-	-	-
MARGULES	3	0.7762	0.1716	-	-	-
MARGULES	4	0.9790	0.3201	0.6223	-	-
ALPHA	2	3.4733	0.9561	-	-	-
ALPHA	3	19.2908	22.9711	-	0.3027	-23.2312

			DEVIATION IN VAPOR PHASE COMPOSITION				
X	Y	PRESS.	LAAR 3	MARG 3	MARG 4	ALPHA2	ALPHA3
0.0330	0.2330	195.70	0.2008	0.0478	0.1219	-0.1070	0.0489
0.1250	0.4460	270.00	0.0581	0.0457	0.0682	-0.1059	-0.0216
0.2670	0.5110	306.50	0.0107	0.0469	0.0261	-0.0039	-0.0066
0.4590	0.5800	330.80	0.0206	0.0281	0.0147	0.0493	0.0152
0.5970	0.6640	343.10	0.0193	0.0103	0.0163	0.0295	0.0061
0.6820	0.7380	349.40	0.0037	-0.0047	0.0069	-0.0060	-0.0141
0.8650	0.8750	354.20	0.0084	0.0183	0.0151	-0.0376	-0.0060
0.8910	0.8900	355.40	0.0151	0.0263	0.0206	-0.0311	0.0036
0.9280	0.9280	355.00	0.0086	0.0196	0.0119	-0.0329	0.0022
0.9490	0.9490	354.60	0.0058	0.0152	0.0078	-0.0291	0.0026
		MEAN	0.0351	0.0263	0.0310	0.0432	0.0127

ACETONE(1)-ACETONITRILE(2)

BROWN I.,SMITH F.:AUSTR.J.CHEM.13,30(1960).

ANTOINE VAPOR PRESSURE CONSTANTS

	A	B	C
(1)	7.23967	1279.870	237.500
(2)	7.24299	1397.929	238.894

T =   45.00

EQUATION	ORDER	A12	A21	D	A122	A211
VAN LAAR	3	-0.0009	0.0010	-	-	-
MARGULES	3	0.0482	-0.0447	-	-	-
MARGULES	4	0.0317	-0.0664	-0.0611	-	-
ALPHA	2	1.4209	-0.5385	-	-	-
ALPHA	3	9.9438	3.2436	-	-8.3533	-3.8033

			DEVIATION IN VAPOR PHASE COMPOSITION				
X	Y	PRESS.	LAAR 3	MARG 3	MARG 4	ALPHA2	ALPHA3
0.0520	0.1200	225.30	-0.0015	0.0072	0.0047	-0.0030	0.0015
0.0950	0.2060	239.90	-0.0014	0.0081	0.0059	-0.0042	0.0005
0.1920	0.3670	268.50	0.0008	0.0039	0.0039	-0.0042	-0.0023
0.3050	0.5100	302.70	0.0073	0.0019	0.0030	0.0007	-0.0002
0.4030	0.6110	331.70	0.0094	0.0018	0.0023	0.0034	0.0017
0.4810	0.6820	355.20	-0.0022	0.0014	0.0010	0.0028	0.0012
0.6060	0.7810	393.20	-0.0016	0.0018	0.0018	-0.0002	-0.0008
0.7060	0.8490	423.60	0.0047	0.0026	0.0016	-0.0030	-0.0025
0.8070	0.9040	454.10	0.0069	0.0076	0.0077	0.0001	0.0013
0.8960	0.9510	481.40	0.0037	0.0057	0.0064	-0.0007	0.0005
		MEAN	0.0040	0.0042	0.0037	0.0022	0.0013

ACETONE(1)-BENZENE(2)

BROWN I,SMITH F.:AUSTR.J.CHEM.10,423(1957).

ANTOINE VAPOR PRESSURE CONSTANTS
```
          A         B          C
(1) 7.23967    1279.870    237.500
(2) 6.90565    1211.033    220.790
```

T =   45.00

EQUATION	ORDER	A12	A21	D	A122	A211
VAN LAAR	3	0.2172	0.1411	-	-	-
MARGULES	3	0.2060	0.1319	-	-	-
MARGULES	4	0.2191	0.1480	0.0464	-	-
ALPHA	2	2.0693	-0.2027	-	-	-
ALPHA	3	3.8333	1.1299	-	-1.1302	-1.4595

			DEVIATION IN VAPOR PHASE COMPOSITION				
X	Y	PRESS.	LAAR 3	MARG 3	MARG 4	ALPHA2	ALPHA3
0.0470	0.1444	250.73	0.0036	0.0016	0.0040	-0.0155	0.0012
0.0963	0.2574	275.02	0.0022	0.0008	0.0028	-0.0196	-0.0012
0.2207	0.4417	324.25	0.0034	0.0039	0.0035	-0.0052	0.0007
0.2936	0.5204	348.40	0.0014	0.0022	0.0013	0.0007	-0.0004
0.4011	0.6139	379.88	0.0004	0.0008	0.0003	0.0062	-0.0002
0.4759	0.6697	399.73	0.0004	0.0004	0.0005	0.0072	0.0003
0.6125	0.7614	432.95	0.0010	0.0004	0.0014	0.0034	0.0004
0.7045	0.8201	453.99	0.0004	0.0000	0.0008	-0.0025	-0.0011
0.8081	0.8805	475.39	0.0036	0.0037	0.0036	-0.0049	0.0009
0.9084	0.9418	495.32	0.0029	0.0034	0.0026	-0.0065	0.0000
0.9529	0.9699	503.96	0.0017	0.0021	0.0014	-0.0050	-0.0005
		MEAN	0.0019	0.0017	0.0020	0.0070	0.0006

ACETONE(1)-BENZENE(2)

OTHMER D.F.:IND.ENG.CHEM.35,614(1943).

ANTOINE VAPOR PRESSURE CONSTANTS
```
          A          B          C
(1) 7.23967    1279.870    237.500
(2) 6.90565    1211.033    220.790
```

P = 760.00

EQUATION	ORDER	A12	A21	D	A122	A211
VAN LAAR	3	0.2874	0.1426	-	-	-
MARGULES	3	0.2607	0.1043	-	-	-
MARGULES	4	0.1865	0.0140	-0.2632	-	-
ALPHA	2	1.9428	-0.0754	-	-	-
ALPHA	3	-0.3911	-0.7358	-	2.5547	0.3771

			DEVIATION IN VAPOR PHASE COMPOSITION				
X	Y	TEMP.	LAAR 3	MARG 3	MARG 4	ALPHA2	ALPHA3
0.0200	0.0630	79.50	0.0113	0.0080	-0.0008	-0.0070	-0.0032
0.0500	0.1400	78.30	0.0217	0.0175	0.0040	-0.0093	-0.0019
0.1000	0.2430	76.40	0.0254	0.0231	0.0123	-0.0077	0.0020
0.2000	0.4000	72.80	0.0108	0.0116	0.0125	-0.0067	0.0008
0.3000	0.5120	69.60	0.0000	0.0004	0.0057	-0.0034	-0.0016
0.4000	0.5940	66.70	0.0018	-0.0001	0.0028	0.0042	0.0006
0.5000	0.6650	64.30	0.0057	0.0022	0.0001	0.0070	0.0002
0.6000	0.7300	62.40	0.0108	0.0072	0.0016	0.0063	0.0001
0.7000	0.7950	60.70	0.0130	0.0111	0.0063	0.0009	-0.0001
0.8000	0.8630	59.60	0.0102	0.0110	0.0107	-0.0077	-0.0005
0.9000	0.9320	58.80	0.0051	0.0074	0.0114	-0.0118	0.0004
		MEAN	0.0105	0.0090	0.0062	0.0065	0.0010

198

ACETONE(1)-BENZENE(2)

REINDERS W.,DE MINJER C.H.:REC.TRAV.CHIM.PAYS BAS 59,369(1940).

ANTOINE VAPOR PRESSURE CONSTANTS

	A	B	C
(1)	7.23967	1279.870	237.500
(2)	6.90565	1211.033	220.790

P =   760.00

EQUATION	ORDER	A12	A21	D	A122	A211
VAN LAAR	3	0.4706	0.1426	-	-	-
MARGULES	3	0.3300	0.0609	-	-	-
MARGULES	4	0.1918	-0.0971	-0.4713	-	-
ALPHA	2	1.6464	-0.1881	-	-	-
ALPHA	3	7.1604	3.2012	-	-4.6868	-3.4647

			DEVIATION IN VAPOR PHASE COMPOSITION				
X	Y	TEMP.	LAAR 3	MARG 3	MARG 4	ALPHA2	ALPHA3
0.0281	0.0816	78.80	0.0504	0.0275	0.0057	-0.0113	0.0043
0.0842	0.2148	76.20	0.0536	0.0383	0.0142	-0.0250	-0.0036
0.1388	0.3015	73.70	0.0449	0.0428	0.0310	-0.0157	0.0001
0.2422	0.4296	69.60	0.0218	0.0294	0.0361	-0.0002	0.0020
0.3400	0.5299	66.70	0.0040	0.0086	0.0182	0.0046	-0.0006
0.4466	0.6199	64.20	-0.0025	-0.0046	-0.0023	0.0075	0.0009
0.4634	0.6359	63.90	-0.0058	-0.0087	-0.0079	0.0046	-0.0017
0.5108	0.6704	63.90	-0.0051	-0.0100	-0.0135	0.0055	0.0006
0.5706	0.7159	0.00	0.0333	0.0278	0.0206	0.0017	-0.0007
0.6796	0.7912	0.00	0.0275	0.0242	0.0159	-0.0033	-0.0001
0.7443	0.8333	0.00	0.0247	0.0242	0.0195	-0.0054	0.0006
0.8597	0.9071	0.00	0.0176	0.0215	0.0256	-0.0071	0.0006
0.9377	0.9585	0.00	0.0088	0.0124	0.0176	-0.0058	-0.0011
		MEAN	0.0231	0.0215	0.0176	0.0075	0.0013

ACETONE(1)-BENZENE(2)

SODAY F.J.,BENNETT G.W.:J.CHEM.EDUC.7,1336(1930).

ANTOINE VAPOR PRESSURE CONSTANTS

	A	B	C
(1)	7.23967	1279.870	237.500
(2)	6.90565	1211.033	220.790

P = 732.00

EQUATION	ORDER	A12	A21	D	A122	A211
VAN LAAR	3	0.2008	0.5210	-	-	-
MARGULES	3	0.1923	0.3748	-	-	-
MARGULES	4	0.3321	0.5751	0.6023	-	-
ALPHA	2	2.8518	-0.1858	-	-	-
ALPHA	3	5.2010	1.4878	-	-1.3355	-1.8830

			DEVIATION IN VAPOR PHASE COMPOSITION				
X	Y	TEMP.	LAAR 3	MARG 3	MARG 4	ALPHA2	ALPHA3
0.0120	0.0530	77.70	-0.0135	-0.0141	-0.0019	-0.0086	0.0016
0.0330	0.1360	75.50	-0.0328	-0.0338	-0.0110	-0.0217	-0.0012
0.0560	0.1920	73.20	-0.0260	-0.0267	-0.0019	-0.0108	0.0143
0.0850	0.2630	72.00	-0.0271	-0.0271	-0.0069	-0.0089	0.0162
0.1430	0.4200	68.80	-0.0679	-0.0664	-0.0621	-0.0488	-0.0312
0.3100	0.5640	63.75	0.0095	0.0090	-0.0050	0.0219	0.0171
0.4040	0.6500	61.75	0.0050	0.0003	-0.0047	0.0143	0.0051
0.4780	0.7190	60.40	-0.0130	-0.0209	-0.0164	-0.0046	-0.0135
0.5540	0.7580	59.35	-0.0081	-0.0185	-0.0061	0.0008	-0.0053
0.6690	0.8140	58.15	-0.0090	-0.0199	-0.0043	0.0037	0.0047
0.7670	0.8670	57.20	-0.0207	-0.0270	-0.0190	-0.0023	0.0050
0.8600	0.9240	56.60	-0.0366	-0.0343	-0.0404	-0.0130	-0.0022
0.9240	0.9590	56.10	-0.0349	-0.0265	-0.0408	-0.0120	-0.0025
		MEAN	0.0234	0.0250	0.0170	0.0132	0.0092

ACETONE(1)-BENZENE(2)

TALLMADGE J.A.,CANJAR L.N.:IND.ENG.CHEM.46,1279(1954).

ANTOINE VAPOR PRESSURE CONSTANTS
```
          A          B          C
(1) 7.23967    1279.870    237.500
(2) 6.90565    1211.033    220.790
```

P = 738.00

EQUATION	ORDER	A12	A21	D	A122	A211
VAN LAAR	3	0.1936	0.1069	-	-	-
MARGULES	3	0.1786	0.0900	-	-	-
MARGULES	4	0.1613	0.0667	-0.0663	-	-
ALPHA	2	1.8038	-0.2065	-	-	-
ALPHA	3	2.1994	0.2723	-	-0.1604	-0.5869

			DEVIATION IN VAPOR PHASE COMPOSITION				
X	Y	TEMP.	LAAR 3	MARG 3	MARG 4	ALPHA2	ALPHA3
0.0148	0.0465	78.02	0.0003	-0.0010	-0.0025	-0.0063	-0.0034
0.0535	0.1468	75.85	0.0020	-0.0002	-0.0030	-0.0128	-0.0059
0.1040	0.2468	73.45	0.0052	0.0039	0.0017	-0.0100	-0.0028
0.1791	0.3510	70.85	0.0172	0.0175	0.0174	0.0088	0.0126
0.2374	0.4438	68.31	-0.0032	-0.0024	-0.0015	-0.0063	-0.0057
0.3403	0.5481	65.40	-0.0024	-0.0020	-0.0009	0.0004	-0.0029
0.4620	0.6461	62.89	0.0025	0.0017	0.0014	0.0055	0.0010
0.6014	0.7446	60.31	0.0066	0.0052	0.0036	0.0030	0.0013
0.7590	0.8475	58.22	0.0078	0.0076	0.0069	-0.0048	-0.0010
0.9276	0.9517	56.25	0.0063	0.0072	0.0082	-0.0046	0.0003
		MEAN	0.0054	0.0049	0.0047	0.0063	0.0037

N

ACETONE(1)-BUTYL ALCOHOL(2)

FORDYCE C.T.,SIMONSON D.R.:IND.ENG.CHEM.41,104(1949).

ANTOINE VAPOR PRESSURE CONSTANTS
	A	B	C
(1)	7.23967	1279.870	237.500
(2)	7.54472	1405.873	183.908

T =   25.00

EQUATION	ORDER	A12	A21	D	A122	A211
VAN LAAR	3	0.3737	0.2885	-	-	-
MARGULES	3	0.3540	0.3030	-	-	-
MARGULES	4	0.6877	0.6070	1.1831	-	-
ALPHA	2	63.6032	0.0301	-	-	-
ALPHA	3	117.1442	3.6007	-	-23.5516	-3.6365

			DEVIATION IN VAPOR PHASE COMPOSITION				
X	Y	PRESS.	LAAR 3	MARG 3	MARG 4	ALPHA2	ALPHA3
0.1600	0.9140	73.00	0.0055	0.0054	0.0060	-0.0024	0.0000
0.4350	0.9610	116.00	0.0053	0.0059	0.0028	0.0046	-0.0003
0.6500	0.9750	149.00	0.0057	0.0059	0.0083	0.0019	0.0008
0.7900	0.9840	164.00	0.0046	0.0045	0.0056	-0.0026	-0.0008
0.9050	0.9890	182.00	0.0058	0.0057	0.0046	-0.0041	-0.0002
0.9510	0.9910	190.00	0.0063	0.0062	0.0049	-0.0037	0.0003
		MEAN	0.0055	0.0056	0.0053	0.0032	0.0004

ACETONE(1)-BUTYL ALCOHOL(2)

MICHALSKI H.,MICHALOWSKI S.,SERWINSKI M.,STRUMILLO C.:ZESZYTY
NAUK.POL.LODZ.,NR .36,73(1961).

ANTOINE VAPOR PRESSURE CONSTANTS

	A	B	C
(1)	7.23967	1279.870	237.500
(2)	7.54472	1405.873	183.908

P = 746.00

EQUATION	ORDER	A12	A21	D	A122	A211
VAN LAAR	3	0.3383	0.1922	-	-	-
MARGULES	3	0.3205	0.1579	-	-	-
MARGULES	4	0.3418	0.2203	0.1356	-	-
ALPHA	2	11.4141	-0.8173	-	-	-
ALPHA	3	-29.0694	-4.1153	-	40.4781	3.0629

			DEVIATION IN VAPOR PHASE COMPOSITION				
X	Y	TEMP.	LAAR 3	MARG 3	MARG 4	ALPHA2	ALPHA3
0.0100	0.1050	115.00	0.0065	0.0029	0.0072	0.0063	0.0063
0.0250	0.2250	112.00	0.0162	0.0109	0.0176	0.0160	0.0161
0.0500	0.4050	107.10	-0.0098	-0.0146	-0.0083	-0.0111	-0.0108
0.0700	0.4720	103.80	0.0103	0.0069	0.0116	0.0091	0.0097
0.1540	0.6980	91.70	-0.0060	-0.0060	-0.0060	-0.0086	-0.0075
0.1980	0.7500	86.90	0.0013	0.0020	0.0013	-0.0006	0.0006
0.2150	0.7730	85.40	-0.0038	-0.0031	-0.0039	-0.0050	-0.0038
0.2650	0.8090	81.70	0.0021	0.0025	0.0019	0.0031	0.0040
0.3120	0.8370	78.80	0.0040	0.0039	0.0038	0.0065	-0.0001
0.5350	0.9300	68.60	-0.0047	-0.0059	-0.0040	-0.0020	0.0002
0.6070	0.9430	66.20	-0.0008	-0.0018	-0.0001	0.0009	0.0028
0.6590	0.9510	64.50	0.0018	0.0011	0.0025	0.0025	0.0041
0.7630	0.9720	61.60	-0.0014	-0.0015	-0.0011	-0.0026	-0.0017
0.8190	0.9780	60.00	0.0008	0.0010	0.0009	-0.0013	-0.0012
0.8970	0.9880	58.10	0.0008	0.0012	0.0006	-0.0019	-0.0044
		MEAN	0.0047	0.0044	0.0047	0.0052	0.0049

ACETONE(1)-CARBON TETRACHLORIDE(2)

ACHARYA M.V.R.,VENKATA RAO C.:TRANS.INDIAN INST.CHEM.ENGRS.
6,129(1953-1954).

ANTOINE VAPOR PRESSURE CONSTANTS
         A   B   C
(1) 7.23967   1279.870   237.500
(2) 6.93390   1242.430   230.000

P =   760.00

EQUATION	ORDER	A12	A21	D	A122	A211
VAN LAAR	3	0.4144	0.2873	-	-	-
MARGULES	3	0.3990	0.2751	-	-	-
MARGULES	4	0.4074	0.2839	0.0283	-	-
ALPHA	2	2.5679	0.4235	-	-	-
ALPHA	3	2.6055	1.5607	-	0.9879	-1.4534

DEVIATION IN VAPOR PHASE COMPOSITION

X	Y	TEMP.	LAAR 3	MARG 3	MARG 4	ALPHA2	ALPHA3
0.0250	0.0900	73.80	0.0169	0.0143	0.0157	-0.0084	0.0096
0.0400	0.1500	72.50	0.0076	0.0045	0.0063	-0.0257	-0.0022
0.0900	0.2700	68.80	0.0117	0.0094	0.0108	-0.0288	-0.0015
0.1700	0.3900	65.00	0.0133	0.0132	0.0134	-0.0157	0.0015
0.2600	0.4860	62.00	0.0043	0.0055	0.0049	-0.0074	-0.0036
0.3600	0.5550	60.10	0.0061	0.0073	0.0066	0.0084	0.0022
0.4400	0.6080	58.90	0.0016	0.0020	0.0018	0.0095	-0.0005
0.5150	0.6500	57.70	0.0028	0.0025	0.0027	0.0119	0.0014
0.6000	0.7000	57.20	0.0019	0.0011	0.0016	0.0080	0.0002
0.6600	0.7350	56.80	0.0030	0.0021	0.0027	0.0047	0.0004
0.7350	0.7900	56.40	-0.0041	-0.0047	-0.0042	-0.0096	-0.0082
0.7900	0.8100	56.20	0.0138	0.0136	0.0138	0.0027	0.0086
0.8300	0.8450	56.10	0.0081	0.0083	0.0082	-0.0066	0.0024
0.8600	0.8750	56.00	0.0012	0.0017	0.0014	-0.0154	-0.0047
0.9000	0.9050	56.00	0.0037	0.0045	0.0040	-0.0138	-0.0019
0.9650	0.9600	56.00	0.0062	0.0069	0.0064	-0.0048	0.0031
		MEAN	0.0066	0.0064	0.0065	0.0113	0.0032

ACETONE(1)-CARBON TETRACHLORIDE(2)

BACHMAN K.C.,SIMONS E.L..IND.ENG.CHEM.44,202(1952).

ANTOINE VAPOR PRESSURE CONSTANTS
	A	B	C
(1)	7.23967	1279.870	237.500
(2)	6.93390	1242.430	230.000

P = 300.00

EQUATION	ORDER	A12	A21	D	A122	A211
VAN LAAR	3	0.4956	0.2845	-	-	-
MARGULES	3	0.4522	0.2595	-	-	-
MARGULES	4	0.4897	0.2943	0.1201	-	-
ALPHA	2	2.5991	0.4505	-	-	-
ALPHA	3	3.6911	2.9511	-	0.9962	-2.9163

DEVIATION IN VAPOR PHASE COMPOSITION

X	Y	TEMP.	LAAR 3	MARG 3	MARG 4	ALPHA2	ALPHA3
0.0510	0.2020	43.95	0.0125	0.0032	0.0119	-0.0479	0.0005
0.1120	0.3300	40.58	0.0140	0.0094	0.0146	-0.0457	0.0004
0.1660	0.4015	38.60	0.0125	0.0123	0.0135	-0.0318	0.0010
0.2345	0.4700	36.80	0.0065	0.0095	0.0073	-0.0169	-0.0014
0.3110	0.5275	35.46	0.0024	0.0063	0.0027	-0.0026	-0.0018
0.3825	0.5745	34.33	-0.0015	0.0015	-0.0015	0.0047	-0.0034
0.3990	0.5755	34.20	0.0070	0.0097	0.0070	0.0150	0.0055
0.4935	0.6370	33.36	-0.0010	-0.0005	-0.0010	0.0117	-0.0015
0.4850	0.6305	33.33	0.0007	0.0014	0.0007	0.0133	0.0002
0.5150	0.6430	33.12	0.0053	0.0053	0.0053	0.0178	0.0048
0.5630	0.6785	32.74	-0.0025	-0.0033	-0.0022	0.0085	-0.0033
0.6555	0.7335	32.21	-0.0013	-0.0029	-0.0006	0.0022	-0.0034
0.7010	0.7565	31.89	0.0051	0.0037	0.0059	0.0035	0.0021
0.7165	0.7670	31.90	0.0050	0.0037	0.0058	0.0014	0.0017
0.7675	0.8030	31.53	0.0043	0.0037	0.0050	-0.0056	0.0001
0.7880	0.8190	31.52	0.0031	0.0029	0.0036	-0.0093	-0.0013
0.8650	0.8760	31.33	0.0050	0.0063	0.0050	-0.0143	0.0004
0.9085	0.9135	31.28	0.0035	0.0053	0.0031	-0.0164	-0.0005
0.9340	0.9360	31.27	0.0031	0.0049	0.0026	-0.0149	-0.0002
0.9470	0.9470	31.27	0.0036	0.0053	0.0031	-0.0126	0.0008
0.9510	0.9515	31.24	0.0027	0.0044	0.0022	-0.0128	0.0001
0.9550	0.9555	31.25	0.0024	0.0039	0.0018	-0.0124	-0.0001
0.9585	0.9585	31.22	0.0025	0.0040	0.0020	-0.0115	0.0002
0.9615	0.9615	31.25	0.0023	0.0037	0.0018	-0.0111	0.0001
0.9680	0.9680	31.20	0.0018	0.0030	0.0013	-0.0100	-0.0001
0.9795	0.9790	31.23	0.0015	0.0024	0.0011	-0.0068	0.0002
0.9940	0.9940	31.25	0.0002	0.0005	0.0001	-0.0025	-0.0002
		MEAN	0.0042	0.0046	0.0042	0.0135	0.0013

ACETONE(1)-CARBON TETRACHLORIDE(2)

BACHMAN K.C.,SIMONS E.L.:IND.ENG.CHEM.44,202(1952).

ANTOINE VAPOR PRESSURE CONSTANTS

	A	B	C
(1)	7.23967	1279.870	237.500
(2)	6.93390	1242.430	230.000

P =   450.00

EQUATION	ORDER	A12	A21	D	A122	A211
VAN LAAR	3	0.4671	0.2918	-	-	-
MARGULES	3	0.4362	0.2761	-	-	-
MARGULES	4	0.4669	0.3012	0.0965	-	-
ALPHA	2	2.5564	0.4118	-	-	-
ALPHA	3	3.0016	2.2724	-	1.2403	-2.2221

			DEVIATION IN VAPOR PHASE COMPOSITION				
X	Y	TEMP.	LAAR 3	MARG 3	MARG 4	ALPHA2	ALPHA3
0.0490	0.1890	55.29	0.0112	0.0046	0.0116	-0.0413	-0.0008
0.0875	0.2795	52.80	0.0137	0.0088	0.0144	-0.0438	-0.0004
0.1625	0.3930	49.50	0.0131	0.0128	0.0137	-0.0294	0.0006
0.2200	0.4530	47.96	0.0101	0.0122	0.0105	-0.0166	0.0005
0.2970	0.5150	46.26	0.0064	0.0096	0.0064	-0.0021	0.0004
0.3830	0.5735	44.88	0.0014	0.0042	0.0012	0.0067	-0.0016
0.4470	0.6100	44.06	0.0015	0.0032	0.0014	0.0123	-0.0002
0.4930	0.6355	43.65	0.0018	0.0027	0.0018	0.0144	0.0006
0.5650	0.6770	43.05	0.0012	0.0009	0.0015	0.0130	-0.0000
0.6040	0.6990	42.68	0.0020	0.0013	0.0025	0.0118	0.0005
0.6525	0.7280	42.42	0.0023	0.0013	0.0030	0.0085	0.0002
0.7025	0.7590	42.11	0.0030	0.0020	0.0038	0.0043	0.0002
0.7410	0.7840	41.92	0.0037	0.0028	0.0044	0.0007	0.0003
0.7630	0.8000	41.82	0.0029	0.0022	0.0035	-0.0027	-0.0007
0.8045	0.8290	41.73	0.0037	0.0035	0.0041	-0.0064	-0.0003
0.8890	0.8945	41.58	0.0045	0.0054	0.0042	-0.0118	0.0007
0.8955	0.9015	41.54	0.0030	0.0039	0.0026	-0.0134	-0.0007
0.9045	0.9090	41.56	0.0031	0.0041	0.0027	-0.0133	-0.0005
0.9125	0.9165	41.57	0.0024	0.0035	0.0020	-0.0138	-0.0009
0.9200	0.9225	41.53	0.0030	0.0040	0.0025	-0.0130	-0.0002
0.9260	0.9270	41.53	0.0037	0.0048	0.0032	-0.0119	0.0007
0.9360	0.9360	41.49	0.0036	0.0047	0.0031	-0.0113	0.0009
0.9450	0.9450	41.46	0.0027	0.0038	0.0022	-0.0112	0.0003
0.9685	0.9675	41.50	0.0020	0.0028	0.0016	-0.0079	0.0004
		MEAN	0.0044	0.0046	0.0045	0.0134	0.0005

ACETONE(1)-CARBON TETRACHLORIDE(2)

BACHMAN K.C.,SIMONS E.L.:IND.ENG.CHEM.44,202(1952).

ANTOINE VAPOR PRESSURE CONSTANTS

	A	B	C
(1)	7.23967	1279.870	237.500
(2)	6.93390	1242.430	230.000

P = 760.00

EQUATION	ORDER	A12	A21	D	A122	A211
VAN LAAR	3	0.4331	0.2898	-	-	-
MARGULES	3	0.4140	0.2755	-	-	-
MARGULES	4	0.4276	0.2895	0.0468	-	-
ALPHA	2	2.6473	0.4719	-	-	-
ALPHA	3	2.6345	1.7917	-	1.2238	-1.7031

DEVIATION IN VAPOR PHASE COMPOSITION

X	Y	TEMP.	LAAR 3	MARG 3	MARG 4	ALPHA2	ALPHA3
0.0590	0.2025	70.80	0.0142	0.0105	0.0134	-0.0269	0.0039
0.0870	0.2710	68.74	0.0099	0.0070	0.0094	-0.0328	-0.0015
0.1790	0.4075	64.45	0.0092	0.0096	0.0096	-0.0180	-0.0004
0.2640	0.4895	61.91	0.0049	0.0066	0.0055	-0.0049	-0.0013
0.3740	0.5655	59.83	0.0035	0.0049	0.0038	0.0089	0.0007
0.4510	0.6125	58.74	0.0018	0.0023	0.0019	0.0118	0.0002
0.5255	0.6550	57.94	0.0017	0.0013	0.0016	0.0119	0.0003
0.6165	0.7065	57.18	0.0031	0.0020	0.0030	0.0086	0.0008
0.6960	0.7560	56.67	0.0026	0.0016	0.0026	0.0007	-0.0010
0.7620	0.7985	56.36	0.0041	0.0036	0.0042	-0.0052	-0.0007
0.8295	0.8460	56.15	0.0055	0.0058	0.0057	-0.0106	0.0001
0.8950	0.8980	56.01	0.0058	0.0068	0.0059	-0.0135	0.0007
0.9140	0.9150	56.02	0.0050	0.0060	0.0051	-0.0139	0.0003
0.9530	0.9520	55.99	0.0028	0.0038	0.0029	-0.0118	-0.0004
		MEAN	0.0053	0.0051	0.0053	0.0128	0.0009

ACETONE(1)-CARBON TETRACHLORIDE(2)

BROWN I.,SMITH F.:AUSTR.J.CHEM.10,423(1957).

ANTOINE VAPOR PRESSURE CONSTANTS

	A	B	C
(1)	7.23967	1279.870	237.500
(2)	6.93390	1242.430	230.000

T = 45.00

EQUATION	ORDER	A12	A21	D	A122	A211
VAN LAAR	3	0.4320	0.2807	-	-	-
MARGULES	3	0.4083	0.2657	-	-	-
MARGULES	4	0.4547	0.3171	0.1596	-	-
ALPHA	2	2.7598	0.4969	-	-	-
ALPHA	3	4.9574	3.4350	-	-0.2406	-3.3337

			DEVIATION IN VAPOR PHASE COMPOSITION				
X	Y	PRESS.	LAAR 3	MARG 3	MARG 4	ALPHA2	ALPHA3
0.0556	0.2165	315.32	-0.0066	-0.0115	-0.0012	-0.0453	-0.0023
0.0903	0.2910	339.70	-0.0015	-0.0052	0.0030	-0.0409	0.0002
0.2152	0.4495	397.77	0.0047	0.0059	0.0043	-0.0100	0.0012
0.2929	0.5137	422.46	0.0030	0.0051	0.0014	0.0026	-0.0001
0.3970	0.5832	448.88	0.0007	0.0023	-0.0003	0.0111	-0.0003
0.4769	0.6309	463.92	-0.0004	0.0002	0.0001	0.0121	0.0002
0.5300	0.6621	472.84	-0.0010	-0.0012	0.0004	0.0101	0.0002
0.6047	0.7081	485.16	-0.0033	-0.0041	-0.0008	0.0028	-0.0017
0.7128	0.7718	498.07	0.0004	-0.0004	0.0027	-0.0051	0.0009
0.8088	0.8360	506.89	0.0028	0.0029	0.0029	-0.0143	0.0010
0.9090	0.9141	512.32	0.0035	0.0046	0.0011	-0.0182	-0.0004
0.9636	0.9636	513.20	0.0020	0.0029	0.0000	-0.0119	-0.0010
		MEAN	0.0025	0.0038	0.0015	0.0154	0.0008

ACETONE(1)-CHLOROBENZENE(2)

OTHMER D.F.:IND.ENG.CHEM.35,614(1943).

ANTOINE VAPOR PRESSURE CONSTANTS

	A	B	C
(1)	7.23967	1279.870	237.500
(2)	6.97429	1431.323	217.975

P = 760.00

EQUATION	ORDER	A12	A21	D	A122	A211
VAN LAAR	3	0.5640	0.1478	-	-	-
MARGULES	3	0.3810	0.0471	-	-	-
MARGULES	4	0.3271	-0.0387	-0.2457	-	-
ALPHA	2	11.6113	-0.2573	-	-	-
ALPHA	3	5.5711	-0.9427	-	5.5604	0.5131

			DEVIATION IN VAPOR PHASE COMPOSITION				
X	Y	TEMP.	LAAR 3	MARG 3	MARG 4	ALPHA2	ALPHA3
0.0200	0.1910	125.50	0.1435	0.0796	0.0602	0.0115	0.0060
0.0500	0.3820	108.00	0.1526	0.1094	0.0919	0.0088	0.0022
0.1000	0.5650	107.00	0.0832	0.0742	0.0672	0.0013	-0.0030
0.2000	0.7290	93.50	0.0326	0.0419	0.0443	0.0016	0.0014
0.3000	0.8120	84.10	0.0153	0.0211	0.0240	-0.0029	-0.0014
0.4000	0.8560	77.50	0.0175	0.0176	0.0183	-0.0005	0.0011
0.5000	0.8870	72.00	0.0216	0.0187	0.0172	-0.0005	0.0004
0.6000	0.9100	68.20	0.0255	0.0225	0.0200	-0.0008	-0.0011
0.7000	0.9260	65.50	0.0308	0.0293	0.0275	0.0013	-0.0000
0.8000	0.9420	62.80	0.0321	0.0325	0.0322	0.0016	0.0001
0.8500	0.9510	62.00	0.0305	0.0314	0.0318	0.0011	0.0003
0.9000	0.9620	61.00	0.0262	0.0274	0.0280	-0.0000	0.0004
0.9500	0.9780	58.10	0.0165	0.0173	0.0178	-0.0025	-0.0006
		MEAN	0.0483	0.0402	0.0370	0.0027	0.0014

ACETONE(1)-CHLOROFORM(2)

KARR A.E.,SCHEIBEL E.G.,BOWES W.M.,OTHMER D.F.:
        IND.ENG.CHEM.43,961(1951).

ANTOINE VAPOR PRESSURE CONSTANTS
          A   B   C
(1) 7.23967   1279.870   237.500
(2) 6.90328   1163.000   227.000

P =   760.00

EQUATION	ORDER	A12	A21	D	A122	A211
VAN LAAR	3	-0.3130	-0.3109	-	-	-
MARGULES	3	-0.3130	-0.3108	-	-	-
MARGULES	4	-0.2829	-0.2851	0.0824	-	-
ALPHA	2	-0.3854	-0.6440	-	-	-
ALPHA	3	-1.0293	-1.4093	-	0.6384	0.9233

| | | | DEVIATION IN VAPOR PHASE COMPOSITION | | | | |
X	Y	TEMP.	LAAR 3	MARG 3	MARG 4	ALPHA2	ALPHA3
0.0563	0.0387	62.06	-0.0016	-0.0016	-0.0001	-0.0008	-0.0016
0.1491	0.1126	63.22	0.0020	0.0020	0.0033	0.0027	-0.0001
0.2224	0.1825	64.07	0.0078	0.0078	0.0077	0.0070	0.0037
0.2747	0.2495	64.35	0.0019	0.0019	0.0007	-0.0005	-0.0031
0.5211	0.5788	63.66	0.0001	0.0001	-0.0002	-0.0070	0.0005
0.6064	0.6887	62.63	-0.0014	-0.0014	-0.0007	-0.0067	-0.0008
0.6448	0.7311	62.03	0.0012	0.0012	0.0021	-0.0029	0.0009
0.7474	0.8337	60.39	0.0039	0.0039	0.0043	0.0037	-0.0005
0.8391	0.9045	58.77	0.0074	0.0074	0.0067	0.0096	0.0002
		MEAN	0.0030	0.0030	0.0029	0.0045	0.0013

ACETONE(1)-CHLOROFORM(2)

REINDERS W.,DE MINJER C.H.:REC.TRAV.CHIM.PAYS BAS 59,369(1940).

ANTOINE VAPOR PRESSURE CONSTANTS

	A	B	C
(1)	7.23967	1279.870	237.500
(2)	6.93708	1171.200	227.000

P =   760.00

EQUATION	ORDER	A12	A21	D	A122	A211
VAN LAAR	3	-0.2789	-0.2877	-	-	-
MARGULES	3	-0.2787	-0.2877	-	-	-
MARGULES	4	-0.2406	-0.2504	0.1050	-	-
ALPHA	2	-0.3548	-0.5956	-	-	-
ALPHA	3	-0.8290	-1.2523	-	0.4246	0.7908

			DEVIATION IN VAPOR PHASE COMPOSITION				
X	Y	TEMP.	LAAR 3	MARG 3	MARG 4	ALPHA2	ALPHA3
0.1013	0.0740	62.70	0.0012	0.0012	0.0036	0.0015	-0.0030
0.1792	0.1428	63.55	0.0053	0.0053	0.0066	0.0048	-0.0011
0.2585	0.2221	63.95	0.0128	0.0128	0.0123	0.0108	0.0062
0.3022	0.2814	64.40	0.0061	0.0061	0.0048	0.0030	0.0003
0.3697	0.3724	64.55	0.0010	0.0010	-0.0007	-0.0037	-0.0023
0.4418	0.4695	64.25	-0.0007	-0.0007	-0.0018	-0.0067	-0.0009
0.5268	0.5862	63.55	-0.0053	-0.0053	-0.0050	-0.0118	-0.0032
0.6318	0.7070	62.30	0.0032	0.0032	0.0045	-0.0018	0.0041
0.6683	0.7526	61.80	-0.0014	-0.0014	-0.0002	-0.0056	-0.0019
0.7020	0.7852	61.27	0.0016	0.0016	0.0025	-0.0016	-0.0003
0.7315	0.8123	60.80	0.0038	0.0038	0.0043	0.0014	0.0005
0.7605	0.8376	60.10	0.0055	0.0055	0.0056	0.0040	0.0009
0.8137	0.8793	59.30	0.0088	0.0088	0.0079	0.0085	0.0023
0.8946	0.9411	57.95	0.0039	0.0039	0.0022	0.0048	-0.0029
0.9433	0.9699	57.10	0.0030	0.0030	0.0016	0.0039	-0.0017
0.9652	0.9822	56.70	0.0019	0.0019	0.0008	0.0025	-0.0014
		MEAN	0.0041	0.0041	0.0040	0.0048	0.0021

ACETONE(1)-CHLOROFORM(2)

ROSANOFF M.A.,EASELEY C.W.:J.AM.CHEM.SOC.31,953(1909).

ANTOINE VAPOR PRESSURE CONSTANTS
```
           A          B          C
(1) 7.23967    1279.870    237.500
(2) 6.90328    1163.000    227.000
```

P = 760.00

EQUATION	ORDER	A12	A21	D	A122	A211
VAN LAAR	3	-0.3090	-0.2940	-	-	-
MARGULES	3	-0.3087	-0.2939	-	-	-
MARGULES	4	-0.3320	-0.3142	-0.0678	-	-
ALPHA	2	-0.3829	-0.6183	-	-0.4525	0.2211
ALPHA	3	-0.0682	-0.7269	-		

			DEVIATION IN VAPOR PHASE COMPOSITION				
X	Y	TEMP.	LAAR 3	MARG 3	MARG 4	ALPHA2	ALPHA3
0.1108	0.0650	62.80	0.0160	0.0161	0.0148	0.0161	0.0079
0.1375	0.1000	63.10	0.0050	0.0051	0.0041	0.0045	-0.0037
0.2108	0.1760	63.80	0.0036	0.0036	0.0037	0.0005	-0.0044
0.2660	0.2370	64.40	0.0065	0.0064	0.0075	0.0007	0.0003
0.4771	0.5170	63.90	0.0049	0.0048	0.0056	-0.0084	0.0046
0.5750	0.6480	62.80	0.0013	0.0013	0.0008	-0.0106	-0.0005
0.6633	0.7505	61.60	0.0018	0.0018	0.0009	-0.0063	-0.0027
0.7388	0.8235	60.40	0.0048	0.0048	0.0042	0.0006	-0.0017
0.7955	0.8688	59.40	0.0085	0.0085	0.0084	0.0071	0.0013
0.8590	0.9165	58.30	0.0075	0.0075	0.0081	0.0083	0.0008
0.9145	0.9522	57.50	0.0060	0.0060	0.0068	0.0075	0.0010
		MEAN	0.0060	0.0060	0.0059	0.0064	0.0026

ACETONE(1)-CHLOROFORM(2)

SODAY F.J.,BENNETT G.W.:J.CHEM.EDUC.7,1336(1930).

ANTOINE VAPOR PRESSURE CONSTANTS
	A	B	C
(1)	7.23967	1279.870	237.500
(2)	6.90328	1163.000	227.000

P =   732.00

EQUATION	ORDER	A12	A21	D	A122	A211
VAN LAAR	3	-0.2833	-0.2802	-	-	-
MARGULES	3	-0.2833	-0.2802	-	-	-
MARGULES	4	-0.2527	-0.2475	0.0981	-	-
ALPHA	2	-0.3324	-0.6118	-	-	-
ALPHA	3	-0.9993	-1.3364	-	0.6650	0.8461

			DEVIATION IN VAPOR PHASE COMPOSITION				
X	Y	TEMP.	LAAR 3	MARG 3	MARG 4	ALPHA2	ALPHA3
0.0790	0.0600	60.30	-0.0029	-0.0029	-0.0012	-0.0011	-0.0022
0.1070	0.0790	60.75	0.0017	0.0017	0.0033	0.0037	0.0020
0.1430	0.1160	61.15	-0.0024	-0.0024	-0.0012	-0.0004	-0.0028
0.1860	0.1600	61.75	-0.0033	-0.0033	-0.0030	-0.0017	-0.0049
0.2660	0.2350	62.20	0.0116	0.0116	0.0103	0.0116	0.0082
0.3940	0.3940	62.40	0.0157	0.0157	0.0141	0.0122	0.0131
0.4620	0.5200	62.00	-0.0197	-0.0197	-0.0202	-0.0243	-0.0204
0.5360	0.5980	61.25	-0.0010	-0.0010	-0.0000	-0.0056	0.0004
0.6180	0.6990	60.30	-0.0015	-0.0015	0.0003	-0.0047	0.0004
0.7150	0.7920	58.90	0.0099	0.0099	0.0113	0.0095	0.0098
0.7700	0.8480	58.20	0.0048	0.0048	0.0053	0.0059	0.0028
0.8210	0.9010	57.20	-0.0066	-0.0066	-0.0070	-0.0045	-0.0101
0.9150	0.9540	56.00	0.0033	0.0033	0.0019	0.0056	-0.0007
		MEAN	0.0065	0.0065	0.0061	0.0070	0.0060

ACETONE(1)-CHLOROFORM(2)

TALLMADGE J.A.,CANJAR L.N.:IND.ENG.CHEM.46,1279(1954).

ANTOINE VAPOR PRESSURE CONSTANTS
```
          A         B         C
(1) 7.23967   1279.870   237.500
(2) 6.90328   1163.000   227.000
```

P =   738.00

EQUATION	ORDER	A12	A21	D	A122	A211
VAN LAAR	3	-0.3431	-0.2983	-	-	-
MARGULES	3	-0.3410	-0.2957	-	-	-
MARGULES	4	-0.3156	-0.2679	0.0715	-	-
ALPHA	2	-0.3606	-0.6671	-	-	-
ALPHA	3	-0.6665	-1.1464	-	0.2470	0.6076

			DEVIATION IN VAPOR PHASE COMPOSITION				
X	Y	TEMP.	LAAR 3	MARG 3	MARG 4	ALPHA2	ALPHA3
0.1355	0.1008	62.14	-0.0013	-0.0012	0.0002	0.0052	-0.0006
0.2714	0.2454	63.41	0.0023	0.0023	0.0021	0.0057	0.0014
0.3576	0.3604	63.63	0.0003	0.0002	-0.0005	-0.0007	-0.0007
0.4171	0.4426	63.63	0.0001	0.0001	-0.0003	-0.0034	-0.0003
0.4742	0.5230	63.39	-0.0014	-0.0013	-0.0012	-0.0063	-0.0011
0.5847	0.6650	62.16	0.0011	0.0012	0.0022	-0.0025	0.0024
0.6911	0.7877	60.36	-0.0020	-0.0019	-0.0012	-0.0010	-0.0016
0.8051	0.8848	58.76	0.0023	0.0023	0.0017	0.0076	0.0005
		MEAN	0.0014	0.0013	0.0012	0.0041	0.0011

ACETONE(1)-CHLOROFORM(2)

TYRER D.:J.CHEM.SOC.101,1104(1912).

ANTOINE VAPOR PRESSURE CONSTANTS
	A	B	C
(1)	7.23967	1279.870	237.500
(2)	6.90328	1163.000	227.000

P = 750.00

EQUATION	ORDER	A12	A21	D	A122	A211
VAN LAAR	3	-0.3239	-0.3140	-	-	-
MARGULES	3	-0.3240	-0.3137	-	-	-
MARGULES	4	-0.2730	-0.2676	0.1398	-	-
ALPHA	2	-0.4370	-0.6264	-	-	-
ALPHA	3	2.4292	0.6202	-	-3.6805	-1.0834

			DEVIATION IN VAPOR PHASE COMPOSITION				
X	Y	TEMP.	LAAR 3	MARG 3	MARG 4	ALPHA2	ALPHA3
0.1860	0.1031	63.02	0.0466	0.0466	0.0479	0.0397	-0.0006
0.3395	0.3181	63.84	0.0149	0.0149	0.0121	-0.0010	0.0029
0.4683	0.5154	63.41	-0.0069	-0.0068	-0.0083	-0.0264	-0.0040
0.5781	0.6521	62.19	0.0018	0.0018	0.0028	-0.0155	0.0011
0.6727	0.7566	61.03	0.0080	0.0080	0.0095	-0.0039	-0.0002
0.7551	0.8315	59.91	0.0142	0.0142	0.0146	0.0078	0.0013
0.8274	0.8898	58.83	0.0144	0.0144	0.0133	0.0122	0.0008
0.8915	0.9359	57.79	0.0106	0.0106	0.0088	0.0107	-0.0005
0.9487	0.9735	57.00	0.0039	0.0039	0.0023	0.0045	-0.0024
		MEAN	0.0135	0.0135	0.0133	0.0135	0.0015

ACETONE(1)-CHLOROFORM(2)

ZAWIDZKI J.:Z.PHYS.CHEM.35,129(1900).

ANTOINE VAPOR PRESSURE CONSTANTS

	A	B	C
(1)	7.23967	1279.870	237.500
(2)	6.90328	1163.000	227.000

T = 35.17

EQUATION	ORDER	A12	A21	D	A122	A211
VAN LAAR	3	-0.3986	-0.3362	-	-	-
MARGULES	3	-0.3947	-0.3322	-	-	-
MARGULES	4	-0.3607	-0.2998	0.0915	-	-
ALPHA	2	-0.4202	-0.6957	-	-	-
ALPHA	3	-0.7276	-1.2576	-	0.1925	0.7171

			DEVIATION IN VAPOR PHASE COMPOSITION				
X	Y	PRESS.	LAAR 3	MARG 3	MARG 4	ALPHA2	ALPHA3
0.0821	0.0500	279.50	-0.0015	-0.0013	0.0005	0.0051	-0.0032
0.0825	0.0465	280.10	0.0023	0.0025	0.0043	0.0089	0.0005
0.1953	0.1464	262.60	0.0017	0.0018	0.0026	0.0104	-0.0030
0.2003	0.1434	261.90	0.0100	0.0100	0.0108	0.0185	0.0052
0.3365	0.3171	249.20	0.0044	0.0043	0.0030	0.0061	-0.0002
0.3390	0.3183	248.80	0.0067	0.0066	0.0052	0.0082	0.0021
0.4188	0.4368	248.40	0.0012	0.0011	-0.0001	-0.0022	-0.0010
0.4469	0.4763	249.40	0.0021	0.0021	0.0012	-0.0026	0.0009
0.4484	0.4796	250.80	0.0010	0.0010	0.0001	-0.0038	-0.0002
0.4857	0.5343	252.80	-0.0005	-0.0004	-0.0008	-0.0064	-0.0006
0.4872	0.5365	252.70	-0.0006	-0.0005	-0.0009	-0.0065	-0.0006
0.4917	0.5443	252.90	-0.0020	-0.0019	-0.0023	-0.0081	-0.0020
0.5061	0.5625	255.40	-0.0000	0.0001	-0.0001	-0.0063	0.0004
0.5070	0.5640	255.70	-0.0003	-0.0002	-0.0003	-0.0065	0.0001
0.5752	0.6583	263.10	-0.0029	-0.0028	-0.0020	-0.0086	-0.0017
0.5768	0.6624	263.20	-0.0050	-0.0048	-0.0041	-0.0106	-0.0037
0.5950	0.6817	266.90	-0.0012	-0.0011	-0.0002	-0.0063	-0.0000
0.6034	0.6868	267.40	0.0041	0.0043	0.0052	-0.0007	0.0052
0.6336	0.7271	272.20	-0.0000	0.0001	0.0011	-0.0036	0.0007
0.6394	0.7354	273.10	-0.0016	-0.0015	-0.0005	-0.0049	-0.0010
0.6432	0.7398	273.50	-0.0017	-0.0015	-0.0005	-0.0048	-0.0011
0.7090	0.8062	285.70	0.0019	0.0020	0.0026	0.0022	0.0008
0.7116	0.8083	286.10	0.0024	0.0025	0.0030	0.0028	0.0012
0.7296	0.8273	290.50	0.0006	0.0006	0.0010	0.0020	-0.0012
0.7343	0.8262	291.60	0.0061	0.0061	0.0064	0.0077	0.0041
0.8147	0.8961	307.30	0.0026	0.0025	0.0018	0.0074	-0.0015
0.8182	0.8971	308.00	0.0042	0.0040	0.0033	0.0090	0.0000
0.8768	0.9362	319.70	0.0038	0.0036	0.0023	0.0092	-0.0008
0.8797	0.9377	320.10	0.0041	0.0039	0.0026	0.0094	-0.0005
0.9397	0.9715	332.10	0.0025	0.0023	0.0011	0.0062	-0.0009
0.9412	0.9724	332.40	0.0023	0.0021	0.0010	0.0060	-0.0010
		MEAN	0.0026	0.0026	0.0023	0.0065	0.0015

ACETONE(1)-ETHYL ALCOHOL(2)

AMER H.H.,PAXTON R.R.,VAN WINKLE M.:IND.ENG.CHEM.48,142(1956).

ANTOINE VAPOR PRESSURE CONSTANTS
```
          A          B          C
(1) 7.23967    1279.870    237.500
(2) 8.16290    1623.220    228.980
```

P = 760.00

EQUATION	ORDER	A12	A21	D	A122	A211
VAN LAAR	3	0.3973	0.1492	-	-	-
MARGULES	3	0.3197	0.0828	-	-	-
MARGULES	4	0.2656	0.0083	-0.2119	-	-
ALPHA	2	2.2137	-0.0272	-	-	-
ALPHA	3	3.2885	1.0105	-	-0.5695	-1.2058

			DEVIATION IN VAPOR PHASE COMPOSITION				
X	Y	TEMP.	LAAR 3	MARG 3	MARG 4	ALPHA2	ALPHA3
0.0330	0.1110	76.40	0.0207	0.0083	-0.0009	-0.0141	-0.0034
0.0780	0.2160	74.00	0.0295	0.0201	0.0108	-0.0110	0.0028
0.1490	0.3450	70.80	0.0165	0.0165	0.0139	-0.0086	0.0004
0.1950	0.4100	69.10	0.0079	0.0111	0.0122	-0.0061	-0.0018
0.3160	0.5340	65.60	0.0025	0.0051	0.0091	0.0055	0.0008
0.4140	0.6140	63.40	0.0037	0.0018	0.0028	0.0074	0.0003
0.5320	0.6970	61.30	0.0093	0.0038	0.0001	0.0044	-0.0006
0.6910	0.7960	59.00	0.0194	0.0162	0.0117	-0.0027	0.0003
0.8520	0.8960	57.30	0.0196	0.0226	0.0244	-0.0093	-0.0001
		MEAN	0.0144	0.0117	0.0095	0.0077	0.0011

o

ACETONE(1)-ETHYL ALCOHOL(2)

DUTTEY,PRIVATE COMMUNICATION TO T.H.CHILTON,1935.(CHU J.C.ET AL.:
          DISTILLATION EQUILIBRIUM DATA,1950).

ANTOINE VAPOR PRESSURE CONSTANTS
        A   B   C
(1) 7.23967   1279.870   237.500
(2) 8.16290   1623.220   228.980

P =   760.00

EQUATION	ORDER	A12	A21	D	A122	A211
VAN LAAR	3	0.4173	0.1587	-	-	-
MARGULES	3	0.3329	0.0887	-	-	-
MARGULES	4	0.2846	0.0247	-0.1750	-	-
ALPHA	2	2.2203	-0.0361	-	-	-
ALPHA	3	1.7152	0.3136	-	0.9500	-0.5975

			DEVIATION IN VAPOR PHASE COMPOSITION				
X	Y	TEMP.	LAAR 3	MARG 3	MARG 4	ALPHA2	ALPHA3
0.0500	0.1550	75.40	0.0315	0.0170	0.0074	-0.0141	-0.0018
0.1000	0.2620	73.00	0.0311	0.0230	0.0155	-0.0114	0.0022
0.1500	0.3480	71.00	0.0197	0.0184	0.0151	-0.0092	0.0013
0.2000	0.4170	69.00	0.0106	0.0131	0.0132	-0.0055	0.0007
0.2500	0.4780	67.30	0.0008	0.0044	0.0065	-0.0053	-0.0034
0.3000	0.5240	65.90	0.0006	0.0034	0.0062	0.0013	-0.0006
0.3500	0.5660	64.70	0.0009	0.0019	0.0042	0.0053	0.0006
0.4000	0.6050	63.60	0.0019	0.0007	0.0019	0.0070	0.0005
0.5000	0.6740	61.80	0.0079	0.0032	0.0012	0.0084	0.0013
0.6000	0.7390	60.40	0.0131	0.0079	0.0040	0.0042	0.0001
0.7000	0.8020	59.10	0.0168	0.0144	0.0113	-0.0025	-0.0009
0.8000	0.8650	58.00	0.0173	0.0191	0.0192	-0.0089	-0.0009
0.9000	0.9290	57.00	0.0136	0.0176	0.0203	-0.0099	0.0011
		MEAN	0.0127	0.0111	0.0097	0.0072	0.0012

ACETONE(1)-ETHYL ALCOHOL(2)

GORDON A.R.,HINES W.G.:CAN.J.RES.24B,254(1946).

ANTOINE VAPOR PRESSURE CONSTANTS
```
          A          B          C
(1) 7.23967    1279.870    237.500
(2) 8.16290    1623.220    228.980
```

T =   32.00

EQUATION	ORDER	A12	A21	D	A122	A211
VAN LAAR	3	0.3349	0.2645	-	-	-
MARGULES	3	0.3305	0.2591	-	-	-
MARGULES	4	0.3361	0.2681	0.0246	-	-
ALPHA	2	5.2876	0.0265	-	-	-
ALPHA	3	4.4539	0.9235	-	2.2162	-1.3859

			DEVIATION IN VAPOR PHASE COMPOSITION				
X	Y	PRESS.	LAAR 3	MARG 3	MARG 4	ALPHA2	ALPHA3
0.0250	0.1605	102.70	-0.0034	-0.0045	-0.0031	-0.0243	-0.0031
0.0500	0.2700	116.20	-0.0027	-0.0039	-0.0024	-0.0295	-0.0026
0.0750	0.3505	128.30	-0.0014	-0.0023	-0.0011	-0.0277	-0.0016
0.1000	0.4130	139.70	-0.0004	-0.0011	-0.0002	-0.0234	-0.0009
0.1500	0.5020	159.50	0.0032	0.0031	0.0033	-0.0107	0.0024
0.2000	0.5690	175.90	0.0015	0.0017	0.0015	-0.0037	0.0007
0.2500	0.6190	190.10	0.0010	0.0013	0.0009	0.0028	0.0004
0.3000	0.6600	202.40	-0.0003	0.0000	-0.0004	0.0066	-0.0007
0.4000	0.7220	223.80	-0.0000	0.0001	-0.0000	0.0115	-0.0000
0.5000	0.7730	242.00	-0.0008	-0.0010	-0.0007	0.0094	-0.0008
0.6000	0.8170	257.40	0.0000	-0.0003	0.0003	0.0044	-0.0002
0.7000	0.8590	271.90	0.0012	0.0010	0.0014	-0.0034	0.0005
0.8000	0.9020	285.30	0.0020	0.0020	0.0021	-0.0124	0.0009
0.9000	0.9500	297.90	0.0002	0.0004	0.0000	-0.0193	-0.0009
		MEAN	0.0013	0.0016	0.0012	0.0135	0.0011

ACETONE(1)-ETHYL ALCOHOL(2)

GORDON A.R.,HINES W.G.:CAN.J.RES.24B,254(1946).

ANTOINE VAPOR PRESSURE CONSTANTS
	A	B	C
(1)	7.23967	1279.870	237.500
(2)	8.16290	1623.220	228.980

T =   40.00

EQUATION	ORDER	A12	A21	D	A122	A211
VAN LAAR	3	0.3448	0.2430	-	-	-
MARGULES	3	0.3368	0.2288	-	-	-
MARGULES	4	0.3209	0.2047	-0.0672	-	-
ALPHA	2	4.5362	0.0766	-	-	-
ALPHA	3	2.7849	0.6928	-	2.9110	-1.1671

			DEVIATION IN VAPOR PHASE COMPOSITION				
X	Y	PRESS.	LAAR 3	MARG 3	MARG 4	ALPHA2	ALPHA3
0.0250	0.1435	152.20	0.0014	-0.0003	-0.0038	-0.0216	-0.0030
0.0500	0.2460	168.50	0.0019	0.0001	-0.0039	-0.0282	-0.0036
0.0750	0.3205	183.30	0.0047	0.0033	-0.0001	-0.0253	-0.0005
0.1000	0.3795	197.40	0.0062	0.0053	0.0029	-0.0203	0.0017
0.1500	0.4700	223.30	0.0052	0.0051	0.0045	-0.0114	0.0025
0.2000	0.5380	244.90	0.0013	0.0017	0.0023	-0.0053	0.0002
0.2500	0.5900	263.90	-0.0011	-0.0007	0.0005	0.0001	-0.0012
0.3000	0.6310	280.60	-0.0016	-0.0014	-0.0002	0.0053	-0.0013
0.4000	0.6940	309.00	0.0006	0.0002	0.0005	0.0125	0.0005
0.5000	0.7470	334.00	0.0019	0.0009	0.0000	0.0119	0.0004
0.6000	0.7950	355.50	0.0036	0.0024	0.0008	0.0064	0.0004
0.7000	0.8430	375.10	0.0038	0.0031	0.0018	-0.0036	-0.0002
0.8000	0.8930	393.00	0.0026	0.0027	0.0025	-0.0151	-0.0006
0.9000	0.9450	409.40	0.0013	0.0020	0.0030	-0.0205	0.0005
		MEAN	0.0027	0.0021	0.0019	0.0134	0.0012

ACETONE(1)-ETHYL ALCOHOL(2)

GORDON A.R.,HINES W.G.:CAN.J.RES.24B,254(1946).

ANTOINE VAPOR PRESSURE CONSTANTS
	A	B	C
(1)	7.23967	1279.870	237.500
(2)	8.16290	1623.220	228.960

T =   48.00

EQUATION	ORDER	A12	A21	D	A122	A211
VAN LAAR	3	0.3056	0.2260	-	-	-
MARGULES	3	0.2993	0.2177	-	-	-
MARGULES	4	0.3030	0.2232	0.0153	-	-
ALPHA	2	3.7338	-0.0101	-	-	-
ALPHA	3	3.5795	0.9552	-	1.2027	-1.3754

			DEVIATION IN VAPOR PHASE COMPOSITION				
X	Y	PRESS.	LAAR 3	MARG 3	MARG 4	ALPHA2	ALPHA3
0.0250	0.1210	223.30	0.0025	0.0013	0.0020	-0.0146	0.0027
0.0500	0.2155	244.10	0.0009	-0.0004	0.0004	-0.0223	0.0011
0.0750	0.2890	263.60	0.0003	-0.0009	-0.0001	-0.0235	0.0003
0.1000	0.3460	281.60	0.0021	0.0013	0.0019	-0.0194	0.0020
0.1500	0.4370	314.20	0.0012	0.0011	0.0012	-0.0125	0.0011
0.2000	0.5070	341.50	-0.0019	-0.0016	-0.0017	-0.0073	-0.0021
0.2500	0.5600	365.50	-0.0021	-0.0017	-0.0020	-0.0005	-0.0023
0.3000	0.6020	387.00	-0.0005	-0.0001	-0.0004	0.0064	-0.0005
0.4000	0.6700	424.90	0.0024	0.0024	0.0024	0.0144	0.0026
0.5000	0.7310	457.40	0.0005	0.0001	0.0003	0.0114	0.0006
0.6000	0.7850	485.00	0.0001	-0.0005	-0.0001	0.0050	0.0000
0.7000	0.8380	510.10	-0.0011	-0.0016	-0.0013	-0.0052	-0.0014
0.8000	0.8890	533.00	-0.0001	-0.0001	-0.0000	-0.0133	-0.0003
0.9000	0.9420	553.60	0.0009	0.0013	0.0011	-0.0162	0.0010
		MEAN	0.0012	0.0010	0.0011	0.0123	0.0013

ACETONE(1)-ETHYL ALCOHOL(2)

HELLWIG L.R.,VAN WINKLE M.:IND.ENG.CHEM.45,624(1953).

ANTOINE VAPOR PRESSURE CONSTANTS

	A	B	C
(1)	7.23967	1279.870	237.500
(2)	8.16290	1623.220	228.980

P = 760.00

EQUATION	ORDER	A12	A21	D	A122	A211
VAN LAAR	3	0.4035	0.1400	-	-	-
MARGULES	3	0.3116	0.0662	-	-	-
MARGULES	4	0.2555	-0.0112	-0.2140	-	-
ALPHA	2	2.0893	-0.0941	-	-	-
ALPHA	3	1.2455	0.1344	-	1.2824	-0.5060

			DEVIATION IN VAPOR PHASE COMPOSITION				
X	Y	TEMP.	LAAR 3	MARG 3	MARG 4	ALPHA2	ALPHA3
0.0580	0.1730	75.10	0.0270	0.0132	0.0026	-0.0178	-0.0045
0.1210	0.2970	72.20	0.0207	0.0167	0.0107	-0.0139	-0.0007
0.1750	0.3760	70.10	0.0137	0.0158	0.0149	-0.0060	0.0030
0.2760	0.4940	66.70	0.0023	0.0064	0.0100	0.0018	0.0016
0.3390	0.5550	65.20	-0.0021	-0.0006	0.0027	0.0028	-0.0018
0.4440	0.6360	62.90	0.0034	-0.0002	-0.0006	0.0070	-0.0013
0.5800	0.7260	60.70	0.0145	0.0084	0.0034	0.0068	0.0009
0.7360	0.8270	58.70	0.0184	0.0172	0.0143	-0.0041	0.0004
0.8750	0.9190	57.20	0.0112	0.0151	0.0177	-0.0132	-0.0005
		MEAN	0.0126	0.0104	0.0085	0.0081	0.0016

ACETONE(1)-HEXANE(2)

SCHAEFER K.,RALL W.:Z.ELEKTROCHEM.62(10),1090(1958).

ANTOINE VAPOR PRESSURE CONSTANTS
	A	B	C
(1)	7.23967	1279.870	237.500
(2)	6.87776	1171.530	224.366

T =   45.00

EQUATION	ORDER	A12	A21	D	A122	A211
VAN LAAR	3	0.6496	0.6574	-	-	-
MARGULES	3	0.6497	0.6572	-	-	-
MARGULES	4	0.7098	0.7105	0.1868	-	-
ALPHA	2	6.2395	3.3680	-	-	-
ALPHA	3	4.3377	3.3307	-	2.3871	-0.7230

			DEVIATION IN VAPOR PHASE COMPOSITION				
X	Y	PRESS.	LAAR 3	MARG 3	MARG 4	ALPHA2	ALPHA3
0.0651	0.2828	444.60	-0.0030	-0.0030	0.0120	-0.0021	0.0078
0.1592	0.4442	545.80	-0.0001	-0.0001	0.0029	-0.0092	-0.0018
0.2549	0.5163	590.20	0.0042	0.0042	-0.0001	-0.0065	-0.0029
0.3478	0.5560	617.30	0.0058	0.0058	0.0002	-0.0014	-0.0009
0.4429	0.5866	632.60	0.0031	0.0030	-0.0001	0.0016	-0.0004
0.5210	0.6068	639.60	0.0017	0.0017	0.0016	0.0048	0.0015
0.5907	0.6258	633.80	-0.0001	-0.0001	0.0023	0.0060	0.0021
0.6202	0.6339	637.10	-0.0002	-0.0002	0.0031	0.0066	0.0028
0.7168	0.6662	631.00	0.0007	0.0006	0.0045	0.0061	0.0039
0.7923	0.7034	627.80	0.0024	0.0024	0.0037	0.0014	0.0025
0.8022	0.7292	623.30	-0.0170	-0.0170	-0.0162	-0.0192	-0.0175
0.8692	0.7583	603.40	0.0095	0.0095	0.0056	-0.0038	0.0031
0.9288	0.8255	583.20	0.0185	0.0185	0.0107	-0.0052	0.0069
0.9658	0.9003	543.30	0.0124	0.0125	0.0051	-0.0107	0.0014
		MEAN	0.0056	0.0056	0.0049	0.0060	0.0040

ACETONE(1)-METHYL ALCOHOL(2)

AMER H.H.,PAXTON R.R.,VAN WINKLE M.:IND.ENG.CHEM.48,142(1956).

ANTOINE VAPOR PRESSURE CONSTANTS

	A	B	C
(1)	7.23967	1279.870	237.500
(2)	8.07246	1574.990	238.860

P = 760.00

EQUATION	ORDER	A12	A21	D	A122	A211
VAN LAAR	3	0.2900	0.2520	-	-	-
MARGULES	3	0.2885	0.2501	-	-	-
MARGULES	4	0.2802	0.2411	-0.0275	-	-
ALPHA	2	1.3848	0.4605	-	-	-
ALPHA	3	1.4753	0.6220	-	-0.0259	-0.1861

			DEVIATION IN VAPOR PHASE COMPOSITION				
X	Y	TEMP.	LAAR 3	MARG 3	MARG 4	ALPHA2	ALPHA3
0.0360	0.0820	63.50	0.0020	0.0018	0.0008	-0.0030	-0.0014
0.0810	0.1610	62.20	0.0086	0.0084	0.0073	0.0009	0.0029
0.1410	0.2510	60.70	0.0095	0.0094	0.0088	0.0013	0.0030
0.2060	0.3360	59.40	0.0029	0.0029	0.0030	-0.0038	-0.0031
0.2930	0.4230	58.10	0.0001	0.0002	0.0007	-0.0035	-0.0039
0.3940	0.5000	56.90	0.0039	0.0039	0.0044	0.0030	0.0019
0.5130	0.5800	56.20	0.0074	0.0074	0.0072	0.0079	0.0068
0.5840	0.6390	55.90	-0.0033	-0.0034	-0.0039	-0.0035	-0.0042
0.6830	0.7050	55.80	-0.0002	-0.0003	-0.0010	-0.0027	-0.0027
0.7420	0.7450	55.80	0.0038	0.0037	0.0033	-0.0008	-0.0002
0.8230	0.8060	55.80	0.0091	0.0092	0.0093	0.0016	0.0027
0.8610	0.8430	55.80	0.0064	0.0065	0.0070	-0.0020	-0.0008
		MEAN	0.0048	0.0048	0.0047	0.0028	0.0028

ACETONE(1)-METHYL ALCOHOL(2)

GRISWOLD J.,BUFORD C.B.: IND.ENG.CHEM. 41,2347(1949).

ANTOINE VAPOR PRESSURE CONSTANTS

	A	B	C
(1)	7.23967	1279.870	237.500
(2)	8.07246	1574.990	238.860

P = 760.00

EQUATION	ORDER	A12	A21	D	A122	A211
VAN LAAR	3	0.1753	0.3246	-	-	-
MARGULES	3	0.1523	0.2911	-	-	-
MARGULES	4	0.1616	0.3029	0.0344	-	-
ALPHA	2	1.2380	0.2393	-	-	-
ALPHA	3	2.3944	0.1628	-	-1.5978	0.4278

			DEVIATION IN VAPOR PHASE COMPOSITION				
X	Y	TEMP.	LAAR 3	MARG 3	MARG 4	ALPHA2	ALPHA3
0.0600	0.0970	62.80	0.0136	0.0104	0.0116	0.0228	0.0070
0.1500	0.2500	60.80	0.0000	-0.0020	-0.0016	0.0090	-0.0070
0.2950	0.4200	58.60	0.0039	0.0053	0.0046	0.0026	0.0020
0.3980	0.5170	57.20	0.0043	0.0060	0.0056	-0.0040	0.0039
0.5250	0.6250	56.30	-0.0043	-0.0043	-0.0038	-0.0157	-0.0049
0.7320	0.7540	55.80	0.0050	0.0034	0.0041	0.0018	0.0017
0.8000	0.8000	55.70	0.0054	0.0048	0.0049	0.0073	0.0012
0.8920	0.8750	55.90	0.0024	0.0042	0.0033	0.0103	-0.0017
		MEAN	0.0049	0.0051	0.0049	0.0092	0.0037

ACETONE(1)-METHYL ALCOHOL(2)

HARPER B.G.,MOORE J.C.: IND.ENG.CHEM. 49,411(1957).

ANTOINE VAPOR PRESSURE CONSTANTS
```
         A          B          C
(1) 7.23967    1279.870    237.500
(2) 8.07246    1574.990    238.860
```

P =   752.00

EQUATION	ORDER	A12	A21	D	A122	A211
VAN LAAR	3	0.2525	0.3365	-	-	-
MARGULES	3	0.2434	0.3320	-	-	-
MARGULES	4	0.1973	0.2807	-0.1650	-	-
ALPHA	2	1.5257	0.3644	-	-	-
ALPHA	3	-0.7745	-1.1155	-	2.0965	1.2779

			DEVIATION IN VAPOR PHASE COMPOSITION				
X	Y	TEMP.	LAAR 3	MARG 3	MARG 4	ALPHA2	ALPHA3
0.0584	0.1175	62.40	0.0063	0.0049	-0.0012	0.0114	0.0058
0.0756	0.1532	61.93	0.0022	0.0008	-0.0052	0.0078	0.0021
0.1360	0.2515	60.54	0.0010	0.0002	-0.0026	0.0065	0.0035
0.1672	0.2950	59.88	0.0007	0.0004	-0.0003	0.0055	0.0046
0.2009	0.3531	59.25	-0.0151	-0.0149	-0.0136	-0.0111	-0.0099
0.2287	0.3810	58.81	-0.0110	-0.0106	-0.0079	-0.0076	-0.0050
0.3629	0.5009	57.12	-0.0045	-0.0032	0.0005	-0.0029	0.0023
0.3977	0.5260	56.78	-0.0023	-0.0010	0.0019	-0.0007	0.0038
0.5838	0.6531	55.61	-0.0037	-0.0034	-0.0068	0.0010	-0.0026
0.6113	0.6662	55.45	0.0004	0.0006	-0.0034	0.0058	0.0013
0.7463	0.7593	55.10	-0.0061	-0.0064	-0.0095	0.0031	-0.0009
0.9166	0.9075	55.40	-0.0135	-0.0132	-0.0087	-0.0047	0.0010
0.9211	0.9134	55.42	-0.0146	-0.0143	-0.0098	-0.0061	-0.0002
		MEAN	0.0063	0.0057	0.0055	0.0057	0.0033

ACETONE(1)-METHYL ALCOHOL(2)

OTHMER D.F.: IND.ENG.CHEM. 20,743(1928).

ANTOINE VAPOR PRESSURE CONSTANTS
```
         A          B          C
(1) 7.23967    1279.870    237.500
(2) 8.07246    1574.990    238.860
```

P =   755.00

EQUATION	ORDER	A12	A21	D	A122	A211
VAN LAAR	3	0.2742	0.3681	-	-	-
MARGULES	3	0.2583	0.3649	-	-	-
MARGULES	4	0.1162	0.2213	-0.4786	-	-
ALPHA	2	1.9202	0.4646	-	-	-
ALPHA	3	-1.2089	-0.8650	-	3.2612	0.8876

			DEVIATION IN VAPOR PHASE COMPOSITION				
X	Y	TEMP.	LAAR 3	MARG 3	MARG 4	ALPHA2	ALPHA3
0.0650	0.1630	61.90	-0.0212	-0.0239	-0.0429	-0.0043	0.0034
0.2380	0.4200	58.30	-0.0318	-0.0314	-0.0226	-0.0108	0.0018
0.3550	0.5280	56.80	-0.0321	-0.0301	-0.0175	-0.0140	-0.0091
0.4940	0.5960	55.60	-0.0019	0.0002	0.0006	0.0142	0.0061
0.5880	0.6480	55.40	0.0032	0.0046	-0.0038	0.0195	0.0056
0.7640	0.7860	54.90	-0.0257	-0.0256	-0.0323	-0.0096	-0.0140
0.8570	0.8480	54.70	-0.0165	-0.0165	-0.0106	-0.0027	0.0067
0.9000	0.8900	55.00	-0.0185	-0.0184	-0.0071	-0.0068	0.0072
0.9400	0.9440	55.20	-0.0285	-0.0283	-0.0153	-0.0200	-0.0059
		MEAN	0.0199	0.0199	0.0170	0.0113	0.0067

ACETONE(1)-METHYL ALCOHOL(2)

UCHIDA S.;OGAWA S.,YAMAGUSHI M.: JAPAN SCI.REV.ENG.SCI.
      1,NO.2,41(1950).

ANTOINE VAPOR PRESSURE CONSTANTS

	A	B	C
(1)	7.23967	1279.870	237.500
(2)	8.07246	1574.990	238.860

P =   760.00

EQUATION	ORDER	A12	A21	D	A122	A211
VAN LAAR	3	0.2516	0.2760	-	-	-
MARGULES	3	0.2512	0.2750	-	-	-
MARGULES	4	0.2674	0.2917	0.0523	-	-
ALPHA	2	1.4023	0.3750	-	-0.2241	-0.3519
ALPHA	3	1.7362	0.7196	-	-0.2241	-0.3519

			DEVIATION IN VAPOR PHASE COMPOSITION				
X	Y	TEMP.	LAAR 3	MARG 3	MARG 4	ALPHA2	ALPHA3
0.0200	0.0470	64.00	-0.0018	-0.0019	-0.0005	-0.0011	0.0006
0.0500	0.1080	63.00	-0.0017	-0.0017	0.0005	-0.0005	0.0027
0.1000	0.1960	61.60	-0.0022	-0.0022	-0.0002	-0.0010	0.0025
0.1500	0.2700	60.60	-0.0030	-0.0031	-0.0021	-0.0020	0.0007
0.2000	0.3350	59.50	-0.0052	-0.0052	-0.0053	-0.0046	-0.0031
0.2500	0.3880	58.70	-0.0038	-0.0038	-0.0046	-0.0033	-0.0028
0.3000	0.4320	58.10	0.0002	0.0002	-0.0010	0.0009	0.0005
0.3500	0.4760	57.40	-0.0002	-0.0002	-0.0014	0.0004	-0.0007
0.4000	0.5140	56.90	0.0015	0.0014	0.0005	0.0023	0.0009
0.4500	0.5490	56.50	0.0034	0.0034	0.0029	0.0046	0.0031
0.5000	0.5880	56.20	-0.0006	-0.0007	-0.0006	0.0009	-0.0004
0.5500	0.6210	56.00	0.0002	0.0001	0.0007	0.0020	0.0010
0.6000	0.6550	55.80	-0.0005	-0.0006	0.0004	0.0015	0.0009
0.6500	0.6910	55.60	-0.0029	-0.0030	-0.0018	-0.0010	-0.0011
0.7000	0.7260	55.50	-0.0035	-0.0036	-0.0025	-0.0018	-0.0015
0.7500	0.7630	55.40	-0.0045	-0.0045	-0.0038	-0.0032	-0.0025
0.8000	0.8002	55.40	-0.0033	-0.0033	-0.0033	-0.0026	-0.0017
0.8010	0.8010	55.40	-0.0033	-0.0033	-0.0033	-0.0026	-0.0017
0.8500	0.8360	55.40	0.0028	0.0028	0.0021	0.0027	0.0038
0.8700	0.8530	55.50	0.0037	0.0037	0.0027	0.0034	0.0044
0.9000	0.8850	55.60	0.0002	0.0003	-0.0010	-0.0003	0.0005
0.9500	0.9410	55.80	-0.0028	-0.0027	-0.0041	-0.0035	-0.0031
0.9800	0.9770	56.00	-0.0030	-0.0029	-0.0037	-0.0034	-0.0033
		MEAN	0.0024	0.0024	0.0021	0.0022	0.0019

ACETONE(1)-ISOPROPYL ALCOHOL(2)

PARKS G.S.,CHAFFEE C.S.:J.PHYS.CHEM.31,439(1927).

ANTOINE VAPOR PRESSURE CONSTANTS
          A   B   C
 (1) 7.23967   1279.870   237.500
 (2) 7.75634   1366.142   197.970

T =   25.00

EQUATION	ORDER	A12	A21	D	A122	A211
VAN LAAR	3	1.6725	0.2758	-	-	-
MARGULES	3	0.6055	0.0547	-	-	-
MARGULES	4	0.5092	-0.0429	-0.2722	-	-
ALPHA	2	7.4672	0.3071	-	-	-
ALPHA	3	4.1180	0.4905	-	4.6384	-0.7120

			DEVIATION IN VAPOR PHASE COMPOSITION				
X	Y	PRESS.	LAAR 3	MARG 3	MARG 4	ALPHA2	ALPHA3
0.1750	0.5990	100.00	0.0640	0.0736	0.0675	-0.0086	0.0020
0.3390	0.7350	139.60	-0.0265	0.0058	0.0093	-0.0012	-0.0039
0.5140	0.7980	167.20	-0.0051	0.0058	0.0053	0.0107	0.0049
0.6690	0.8550	190.00	0.0120	0.0182	0.0161	-0.0016	-0.0032
0.8390	0.9100	221.60	0.0299	0.0401	0.0419	-0.0088	0.0008
		MEAN	0.0275	0.0287	0.0280	0.0062	0.0029

ACETONE(1)-WATER(2)

BEARE W.G.,MCVICAR G.A.,FERGUSSON J.B.:J.PHYS.CHEM.34,1310(1930).

ANTOINE VAPOR PRESSURE CONSTANTS

	A	B	C
(1)	7.23967	1279.870	237.500
(2)	7.96681	1668.210	228.000

T = 25.00

EQUATION	ORDER	A12	A21	D	A122	A211
VAN LAAR	3	0.8660	0.7458	-	-	-
MARGULES	3	0.8617	0.7424	-	-	-
MARGULES	4	0.9049	0.8976	0.3491	-	-
ALPHA	2	76.1503	5.2692	-	-	-
ALPHA	3	-845.3442	-7.7263	-	915.6645	-60.6552

			DEVIATION IN VAPOR PHASE COMPOSITION				
X	Y	PRESS.	LAAR 3	MARG 3	MARG 4	ALPHA2	ALPHA3
0.0194	0.5234	50.10	0.0420	0.0401	0.0587	0.0526	0.0334
0.0289	0.6212	61.80	0.0308	0.0293	0.0440	0.0382	0.0200
0.0449	0.7168	81.30	0.0167	0.0157	0.0252	0.0202	0.0022
0.0556	0.7591	91.90	0.0075	0.0067	0.0137	0.0094	-0.0109
0.0939	0.8351	126.10	-0.0054	-0.0056	-0.0040	-0.0060	0.0049
0.0951	0.8416	126.60	-0.0107	-0.0109	-0.0094	-0.0113	-0.0015
0.1310	0.8618	144.30	-0.0042	-0.0041	-0.0051	-0.0048	-0.0040
0.1470	0.8768	150.60	-0.0115	-0.0114	-0.0129	-0.0118	-0.0117
0.1791	0.8782	159.80	-0.0018	-0.0016	-0.0038	-0.0012	-0.0015
0.2654	0.8856	176.10	0.0056	0.0060	0.0042	0.0100	0.0096
0.3538	0.8954	184.40	0.0017	0.0020	0.0026	0.0102	0.0099
0.5808	0.9158	199.10	-0.0114	-0.0115	-0.0043	0.0025	0.0018
0.7852	0.9421	213.50	-0.0162	-0.0164	-0.0146	-0.0170	-0.0196
		MEAN	0.0127	0.0124	0.0156	0.0150	0.0101

ACETONE(1)-WATER(2)

OTHMER D.F.,CHUDGAR M.M.,LEVY S.L.:IND.ENG.CHEM. 44,1872(1952).

ANTOINE VAPOR PRESSURE CONSTANTS
```
         A          B          C
(1) 7.23967    1279.870    237.500
(2) 7.96681    1668.210    228.000
```

P =   760.00

EQUATION	ORDER	A12	A21	D	A122	A211
VAN LAAR	3	0.9095	0.6161	-	-	-
MARGULES	3	0.8699	0.5908	-	-	-
MARGULES	4	0.9642	0.7029	0.3411	-	-
ALPHA	2	32.5221	4.2850	-	-	-
ALPHA	3	3.2008	8.3204	-	46.9552	-9.4355

			DEVIATION IN VAPOR PHASE COMPOSITION				
X	Y	TEMP.	LAAR 3	MARG 3	MARG 4	ALPHA2	ALPHA3
0.0100	0.3350	87.80	-0.0913	-0.1062	-0.0693	-0.0917	-0.0160
0.0230	0.4620	83.00	-0.0396	-0.0570	-0.0142	-0.0494	0.0307
0.0410	0.5850	76.50	-0.0226	-0.0371	-0.0013	-0.0457	0.0189
0.1200	0.7560	66.20	-0.0062	-0.0092	-0.0004	-0.0287	-0.0097
0.2640	0.8020	61.80	0.0055	0.0079	0.0034	0.0056	0.0007
0.3000	0.8090	61.10	0.0038	0.0063	0.0015	0.0077	0.0005
0.4440	0.8320	60.00	-0.0059	-0.0048	-0.0058	0.0080	-0.0023
0.5060	0.8370	59.70	-0.0046	-0.0044	-0.0027	0.0096	0.0003
0.5380	0.8400	59.50	-0.0035	-0.0038	-0.0008	0.0095	0.0013
0.6090	0.8470	58.90	0.0009	-0.0001	0.0048	0.0085	0.0042
0.6610	0.8600	58.50	-0.0015	-0.0028	0.0025	-0.0005	-0.0001
0.7930	0.9000	57.40	-0.0036	-0.0040	-0.0029	-0.0293	-0.0057
0.8500	0.9170	57.10	0.0015	0.0020	-0.0004	-0.0393	0.0027
		MEAN	0.0147	0.0189	0.0085	0.0257	0.0072

ACETONE(1)-WATER(2)

YORK R.,HOLMES R.C.: IND.ENG.CHEM. 34,345(1942).

ANTOINE VAPOR PRESSURE CONSTANTS
```
            A          B          C
(1)  7.23967    1279.870    237.500
(2)  7.96681    1668.210    228.000
```

P = 760.00

EQUATION	ORDER	A12	A21	D	A122	A211
VAN LAAR	3	0.8569	0.6949	-	-	-
MARGULES	3	0.8484	0.6867	-	-	-
MARGULES	4	0.8806	0.7450	0.1722	-	-
ALPHA	2	34.0904	4.1338	-	-	-
ALPHA	3	2.4809	4.9334	-	37.5587	-5.4350

			DEVIATION IN VAPOR PHASE COMPOSITION				
X	Y	TEMP.	LAAR 3	MARG 3	MARG 4	ALPHA2	ALPHA3
0.0030	0.0440	100.00	0.0289	0.0276	0.0326	0.0502	0.0639
0.0110	0.2040	100.00	0.0145	0.0116	0.0229	0.0657	0.0939
0.0250	0.4220	84.70	-0.0081	-0.0117	0.0017	0.0211	0.0505
0.0540	0.6150	75.00	-0.0105	-0.0128	-0.0042	-0.0069	0.0128
0.0820	0.6780	75.10	-0.0008	-0.0020	0.0025	0.0050	0.0172
0.1040	0.7300	68.30	-0.0021	-0.0027	-0.0006	-0.0108	-0.0029
0.1560	0.7910	64.60	-0.0114	-0.0111	-0.0122	-0.0211	-0.0199
0.1940	0.7950	64.00	0.0026	0.0032	0.0009	-0.0032	-0.0050
0.3300	0.8250	63.80	-0.0022	-0.0015	-0.0036	0.0074	0.0002
0.4450	0.8330	62.40	0.0002	0.0004	0.0010	0.0161	0.0077
0.6220	0.8550	63.30	-0.0082	-0.0088	-0.0046	0.0098	0.0044
0.7950	0.8960	60.40	-0.0079	-0.0082	-0.0068	-0.0175	-0.0065
0.9410	0.9590	60.00	-0.0028	-0.0024	-0.0056	-0.0514	0.0019
		MEAN	0.0077	0.0080	0.0076	0.0220	0.0221

ALLYL ALCOHOL(1)-WATER(2)

GRABNER R.W.,CLUMP C.W.:J.CHEM.ENG.DATA 10(1),13(1965.).

ANTOINE VAPOR PRESSURE CONSTANTS
```
        A          B         C
(1) 9.49445    2489.345   282.468
(2) 7.96681    1668.210   228.000
```

P =   760.00

EQUATION	ORDER	A12	A21	D	A122	A211
VAN LAAR	3	1.0031	0.4570	-	-	-
MARGULES	3	0.8836	0.3664	-	-	-
MARGULES	4	0.8910	0.3736	0.0286	-	-
ALPHA	2	4.3635	4.1092	-	-	-
ALPHA	3	6.5534	22.9740	-	7.3821	-21.9859

			DEVIATION IN VAPOR PHASE COMPOSITION				
X	Y	TEMP.	LAAR 3	MARG 3	MARG 4	ALPHA2	ALPHA3
0.0025	0.0309	99.18	-0.0006	-0.0075	-0.0071	-0.0178	0.0032
0.0049	0.0554	98.68	0.0011	-0.0112	-0.0105	-0.0303	0.0062
0.0113	0.0994	97.54	0.0156	-0.0063	-0.0050	-0.0445	0.0185
0.0193	0.1446	96.14	0.0263	-0.0013	0.0004	-0.0568	0.0215
0.0267	0.1780	95.18	0.0329	0.0035	0.0054	-0.0632	0.0199
0.0397	0.2260	94.04	0.0370	0.0089	0.0109	-0.0703	0.0109
0.0556	0.2634	93.06	0.0430	0.0200	0.0219	-0.0663	0.0051
0.0622	0.2793	92.48	0.0406	0.0201	0.0218	-0.0673	-0.0008
0.1058	0.3456	90.58	0.0287	0.0243	0.0251	-0.0577	-0.0239
0.1680	0.3658	89.96	0.0397	0.0498	0.0496	-0.0097	-0.0106
0.4216	0.4336	89.14	0.0245	0.0261	0.0255	0.0509	0.0048
0.5517	0.4750	89.06	0.0301	0.0188	0.0192	0.0519	0.0136
0.6921	0.5747	90.04	0.0134	-0.0005	0.0005	0.0034	-0.0072
0.7658	0.6338	90.88	0.0163	0.0084	0.0093	-0.0193	-0.0060
0.8340	0.7058	92.18	0.0175	0.0191	0.0193	-0.0440	-0.0030
0.8813	0.7690	93.16	0.0163	0.0247	0.0243	-0.0598	0.0011
0.9402	0.8696	94.96	0.0099	0.0220	0.0212	-0.0666	0.0061
0.9824	0.9596	96.58	0.0019	0.0080	0.0076	-0.0369	0.0041
		MEAN	0.0219	0.0156	0.0158	0.0454	0.0093

233

P

METHYL ACETATE(1)-BENZENE(2)

NAGATA I.:J.CHEM.ENG.DATA 7(3),360(1962).

ANTOINE VAPOR PRESSURE CONSTANTS

	A	B	C
(1)	7.06524	1157.622	219.724
(2)	6.90565	1211.033	220.790

P = 760.00

EQUATION	ORDER	A12	A21	D	A122	A211
VAN LAAR	3	0.2144	0.0675	-	-	-
MARGULES	3	0.1592	0.0249	-	-	-
MARGULES	4	0.1295	-0.0149	-0.1146	-	-
ALPHA	2	1.4993	-0.2628	-	-	-
ALPHA	3	1.0273	-0.1683	-	0.6835	-0.2280

			DEVIATION IN VAPOR PHASE COMPOSITION				
X	Y	TEMP.	LAAR 3	MARG 3	MARG 4	ALPHA2	ALPHA3
0.0550	0.1330	76.90	0.0153	0.0088	0.0041	-0.0081	-0.0013
0.0760	0.1750	75.90	0.0158	0.0106	0.0061	-0.0082	-0.0006
0.1390	0.2820	73.50	0.0126	0.0123	0.0105	-0.0046	0.0022
0.1890	0.3560	71.80	0.0059	0.0081	0.0084	-0.0039	0.0005
0.2120	0.3870	71.00	0.0031	0.0058	0.0069	-0.0037	-0.0005
0.3380	0.5280	67.50	-0.0030	-0.0022	-0.0001	-0.0005	-0.0037
0.3620	0.5450	66.80	0.0030	0.0030	0.0048	0.0059	0.0018
0.5050	0.6650	63.80	0.0079	0.0043	0.0029	0.0071	0.0012
0.6200	0.7490	61.80	0.0124	0.0090	0.0062	0.0047	0.0014
0.7350	0.8320	60.10	0.0095	0.0087	0.0070	-0.0043	-0.0026
0.8630	0.9140	58.40	0.0082	0.0103	0.0115	-0.0064	-0.0001
0.8950	0.9330	57.90	0.0081	0.0105	0.0120	-0.0049	0.0015
		MEAN	0.0087	0.0078	0.0067	0.0052	0.0015

METHYL ACETATE(1)-CHLOROFORM(2)

NAGATA I.:J.CHEM.ENG.DATA 7(3),360(1962).

ANTOINE VAPOR PRESSURE CONSTANTS

	A	B	C
(1)	7.06524	1157.622	219.724
(2)	6.93708	1171.200	227.000

P = 760.00

EQUATION	ORDER	A12	A21	D	A122	A211
VAN LAAR	3	-0.3242	-0.2593	-	-	-
MARGULES	3	-0.3192	-0.2542	-	-	-
MARGULES	4	-0.3137	-0.2487	0.0160	-	-
ALPHA	2	-0.3531	-0.6047	-	-	-
ALPHA	3	-0.3368	-0.9582	-	-0.1432	0.5058

			DEVIATION IN VAPOR PHASE COMPOSITION				
X	Y	TEMP.	LAAR 3	MARG 3	MARG 4	ALPHA2	ALPHA3
0.0640	0.0400	62.20	0.0012	0.0014	0.0017	0.0055	-0.0009
0.1590	0.1170	63.50	0.0059	0.0061	0.0063	0.0112	0.0021
0.1710	0.1300	63.70	0.0049	0.0051	0.0052	0.0100	0.0011
0.2240	0.1910	64.20	0.0012	0.0012	0.0012	0.0041	-0.0025
0.2630	0.2360	64.60	0.0023	0.0022	0.0020	0.0029	-0.0010
0.3350	0.3270	64.70	0.0028	0.0026	0.0024	-0.0011	0.0009
0.4060	0.4250	64.70	0.0000	-0.0001	-0.0003	-0.0078	-0.0005
0.4630	0.5020	64.20	-0.0001	-0.0001	-0.0002	-0.0095	0.0002
0.5320	0.5920	63.70	0.0004	0.0006	0.0007	-0.0088	0.0008
0.5630	0.6310	63.20	0.0003	0.0005	0.0007	-0.0081	0.0004
0.6400	0.7190	62.40	0.0027	0.0030	0.0033	-0.0022	0.0013
0.7060	0.7910	61.40	-0.0002	-0.0000	0.0001	-0.0013	-0.0034
0.7820	0.8540	60.30	0.0059	0.0059	0.0059	0.0087	0.0009
0.8510	0.9070	59.20	0.0061	0.0059	0.0057	0.0107	0.0005
0.9200	0.9530	58.10	0.0048	0.0045	0.0043	0.0088	0.0004
		MEAN	0.0026	0.0026	0.0027	0.0067	0.0011

235

METHYL ACETATE(1)-CYCLOHEXANE(2)

NAGATA I.:J.CHEM.ENG.DATA 7(4),461(1962).

ANTOINE VAPOR PRESSURE CONSTANTS
	A	B	C
(1)	7.06524	1157.622	219.724
(2)	6.84498	1203.526	222.863

P =   760.00

EQUATION	ORDER	A12	A21	D	A122	A211
VAN LAAR	3	0.5746	0.4802	-	-	-
MARGULES	3	0.5694	0.4747	-	-	-
MARGULES	4	0.5645	0.4692	-0.0161	-	-
ALPHA	2	5.3454	1.4720	-	-	-
ALPHA	3	2.9881	1.7110	-	3.2758	-0.9775

			DEVIATION IN VAPOR PHASE COMPOSITION				
X	Y	TEMP.	LAAR 3	MARG 3	MARG 4	ALPHA2	ALPHA3
0.0330	0.1820	74.30	0.0173	0.0158	0.0145	-0.0148	0.0015
0.0850	0.3500	68.40	0.0169	0.0158	0.0147	-0.0228	-0.0042
0.1420	0.4430	64.90	0.0233	0.0229	0.0225	-0.0097	0.0040
0.2830	0.5750	59.70	0.0097	0.0101	0.0104	-0.0012	0.0004
0.3130	0.5940	59.00	0.0062	0.0065	0.0068	-0.0010	-0.0013
0.3730	0.6250	57.90	0.0015	0.0018	0.0020	0.0006	-0.0026
0.4780	0.6640	56.80	0.0015	0.0014	0.0014	0.0068	0.0007
0.5070	0.6730	56.70	0.0028	0.0026	0.0025	0.0086	0.0024
0.6160	0.7140	56.00	0.0025	0.0022	0.0018	0.0057	0.0009
0.6880	0.7440	55.80	0.0038	0.0034	0.0031	0.0011	-0.0007
0.7220	0.7590	55.70	0.0055	0.0052	0.0050	-0.0011	-0.0008
0.7810	0.7890	55.50	0.0086	0.0085	0.0084	-0.0063	-0.0012
0.8350	0.8200	55.55	0.0136	0.0138	0.0140	-0.0097	0.0009
0.9400	0.9140	55.80	0.0126	0.0131	0.0136	-0.0173	0.0005
		MEAN	0.0090	0.0088	0.0086	0.0076	0.0016

PROPYL ALCOHOL(1)-BENZENE(2)

BROWN I.,SMITH F.:AUSTR.J.CHEM.12,407(1959).

ANTOINE VAPOR PRESSURE CONSTANTS
	A	B	C
(1)	7.85418	1497.910	204.112
(2)	6.90565	1211.033	220.790

T =   45.00

EQUATION	ORDER	A12	A21	D	A122	A211
VAN LAAR	3	0.8102	0.5222	-	-	-
MARGULES	3	0.7621	0.4990	-	-	-
MARGULES	4	0.8637	0.5969	0.3346	-	-
ALPHA	2	1.7175	11.6469	-	-	-
ALPHA	3	1.4091	13.5290	-	0.7408	-2.9618

			DEVIATION IN VAPOR PHASE COMPOSITION				
X	Y	PRESS.	LAAR 3	MARG 3	MARG 4	ALPHA2	ALPHA3
0.0470	0.0843	235.26	-0.0136	-0.0178	-0.0080	-0.0068	-0.0007
0.0977	0.1194	238.35	-0.0054	-0.0085	-0.0007	-0.0050	0.0001
0.2144	0.1560	237.55	0.0045	0.0061	0.0038	-0.0011	0.0008
0.2973	0.1728	234.89	0.0039	0.0069	0.0013	0.0002	0.0003
0.4061	0.1935	229.51	-0.0003	0.0023	-0.0022	0.0007	-0.0007
0.4807	0.2080	224.13	-0.0023	-0.0008	-0.0021	0.0017	-0.0003
0.5252	0.2185	220.42	-0.0037	-0.0031	-0.0021	0.0016	-0.0006
0.6053	0.2402	211.04	-0.0043	-0.0051	-0.0001	0.0020	0.0001
0.7033	0.2790	194.66	-0.0031	-0.0048	0.0024	0.0012	0.0008
0.7982	0.3437	170.06	-0.0001	-0.0009	0.0017	-0.0028	0.0004
0.9140	0.5252	122.08	0.0075	0.0126	-0.0093	-0.0136	-0.0015
		MEAN	0.0044	0.0063	0.0031	0.0033	0.0006

PROPYL ALCOHOL(1)-ETHYLBENZENE(2)

MALYUSOV V.A.,MALAFEEV N.A.,ZHAVORONKOV N.M.:ZH.FIZ.KHIM.
31,699(1957).

ANTOINE VAPOR PRESSURE CONSTANTS
	A	B	C
(1)	7.85418	1497.910	204.112
(2)	6.95719	1424.255	213.206

P =   50.00

EQUATION	ORDER	A12	A21	D	A122	A211
VAN LAAR	3	0.8383	0.4396	-	-	-
MARGULES	3	0.7589	0.3554	-	-	-
MARGULES	4	0.6098	0.2098	-0.4751	-	-
ALPHA	2	7.3384	2.5879	-	-	-
ALPHA	3	-0.0361	2.8436	-	9.2841	-2.7181

			DEVIATION IN VAPOR PHASE COMPOSITION				
X	Y	TEMP.	LAAR 3	MARG 3	MARG 4	ALPHA2	ALPHA3
0.0350	0.1940	52.00	0.1449	0.1183	0.0673	0.0178	0.0473
0.0770	0.3550	47.60	0.1129	0.0965	0.0578	-0.0040	0.0260
0.1010	0.4180	46.00	0.0867	0.0761	0.0476	-0.0144	0.0124
0.1490	0.5010	43.40	0.0463	0.0438	0.0328	-0.0231	-0.0040
0.1640	0.5350	42.60	0.0207	0.0201	0.0133	-0.0394	-0.0228
0.1720	0.5270	42.60	0.0334	0.0335	0.0288	-0.0227	-0.0074
0.2240	0.5720	41.40	0.0098	0.0130	0.0185	-0.0218	-0.0140
0.2590	0.5760	40.60	0.0164	0.0200	0.0295	-0.0020	0.0012
0.3140	0.6040	40.00	0.0036	0.0062	0.0181	-0.0002	-0.0031
0.3630	0.6130	39.60	0.0075	0.0082	0.0192	0.0121	0.0046
0.4480	0.6380	39.00	0.0068	0.0034	0.0086	0.0170	0.0034
0.4980	0.6430	38.60	0.0181	0.0126	0.0132	0.0270	0.0112
0.5820	0.6650	38.40	0.0289	0.0216	0.0152	0.0282	0.0117
0.6200	0.6930	38.40	0.0182	0.0111	0.0025	0.0106	-0.0046
0.6570	0.7000	38.20	0.0293	0.0230	0.0135	0.0139	0.0015
0.7160	0.7380	38.20	0.0241	0.0205	0.0122	-0.0064	-0.0106
0.7450	0.7450	38.00	0.0347	0.0328	0.0264	-0.0037	-0.0018
0.8250	0.8030	38.20	0.0326	0.0365	0.0385	-0.0285	-0.0019
0.8710	0.8500	38.40	0.0228	0.0294	0.0364	-0.0484	-0.0041
0.9390	0.9180	38.80	0.0172	0.0243	0.0336	-0.0514	0.0078
0.9620	0.9510	39.00	0.0076	0.0131	0.0204	-0.0483	0.0035
		MEAN	0.0344	0.0316	0.0264	0.0210	0.0098

PROPYL ALCOHOL(1)-ETHYLBENZENE(2)

MALYUSOV V.A.,MALAFEEV N.A.,ZHAVORONKOV N.M.:ZH.FIZ.KHIM.
        31,699(1957).

ANTOINE VAPOR PRESSURE CONSTANTS
            A   B   C
    (1) 7.85418   1497.910   204.112
    (2) 6.95719   1424.255   213.206

P = 200.00

EQUATION	ORDER	A12	A21	D	A122	A211
VAN LAAR	3	0.8141	0.4884	-	-	-
MARGULES	3	0.7799	0.4316	-	-	-
MARGULES	4	0.6350	0.2982	-0.4855	-	-
ALPHA	2	5.8763	1.3983	-	-	-
ALPHA	3	0.0368	1.4436	-	7.8565	-1.3741

			DEVIATION IN VAPOR PHASE COMPOSITION				
X	Y	TEMP.	LAAR 3	MARG 3	MARG 4	ALPHA2	ALPHA3
0.0260	0.1120	86.20	0.2047	0.1932	0.1407	0.0356	0.0674
0.0640	0.3240	79.60	0.1572	0.1508	0.1084	-0.0343	0.0085
0.1120	0.4580	74.00	0.1061	0.1056	0.0855	-0.0539	-0.0146
0.2880	0.6220	68.80	0.0322	0.0351	0.0488	-0.0228	-0.0132
0.3900	0.6430	67.60	0.0347	0.0337	0.0450	0.0118	0.0077
0.4900	0.6800	66.60	0.0218	0.0170	0.0194	0.0150	0.0011
0.6220	0.7080	65.60	0.0362	0.0295	0.0208	0.0312	0.0121
0.7150	0.7620	65.00	0.0228	0.0183	0.0087	0.0084	-0.0053
0.7620	0.7890	64.80	0.0208	0.0184	0.0116	-0.0010	-0.0076
0.8120	0.8130	64.60	0.0271	0.0274	0.0255	-0.0032	0.0012
0.8380	0.8380	64.50	0.0195	0.0212	0.0223	-0.0150	-0.0036
0.8800	0.8720	64.60	0.0166	0.0203	0.0258	-0.0234	-0.0003
0.9200	0.9070	64.80	0.0148	0.0192	0.0273	-0.0261	0.0053
0.9450	0.9380	65.00	0.0064	0.0106	0.0184	-0.0306	0.0014
0.9750	0.9710	65.60	0.0027	0.0054	0.0103	-0.0212	0.0017
		MEAN	0.0482	0.0470	0.0412	0.0222	0.0101

PROPYL ALCOHOL(1)-STYRENE(2)

MALYUSOV V.A.,MALAFEEV N.A.,ZHAVORONKOV N.M.:ZH.FIZ.KHIM.
          31,699(1957).

ANTOINE VAPOR PRESSURE CONSTANTS
          A   B   C
(1) 7.85418   1497.910   204.112
(2) 6.92409   1420.000   206.000

P =   50.00

EQUATION	ORDER	A12	A21	D	A122	A211
VAN LAAR	3	0.8212	0.5159	-	-	-
MARGULES	3	0.7738	0.4749	-	-	-
MARGULES	4	0.7176	0.4168	-0.1833	-	-
ALPHA	2	12.6417	2.5093	-	-	-
ALPHA	3	0.6087	2.7627	-	15.2691	-2.8598

DEVIATION IN VAPOR PHASE COMPOSITION

X	Y	TEMP.	LAAR 3	MARG 3	MARG 4	ALPHA2	ALPHA3
0.0410	0.3450	56.20	0.1207	0.1033	0.0814	-0.0078	0.0311
0.0870	0.5360	50.00	0.0615	0.0525	0.0397	-0.0408	-0.0084
0.1310	0.5940	46.81	0.0568	0.0533	0.0475	-0.0178	0.0059
0.1530	0.6200	45.80	0.0473	0.0457	0.0423	-0.0155	0.0043
0.1990	0.6660	44.00	0.0244	0.0255	0.0259	-0.0177	-0.0048
0.2300	0.6850	43.00	0.0160	0.0180	0.0200	-0.0147	-0.0058
0.2700	0.6970	42.60	0.0150	0.0176	0.0207	-0.0042	0.0004
0.2780	0.7050	42.40	0.0088	0.0114	0.0147	-0.0083	-0.0045
0.3640	0.7250	41.20	0.0057	0.0076	0.0108	0.0050	0.0019
0.3680	0.7270	41.20	0.0045	0.0062	0.0094	0.0043	0.0008
0.4090	0.7450	40.80	-0.0060	-0.0051	-0.0027	-0.0019	-0.0078
0.4510	0.7400	40.60	0.0071	0.0071	0.0084	0.0137	0.0056
0.4680	0.7540	40.60	-0.0034	-0.0037	-0.0030	0.0036	-0.0053
0.5080	0.7520	40.40	0.0073	0.0061	0.0057	0.0142	0.0039
0.5140	0.7640	40.40	-0.0033	-0.0046	-0.0052	0.0034	-0.0070
0.5670	0.7540	40.00	0.0198	0.0177	0.0158	0.0237	0.0124
0.6160	0.7670	40.00	0.0210	0.0186	0.0158	0.0197	0.0087
0.6250	0.7770	39.80	0.0137	0.0113	0.0084	0.0114	0.0005
0.6780	0.7880	39.60	0.0208	0.0187	0.0156	0.0101	0.0017
0.7480	0.8200	39.40	0.0173	0.0164	0.0145	-0.0080	-0.0088
0.8000	0.8330	39.00	0.0292	0.0297	0.0294	-0.0087	0.0009
0.8340	0.8610	38.80	0.0196	0.0210	0.0221	-0.0267	-0.0077
0.8840	0.8930	38.60	0.0181	0.0205	0.0233	-0.0392	-0.0034
0.9000	0.9000	38.50	0.0217	0.0244	0.0275	-0.0379	0.0032
0.9420	0.9380	38.80	0.0140	0.0167	0.0199	-0.0447	0.0051
0.9800	0.9780	39.00	0.0046	0.0059	0.0076	-0.0313	0.0031
		MEAN	0.0226	0.0219	0.0207	0.0167	0.0059

PROPYL ALCOHOL(1)-WATER(2)

DOROSHEVSKY A.,POLANSKY E.:Z.PHYS.CHEM.73,192(1910).

ANTOINE VAPOR PRESSURE CONSTANTS
	A	B	C
(1)	7.85418	1497.910	204.112
(2)	7.96681	1668.210	228.000

P = 760.00

EQUATION	ORDER	A12	A21	D	A122	A211
VAN LAAR	3	1.1204	0.4927	-	-	-
MARGULES	3	0.9480	0.4039	-	-	-
MARGULES	4	1.0948	0.5676	0.5549	-	-
ALPHA	2	4.7854	5.2102	-	-	-
ALPHA	3	18.1466	72.1386	-	24.1692	-71.6730

			DEVIATION IN VAPOR PHASE COMPOSITION				
X	Y	TEMP.	LAAR 3	MARG 3	MARG 4	ALPHA2	ALPHA3
0.0100	0.1100	95.00	0.0062	-0.0252	0.0019	-0.0578	0.0909
0.0200	0.2160	92.00	-0.0247	-0.0677	-0.0291	-0.1208	0.0460
0.0400	0.3200	90.50	-0.0391	-0.0832	-0.0404	-0.1583	-0.0098
0.0600	0.3510	89.30	-0.0216	-0.0570	-0.0194	-0.1400	-0.0194
0.1000	0.3720	88.50	0.0034	-0.0117	0.0100	-0.0926	-0.0184
0.2000	0.3920	88.10	0.0099	0.0243	0.0166	-0.0205	-0.0105
0.3000	0.4040	87.90	0.0038	0.0219	0.0057	0.0171	-0.0006
0.4000	0.4240	87.80	-0.0036	0.0062	-0.0042	0.0316	0.0034
0.4320	0.4320	87.80	-0.0051	0.0011	-0.0056	0.0332	0.0041
0.5000	0.4520	87.90	-0.0062	-0.0079	-0.0055	0.0328	0.0050
0.6000	0.4920	88.30	-0.0047	-0.0149	-0.0000	0.0224	0.0040
0.7000	0.5510	89.00	-0.0012	-0.0129	0.0061	-0.0009	0.0002
0.8000	0.6410	90.50	0.0011	-0.0019	0.0050	-0.0387	-0.0055
0.8500	0.7040	91.50	0.0001	0.0045	-0.0011	-0.0620	-0.0075
0.9000	0.7780	92.80	0.0027	0.0142	-0.0047	-0.0775	-0.0005
0.9600	0.9000	95.00	-0.0009	0.0113	-0.0114	-0.0735	0.0098
		MEAN	0.0084	0.0229	0.0104	0.0612	0.0147

PROPYL ALCOHOL(1)-WATER(2)

MURTI P.S.,VAN WINKLE M.:CHEM.ENG.DATA SERIES 3,72(1958).

ANTOINE VAPOR PRESSURE CONSTANTS

	A	B	C
(1)	7.85418	1497.910	204.112
(2)	7.96681	1668.210	228.000

T = 40.00

EQUATION	ORDER	A12	A21	D	A122	A211
VAN LAAR	3	1.0164	0.4928	-	-	-
MARGULES	3	0.8897	0.4154	-	-	-
MARGULES	4	0.9680	0.4936	0.2739	-	-
ALPHA	2	4.1759	5.1532	-	-	-
ALPHA	3	17.3307	96.0316	-	35.2253	-98.3424

				DEVIATION IN VAPOR PHASE COMPOSITION				
X	Y	PRESS.	LAAR 3	MARG 3	MARG 4	ALPHA2	ALPHA3
0.0805	0.3410	85.75	-0.0392	-0.0575	-0.0424	-0.1105	-0.0053
0.1295	0.3555	86.50	-0.0132	-0.0172	-0.0108	-0.0629	-0.0049
0.1525	0.3615	86.50	-0.0092	-0.0080	-0.0053	-0.0477	-0.0058
0.3050	0.3870	86.50	-0.0072	0.0044	-0.0037	0.0127	-0.0028
0.3980	0.3995	86.75	-0.0049	0.0014	-0.0046	0.0328	0.0041
0.4700	0.4225	86.50	-0.0110	-0.0105	-0.0124	0.0319	-0.0007
0.5755	0.4540	86.00	-0.0049	-0.0119	-0.0071	0.0326	0.0029
0.6660	0.4995	83.50	-0.0020	-0.0112	-0.0029	0.0191	0.0009
0.7385	0.5405	81.50	0.0109	0.0042	0.0117	0.0109	0.0103
0.8440	0.6625	72.50	0.0041	0.0087	0.0066	-0.0370	0.0103
0.8500	0.7260	69.50	-0.0510	-0.0456	-0.0486	-0.0946	-0.0434
0.8975	0.7535	69.00	-0.0029	0.0086	-0.0007	-0.0638	0.0221
		MEAN	0.0134	0.0158	0.0131	0.0464	0.0095

PROPYL ALCOHOL(1)-WATER(2)

MURTI P.S.,VAN WINKLE M.:CHEM.ENG.DATA SERIES 3,72(1958).

ANTOINE VAPOR PRESSURE CONSTANTS

	A	B	C
(1)	7.85418	1497.910	204.112
(2)	7.96681	1668.210	228.000

T = 60.00

EQUATION	ORDER	A12	A21	D	A122	A211
VAN LAAR	3	0.9992	0.4986	-	-	-
MARGULES	3	0.8592	0.4464	-	-	-
MARGULES	4	1.0495	0.6336	0.6507	-	-
ALPHA	2	4.1879	4.7192	-	-	-
ALPHA	3	50.7797	169.0829	-	39.7807	-166.1106

			DEVIATION IN VAPOR PHASE COMPOSITION				
X	Y	PRESS.	LAAR 3	MARG 3	MARG 4	ALPHA2	ALPHA3
0.0390	0.2800	202.50	-0.0528	-0.0871	-0.0360	-0.1331	0.0436
0.0650	0.3575	223.00	-0.0668	-0.0959	-0.0508	-0.1502	-0.0186
0.1545	0.3670	228.50	0.0034	0.0014	0.0085	-0.0427	-0.0035
0.1790	0.3735	230.50	0.0050	0.0081	0.0073	-0.0294	-0.0048
0.1960	0.3750	228.50	0.0078	0.0137	0.0084	-0.0189	-0.0027
0.2620	0.3920	229.00	0.0022	0.0148	-0.0018	0.0017	-0.0055
0.3000	0.3940	229.50	0.0055	0.0192	0.0003	0.0169	0.0011
0.4090	0.4120	231.00	0.0055	0.0161	0.0029	0.0388	0.0111
0.4260	0.4175	231.00	0.0037	0.0133	0.0021	0.0388	0.0106
0.4895	0.4455	230.00	-0.0077	-0.0024	-0.0045	0.0307	0.0030
0.5660	0.5030	229.00	-0.0379	-0.0377	-0.0282	-0.0028	-0.0250
0.7050	0.5530	216.00	-0.0098	-0.0140	0.0049	-0.0005	-0.0006
0.7350	0.5575	215.00	0.0091	0.0053	0.0223	0.0095	0.0163
0.7960	0.6210	204.50	0.0033	0.0021	0.0094	-0.0176	0.0050
0.8800	0.7460	184.50	-0.0121	-0.0065	-0.0239	-0.0646	-0.0194
0.8940	0.7600	178.50	-0.0032	0.0035	-0.0183	-0.0600	-0.0119
0.9250	0.7850	182.50	0.0287	0.0372	0.0076	-0.0337	0.0175
0.9500	0.8500	169.50	0.0170	0.0255	-0.0058	-0.0427	0.0051
		MEAN	0.0156	0.0224	0.0135	0.0407	0.0114

PROPYL ALCOHOL(1)-WATER(2)

MURTI P.S.,VAN WINKLE M.:CHEM.ENG.DATA SERIES 3,72(1958).

ANTOINE VAPOR PRESSURE CONSTANTS

	A	B	C
(1)	7.85418	1497.910	204.112
(2)	7.96681	1668.210	228.000

P =   760.00

EQUATION	ORDER	A12	A21	D	A122	A211
VAN LAAR	3	1.1525	0.4883	-	-	-
MARGULES	3	0.9663	0.3989	-	-	-
MARGULES	4	1.1111	0.5762	0.6129	-	-
ALPHA	2	6.1786	6.6495	-	-	-
ALPHA	3	46.5835	184.5992	-	57.2624	-185.5050

			DEVIATION IN VAPOR PHASE COMPOSITION				
X	Y	TEMP.	LAAR 3	MARG 3	MARG 4	ALPHA2	ALPHA3
0.0390	0.2810	92.35	0.0069	-0.0408	0.0006	-0.0983	0.0548
0.0720	0.3600	88.85	-0.0049	-0.0346	-0.0046	-0.0989	-0.0075
0.0750	0.3750	89.05	-0.0162	-0.0440	-0.0154	-0.1085	-0.0215
0.1790	0.3880	87.95	0.0116	0.0259	0.0180	-0.0112	-0.0114
0.2000	0.3790	88.00	0.0217	0.0397	0.0276	0.0104	0.0014
0.4250	0.4260	87.50	-0.0044	0.0037	-0.0037	0.0419	0.0031
0.4820	0.4380	87.80	-0.0013	-0.0007	0.0009	0.0437	0.0076
0.7120	0.5600	89.20	-0.0018	-0.0152	0.0087	-0.0149	-0.0076
0.8500	0.6850	91.70	0.0198	0.0226	0.0192	-0.0636	0.0026
0.9400	0.8550	95.00	0.0018	0.0152	-0.0106	-0.1074	0.0002
		MEAN	0.0090	0.0242	0.0109	0.0599	0.0118

ISOPROPYL ALCOHOL(1)-BENZENE(2)

BROWN I.,FOCK W.,SMITH F.:AUSTR.J.CHEM.9,364(1956).

ANTOINE VAPOR PRESSURE CONSTANTS

	A	B	C
(1)	7.75634	1366.142	197.970
(2)	6.90565	1211.033	220.790

T =   45.00

EQUATION	ORDER	A12	A21	D	A122	A211
VAN LAAR	3	0.7692	0.5500	-	-	-
MARGULES	3	0.7393	0.5381	-	-	-
MARGULES	4	0.8403	0.6335	0.3296	-	-
ALPHA	2	2.8635	6.7097	-	-	-
ALPHA	3	2.7848	10.9615	-	1.6603	-5.3212

			DEVIATION IN VAPOR PHASE COMPOSITION				
X	Y	PRESS.	LAAR 3	MARG 3	MARG 4	ALPHA2	ALPHA3
0.0472	0.1467	252.50	-0.0230	-0.0275	-0.0112	-0.0237	-0.0015
0.0980	0.2066	264.13	-0.0093	-0.0127	-0.0004	-0.0164	0.0005
0.2047	0.2663	272.06	0.0057	0.0067	0.0042	-0.0041	0.0002
0.2960	0.2953	273.40	0.0075	0.0103	0.0022	0.0028	0.0010
0.3862	0.3211	272.22	0.0028	0.0057	-0.0016	0.0047	-0.0002
0.4753	0.3463	269.49	-0.0025	-0.0005	-0.0030	0.0048	-0.0008
0.5504	0.3692	264.92	-0.0052	-0.0044	-0.0017	0.0045	-0.0004
0.6198	0.3951	259.35	-0.0069	-0.0072	-0.0003	0.0025	-0.0005
0.7096	0.4378	247.70	-0.0050	-0.0060	0.0026	-0.0006	0.0006
0.8073	0.5107	227.14	0.0017	0.0012	0.0028	-0.0074	0.0013
0.9120	0.6658	189.28	0.0114	0.0134	-0.0049	-0.0201	-0.0009
0.9655	0.8252	159.80	0.0113	0.0139	-0.0087	-0.0208	-0.0025
		MEAN	0.0077	0.0091	0.0036	0.0094	0.0009

ISO-PROPYL ALCOHOL(1)-CARBON TETRACHLORIDE(2)

YUAN K.S.,LU B.C.Y.:J.CHEM.ENG.DATA 8(4),549(1963).

ANTOINE VAPOR PRESSURE CONSTANTS

	A	B	C
(1)	7.75634	1366.142	197.970
(2)	6.93390	1242.430	230.000

P = 760.00

EQUATION	ORDER	A12	A21	D	A122	A211
VAN LAAR	3	0.5841	0.5672	-	-	-
MARGULES	3	0.5837	0.5673	-	-	-
MARGULES	4	0.6369	0.6128	0.1530	-	-
ALPHA	2	2.5341	4.1340	-	-	-
ALPHA	3	2.1963	6.1672	-	1.3642	-2.7375

			DEVIATION IN VAPOR PHASE COMPOSITION				
X	Y	TEMP.	LAAR 3	MARG 3	MARG 4	ALPHA2	ALPHA3
0.0110	0.0600	74.79	-0.0288	-0.0289	-0.0253	-0.0240	-0.0152
0.0540	0.1580	71.40	-0.0345	-0.0346	-0.0262	-0.0212	-0.0018
0.1460	0.2650	69.59	-0.0222	-0.0222	-0.0184	-0.0128	-0.0007
0.1530	0.2580	69.28	-0.0092	-0.0092	-0.0060	0.0002	0.0115
0.1760	0.2890	69.28	-0.0201	-0.0201	-0.0184	-0.0127	-0.0039
0.2110	0.3050	69.08	-0.0104	-0.0104	-0.0108	-0.0051	0.0003
0.2110	0.2920	69.00	0.0024	0.0024	0.0020	0.0079	0.0133
0.2420	0.3250	68.88	-0.0112	-0.0112	-0.0131	-0.0071	-0.0043
0.2580	0.3320	68.90	-0.0091	-0.0090	-0.0116	-0.0056	-0.0040
0.2590	0.3300	68.91	-0.0065	-0.0065	-0.0090	-0.0031	-0.0016
0.3070	0.3710	68.75	-0.0240	-0.0240	-0.0277	-0.0212	-0.0228
0.3190	0.3510	68.76	0.0013	0.0014	-0.0025	0.0041	0.0018
0.3390	0.3590	68.56	0.0012	0.0013	-0.0027	0.0046	0.0012
0.4130	0.3810	68.31	0.0061	0.0062	0.0029	0.0116	0.0056
0.4880	0.4110	68.86	0.0027	0.0027	0.0013	0.0095	0.0023
0.5060	0.4120	68.46	0.0064	0.0065	0.0056	0.0153	0.0080
0.5380	0.4420	69.43	-0.0101	-0.0100	-0.0099	-0.0025	-0.0097
0.5780	0.4160	69.15	0.0291	0.0291	0.0304	0.0394	0.0328
0.6230	0.4690	69.70	-0.0050	-0.0050	-0.0027	0.0057	0.0002
0.6570	0.5000	69.54	-0.0219	-0.0218	-0.0191	-0.0093	-0.0136
0.6830	0.5030	70.12	-0.0107	-0.0107	-0.0078	0.0010	-0.0021
0.7060	0.5170	70.51	-0.0114	-0.0114	-0.0086	-0.0002	-0.0021
0.7110	0.5270	70.63	-0.0183	-0.0182	-0.0155	-0.0073	-0.0089
0.7200	0.5370	70.87	-0.0223	-0.0223	-0.0197	-0.0119	-0.0129
0.7640	0.5620	71.62	-0.0162	-0.0162	-0.0146	-0.0076	-0.0055
0.7750	0.5670	71.78	-0.0124	-0.0124	-0.0113	-0.0043	-0.0014
0.8100	0.6010	72.54	-0.0141	-0.0142	-0.0148	-0.0088	-0.0028
0.8420	0.6550	73.71	-0.0313	-0.0313	-0.0341	-0.0302	-0.0212
0.8600	0.6040	74.27	0.0434	0.0434	0.0392	0.0423	0.0530
0.8620	0.6900	74.90	-0.0383	-0.0383	-0.0427	-0.0412	-0.0303
0.8680	0.6040	74.48	0.0548	0.0547	0.0499	0.0527	0.0641
0.8960	0.7170	75.71	-0.0122	-0.0122	-0.0194	-0.0187	-0.0048
0.9420	0.8620	79.15	-0.0558	-0.0559	-0.0655	-0.0701	-0.0544
		MEAN	0.0183	0.0183	0.0178	0.0157	0.0127

ISOPROPYL ALCOHOL(1)-ETHYL ACETATE(2)

MURTI P.S.,VAN WINKLE M.:CHEM.ENG.DATA SERIES 3,72(1958).

ANTOINE VAPOR PRESSURE CONSTANTS
	A	B	C
(1)	7.75634	1366.142	197.970
(2)	7.10233	1245.239	217.911

T = 40.00

EQUATION	ORDER	A12	A21	D	A122	A211
VAN LAAR	3	0.3585	0.3718	-	-	-
MARGULES	3	0.3584	0.3717	-	-	-
MARGULES	4	0.3586	0.3718	0.0006	-	-
ALPHA	2	0.7063	2.9425	-	-	-
ALPHA	3	0.6151	1.7685	-	-0.2402	1.4010

			DEVIATION IN VAPOR PHASE COMPOSITION				
X	Y	PRESS.	LAAR 3	MARG 3	MARG 4	ALPHA2	ALPHA3
0.0915	0.1014	192.00	-0.0027	-0.0027	-0.0027	0.0139	0.0032
0.1860	0.1850	194.00	-0.0092	-0.0092	-0.0092	0.0037	-0.0030
0.3595	0.2880	191.00	-0.0053	-0.0053	-0.0053	-0.0043	-0.0022
0.4600	0.3335	183.50	0.0025	0.0025	0.0025	-0.0002	0.0040
0.5385	0.3775	178.00	0.0004	0.0004	0.0004	-0.0030	0.0013
0.6150	0.4220	174.00	0.0001	0.0001	0.0001	-0.0023	0.0010
0.7100	0.4960	164.00	-0.0084	-0.0084	-0.0084	-0.0075	-0.0069
0.7300	0.5050	161.50	-0.0012	-0.0012	-0.0012	0.0006	0.0005
0.8010	0.5770	152.50	-0.0049	-0.0049	-0.0049	0.0006	-0.0023
0.8500	0.6270	148.50	0.0065	0.0065	0.0065	0.0146	0.0096
0.8845	0.6935	137.00	-0.0061	-0.0061	-0.0061	0.0034	-0.0027
0.9300	0.7825	126.00	-0.0037	-0.0036	-0.0037	0.0061	-0.0004
0.9500	0.8340	120.50	-0.0046	-0.0046	-0.0046	0.0043	-0.0017
0.9620	0.8653	118.00	-0.0015	-0.0015	-0.0015	0.0063	0.0010
		MEAN	0.0041	0.0041	0.0041	0.0051	0.0028

247

ISOPROPYL ALCOHOL(1)-ETHYL ACETATE(2)

MURTI P.S.,VAN WINKLE M.:CHEM.ENG.DATA SERIES 3,72(1958).

ANTOINE VAPOR PRESSURE CONSTANTS
```
          A         B          C
(1) 7.75634   1366.142   197.970
(2) 7.10232   1245.239   217.911
```

T =   60.00

EQUATION	ORDER	A12	A21	D	A122	A211
VAN LAAR	3	0.2889	0.3194	-	-	-
MARGULES	3	0.2878	0.3187	-	-	-
MARGULES	4	0.2812	0.3137	-0.0201	-	-
ALPHA	2	0.6411	1.7761	-	-	-
ALPHA	3	0.6157	1.3275	-	-0.1255	0.5599

DEVIATION IN VAPOR PHASE COMPOSITION

X	Y	PRESS.	LAAR 3	MARG 3	MARG 4	ALPHA2	ALPHA3
0.0775	0.1068	425.00	-0.0137	-0.0138	-0.0144	-0.0016	-0.0069
0.0805	0.1000	432.00	-0.0037	-0.0038	-0.0044	0.0085	0.0032
0.1650	0.1795	430.50	-0.0016	-0.0016	-0.0017	0.0105	0.0064
0.2475	0.2555	434.00	-0.0107	-0.0106	-0.0102	-0.0025	-0.0040
0.3200	0.3023	432.00	-0.0056	-0.0054	-0.0048	-0.0012	-0.0006
0.4095	0.3578	430.50	-0.0023	-0.0022	-0.0016	-0.0015	0.0007
0.5085	0.4168	420.00	0.0012	0.0012	0.0014	0.0001	0.0029
0.5680	0.4587	413.00	-0.0024	-0.0024	-0.0025	-0.0034	-0.0009
0.5725	0.4631	411.50	-0.0038	-0.0038	-0.0040	-0.0048	-0.0023
0.6400	0.5119	404.50	-0.0062	-0.0062	-0.0066	-0.0060	-0.0043
0.6865	0.5387	394.00	0.0020	0.0020	0.0015	0.0036	0.0044
0.7335	0.5787	385.00	0.0011	0.0011	0.0006	0.0045	0.0042
0.8245	0.6795	358.50	-0.0072	-0.0072	-0.0073	0.0001	-0.0025
0.8410	0.7020	355.00	-0.0095	-0.0095	-0.0095	-0.0016	-0.0046
0.8705	0.7372	343.00	-0.0050	-0.0050	-0.0047	0.0037	0.0002
0.9065	0.7824	330.50	0.0060	0.0061	0.0067	0.0151	0.0112
0.9145	0.8100	329.50	-0.0077	-0.0076	-0.0070	0.0013	-0.0025
0.9260	0.8415	321.50	-0.0181	-0.0180	-0.0173	-0.0094	-0.0131
0.9545	0.8746	312.50	0.0073	0.0074	0.0082	0.0144	0.0113
		MEAN	0.0061	0.0061	0.0060	0.0049	0.0045

ISOPROPYL ALCOHOL(1)-ETHYL ACETATE(2)

MURTI P.S.,VAN WINKLE M.:CHEM.ENG.DATA SERIES 3,72(1958).

ANTOINE VAPOR PRESSURE CONSTANTS

	A	B	C
(1)	7.75634	1366.142	197.970
(2)	7.10232	1245.239	217.911

P = 760.00

EQUATION	ORDER	A12	A21	D	A122	A211
VAN LAAR	3	0.2230	0.2059	-	-	-
MARGULES	3	0.2222	0.2059	-	-	-
MARGULES	4	0.2577	0.2351	0.1043	-	-
ALPHA	2	0.3151	0.9263	-	-	-
ALPHA	3	2.1481	3.1514	-	-1.7408	-2.0124

			DEVIATION IN VAPOR PHASE COMPOSITION				
X	Y	TEMP.	LAAR 3	MARG 3	MARG 4	ALPHA2	ALPHA3
0.0960	0.1114	76.85	0.0029	0.0028	0.0059	-0.0000	0.0002
0.3850	0.3538	75.92	0.0041	0.0042	0.0015	0.0014	-0.0017
0.5390	0.4662	76.40	0.0067	0.0068	0.0068	0.0056	0.0064
0.5985	0.5240	76.85	-0.0038	-0.0038	-0.0026	-0.0047	-0.0027
0.6555	0.5750	77.25	-0.0070	-0.0070	-0.0050	-0.0074	-0.0050
0.7710	0.6750	78.70	0.0039	0.0039	0.0050	0.0030	0.0036
0.8295	0.7410	79.38	0.0034	0.0034	0.0028	0.0025	0.0007
0.8815	0.8068	80.30	0.0038	0.0038	0.0014	0.0027	-0.0016
		MEAN	0.0045	0.0045	0.0039	0.0034	0.0027

Q

ISO-PROPYL ALCOHOL(1)-METHYLCYCLOHEXANE(2)

NAGATA I.:J.CHEM.ENG.DATA 10,106(1965).

ANTOINE VAPOR PRESSURE CONSTANTS

	A	B	C
(1)	7.75634	1366.142	197.970
(2)	6.82689	1272.864	221.630

P = 500.00

EQUATION	ORDER	A12	A21	D	A122	A211
VAN LAAR	3	0.8758	0.6457	-	-	-
MARGULES	3	0.8410	0.6397	-	-	-
MARGULES	4	1.0101	0.7924	0.5428	-	-
ALPHA	2	8.3341	5.1065	-	16.0382	-
ALPHA	3	9.4219	21.9649	-	16.0382	-19.1297

			DEVIATION IN VAPOR PHASE COMPOSITION				
X	Y	TEMP.	LAAR 3	MARG 3	MARG 4	ALPHA2	ALPHA3
0.0300	0.3440	75.40	-0.0981	-0.1093	-0.0521	-0.1481	-0.0218
0.0770	0.4340	71.00	-0.0354	-0.0442	-0.0021	-0.0917	-0.0035
0.1800	0.4910	64.80	0.0101	0.0097	0.0105	-0.0184	0.0060
0.2500	0.5110	68.00	0.0254	0.0282	0.0154	0.0039	0.0072
0.3310	0.5400	67.00	0.0102	0.0146	-0.0023	0.0074	-0.0029
0.4270	0.5580	66.80	0.0041	0.0084	-0.0024	0.0170	-0.0002
0.5090	0.5810	66.70	-0.0074	-0.0042	-0.0054	0.0136	-0.0043
0.5810	0.6000	66.60	-0.0121	-0.0102	-0.0031	0.0109	-0.0042
0.6680	0.6230	66.50	-0.0086	-0.0082	0.0049	0.0091	0.0014
0.7350	0.6500	66.50	-0.0045	-0.0048	0.0069	0.0018	0.0031
0.7880	0.6820	67.00	-0.0016	-0.0021	0.0038	-0.0102	0.0005
0.8430	0.7190	67.50	0.0098	0.0094	0.0048	-0.0187	0.0037
0.8870	0.7680	68.10	0.0121	0.0122	-0.0028	-0.0342	-0.0017
0.9350	0.8350	69.20	0.0194	0.0200	-0.0036	-0.0416	-0.0011
0.9650	0.8990	70.30	0.0149	0.0155	-0.0062	-0.0417	-0.0044
		MEAN	0.0182	0.0201	0.0084	0.0312	0.0044

ISOPROPYL ALCOHOL(1)-PROPYL ALCOHOL(2)

BALLARD L.H.,VAN WINKLE M.:IND.ENG.CHEM.44,2450(1952).

ANTOINE VAPOR PRESSURE CONSTANTS
	A	B	C
(1)	7.75634	1366.142	197.970
(2)	7.85418	1497.910	204.112

P =   760.00

EQUATION	ORDER	A12	A21	D	A122	A211
VAN LAAR	3	0.0031	-0.0003	-	-	-
MARGULES	3	0.0415	-0.0051	-	-	-
MARGULES	4	0.0354	-0.0131	-0.0219	-	-
ALPHA	2	0.8067	-0.4086	-	-	-
ALPHA	3	17.7119	9.6557	-	-16.5593	-10.2035

			DEVIATION IN VAPOR PHASE COMPOSITION				
X	Y	TEMP.	LAAR 3	MARG 3	MARG 4	ALPHA2	ALPHA3
0.0575	0.1100	96.10	-0.0051	-0.0063	-0.0071	-0.0109	-0.0031
0.0610	0.1110	95.90	0.0036	-0.0014	-0.0022	-0.0062	0.0016
0.1455	0.2325	94.20	0.0007	0.0067	0.0063	0.0017	0.0068
0.2285	0.3510	92.80	-0.0061	-0.0022	-0.0021	-0.0046	-0.0035
0.3095	0.4435	91.40	0.0003	0.0002	0.0005	0.0006	-0.0009
0.3125	0.4500	91.40	-0.0027	-0.0030	-0.0026	-0.0025	-0.0040
0.4200	0.5545	90.00	0.0092	0.0046	0.0046	0.0075	0.0047
0.4355	0.5725	89.70	0.0069	0.0018	0.0018	0.0048	0.0020
0.5190	0.6600	88.50	-0.0012	-0.0078	-0.0081	-0.0047	-0.0071
0.6310	0.7480	87.00	0.0062	0.0001	-0.0005	0.0014	0.0008
0.7305	0.8225	85.80	0.0073	0.0031	0.0028	0.0022	0.0036
0.7675	0.8495	85.30	0.0065	0.0031	0.0030	0.0015	0.0036
0.8585	0.9175	84.10	-0.0012	-0.0024	-0.0022	-0.0051	-0.0016
0.9100	0.9525	83.40	-0.0044	-0.0048	-0.0044	-0.0073	-0.0035
		MEAN	0.0044	0.0034	0.0034	0.0044	0.0033

251

ISOPROPYL ALCOHOL(1)-TOLUENE(2)

KIREEV V.A.,SHEINKER YU.N.,PERESLENI E.M.:ZH.FIZ.KHIM.
26,352(1952).

ANTOINE VAPOR PRESSURE CONSTANTS
	A	B	C
(1)	7.75634	1366.142	197.970
(2)	6.95334	1343.943	219.377

P = 760.00

EQUATION	ORDER	A12	A21	D	A122	A211
VAN LAAR	3	0.5680	0.4939	-	-	-
MARGULES	3	0.5661	0.4892	-	-	-
MARGULES	4	0.5303	0.4472	-0.1261	-	-
ALPHA	2	6.7202	1.5774	-	-	-
ALPHA	3	2.0105	1.7760	-	6.0677	-1.4378

DEVIATION IN VAPOR PHASE COMPOSITION

X	Y	TEMP.	LAAR 3	MARG 3	MARG 4	ALPHA2	ALPHA3
0.0300	0.1790	104.60	0.0431	0.0425	0.0320	0.0027	0.0239
0.0780	0.3770	96.60	0.0174	0.0170	0.0084	-0.0255	-0.0020
0.1030	0.4340	94.40	0.0144	0.0141	0.0079	-0.0242	-0.0035
0.1490	0.5020	91.00	0.0143	0.0143	0.0119	-0.0142	0.0003
0.2030	0.5620	88.50	0.0065	0.0066	0.0073	-0.0112	-0.0037
0.2560	0.5980	86.60	0.0059	0.0060	0.0082	-0.0028	-0.0011
0.3120	0.6220	85.40	0.0103	0.0103	0.0129	0.0089	0.0056
0.3970	0.6630	84.00	0.0025	0.0023	0.0039	0.0086	0.0000
0.4690	0.6900	83.20	-0.0006	-0.0009	-0.0009	0.0087	-0.0025
0.5740	0.7210	82.20	0.0025	0.0020	-0.0002	0.0110	-0.0005
0.6760	0.7550	81.80	0.0067	0.0063	0.0035	0.0074	0.0007
0.7040	0.7650	81.50	0.0086	0.0083	0.0057	0.0061	0.0019
0.7420	0.7810	81.50	0.0106	0.0104	0.0085	0.0027	0.0027
0.7800	0.8070	81.40	0.0046	0.0046	0.0037	-0.0094	-0.0041
0.8150	0.8240	81.20	0.0082	0.0083	0.0086	-0.0118	-0.0006
0.8580	0.8540	81.20	0.0074	0.0077	0.0095	-0.0202	-0.0009
0.9330	0.9180	81.60	0.0068	0.0072	0.0107	-0.0270	0.0021
		MEAN	0.0100	0.0099	0.0084	0.0119	0.0033

252

ISO-PROPYL ALCOHOL(1)-WATER(2)

BRUNJES A.S.,BOGART M.J.P.:IND.ENG.CHEM.35,255(1943).

ANTOINE VAPOR PRESSURE CONSTANTS

	A	B	C
(1)	7.75634	1366.142	197.970
(2)	7.96681	1668.210	228.000

P = 760.00

EQUATION	ORDER	A12	A21	D	A122	A211
VAN LAAR	3	1.0302	0.4987	-	-	-
MARGULES	3	0.8755	0.4425	-	-	-
MARGULES	4	1.0379	0.5628	0.4678	-	-
ALPHA	2	9.2055	3.9823	-	-	-
ALPHA	3	13.6161	33.3698	-	27.8313	-33.6596

DEVIATION IN VAPOR PHASE COMPOSITION

X	Y	TEMP.	LAAR 3	MARG 3	MARG 4	ALPHA2	ALPHA3
0.0016	0.0364	98.87	-0.0044	-0.0136	-0.0038	-0.0204	0.0241
0.0083	0.1473	95.30	-0.0080	-0.0426	-0.0052	-0.0715	0.0684
0.0136	0.2244	93.19	-0.0220	-0.0671	-0.0178	-0.1079	0.0589
0.0204	0.2308	90.80	0.0338	-0.0177	0.0395	-0.0691	0.1082
0.0254	0.3399	89.04	-0.0390	-0.0923	-0.0324	-0.1490	0.0276
0.0488	0.4660	83.80	-0.0575	-0.1048	-0.0482	-0.1707	-0.0253
0.0843	0.5024	82.63	-0.0197	-0.0486	-0.0092	-0.1085	-0.0157
0.1232	0.5378	81.41	-0.0191	-0.0310	-0.0093	-0.0769	-0.0259
0.1629	0.5298	81.39	0.0067	0.0070	0.0145	-0.0230	-0.0015
0.1986	0.5444	81.19	0.0008	0.0085	0.0065	-0.0075	-0.0049
0.2387	0.5559	81.11	-0.0043	0.0084	-0.0012	0.0069	-0.0057
0.3314	0.5654	80.77	-0.0023	0.0129	-0.0035	0.0392	0.0066
0.4597	0.5939	80.44	-0.0063	0.0016	-0.0082	0.0484	0.0092
0.5838	0.6358	80.14	-0.0063	-0.0072	-0.0047	0.0351	0.0056
0.6496	0.6659	80.04	-0.0050	-0.0086	-0.0015	0.0200	0.0019
0.6813	0.6813	80.16	-0.0027	-0.0067	0.0013	0.0124	0.0015
0.6838	0.6846	80.05	-0.0045	-0.0086	-0.0005	0.0097	-0.0006
0.6857	0.6824	80.04	-0.0012	-0.0053	0.0029	0.0124	0.0026
0.6905	0.6879	80.03	-0.0038	-0.0079	0.0003	0.0082	-0.0005
0.6971	0.6931	80.05	-0.0049	-0.0090	-0.0008	0.0047	-0.0022
0.7333	0.7142	80.07	-0.0023	-0.0059	0.0016	-0.0063	-0.0027
0.7702	0.7401	80.14	-0.0011	-0.0034	0.0018	-0.0204	-0.0042
0.8100	0.7698	80.30	0.0021	0.0019	0.0032	-0.0348	-0.0028
0.8520	0.8126	80.51	-0.0014	0.0010	-0.0030	-0.0567	-0.0058
0.8872	0.8500	80.75	-0.0016	0.0028	-0.0056	-0.0704	-0.0032
0.9153	0.8801	81.01	0.0010	0.0065	-0.0045	-0.0746	0.0037
0.9319	0.9011	81.21	0.0008	0.0065	-0.0053	-0.0758	0.0065
		MEAN	0.0097	0.0199	0.0088	0.0496	0.0158

ISOPROPANOL(1)-WATER(2)

DUNLOP J.G.:M.S.THESIS,BROOKLYN POLYTECHN.INST.,1948.

ANTOINE VAPOR PRESSURE CONSTANTS

	A	B	C
(1)	7.75634	1366.142	197.970
(2)	7.96681	1668.210	228.000

P =   760.00

EQUATION	ORDER	A12	A21	D	A122	A211
VAN LAAR	3	1.0318	0.4982	-	-	-
MARGULES	3	0.8759	0.4415	-	-	-
MARGULES	4	1.0388	0.5621	0.4689	-	-
ALPHA	2	9.1947	3.9777	-	-	-
ALPHA	3	13.7611	33.6889	-	27.9989	-33.9909

DEVIATION IN VAPOR PHASE COMPOSITION

X	Y	TEMP.	LAAR 3	MARG 3	MARG 4	ALPHA2	ALPHA3
0.0020	0.0360	98.90	0.0037	-0.0078	0.0043	-0.0162	0.0381
0.0080	0.1470	95.30	-0.0113	-0.0455	-0.0088	-0.0737	0.0647
0.0140	0.2240	93.20	-0.0169	-0.0630	-0.0129	-0.1048	0.0643
0.0200	0.2310	90.80	0.0310	-0.0208	0.0364	-0.0719	0.1061
0.0250	0.3400	89.00	-0.0412	-0.0948	-0.0349	-0.1514	0.0262
0.0490	0.4660	83.80	-0.0564	-0.1039	-0.0472	-0.1701	-0.0245
0.0840	0.5020	82.60	-0.0194	-0.0487	-0.0090	-0.1089	-0.0154
0.1230	0.5380	81.40	-0.0193	-0.0314	-0.0096	-0.0775	-0.0262
0.1630	0.5300	81.40	0.0065	0.0068	0.0143	-0.0232	-0.0017
0.1990	0.5440	81.20	0.0012	0.0090	0.0069	-0.0070	-0.0045
0.2390	0.5560	81.10	-0.0046	0.0082	-0.0014	0.0069	-0.0058
0.3310	0.5650	80.80	-0.0022	0.0131	-0.0034	0.0393	0.0067
0.4600	0.5940	80.40	-0.0065	0.0013	-0.0085	0.0483	0.0091
0.5840	0.6360	80.10	-0.0065	-0.0075	-0.0049	0.0349	0.0054
0.6500	0.6660	80.00	-0.0050	-0.0086	-0.0015	0.0200	0.0019
0.6810	0.6810	80.20	-0.0026	-0.0067	0.0014	0.0126	0.0016
0.6840	0.6850	80.10	-0.0048	-0.0089	-0.0008	0.0094	-0.0009
0.6860	0.6820	80.00	-0.0006	-0.0047	0.0034	0.0129	0.0031
0.6910	0.6880	80.00	-0.0036	-0.0077	0.0004	0.0082	-0.0003
0.6970	0.6930	80.10	-0.0049	-0.0090	-0.0008	0.0048	-0.0022
0.7330	0.7140	80.10	-0.0023	-0.0059	0.0016	-0.0062	-0.0027
0.7700	0.7400	80.10	-0.0011	-0.0034	0.0018	-0.0204	-0.0042
0.8100	0.7700	80.30	0.0019	0.0018	0.0031	-0.0350	-0.0029
0.8520	0.8130	80.50	-0.0017	0.0008	-0.0032	-0.0571	-0.0061
0.8870	0.8500	80.80	-0.0018	0.0028	-0.0057	-0.0705	-0.0033
0.9150	0.8800	81.00	0.0008	0.0064	-0.0047	-0.0748	0.0035
0.9320	0.9010	81.20	0.0011	0.0069	-0.0049	-0.0755	0.0069
		MEAN	0.0096	0.0198	0.0087	0.0497	0.0162

254

ISO-PROPYL ALCOHOL(1)-WATER(2)

LEBO R.B.:J.AM.CHEM.SOC.43,1005(1921).

ANTOINE VAPOR PRESSURE CONSTANTS
	A	B	C
(1)	7.75634	1366.142	197.970
(2)	7.96681	1668.210	228.000

P = 760.00

EQUATION	ORDER	A12	A21	D	A122	A211
VAN LAAR	3	1.0123	0.5663	-	-	-
MARGULES	3	0.9099	0.5451	-	-	-
MARGULES	4	1.1425	0.8360	0.9346	-	-
ALPHA	2	13.3668	5.7678	-	-	-
ALPHA	3	27.8048	45.6702	-	30.6309	-43.1211

			DEVIATION IN VAPOR PHASE COMPOSITION				
X	Y	TEMP.	LAAR 3	MARG 3	MARG 4	ALPHA2	ALPHA3
0.0100	0.1900	95.00	-0.0324	-0.0585	0.0075	-0.0703	0.0994
0.0200	0.3400	90.00	-0.0813	-0.1159	-0.0308	-0.1349	0.0430
0.0300	0.4300	86.70	-0.1016	-0.1376	-0.0502	-0.1609	-0.0003
0.0600	0.5050	83.50	-0.0589	-0.0864	-0.0191	-0.1136	-0.0139
0.1500	0.5600	81.50	-0.0082	-0.0084	-0.0029	-0.0209	-0.0147
0.3000	0.5800	81.00	0.0006	0.0148	-0.0102	0.0391	0.0042
0.5000	0.6300	80.70	-0.0231	-0.0169	-0.0090	0.0342	0.0034
0.7000	0.7000	80.50	-0.0194	-0.0234	0.0061	-0.0011	0.0045
0.8000	0.7700	81.00	-0.0220	-0.0257	-0.0165	-0.0465	-0.0119
0.9000	0.8300	82.30	0.0181	0.0183	-0.0098	-0.0574	0.0068
		MEAN	0.0366	0.0506	0.0162	0.0679	0.0202

ISO-PROPYL ALCOHOL(1)-WATER(2)

WILSON A.,SIMONS E.L.:IND.ENG.CHEM.44,2214(1952).

ANTOINE VAPOR PRESSURE CONSTANTS
```
          A           B          C
(1) 7.75634    1366.142    197.970
(2) 7.96681    1668.210    228.000
```

P =   95.00

EQUATION	ORDER	A12	A21	D	A122	A211
VAN LAAR	3	0.9581	0.5235	-	-	-
MARGULES	3	0.8540	0.4790	-	-	-
MARGULES	4	0.9798	0.5830	0.3667	-	-
ALPHA	2	9.0253	4.3185	-	-	-
ALPHA	3	8.7767	22.2248	-	19.3332	-21.8498

			DEVIATION IN VAPOR PHASE COMPOSITION				
X	Y	TEMP.	LAAR 3	MARG 3	MARG 4	ALPHA2	ALPHA3
0.0055	0.0600	49.17	0.0240	0.0081	0.0279	-0.0089	0.0645
0.0140	0.1655	47.10	0.0144	-0.0142	0.0219	-0.0485	0.0706
0.0345	0.3105	43.44	0.0089	-0.0258	0.0197	-0.0785	0.0503
0.0510	0.4055	41.19	-0.0215	-0.0528	-0.0102	-0.1091	0.0035
0.0795	0.4820	39.01	-0.0316	-0.0537	-0.0210	-0.1076	-0.0265
0.1395	0.5165	37.85	-0.0029	-0.0083	0.0046	-0.0465	-0.0143
0.1850	0.5285	37.59	0.0061	0.0088	0.0110	-0.0153	-0.0065
0.2610	0.5475	37.14	0.0049	0.0142	0.0060	0.0126	-0.0021
0.3875	0.5705	36.87	0.0019	0.0105	0.0005	0.0365	0.0061
0.5080	0.6030	36.14	-0.0039	-0.0013	-0.0037	0.0345	0.0051
0.5725	0.6250	36.17	-0.0048	-0.0053	-0.0030	0.0270	0.0034
0.6495	0.6565	36.23	-0.0032	-0.0060	0.0000	0.0132	0.0014
0.6505	0.6565	36.38	-0.0026	-0.0055	0.0006	0.0135	0.0018
0.6580	0.6605	36.21	-0.0030	-0.0059	0.0003	0.0113	0.0011
0.6860	0.6740	36.01	-0.0016	-0.0047	0.0018	0.0049	0.0007
0.7350	0.7040	36.07	-0.0015	-0.0041	0.0013	-0.0113	-0.0026
0.7385	0.7055	35.78	-0.0007	-0.0033	0.0020	-0.0118	-0.0020
0.7400	0.7070	36.23	-0.0010	-0.0035	0.0017	-0.0128	-0.0026
0.7715	0.7275	36.33	0.0012	-0.0004	0.0029	-0.0228	-0.0025
0.8195	0.7665	36.39	0.0020	0.0025	0.0013	-0.0420	-0.0038
0.8840	0.8300	37.04	0.0040	0.0077	-0.0007	-0.0660	-0.0006
0.9425	0.9040	37.65	0.0044	0.0093	-8.0021	-0.0732	0.0062
		MEAN	0.0068	0.0116	0.0066	0.0367	0.0126

ISO-PROPYL ALCOHOL(1)-WATER(2)

WILSON A.,SIMONS E.L.:IND.ENG.CHEM.44,2214(1952).

ANTOINE VAPOR PRESSURE CONSTANTS

	A	B	C
(1)	7.75634	1366.142	197.970
(2)	7.96681	1668.210	228.000

P = 190.00

EQUATION	ORDER	A12	A21	D	A122	A211
VAN LAAR	3	0.9755	0.5133	-	-	-
MARGULES	3	0.8628	0.4571	-	-	-
MARGULES	4	0.9728	0.5549	0.3385	-	-
ALPHA	2	8.6679	4.0164	-	-	-
ALPHA	3	8.6564	22.0053	-	19.4587	-21.8039

DEVIATION IN VAPOR PHASE COMPOSITION

X	Y	TEMP.	LAAR 3	MARG 3	MARG 4	ALPHA2	ALPHA3
0.0075	0.0985	62.86	0.0165	-0.0058	0.0162	-0.0327	0.0589
0.0175	0.1915	60.41	0.0280	-0.0066	0.0285	-0.0548	0.0762
0.0300	0.2940	57.66	0.0119	-0.0261	0.0135	-0.0879	0.0491
0.0485	0.4045	54.70	-0.0188	-0.0532	-0.0157	-0.1216	-0.0001
0.0840	0.4840	51.99	-0.0181	-0.0397	-0.0132	-0.1040	-0.0218
0.1500	0.5210	51.12	0.0038	0.0009	0.0087	-0.0414	-0.0116
0.1625	0.5255	50.81	0.0049	0.0044	0.0096	-0.0332	-0.0104
0.2115	0.5385	50.47	0.0070	0.0134	0.0104	-0.0066	-0.0053
0.2725	0.5510	50.41	0.0057	0.0157	0.0072	0.0155	-0.0003
0.3860	0.5725	49.97	0.0017	0.0098	0.0011	0.0366	0.0063
0.4765	0.5955	49.57	-0.0018	0.0012	-0.0022	0.0382	0.0066
0.5710	0.6270	49.32	-0.0033	-0.0053	-0.0021	0.0290	0.0040
0.5890	0.6340	49.34	-0.0032	-0.0059	-0.0017	0.0262	0.0033
0.6645	0.6670	49.33	-0.0009	-0.0051	0.0018	0.0114	0.0012
0.6860	0.6790	49.35	-0.0010	-0.0052	0.0018	0.0051	-0.0005
0.7075	0.6910	49.23	-0.0003	-0.0041	0.0026	-0.0010	-0.0013
0.7530	0.7200	49.39	0.0011	-0.0014	0.0035	-0.0159	-0.0031
0.7580	0.7235	49.40	0.0012	-0.0010	0.0035	-0.0177	-0.0033
0.7940	0.7500	49.55	0.0027	0.0023	0.0040	-0.0307	-0.0036
0.8545	0.8035	49.86	0.0044	0.0078	0.0034	-0.0537	-0.0019
0.8870	0.8390	50.00	0.0037	0.0088	0.0011	-0.0660	-0.0004
0.9520	0.9230	50.62	0.0022	0.0079	-0.0018	-0.0696	0.0068
		MEAN	0.0065	0.0105	0.0070	0.0408	0.0125

ISO-PROPYL ALCOHOL(1)-WATER(2)

WILSON A.,SIMONS E.L.:IND.ENG.CHEM.44,2214(1952).

ANTOINE VAPOR PRESSURE CONSTANTS

	A	B	C
(1)	7.75634	1366.142	197.970
(2)	7.96681	1668.210	228.000

P = 380.00

EQUATION	ORDER	A12	A21	D	A122	A211
VAN LAAR	3	1.0392	0.4984	-	-	-
MARGULES	3	0.8905	0.4292	-	-	-
MARGULES	4	1.0429	0.5698	0.4907	-	-
ALPHA	2	7.6457	3.3104	-	-	-
ALPHA	3	23.6237	51.4803	-	34.2380	-51.7543

			DEVIATION IN VAPOR PHASE COMPOSITION				
X	Y	TEMP.	LAAR 3	MARG 3	MARG 4	ALPHA2	ALPHA3
0.0065	0.0925	79.20	0.0235	-0.0054	0.0249	-0.0403	0.1299
0.0410	0.3905	70.12	-0.0074	-0.0554	-0.0004	-0.1517	0.0476
0.0605	0.4565	67.76	-0.0147	-0.0533	-0.0062	-0.1514	0.0105
0.0770	0.5100	66.25	-0.0371	-0.0672	-0.0281	-0.1612	-0.0279
0.1305	0.5255	65.59	-0.0030	-0.0106	0.0053	-0.0806	-0.0154
0.1765	0.5365	65.31	0.0026	0.0069	0.0088	-0.0398	-0.0117
0.2620	0.5465	65.02	0.0059	0.0196	0.0073	0.0116	-0.0002
0.2680	0.5490	64.98	0.0040	0.0179	0.0052	0.0124	-0.0013
0.3350	0.5625	64.61	-0.0015	0.0119	-0.0027	0.0299	0.0008
0.3915	0.5700	64.60	-0.0000	0.0102	-0.0019	0.0430	0.0072
0.4765	0.5960	64.18	-0.0067	-0.0031	-0.0072	0.0426	0.0044
0.5865	0.6335	63.95	-0.0055	-0.0093	-0.0022	0.0339	0.0043
0.6585	0.6685	63.90	-0.0052	-0.0111	-0.0000	0.0180	0.0005
0.6715	0.6740	63.93	-0.0034	-0.0093	0.0020	0.0161	0.0015
0.6930	0.6875	63.91	-0.0041	-0.0098	0.0014	0.0089	-0.0008
0.7450	0.7200	63.96	-0.0019	-0.0058	0.0029	-0.0068	-0.0023
0.7565	0.7280	63.96	-0.0014	-0.0047	0.0030	-0.0107	-0.0026
0.7980	0.7600	63.99	-0.0002	-0.0008	0.0022	-0.0259	-0.0037
0.8530	0.8100	64.24	0.0009	0.0045	-0.0008	-0.0468	-0.0035
0.8760	0.8330	64.51	0.0020	0.0073	-0.0015	-0.0538	-0.0016
0.8865	0.8435	64.56	0.0030	0.0090	-0.0013	-0.0559	0.0002
0.8935	0.8510	64.64	0.0035	0.0098	-0.0014	-0.0574	0.0012
0.9285	0.8935	64.90	0.0033	0.0107	-0.0034	-0.0621	0.0054
		MEAN	0.0061	0.0154	0.0052	0.0505	0.0124

ISO-PROPYL ALCOHOL(1)-WATER(2)

WILSON A.,SIMONS E.L.:IND.ENG.CHEM.44,2214(1952).

ANTOINE VAPOR PRESSURE CONSTANTS

	A	B	C
(1)	7.75634	1366.142	197.970
(2)	7.96681	1668.210	228.000

P = 760.00

EQUATION	ORDER	A12	A21	D	A122	A211
VAN LAAR	3	1.0117	0.4822	-	-	-
MARGULES	3	0.8667	0.3911	-	-	-
MARGULES	4	0.9993	0.5391	0.4425	-	-
ALPHA	2	8.9649	4.2959	-	-	-
ALPHA	3	15.0650	33.6826	-	24.3568	-33.8657

			DEVIATION IN VAPOR PHASE COMPOSITION				
X	Y	TEMP.	LAAR 3	MARG 3	MARG 4	ALPHA2	ALPHA3
0.0115	0.1630	95.17	0.0100	-0.0281	0.0075	-0.0644	0.0873
0.0160	0.2115	93.40	0.0075	-0.0365	0.0052	-0.0815	0.0836
0.0365	0.3655	88.05	-0.0133	-0.0614	-0.0131	-0.1256	0.0325
0.0570	0.4565	84.57	-0.0345	-0.0745	-0.0319	-0.1419	-0.0142
0.1000	0.5015	82.70	-0.0101	-0.0305	-0.0046	-0.0881	-0.0155
0.1215	0.5120	82.32	-0.0031	-0.0156	0.0027	-0.0655	-0.0133
0.1665	0.5215	81.99	0.0084	0.0078	0.0138	-0.0249	-0.0037
0.1895	0.5375	81.58	-0.0012	0.0023	0.0035	-0.0217	-0.0120
0.1935	0.5320	81.75	0.0053	0.0093	0.0098	-0.0132	-0.0053
0.2450	0.5390	81.62	0.0077	0.0166	0.0103	0.0122	0.0023
0.2835	0.5530	81.23	-0.0006	0.0093	0.0006	0.0168	-0.0018
0.2975	0.5540	81.29	0.0004	0.0103	0.0012	0.0217	0.0007
0.2980	0.5510	81.28	0.0035	0.0134	0.0042	0.0249	0.0039
0.3835	0.5700	80.90	-0.0009	0.0054	-0.0018	0.0353	0.0064
0.4460	0.5920	80.67	-0.0084	-0.0064	-0.0091	0.0303	0.0012
0.5145	0.6075	80.38	-0.0030	-0.0056	-0.0022	0.0311	0.0065
0.5590	0.6255	80.31	-0.0042	-0.0091	-0.0021	0.0231	0.0038
0.6460	0.6645	80.15	-0.0022	-0.0086	0.0023	0.0041	0.0005
0.6605	0.6715	80.16	-0.0012	-0.0074	0.0035	0.0006	0.0005
0.6955	0.6915	80.11	-0.0003	-0.0054	0.0047	-0.0103	-0.0013
0.7650	0.7370	80.23	0.0026	0.0019	0.0065	-0.0347	-0.0031
0.8090	0.7745	80.37	0.0012	0.0043	0.0032	-0.0548	-0.0061
0.8725	0.8340	80.70	0.0026	0.0111	0.0007	-0.0786	-0.0027
0.9535	0.9325	81.48	0.0004	0.0091	-0.0046	-0.0817	0.0067
		MEAN	0.0055	0.0163	0.0062	0.0453	0.0131

ETHYL ACETATE(1)-BENZENE(2)

CARR A.D.,KROPHOLLER H.W.:J.CHEM.ENG.DATA 7,26(1962).

ANTOINE VAPOR PRESSURE CONSTANTS

	A	B	C
(1)	7.10232	1245.239	217.911
(2)	6.90565	1211.033	220.790

P = 760.00

EQUATION	ORDER	A12	A21	D	A122	A211
VAN LAAR	3	0.0516	0.0371	-	-	-
MARGULES	3	0.0502	0.0355	-	-	-
MARGULES	4	0.0469	0.0319	-0.0115	-	-
ALPHA	2	0.2046	0.0124	-	-	-
ALPHA	3	-0.3966	-0.4803	-	0.6163	0.4661

			DEVIATION IN VAPOR PHASE COMPOSITION				
X	Y	TEMP.	LAAR 3	MARG 3	MARG 4	ALPHA2	ALPHA3
0.0160	0.0200	80.01	-0.0003	-0.0003	-0.0005	-0.0008	-0.0006
0.0270	0.0330	79.95	0.0001	-0.0000	-0.0002	-0.0008	-0.0004
0.0600	0.0710	79.77	0.0012	0.0011	0.0008	-0.0003	0.0004
0.1360	0.1570	79.40	0.0009	0.0009	0.0007	-0.0009	0.0002
0.1600	0.1840	79.28	-0.0001	-0.0001	-0.0002	-0.0018	-0.0007
0.2300	0.2560	78.96	0.0015	0.0016	0.0017	0.0004	0.0013
0.3000	0.3290	78.66	-0.0007	-0.0006	-0.0003	-0.0010	-0.0006
0.3590	0.3870	78.43	-0.0005	-0.0004	-0.0002	-0.0002	-0.0004
0.4220	0.4480	78.19	-0.0005	-0.0005	-0.0004	0.0002	-0.0005
0.4410	0.4650	78.12	0.0007	0.0007	0.0008	0.0015	0.0007
0.5280	0.5470	77.84	0.0014	0.0013	0.0012	0.0023	0.0011
0.5870	0.6050	77.67	-0.0009	-0.0010	-0.0013	-0.0002	-0.0014
0.6970	0.7070	77.49	0.0009	0.0008	0.0005	0.0008	0.0002
0.7740	0.7800	77.38	0.0010	0.0009	0.0008	0.0002	0.0004
0.8340	0.8380	77.32	0.0003	0.0003	0.0004	-0.0008	0.0000
0.8410	0.8450	77.31	0.0000	0.0001	0.0002	-0.0011	-0.0002
0.8570	0.8600	77.29	0.0004	0.0005	0.0006	-0.0008	0.0002
0.9120	0.9140	77.23	-0.0004	-0.0002	0.0000	-0.0015	-0.0003
0.9500	0.9510	77.19	-0.0003	-0.0002	0.0001	-0.0011	-0.0002
		MEAN	0.0006	0.0006	0.0006	0.0009	0.0005

ETHYL ACETATE(1)-PROPYL ALCOHOL(2)

MURTI P.S.,VAN WINKLE M.:CHEM.ENG. DATA SERIES 3,72(1958).

ANTOINE VAPOR PRESSURE CONSTANTS

	A	B	C
(1)	7.10232	1245.239	217.911
(2)	7.85418	1497.910	204.112

T =   40.00

EQUATION	ORDER	A12	A21	D	A122	A211
VAN LAAR	3	0.4174	0.2859	-	-	-
MARGULES	3	0.4006	0.2720	-	-	-
MARGULES	4	0.3742	0.2471	-0.0836	-	-
ALPHA	2	5.5945	0.1212	-	-	-
ALPHA	3	4.5809	1.1698	-	2.7992	-1.5167

			DEVIATION IN VAPOR PHASE COMPOSITION				
X	Y	PRESS.	LAAR 3	MARG 3	MARG 4	ALPHA2	ALPHA3
0.1010	0.4285	84.00	0.0193	0.0166	0.0117	-0.0276	0.0007
0.1500	0.5131	95.00	0.0183	0.0175	0.0157	-0.0137	0.0037
0.2012	0.5870	109.00	0.0040	0.0044	0.0048	-0.0135	-0.0067
0.3087	0.6656	127.50	0.0079	0.0090	0.0111	0.0113	0.0033
0.5126	0.7715	154.00	0.0039	0.0036	0.0036	0.0153	0.0019
0.6149	0.8198	161.50	-0.0003	-0.0010	-0.0020	0.0044	-0.0040
0.6481	0.8305	165.00	0.0032	0.0025	0.0014	0.0047	-0.0011
0.7223	0.8584	170.00	0.0076	0.0071	0.0062	0.0009	0.0016
0.7450	0.8700	171.00	0.0060	0.0057	0.0050	-0.0033	-0.0004
0.8201	0.9000	176.50	0.0102	0.0104	0.0105	-0.0073	0.0028
0.8524	0.9175	178.00	0.0079	0.0084	0.0088	-0.0122	0.0006
0.9143	0.9523	185.00	0.0033	0.0040	0.0049	-0.0180	-0.0028
		MEAN	0.0077	0.0075	0.0072	0.0110	0.0025

ETHYL ACETATE(1)-PROPYL ALCOHOL(2)

MURTI P.S.,VAN WINKLE M.:CHEM.ENG. DATA SERIES 3,72(1958).

ANTOINE VAPOR PRESSURE CONSTANTS

	A	B	C
(1)	7.10232	1245.239	217.911
(2)	7.85418	1497.910	204.112

T = 60.00

EQUATION	ORDER	A12	A21	D	A122	A211
VAN LAAR	3	0.2865	0.2889	-	-	-
MARGULES	3	0.2865	0.2889	-	-	-
MARGULES	4	0.2680	0.2613	-0.0784	-	-
ALPHA	2	3.8226	0.0521	-	-.	-
ALPHA	3	1.2166	-0.2401	-	2.9661	-0.0943

DEVIATION IN VAPOR PHASE COMPOSITION

X	Y	PRESS.	LAAR 3	MARG 3	MARG 4	ALPHA2	ALPHA3
0.0480	0.1950	179.00	0.0060	0.0060	0.0020	-0.0058	0.0032
0.0638	0.2425	289.00	0.0074	0.0074	0.0035	-0.0053	0.0045
0.0905	0.3100	205.00	0.0096	0.0096	0.0066	-0.0028	0.0070
0.1175	0.3765	222.00	0.0011	0.0010	-0.0008	-0.0099	-0.0010
0.1596	0.4595	239.50	-0.0094	-0.0094	-0.0096	-0.0171	-0.0104
0.2189	0.5232	262.00	0.0045	0.0045	0.0057	0.0016	0.0045
0.2770	0.5907	285.50	-0.0047	-0.0047	-0.0032	-0.0037	-0.0041
0.3625	0.6472	305.50	0.0051	0.0052	0.0061	0.0101	0.0058
0.4158	0.6891	324.50	-0.0027	-0.0027	-0.0025	0.0034	-0.0023
0.5284	0.7500	343.00	-0.0020	-0.0020	-0.0033	0.0035	-0.0028
0.6262	0.7878	362.50	0.0080	0.0081	0.0061	0.0098	0.0061
0.7023	0.8340	379.00	-0.0018	-0.0018	-0.0034	-0.0045	-0.0039
0.8140	0.8880	393.00	0.0000	0.0000	0.0000	-0.0102	-0.0012
0.9157	0.9440	402.00	0.0010	0.0010	0.0024	-0.0120	0.0017
		MEAN	0.0045	0.0045	0.0039	0.0071	0.0042

ETHYL ACETATE(1)-PROPYL ALCOHOL(2)

MURTI P.S.,VAN WINKLE M.:CHEM.ENG. DATA SERIES 3,72(1958).

ANTOINE VAPOR PRESSURE CONSTANTS

	A	B	C
(1)	7.10232	1245.239	217.911
(2)	7.85418	1497.910	204.112

P = 760.00

EQUATION	ORDER	A12	A21	D	A122	A211
VAN LAAR	3	0.2455	0.2335	-	-	-
MARGULES	3	0.2454	0.2332	-	-	-
MARGULES	4	0.2413	0.2269	-0.0177	-	-
ALPHA	2	2.1354	-0.0735	-	-	-
ALPHA	3	1.5216	-0.1794	-	0.7024	0.0122

			DEVIATION IN VAPOR PHASE COMPOSITION				
X	Y	TEMP.	LAAR 3	MARG 3	MARG 4	ALPHA2	ALPHA3
0.0238	0.0763	96.00	-0.0039	-0.0039	-0.0044	-0.0062	-0.0046
0.0523	0.1331	94.90	0.0142	0.0141	0.0135	0.0104	0.0131
0.1137	0.2731	91.95	0.0044	0.0044	0.0040	-0.0008	0.0021
0.0938	0.2322	93.25	0.0063	0.0063	0.0058	0.0021	0.0051
0.1494	0.3435	90.55	-0.0052	-0.0052	-0.0054	-0.0101	-0.0077
0.1802	0.3884	89.20	-0.0037	-0.0036	-0.0036	-0.0085	-0.0066
0.2440	0.4608	87.50	0.0037	0.0037	0.0040	0.0013	0.0017
0.3221	0.5434	85.47	0.0004	0.0004	0.0007	0.0002	-0.0009
0.3822	0.5870	84.15	0.0081	0.0081	0.0083	0.0092	0.0071
0.4684	0.6627	82.80	-0.0042	-0.0042	-0.0044	-0.0019	-0.0045
0.6377	0.7619	80.38	0.0046	0.0045	0.0040	0.0042	0.0027
0.7320	0.8236	79.50	-0.0007	-0.0007	-0.0011	-0.0042	-0.0038
0.8333	0.8791	78.50	0.0062	0.0062	0.0063	-0.0008	0.0020
0.9074	0.9301	78.00	0.0036	0.0036	0.0040	-0.0038	-0.0002
		MEAN	0.0049	0.0049	0.0050	0.0045	0.0044

ETHYL ACETATE(1)-TOLUENE(2)

CARR A.D.,KROPHOLLER H.W.:J.CHEM.ENG.DATA 7,26(1962).

ANTOINE VAPOR PRESSURE CONSTANTS

	A	B	C
(1)	7.10232	1245.239	217.911
(2)	6.95334	1343.945	219.377

P = 760.00

EQUATION	ORDER	A12	A21	D	A122	A211
VAN LAAR	3	0.0729	0.0781	-	-	-
MARGULES	3	0.0724	0.0783	-	-	-
MARGULES	4	0.0545	0.0537	-0.0678	-	-
ALPHA	2	2.1308	-0.5466	-	-	-
ALPHA	3	-1.4895	-1.7668	-	3.5978	1.1652

			DEVIATION IN VAPOR PHASE COMPOSITION				
X	Y	TEMP.	LAAR 3	MARG 3	MARG 4	ALPHA2	ALPHA3
0.0210	0.0640	108.82	-0.0001	-0.0002	-0.0022	-0.0012	-0.0016
0.0320	0.0970	107.87	-0.0020	-0.0020	-0.0046	-0.0036	-0.0040
0.0480	0.1370	106.94	0.0007	0.0006	-0.0025	-0.0014	-0.0019
0.1070	0.2650	103.46	0.0076	0.0076	0.0052	0.0048	0.0046
0.1750	0.3910	99.80	0.0044	0.0043	0.0040	0.0018	0.0023
0.2700	0.5280	95.51	-0.0009	-0.0008	0.0002	-0.0022	-0.0012
0.2830	0.5450	95.02	-0.0025	-0.0025	-0.0014	-0.0036	-0.0026
0.3650	0.6290	92.09	-0.0008	-0.0007	0.0001	-0.0008	-0.0001
0.4520	0.7030	89.22	0.0001	0.0002	-0.0000	0.0007	0.0005
0.5980	0.8000	85.16	0.0037	0.0038	0.0024	0.0040	0.0023
0.6560	0.8370	83.55	0.0006	0.0007	-0.0006	0.0005	-0.0013
0.7150	0.8670	82.25	0.0025	0.0025	0.0016	0.0018	0.0005
0.7730	0.8990	81.14	-0.0002	-0.0002	-0.0006	-0.0016	-0.0018
0.8350	0.9280	79.91	0.0002	0.0002	0.0006	-0.0016	-0.0005
0.8910	0.9530	78.80	0.0005	0.0005	0.0013	-0.0015	0.0005
0.9220	0.9670	78.39	0.0000	0.0000	0.0009	-0.0017	0.0003
0.9540	0.9800	77.81	0.0007	0.0007	0.0014	-0.0006	0.0011
0.9700	0.9870	77.60	0.0005	0.0005	0.0010	-0.0005	0.0008
		MEAN	0.0016	0.0016	0.0017	0.0019	0.0016

ETHYL ACETATE(1)-P-XYLENE(2)

CARR A.D.,KROPHOLLER H.W.:J.CHEM.ENG.DATA 7,26(1962).

ANTOINE VAPOR PRESSURE CONSTANTS

	A	B	C
(1)	7.10232	1245.239	217.911
(2)	6.99052	1453.430	215.307

P = 760.00

EQUATION	ORDER	A12	A21	D	A122	A211
VAN LAAR	3	0.1795	0.1726	-	-	-
MARGULES	3	0.1792	0.1728	-	-	-
MARGULES	4	0.1991	0.2132	0.1077	-	-
ALPHA	2	7.0924	-0.6765	-	-	-
ALPHA	3	6.0828	-0.6347	-	1.3573	-0.1292

			DEVIATION IN VAPOR PHASE COMPOSITION				
X	Y	TEMP.	LAAR 3	MARG 3	MARG 4	ALPHA2	ALPHA3
0.0060	0.0480	136.54	-0.0016	-0.0016	0.0003	-0.0015	0.0004
0.0190	0.1380	133.12	-0.0025	-0.0025	0.0019	-0.0030	0.0016
0.0330	0.2200	130.03	-0.0033	-0.0034	0.0019	-0.0047	0.0013
0.0440	0.2800	127.88	-0.0083	-0.0083	-0.0031	-0.0104	-0.0038
0.0700	0.3840	123.44	-0.0056	-0.0056	-0.0015	-0.0089	-0.0023
0.1000	0.4750	119.08	-0.0030	-0.0030	-0.0008	-0.0069	-0.0014
0.1240	0.5300	115.97	0.0011	0.0010	0.0019	-0.0029	0.0016
0.1650	0.6120	111.13	-0.0021	-0.0021	-0.0029	-0.0057	-0.0031
0.2350	0.6950	105.05	0.0076	0.0076	0.0058	0.0062	0.0063
0.3100	0.7710	99.62	-0.0014	-0.0014	-0.0027	-0.0009	-0.0025
0.3990	0.8250	95.00	-0.0004	-0.0003	-0.0003	0.0022	-0.0003
0.4400	0.8470	93.17	-0.0024	-0.0024	-0.0017	0.0006	-0.0019
0.5240	0.8780	90.12	0.0006	0.0006	0.0021	0.0038	0.0018
0.6350	0.9160	86.54	-0.0016	-0.0016	0.0000	0.0003	-0.0003
0.6670	0.9250	85.65	-0.0017	-0.0016	-0.0001	-0.0004	-0.0005
0.7650	0.9480	83.10	0.0004	0.0004	0.0010	-0.0007	0.0009
0.8460	0.9670	80.91	0.0002	0.0002	-0.0000	-0.0029	-0.0001
0.9120	0.9810	79.20	0.0006	0.0006	-0.0000	-0.0033	-0.0002
0.9490	0.9890	78.17	0.0005	0.0005	-0.0002	-0.0029	-0.0004
0.9650	0.9920	77.92	0.0008	0.0008	0.0003	-0.0020	0.0001
		MEAN	0.0023	0.0023	0.0014	0.0035	0.0015

R

1-BROMOBUTANE(1)-BUTYL ALCOHOL(2)

SMITH C.P.,ENGEL E.W.:J.AM.CHEM.SOC.51,2660(1929).

ANTOINE VAPOR PRESSURE CONSTANTS

	A	B	C
(1)	6.92254	1298.608	219.700
(2)	7.54472	1405.873	183.908

T =   50.00

EQUATION	ORDER	A12	A21	D	A122	A211
VAN LAAR	3	0.5545	0.8813	-	-	-
MARGULES	3	0.5357	0.8078	-	-	-
MARGULES	4	0.6716	1.0010	0.5268	-	-
ALPHA	2	15.9578	1.9780	-	-	-
ALPHA	3	13.7156	2.1101	-	2.9674	-0.5412

DEVIATION IN VAPOR PHASE COMPOSITION

X	Y	PRESS.	LAAR 3	MARG 3	MARG 4	ALPHA2	ALPHA3
0.0261	0.3707	43.70	-0.1144	-0.1208	-0.0722	-0.0742	-0.0669
0.0617	0.4852	54.20	-0.0419	-0.0475	-0.0063	-0.0017	0.0048
0.1168	0.6142	70.50	-0.0221	-0.0245	-0.0069	0.0043	0.0080
0.2249	0.7274	92.80	-0.0069	-0.0067	-0.0110	0.0012	0.0016
0.2907	0.7641	103.00	-0.0047	-0.0047	-0.0112	-0.0019	-0.0025
0.3632	0.7872	109.50	0.0000	-0.0009	-0.0050	0.0002	-0.0009
0.4210	0.8029	115.20	-0.0004	-0.0022	-0.0027	-0.0004	-0.0017
0.4774	0.8160	119.00	-0.0024	-0.0052	-0.0018	-0.0016	-0.0028
0.4967	0.8133	120.00	0.0035	0.0003	0.0050	0.0047	0.0036
0.6102	0.8338	125.80	-0.0029	-0.0073	0.0027	0.0028	0.0023
0.7185	0.8522	128.60	-0.0101	-0.0136	-0.0053	0.0005	0.0011
0.7406	0.8580	129.60	-0.0131	-0.0161	-0.0093	-0.0018	-0.0010
0.7860	0.8678	130.90	-0.0160	-0.0173	-0.0151	-0.0039	-0.0023
0.8545	0.8807	132.40	-0.0118	-0.0092	-0.0175	-0.0019	0.0011
		MEAN	0.0179	0.0197	0.0123	0.0072	0.0072

1-CHLOROBUTANE(1)-1-BROMOBUTANE(2)

SMITH C.P.,ENGEL E.W.:J.AM.CHEM.SOC.51,2646(1929).

ANTOINE VAPOR PRESSURE CONSTANTS

	A	B	C
(1)	6.93790	1227.433	224.100
(2)	6.92254	1298.608	219.700

T =   50.00

EQUATION	ORDER	A12	A21	D	A122	A211
VAN LAAR	3	-0.0360	-0.0187	-	-	-
MARGULES	3	-0.0319	-0.0155	-	-	-
MARGULES	4	-0.0246	-0.0069	0.0240	-	-
ALPHA	2	1.1640	-0.5997	-	-	-
ALPHA	3	-5.3433	-3.4959	-	6.5665	2.9428

			DEVIATION IN VAPOR PHASE COMPOSITION				
X	Y	PRESS.	LAAR 3	MARG 3	MARG 4	ALPHA2	ALPHA3
0.0417	0.0813	131.60	0.0024	0.0028	0.0037	0.0050	0.0069
0.1076	0.1972	141.50	0.0075	0.0078	0.0087	0.0109	0.0140
0.1774	0.3257	153.50	-0.0072	-0.0071	-0.0069	-0.0049	-0.0021
0.2446	0.4243	166.40	-0.0089	-0.0090	-0.0093	-0.0080	-0.0067
0.2918	0.4807	173.50	-0.0039	-0.0041	-0.0045	-0.0040	-0.0041
0.3666	0.5461	184.00	0.0180	0.0179	0.0175	0.0170	0.0135
0.4519	0.6354	199.70	0.0153	0.0153	0.0153	0.0140	-0.0038
0.4991	0.7035	207.10	-0.0100	-0.0099	-0.0098	-0.0111	0.0070
0.5122	0.7222	209.50	-0.0174	-0.0173	-0.0171	-0.0184	-0.0088
0.5778	0.7579	220.60	0.0000	0.0002	0.0006	-0.0004	-0.0005
0.6720	0.8230	236.10	0.0023	0.0024	0.0028	0.0028	0.0002
0.7502	0.8698	249.70	0.0046	0.0046	0.0048	0.0059	0.0004
0.8620	0.9372	267.70	-0.0014	-0.0015	-0.0018	0.0003	-0.0039
0.9499	0.9808	281.90	-0.0028	-0.0029	-0.0032	-0.0018	-0.0038
		MEAN	0.0073	0.0074	0.0076	0.0075	0.0054

1-CHLOROBUTANE(1)-HEPTANE(2)

SMITH C.P.,ENGEL E.W.:J.AM.CHEM.SOC.51,2646(1929).

ANTOINE VAPOR PRESSURE CONSTANTS
	A	B	C
(1)	6.93790	1227.433	224.100
(2)	6.90240	1268.115	216.900

T = 50.00

EQUATION	ORDER	A12	A21	D	A122	A211
VAN LAAR	3	0.1251	0.1427	-	-	-
MARGULES	3	0.1254	0.1407	-	-	-
MARGULES	4	0.1761	0.1959	0.1684	-	-
ALPHA	2	1.6164	-0.2473	-	-	-
ALPHA	3	9.5189	3.5175	-	-7.5774	-3.5137

DEVIATION IN VAPOR PHASE COMPOSITION

X	Y	PRESS.	LAAR 3	MARG 3	MARG 4	ALPHA2	ALPHA3
0.0496	0.1129	152.40	0.0085	0.0086	0.0166	0.0053	0.0125
0.1128	0.2261	167.20	0.0196	0.0197	0.0257	0.0154	0.0205
0.1580	0.3496	182.80	-0.0302	-0.0302	-0.0276	-0.0340	-0.0319
0.2870	0.4769	201.70	0.0071	0.0070	0.0033	0.0057	0.0025
0.3853	0.5685	217.60	0.0114	0.0112	0.0083	0.0116	0.0084
0.4462	0.6262	229.00	0.0050	0.0048	0.0036	0.0057	0.0036
0.5273	0.6978	239.60	-0.0049	-0.0052	-0.0039	-0.0043	-0.0044
0.5955	0.7429	248.20	-0.0025	-0.0027	0.0001	-0.0023	-0.0009
0.6386	0.7743	255.20	-0.0052	-0.0054	-0.0023	-0.0056	-0.0035
0.6979	0.8080	262.50	-0.0007	-0.0009	0.0019	-0.0020	0.0004
0.7797	0.8543	271.20	0.0043	0.0043	0.0051	0.0017	0.0031
0.8645	0.9100	280.20	0.0017	0.0018	-0.0002	-0.0018	-0.0035
0.9565	0.9602	288.80	0.0106	0.0107	0.0082	0.0084	0.0038
		MEAN	0.0086	0.0086	0.0082	0.0080	0.0076

BUTYL ALCOHOL(1)-1,2,3,4-TETRAHYDRONAPHTALENE(2)

PADGHAM D.N.,CRAN J.:ROY.AUSTRALIAN CHEM.INST.J.PROC.17,189(1950)

ANTOINE VAPOR PRESSURE CONSTANTS

	A	B	C
(1)	7.54472	1405.873	183.908
(2)	6.96965	1662.400	199.000

P = 760.00

EQUATION	ORDER	A12	A21	D	A122	A211
VAN LAAR	3	0.4403	0.7561	-	-	-
MARGULES	3	0.4358	0.6609	-	-	-
MARGULES	4	0.5471	0.9155	0.6404	-	-
ALPHA	2	35.3421	0.0458	-	-	-
ALPHA	3	31.5081	0.8092	-	8.8746	-1.3168

DEVIATION IN VAPOR PHASE COMPOSITION

X	Y	TEMP.	LAAR 3	MARG 3	MARG 4	ALPHA2	ALPHA3
0.0197	0.3400	194.00	0.0568	0.0553	0.1049	0.0771	0.1044
0.0590	0.7000	166.00	-0.0248	-0.0249	0.0002	-0.0183	-0.0017
0.1240	0.8260	147.00	-0.0077	-0.0071	-0.0044	-0.0079	-0.0033
0.1730	0.8700	140.00	-0.0068	-0.0063	-0.0092	-0.0075	-0.0071
0.1880	0.8740	139.50	-0.0016	-0.0012	-0.0048	-0.0019	-0.0023
0.2890	0.8950	135.00	0.0166	0.0161	0.0128	0.0179	0.0148
0.3210	0.9180	130.40	0.0017	0.0009	-0.0012	0.0029	-0.0004
0.4380	0.9470	125.00	-0.0088	-0.0106	-0.0082	-0.0060	-0.0087
0.4850	0.9480	125.00	-0.0051	-0.0072	-0.0033	-0.0015	-0.0035
0.5220	0.9500	124.40	-0.0039	-0.0061	-0.0013	0.0002	-0.0011
0.5610	0.9560	124.00	-0.0070	-0.0093	-0.0038	-0.0023	-0.0029
0.6810	0.9600	122.80	-0.0035	-0.0055	-0.0002	0.0021	0.0045
0.8270	0.9750	121.00	-0.0085	-0.0082	-0.0089	-0.0046	0.0016
0.9080	0.9860	119.80	-0.0096	-0.0078	-0.0132	-0.0102	-0.0023
0.9650	0.9990	117.70	-0.0105	-0.0088	-0.0140	-0.0156	-0.0084
		MEAN	0.0115	0.0117	0.0127	0.0117	0.0111

ETHYL ETHER(1)-ACETONE(2)

CUNAENS E.H.J.:Z.PHYS.CHEM.36,232(1901).

ANTOINE VAPOR PRESSURE CONSTANTS

	A	B	C
(1)	6.89227	1051.305	227.432
(2)	7.23967	1279.870	237.500

T =   0.00

EQUATION	ORDER	A12	A21	D	A122	A211
VAN LAAR	3	0.9844	0.2310	-	-	-
MARGULES	3	0.5113	0.1041	-	-	-
MARGULES	4	0.3402	-0.0685	-0.5258	-	-
ALPHA	2	3.7139	0.6672	-	-	-
ALPHA	3	-0.9240	-0.3659	-	4.8800	0.1505

			DEVIATION IN VAPOR PHASE COMPOSITION				
X	Y	PRESS.	LAAR 3	MARG 3	MARG 4	ALPHA2	ALPHA3
0.1560	0.4460	119.00	0.0247	0.0197	0.0066	-0.0370	-0.0268
0.1920	0.4360	117.00	0.0522	0.0632	0.0601	0.0213	0.0303
0.3640	0.6170	142.00	-0.0350	-0.0142	-0.0037	-0.0094	-0.0107
0.5100	0.6700	150.00	0.0042	0.0106	0.0095	0.0165	0.0073
0.6170	0.7280	167.00	0.0172	0.0185	0.0105	0.0064	-0.0028
0.8350	0.8610	181.00	0.0304	0.0386	0.0423	-0.0212	0.0002
		MEAN	0.0273	0.0275	0.0221	0.0186	0.0130

ETHYL ETHER(1)-ACETONE(2)

SAMESHIMA J.:J.AM.CHEM.SOC.40,1482(1918).

ANTOINE VAPOR PRESSURE CONSTANTS

	A	B	C
(1)	6.89227	1051.305	227.432
(2)	7.23967	1279.870	237.500

T =   30.00

EQUATION	ORDER	A12	A21	D	A122	A211
VAN LAAR	3	0.2752	0.3108	-	-	-
MARGULES	3	0.2740	0.3094	-	-	-
MARGULES	4	0.2707	0.3048	-0.0131	-	-
ALPHA	2	2.9771	0.1278	-	-	-
ALPHA	3	1.7713	-0.0517	-	1.3926	-0.0092

DEVIATION IN VAPOR PHASE COMPOSITION

X	Y	PRESS.	LAAR 3	MARG 3	MARG 4	ALPHA2	ALPHA3
0.0201	0.0700	297.40	0.0086	0.0084	0.0079	0.0042	0.0072
0.0472	0.1697	325.90	-0.0033	-0.0035	-0.0042	-0.0108	-0.0056
0.0663	0.2102	337.80	0.0086	0.0083	0.0077	-0.0000	0.0059
0.1619	0.4065	409.60	-0.0013	-0.0013	-0.0014	-0.0084	-0.0036
0.2953	0.5545	479.70	0.0024	0.0025	0.0027	0.0012	0.0015
0.3493	0.5997	502.70	0.0010	0.0011	0.0013	0.0018	0.0004
0.5042	0.6997	557.80	-0.0001	-0.0002	-0.0003	0.0032	-0.0005
0.6546	0.7782	597.00	0.0009	0.0008	0.0005	0.0019	-0.0003
0.7491	0.8270	616.90	0.0020	0.0019	0.0017	-0.0006	0.0001
0.8673	0.8961	637.00	0.0027	0.0028	0.0030	-0.0047	0.0006
0.9613	0.9667	654.30	0.0001	0.0002	0.0004	-0.0057	-0.0009
		MEAN	0.0028	0.0028	0.0028	0.0039	0.0024

ETHYL ETHER(1)-CARBON DISULPHIDE(2)

HIRSCHBERG J.:BULL.SOC.CHIM.BELG.41,163(1932).

ANTOINE VAPOR PRESSURE CONSTANTS
```
            A          B          C
(1) 6.89227    1051.305    227.432
(2) 6.85145    1122.500    236.460
```

T =   29.20

EQUATION	ORDER	A12	A21	D	A122	A211
VAN LAAR	3	0.3670	0.1966	-	-	-
MARGULES	3	0.3251	0.1640	-	-	-
MARGULES	4	0.3224	0.1612	-0.0077	-	-
ALPHA	2	1.4535	0.4295	-	-	-
ALPHA	3	1.0426	1.4551	-	1.1440	-1.5474

			DEVIATION IN VAPOR PHASE COMPOSITION				
X	Y	PRESS.	LAAR 3	MARG 3	MARG 4	ALPHA2	ALPHA3
0.1600	0.3000	382.50	0.0017	-0.0003	-0.0006	-0.0164	-0.0001
0.1790	0.3200	387.00	0.0022	0.0010	0.0008	-0.0125	0.0011
0.3030	0.4380	410.10	-0.0065	-0.0052	-0.0050	-0.0016	-0.0049
0.3300	0.4500	414.50	0.0021	0.0034	0.0036	0.0099	0.0037
0.5100	0.5800	438.90	0.0020	0.0009	0.0009	0.0138	0.0011
0.5850	0.6350	448.20	0.0015	-0.0001	-0.0002	0.0086	-0.0002
0.6750	0.7050	456.00	-0.0001	-0.0012	-0.0013	-0.0017	-0.0014
0.7200	0.7400	459.00	0.0008	0.0005	0.0004	-0.0057	0.0004
0.8000	0.8100	463.00	-0.0021	-0.0006	-0.0005	-0.0165	0.0004
		MEAN	0.0021	0.0015	0.0015	0.0096	0.0015

272

ETHYL ETHER(1)-CARBON TETRACHLORIDE(2)

TYRER D.:J.CHEM.SOC.101,81(1912).

ANTOINE VAPOR PRESSURE CONSTANTS
	A	B	C
(1)	6.89227	1051.305	227.432
(2)	6.93390	1242.430	230.000

P = 760.00

EQUATION	ORDER	A12	A21	D	A122	A211
VAN LAAR	3	0.0011	-0.0043	-	-	-
MARGULES	3	-0.0717	0.2217	-	-	-
MARGULES	4	-0.0926	0.1992	-0.0619	-	-
ALPHA	2	3.2923	-0.8055	-	-	-
ALPHA	3	-.1003/ 06	-.2304/ 05	-	.1006/ 06	.2345/ 05

			DEVIATION IN VAPOR PHASE COMPOSITION				
X	Y	TEMP.	LAAR 3	MARG 3	MARG 4	ALPHA2	ALPHA3
0.1870	0.4960	73.70	-0.0412	-0.0225	-0.0234	0.0029	0.0025
0.3420	0.6990	69.73	-0.0416	-0.0045	-0.0037	-0.0047	-0.0044
0.4710	0.7970	65.54	-0.0266	0.0024	0.0024	0.0003	0.0002
0.5800	0.8590	61.44	-0.0171	-0.0019	-0.0024	0.0016	0.0013
0.6750	0.9010	57.16	-0.0086	-0.0065	-0.0069	0.0030	0.0028
0.7570	0.9340	53.09	0.0048	-0.0125	-0.0126	0.0009	0.0009
0.8290	0.9600	48.98	0.0040	-0.0165	-0.0161	-0.0020	-0.0014
0.8920	0.9790	44.84	-0.0063	-0.0161	-0.0155	-0.0035	-0.0021
0.9500	0.9920	41.35	-0.0041	-0.0102	-0.0096	-0.0025	0.0001
		MEAN	0.0172	0.0103	0.0103	0.0024	0.0018

ETHYL ETHER(1)-ETHYL ALCOHOL(2)

GORDON A.R.,HORNIBROOK E.J.:CAN.J.RES.24B,263(1946).

ANTOINE VAPOR PRESSURE CONSTANTS

	A	B	C
(1)	6.89227	1051.305	227.432
(2)	8.16290	1623.220	228.980

T = 25.00

EQUATION	ORDER	A12	A21	D	A122	A211
VAN LAAR	3	0.4403	0.6445	-	-	-
MARGULES	3	0.4313	0.6112	-	-	-
MARGULES	4	0.4365	0.6249	0.0337	-	-
ALPHA	2	23.1174	0.2916	-	23.3075	-
ALPHA	3	1.4513	-0.2713	-	23.3075	-0.5563

			DEVIATION IN VAPOR PHASE COMPOSITION				
X	Y	PRESS.	LAAR 3	MARG 3	MARG 4	ALPHA2	ALPHA3
0.0050	0.1105	66.00	0.0000	-0.0019	-0.0008	-0.0030	0.0033
0.0100	0.1985	73.10	0.0002	-0.0027	-0.0010	-0.0046	0.0054
0.0200	0.3305	86.70	0.0002	-0.0034	-0.0014	-0.0062	0.0069
0.0400	0.4950	112.80	0.0000	-0.0031	-0.0014	-0.0065	0.0066
0.0600	0.5940	137.30	-0.0007	-0.0029	-0.0017	-0.0062	0.0048
0.0800	0.6600	160.10	-0.0014	-0.0028	-0.0020	-0.0058	0.0031
0.1000	0.7070	181.30	-0.0018	-0.0026	-0.0022	-0.0051	0.0017
0.1500	0.7815	228.50	-0.0031	-0.0029	-0.0030	-0.0042	-0.0010
0.2000	0.8245	268.80	-0.0036	-0.0030	-0.0034	-0.0029	-0.0023
0.2500	0.8520	302.20	-0.0034	-0.0028	-0.0031	-0.0013	-0.0025
0.3000	0.8700	330.70	-0.0019	-0.0015	-0.0017	0.0013	-0.0011
0.3500	0.8830	355.40	-0.0005	-0.0004	-0.0004	0.0038	0.0005
0.4000	0.8930	376.10	0.0007	0.0005	0.0006	0.0058	0.0020
0.5000	0.9110	410.60	-0.0010	-0.0018	-0.0014	0.0054	0.0014
0.6000	0.9250	438.90	-0.0027	-0.0039	-0.0034	0.0040	0.0009
0.7000	0.9390	462.50	-0.0057	-0.0067	-0.0063	-0.0001	-0.0002
0.8000	0.9560	485.20	-0.0099	-0.0102	-0.0102	-0.0080	-0.0018
0.9000	0.9760	509.20	-0.0109	-0.0101	-0.0105	-0.0166	-0.0000
0.9500	0.9880	522.20	-0.0084	-0.0075	-0.0079	-0.0182	0.0009
		MEAN	0.0030	0.0037	0.0033	0.0057	0.0025

ETHYL ETHER(1)-ETHYL ALCOHOL(2)

KIREEV V.A.,KHACHADUROVA E.M.:ZH.PRIKL.KHIM.7,495(1934).

ANTOINE VAPOR PRESSURE CONSTANTS

	A	B	C
(1)	6.89227	1051.305	227.432
(2)	8.16290	1623.220	228.980

P = 760.00

EQUATION	ORDER	A12	A21	D	A122	A211
VAN LAAR	3	0.4854	1.2895	-	-	-
MARGULES	3	0.4648	0.9087	-	-	-
MARGULES	4	0.5664	1.1821	0.6749	-	-
ALPHA	2	14.7258	-0.3365	-	-	-
ALPHA	3	-2.0202	-1.3478	-	17.1007	0.4463

DEVIATION IN VAPOR PHASE COMPOSITION

X	Y	TEMP.	LAAR 3	MARG 3	MARG 4	ALPHA2	ALPHA3
0.0063	0.0480	76.60	0.0208	0.0181	0.0328	0.0423	0.0442
0.0125	0.0988	75.10	0.0315	0.0272	0.0510	0.0662	0.0692
0.0189	0.1530	73.70	0.0348	0.0297	0.0588	0.0775	0.0813
0.0252	0.2220	72.20	0.0177	0.0123	0.0439	0.0639	0.0682
0.0317	0.2710	70.80	0.0173	0.0119	0.0441	0.0645	0.0691
0.0382	0.3310	69.50	0.0013	-0.0036	0.0278	0.0479	0.0527
0.0447	0.3830	68.20	-0.0103	-0.0147	0.0149	0.0342	0.0390
0.0513	0.4270	66.90	-0.0168	-0.0204	0.0068	0.0247	0.0295
0.0579	0.4660	65.70	-0.0216	-0.0245	-0.0001	0.0165	0.0212
0.0646	0.5330	64.50	-0.0568	-0.0589	-0.0374	-0.0225	-0.0179
0.0989	0.6210	59.40	-0.0194	-0.0183	-0.0106	-0.0027	0.0011
0.1340	0.7040	55.80	-0.0194	-0.0165	-0.0183	-0.0137	-0.0109
0.1720	0.7590	53.10	-0.0145	-0.0112	-0.0180	-0.0148	-0.0129
0.2100	0.7960	50.60	-0.0084	-0.0058	-0.0141	-0.0128	-0.0117
0.2510	0.8230	48.90	-0.0032	-0.0019	-0.0094	-0.0081	-0.0078
0.2930	0.8410	47.40	0.0034	0.0029	-0.0024	-0.0011	-0.0015
0.3370	0.8570	45.90	0.0071	0.0045	0.0023	0.0035	0.0025
0.3830	0.8710	44.60	0.0086	0.0038	0.0051	0.0069	0.0052
0.4320	0.8860	43.50	0.0060	-0.0011	0.0037	0.0069	0.0047
0.4830	0.8990	42.30	0.0030	-0.0064	0.0017	0.0067	0.0042
0.5360	0.9110	41.10	-0.0011	-0.0125	-0.0019	0.0059	0.0032
0.5920	0.9230	40.00	-0.0071	-0.0202	-0.0080	0.0040	0.0014
0.6510	0.9360	39.20	-0.0164	-0.0302	-0.0178	0.0002	-0.0017
0.7130	0.9480	38.40	-0.0266	-0.0392	-0.0288	-0.0033	-0.0033
0.7790	0.9600	37.70	-0.0387	-0.0468	-0.0418	-0.0069	-0.0032
0.8480	0.9730	37.00	-0.0528	-0.0507	-0.0556	-0.0111	-0.0015
0.9220	0.9850	36.40	-0.0601	-0.0409	-0.0581	-0.0114	0.0029
		MEAN	0.0194	0.0198	0.0228	0.0215	0.0212

ETHYL ETHER(1)-ETHYL ALCOHOL(2)

NAGAI J.,ISII N.:J.SOC.CHEM.IND.JAPAN38,8(1935).

ANTOINE VAPOR PRESSURE CONSTANTS
	A	B	C
(1)	6.89227	1051.305	227.432
(2)	8.16290	1623.220	228.980

T =   0.00

EQUATION	ORDER	A12	A21	D	A122	A211
VAN LAAR	3	0.4496	0.5847	-	-	-
MARGULES	3	0.4439	0.5707	-	-	-
MARGULES	4	0.4551	0.5917	0.0570	-	-
ALPHA	2	40.6412	0.4312	-	-	-
ALPHA	3	8.7314	0.0204	-	34.8068	-0.6961

DEVIATION IN VAPOR PHASE COMPOSITION

X	Y	PRESS.	LAAR 3	MARG 3	MARG 4	ALPHA2	ALPHA3
0.0500	0.6787	36.83	0.0034	0.0018	0.0048	-0.0076	0.0032
0.1000	0.8034	57.38	0.0030	0.0025	0.0034	-0.0033	0.0019
0.1500	0.8590	76.36	-0.0006	-0.0006	-0.0006	-0.0041	-0.0019
0.2000	0.8866	90.79	0.0004	0.0006	0.0002	-0.0014	-0.0009
0.2500	0.9039	102.77	0.0012	0.0014	0.0010	0.0006	-0.0000
0.3000	0.9159	112.89	0.0016	0.0018	0.0015	0.0020	0.0006
0.3500	0.9259	121.61	0.0008	0.0009	0.0007	0.0018	-0.0000
0.4000	0.9323	129.24	0.0014	0.0015	0.0014	0.0029	0.0008
0.4500	0.9386	136.36	0.0009	0.0008	0.0009	0.0026	0.0004
0.5000	0.9442	142.97	0.0001	-0.0001	0.0002	0.0019	-0.0003
0.5500	0.9487	148.51	-0.0002	-0.0005	-0.0001	0.0014	-0.0006
0.6000	0.9523	152.88	0.0001	-0.0002	0.0003	0.0013	-0.0003
0.6500	0.9555	156.46	0.0007	0.0004	0.0008	0.0012	0.0000
0.7000	0.9585	159.82	0.0015	0.0012	0.0016	0.0009	0.0005
0.7500	0.9622	163.36	0.0019	0.0018	0.0020	-0.0002	0.0004
0.8000	0.9667	167.46	0.0021	0.0020	0.0021	-0.0022	-0.0000
0.8500	0.9720	172.00	0.0023	0.0023	0.0022	-0.0049	-0.0005
0.9000	0.9783	176.42	0.0026	0.0028	0.0025	-0.0079	-0.0005
0.9500	0.9861	180.90	0.0032	0.0034	0.0030	-0.0100	0.0006
		MEAN	0.0015	0.0014	0.0015	0.0031	0.0007

ETHYL ETHER(1)-ETHYL ALCOHOL(2)

NAGAI J.,ISII N.:J.SOC.CHEM.IND.JAPAN38,8(1935).

ANTOINE VAPOR PRESSURE CONSTANTS

	A	B	C
(1)	6.89227	1051.305	227.432
(2)	8.16290	1623.220	228.980

T = 10.00

EQUATION	ORDER	A12	A21	D	A122	A211
VAN LAAR	3	0.4372	0.5722	-	-	-
MARGULES	3	0.4314	0.5574	-	-	-
MARGULES	4	0.4471	0.5844	0.0745	-	-
ALPHA	2	30.7866	0.3394	-	-	-
ALPHA	3	9.1001	0.1214	-	24.6929	-0.7412

DEVIATION IN VAPOR PHASE COMPOSITION

X	Y	PRESS.	LAAR 3	MARG 3	MARG 4	ALPHA2	ALPHA3
0.0500	0.6116	59.67	0.0115	0.0097	0.0145	-0.0014	0.0144
0.1000	0.7636	93.10	-0.0007	-0.0013	0.0003	-0.0084	-0.0001
0.1500	0.8269	121.41	-0.0024	-0.0024	-0.0023	-0.0067	-0.0029
0.2000	0.8593	142.55	-0.0002	0.0000	-0.0005	-0.0022	-0.0011
0.2500	0.8801	161.22	0.0012	0.0015	0.0008	0.0009	0.0002
0.3000	0.8953	177.03	0.0015	0.0017	0.0012	0.0024	0.0006
0.3500	0.9068	190.45	0.0015	0.0016	0.0013	0.0032	0.0007
0.4000	0.9161	202.70	0.0011	0.0011	0.0010	0.0034	0.0006
0.4500	0.9245	214.81	-0.0001	-0.0002	-0.0000	0.0027	-0.0003
0.5000	0.9311	224.57	-0.0006	-0.0008	-0.0004	0.0023	-0.0006
0.5500	0.9362	232.01	-0.0003	-0.0006	-0.0000	0.0024	-0.0001
0.6000	0.9404	238.40	0.0005	0.0001	0.0008	0.0028	0.0007
0.6500	0.9457	245.06	-0.0001	-0.0004	0.0003	0.0014	0.0002
0.7000	0.9501	251.97	0.0004	0.0001	0.0007	0.0006	0.0005
0.7500	0.9556	259.00	0.0001	-0.0001	0.0003	-0.0015	-0.0002
0.8000	0.9613	265.98	0.0002	0.0001	0.0002	-0.0038	-0.0006
0.8500	0.9675	272.35	0.0008	0.0009	0.0007	-0.0064	-0.0007
0.9000	0.9745	278.70	0.0020	0.0023	0.0018	-0.0088	-0.0000
0.9500	0.9841	285.13	0.0027	0.0030	0.0025	-0.0108	0.0006
		MEAN	0.0015	0.0015	0.0015	0.0038	0.0013

ETHYL ETHER(1)-ETHYL ALCOHOL(2)

NAGAI J.,ISII N.:J.SOC.CHEM.IND.JAPAN38,8(1935).

ANTOINE VAPOR PRESSURE CONSTANTS
```
            A          B          C
(1) 6.89227    1051.305    227.432
(2) 8.16290    1623.220    228.980
```

T =   20.00

EQUATION	ORDER	A12	A21	D	A122	A211
VAN LAAR	3	0.4315	0.5917	-	-	-
MARGULES	3	0.4230	0.5713	-	-	-
MARGULES	4	0.4435	0.6041	0.0916	-	-
ALPHA	2	24.5944	0.3220	-	-	-
ALPHA	3	6.6910	0.0181	-	20.1820	-0.6135

			DEVIATION IN VAPOR PHASE COMPOSITION				
X	Y	PRESS.	LAAR 3	MARG 3	MARG 4	ALPHA2	ALPHA3
0.0500	0.5681	97.87	0.0009	-0.0019	0.0047	-0.0102	0.0058
0.1000	0.7224	145.07	-0.0017	-0.0028	-0.0003	-0.0089	0.0004
0.1500	0.7918	184.51	-0.0009	-0.0011	-0.0008	-0.0051	-0.0003
0.2000	0.8317	218.11	-0.0004	-0.0002	-0.0008	-0.0025	-0.0006
0.2500	0.8575	246.62	0.0000	0.0004	-0.0005	-0.0004	-0.0005
0.3000	0.8760	271.01	-0.0000	0.0003	-0.0005	0.0008	-0.0006
0.3500	0.8893	291.23	0.0004	0.0006	0.0000	0.0022	-0.0001
0.4000	0.8995	308.18	0.0009	0.0009	0.0007	0.0035	0.0006
0.4500	0.9082	323.40	0.0008	0.0006	0.0008	0.0039	0.0008
0.5000	0.9159	337.48	0.0004	-0.0000	0.0005	0.0038	0.0006
0.5500	0.9230	350.50	-0.0004	-0.0009	-0.0001	0.0031	0.0001
0.6000	0.9294	362.37	-0.0010	-0.0015	-0.0006	0.0022	-0.0003
0.6500	0.9353	373.15	-0.0013	-0.0019	-0.0009	0.0012	-0.0005
0.7000	0.9405	382.11	-0.0008	-0.0014	-0.0005	0.0005	0.0000
0.7500	0.9461	390.97	-0.0004	-0.0008	-0.0002	-0.0008	0.0003
0.8000	0.9529	400.45	-0.0004	-0.0005	-0.0004	-0.0033	0.0000
0.8500	0.9610	410.82	-0.0005	-0.0003	-0.0007	-0.0067	-0.0005
0.9000	0.9704	421.05	0.0000	0.0004	-0.0003	-0.0102	-0.0003
0.9500	0.9821	431.50	0.0010	0.0015	0.0007	-0.0120	0.0006
		MEAN	0.0006	0.0009	0.0007	0.0043	0.0007

ETHYL ETHER(1)-ETHYL ALCOHOL(2)

NAGAI J.,ISII N.:J.SOC.CHEM.IND.JAPAN38,8(1935).

ANTOINE VAPOR PRESSURE CONSTANTS

	A	B	C
(1)	6.89227	1051.305	227.432
(2)	8.16290	1623.220	228.980

T = 30.00

EQUATION	ORDER	A12	A21	D	A122	A211
VAN LAAR	3	0.4161	0.5608	-	-	-
MARGULES	3	0.4097	0.5420	-	-	-
MARGULES	4	0.4464	0.5970	0.1551	-	-
ALPHA	2	19.0274	0.2593	-	-	-
ALPHA	3	8.7299	0.3059	-	12.7886	-0.7654

			DEVIATION IN VAPOR PHASE COMPOSITION				
X	Y	PRESS.	LAAR 3	MARG 3	MARG 4	ALPHA2	ALPHA3
0.0500	0.5092	153.96	0.0018	-0.0003	0.0121	-0.0114	0.0105
0.1000	0.6771	223.00	-0.0054	-0.0062	-0.0011	-0.0146	-0.0010
0.1500	0.7538	279.00	-0.0035	-0.0036	-0.0026	-0.0091	-0.0020
0.2000	0.7977	324.95	-0.0008	-0.0006	-0.0016	-0.0037	-0.0009
0.2500	0.8278	364.96	-0.0001	0.0002	-0.0015	-0.0008	-0.0009
0.3000	0.8455	389.48	0.0042	0.0044	0.0028	0.0051	0.0031
0.3500	0.8656	429.04	0.0006	0.0007	-0.0004	0.0028	-0.0004
0.4000	0.8785	454.50	0.0007	0.0006	0.0002	0.0038	-0.0000
0.4500	0.8897	478.02	0.0001	-0.0002	0.0001	0.0039	-0.0002
0.5000	0.8993	498.74	-0.0005	-0.0009	0.0000	0.0037	-0.0003
0.5500	0.9077	517.24	-0.0009	-0.0015	-0.0001	0.0032	-0.0002
0.6000	0.9153	533.60	-0.0012	-0.0019	-0.0001	0.0026	-0.0000
0.6500	0.9223	548.30	-0.0012	-0.0018	-0.0000	0.0018	0.0003
0.7000	0.9305	572.00	-0.0023	-0.0028	-0.0013	-0.0007	-0.0006
0.7500	0.9371	577.52	-0.0013	-0.0017	-0.0007	-0.0018	0.0002
0.8000	0.9451	591.00	-0.0009	-0.0011	-0.0008	-0.0042	0.0001
0.8500	0.9540	604.80	-0.0001	0.0001	-0.0005	-0.0069	0.0002
0.9000	0.9651	618.60	0.0007	0.0011	-0.0003	-0.0102	-0.0002
0.9500	0.9789	632.55	0.0018	0.0023	0.0007	-0.0114	0.0001
		MEAN	0.0015	0.0017	0.0014	0.0054	0.0011

ETHYL ETHER(1)-ETHYL ALCOHOL(2)

NAGAI J.,ISII N.:J.SOC.CHEM.IND.JAPAN38,8(1935).

ANTOINE VAPOR PRESSURE CONSTANTS
```
          A          B          C
(1) 6.89227    1051.305    227.432
(2) 8.16290    1623.220    228.980
```

T =   40.00

EQUATION	ORDER	A12	A21	D	A122	A211
VAN LAAR	3	0.4083	0.5605	-	-	-
MARGULES	3	0.3995	0.5411	-	-	-
MARGULES	4	0.4255	0.5777	0.1045	-	-
ALPHA	2	15.4862	0.2622	-	-	-
ALPHA	3	5.8372	0.0832	-	11.3240	-0.5338

			DEVIATION IN VAPOR PHASE COMPOSITION				
X	Y	PRESS.	LAAR 3	MARG 3	MARG 4	ALPHA2	ALPHA3
0.0500	0.4618	238.00	-0.0008	-0.0038	0.0050	-0.0124	0.0063
0.1000	0.6303	330.30	-0.0036	-0.0050	-0.0009	-0.0124	0.0005
0.1500	0.7150	408.50	-0.0031	-0.0034	-0.0025	-0.0087	-0.0011
0.2000	0.7650	473.20	-0.0012	-0.0010	-0.0016	-0.0042	-0.0006
0.2500	0.7992	529.50	-0.0004	-0.0000	-0.0012	-0.0014	-0.0006
0.3000	0.8234	577.00	0.0006	0.0010	-0.0002	0.0013	0.0001
0.3500	0.8420	617.60	0.0011	0.0013	0.0005	0.0031	0.0006
0.4000	0.8571	653.70	0.0011	0.0012	0.0008	0.0042	0.0007
0.4500	0.8706	687.60	0.0000	-0.0002	-0.0000	0.0039	-0.0001
0.5000	0.8820	717.60	-0.0008	-0.0013	-0.0006	0.0034	-0.0006
0.5500	0.8912	742.00	-0.0007	-0.0013	-0.0003	0.0037	-0.0001
0.6000	0.8998	764.30	-0.0008	-0.0015	-0.0002	0.0033	0.0002
0.6500	0.9083	785.20	-0.0011	-0.0018	-0.0005	0.0023	0.0002
0.7000	0.9167	805.90	-0.0012	-0.0018	-0.0007	0.0009	0.0003
0.7500	0.9257	826.50	-0.0014	-0.0019	-0.0011	-0.0014	0.0000
0.8000	0.9357	846.80	-0.0017	-0.0018	-0.0017	-0.0044	-0.0004
0.8500	0.9466	866.00	-0.0012	-0.0010	-0.0015	-0.0076	-0.0003
0.9000	0.9598	885.00	-0.0005	0.0000	-0.0011	-0.0109	-0.0002
0.9500	0.9762	903.30	0.0007	0.0013	0.0001	-0.0119	0.0005
		MEAN	0.0012	0.0016	0.0011	0.0053	0.0007

ETHYL ETHER(1)-ETHYL ALCOHOL(2)

NAGAI J.,ISII N.:J.SOC.CHEM.IND.JAPAN38,8(1935).

ANTOINE VAPOR PRESSURE CONSTANTS

	A	B	C
(1)	6.89227	1051.305	227.432
(2)	8.16290	1623.220	228.980

T = 50.00

EQUATION	ORDER	A12	A21	D	A122	A211
VAN LAAR	3	0.3948	0.5410	-	-	-
MARGULES	3	0.3864	0.5221	-	-	-
MARGULES	4	0.4185	0.5654	0.1242	-	-
ALPHA	2	12.4716	0.2344	-	-	-
ALPHA	3	5.9311	0.2054	-	8.0910	-0.5693

			DEVIATION IN VAPOR PHASE COMPOSITION				
X	Y	PRESS.	LAAR 3	MARG 3	MARG 4	ALPHA2	ALPHA3
0.0500	0.4203	364.00	-0.0086	-0.0114	-0.0006	-0.0197	0.0007
0.1000	0.5863	486.10	-0.0070	-0.0085	-0.0030	-0.0160	-0.0010
0.1500	0.6723	584.60	-0.0021	-0.0026	-0.0011	-0.0081	0.0010
0.2000	0.7285	674.00	-0.0013	-0.0011	-0.0018	-0.0046	-0.0001
0.2500	0.7666	748.80	-0.0002	0.0002	-0.0014	-0.0013	-0.0002
0.3000	0.7946	813.00	0.0005	0.0008	-0.0008	0.0012	-0.0001
0.3500	0.8161	868.80	0.0010	0.0012	-0.0001	0.0031	0.0003
0.4000	0.8341	919.60	0.0005	0.0005	-0.0001	0.0038	-0.0001
0.4500	0.8496	967.50	-0.0005	-0.0007	-0.0006	0.0037	-0.0007
0.5000	0.8621	1006.80	-0.0005	-0.0010	-0.0003	0.0042	-0.0003
0.5500	0.8732	1041.00	-0.0006	-0.0013	0.0000	0.0042	0.0001
0.6000	0.8835	1070.70	-0.0008	-0.0015	0.0001	0.0038	0.0005
0.6500	0.8939	1100.80	-0.0014	-0.0022	-0.0005	0.0024	0.0004
0.7000	0.9040	1128.30	-0.0017	-0.0023	-0.0009	0.0008	0.0004
0.7500	0.9148	1157.40	-0.0021	-0.0026	-0.0016	-0.0017	0.0001
0.8000	0.9268	1183.60	-0.0026	-0.0028	-0.0026	-0.0051	-0.0006
0.8500	0.9396	1209.00	-0.0021	-0.0019	-0.0026	-0.0083	-0.0006
0.9000	0.9543	1233.30	-0.0008	-0.0002	-0.0017	-0.0108	-0.0001
0.9500	0.9727	1255.20	0.0010	0.0017	0.0000	-0.0108	0.0008
		MEAN	0.0019	0.0023	0.0010	0.0060	0.0004

S

DIETHYLAMINE(1)-WATER(2)

COPP J.L.,EVERETT D.H.:DIS.FARAD.SOC.,15,174(1953)

ANTOINE VAPOR PRESSURE CONSTANTS
	A	B	C
(1)	7.14099	1209.900	229.000
(2)	7.96681	1668.210	228.000

T = 38.35

EQUATION	ORDER	A12	A21	D	A122	A211
VAN LAAR	3	0.7318	0.2895	-	-	-
MARGULES	3	0.5817	0.2337	-	-	-
MARGULES	4	0.7875	0.5609	0.9332	-	-
ALPHA	2	18.2829	0.6212	-	-	-
ALPHA	3	415.3628	72.1570	-	-257.9683	-67.0787

			DEVIATION IN VAPOR PHASE COMPOSITION				
X	Y	PRESS.	LAAR 3	MARG 3	MARG 4	ALPHA2	ALPHA3
0.0500	0.6530	140.30	-0.0403	-0.0818	-0.0177	-0.1691	0.0056
0.1000	0.7150	167.70	-0.0046	-0.0170	0.0071	-0.0689	-0.0055
0.2000	0.7660	197.20	0.0106	0.0186	0.0100	0.0105	0.0014
0.3000	0.8090	227.00	0.0026	0.0121	0.0002	0.0239	0.0011
0.4000	0.8460	257.10	-0.0050	0.0003	-0.0026	0.0187	-0.0002
0.5000	0.8770	287.00	-0.0079	-0.0072	0.0005	0.0086	-0.0005
0.6000	0.9050	316.90	-0.0083	-0.0102	0.0027	-0.0042	-0.0014
0.7000	0.9280	344.50	-0.0042	-0.0062	0.0041	-0.0146	-0.0004
0.8000	0.9470	367.20	0.0031	0.0027	0.0045	-0.0214	0.0022
0.9000	0.9680	388.10	0.0076	0.0086	0.0022	-0.0257	0.0007
0.9500	0.9800	397.90	0.0079	0.0089	0.0023	-0.0219	-0.0020
		MEAN	0.0093	0.0158	0.0049	0.0352	0.0019

DIETHYLAMINE(1)-WATER(2)

COPP J.L.,EVERETT D.H.:DIS.FARAD.SOC.,15,174(1953)

ANTOINE VAPOR PRESSURE CONSTANTS
	A	B	C
(1)	7.14099	1209.900	229.000
(2)	7.96681	1668.210	228.000

T =   49.10

EQUATION	ORDER	A12	A21	D	A122	A211
VAN LAAR	3	0.7987	0.3484	-	-	-
MARGULES	3	0.6560	0.3031	-	-	-
MARGULES	4	0.8855	0.6319	0.9764	-	-
ALPHA	2	18.4247	1.0838	-	-	-
ALPHA	3	401.9419	90.4667	-	-199.7178	-83.4454

			DEVIATION IN VAPOR PHASE COMPOSITION				
X	Y	PRESS.	LAAR 3	MARG 3	MARG 4	ALPHA2	ALPHA3
0.0500	0.6650	255.00	-0.0535	-0.0947	-0.0223	-0.1848	0.0024
0.1000	0.7110	291.00	-0.0040	-0.0176	0.0109	-0.0730	-0.0031
0.2000	0.7550	333.00	0.0123	0.0198	0.0108	0.0088	0.0005
0.3000	0.7920	373.00	0.0048	0.0152	0.0006	0.0260	0.0011
0.4000	0.8260	413.50	-0.0037	0.0032	-0.0022	0.0226	0.0005
0.5000	0.8580	454.00	-0.0094	-0.0073	-0.0002	0.0108	-0.0010
0.6000	0.8870	493.50	-0.0104	-0.0117	0.0026	-0.0032	-0.0017
0.7000	0.9110	528.50	-0.0049	-0.0070	0.0053	-0.0145	0.0006
0.8000	0.9350	558.00	0.0017	0.0010	0.0035	-0.0256	0.0013
0.9000	0.9580	584.00	0.0102	0.0110	0.0029	-0.0297	0.0017
0.9500	0.9740	596.40	0.0100	0.0110	0.0023	-0.0267	-0.0029
		MEAN	0.0114	0.0181	0.0058	0.0387	0.0015

DIETHYLAMINE(1)-WATER(2)

COPP J.L.,EVERETT D.H.:DIS.FARAD.SOC.,15,174(1953)

ANTOINE VAPOR PRESSURE CONSTANTS

	A	B	C
(1)	7.14099	1209.900	229.000
(2)	7.96681	1668.210	228.000

T = 56.80

EQUATION	ORDER	A12	A21	D	A122	A211
VAN LAAR	3	0.8725	0.4123	-	-	-
MARGULES	3	0.7420	0.3648	-	-	-
MARGULES	4	0.9563	0.6585	0.8735	-	-
ALPHA	2	21.2416	1.9588	-	-	-
ALPHA	3	186.6465	58.0292	-	-32.3537	-54.4420

			DEVIATION IN VAPOR PHASE COMPOSITION				
X	Y	PRESS.	LAAR 3	MARG 3	MARG 4	ALPHA2	ALPHA3
0.0500	0.6810	388.50	-0.0560	-0.0942	-0.0262	-0.1772	-0.0052
0.1000	0.7200	438.00	-0.0043	-0.0173	0.0103	-0.0685	-0.0029
0.2000	0.7530	485.50	0.0151	0.0215	0.0146	0.0107	0.0035
0.3000	0.7840	532.30	0.0066	0.0162	0.0031	0.0267	0.0020
0.4000	0.8140	579.50	-0.0035	0.0030	-0.0024	0.0231	-0.0006
0.5000	0.8430	626.50	-0.0100	-0.0081	-0.0019	0.0115	-0.0028
0.6000	0.8700	672.00	-0.0107	-0.0121	0.0015	-0.0025	-0.0030
0.7000	0.8940	709.00	-0.0045	-0.0068	0.0054	-0.0151	0.0001
0.8000	0.9170	740.50	0.0063	0.0055	0.0078	-0.0260	0.0047
0.9000	0.9460	769.00	0.0142	0.0154	0.0064	-0.0355	0.0042
0.9500	0.9860	782.50	-0.0062	-0.0050	-0.0146	-0.0540	-0.0207
		MEAN	0.0125	0.0186	0.0085	0.0410	0.0045

CYCLOPENTANE(1)-BENZENE(2)

MYERS H.S.:IND.ENG.CHEM.48,1104(1956).

ANTOINE VAPOR PRESSURE CONSTANTS
```
         A          B          C
(1) 6.87798    1119.208    230.738
(2) 6.90565    1211.033    220.790
```

P = 760.00

EQUATION	ORDER	A12	A21	D	A122	A211
VAN LAAR	3	0.1491	0.1229	-	-	-
MARGULES	3	0.1478	0.1210	-	-	-
MARGULES	4	0.1423	0.1133	-0.0211	-	-
ALPHA	2	2.2870	-0.4245	-	-	-
ALPHA	3	1.3564	-0.6491	-	1.0121	0.1642

			DEVIATION IN VAPOR PHASE COMPOSITION				
X	Y	TEMP.	LAAR 3	MARG 3	MARG 4	ALPHA2	ALPHA3
0.0060	0.0220	79.70	-0.0013	-0.0014	-0.0016	-0.0026	-0.0021
0.0170	0.0560	78.80	0.0008	0.0007	0.0001	-0.0024	-0.0013
0.0320	0.0990	77.90	0.0034	0.0033	0.0024	-0.0018	0.0002
0.0500	0.1470	76.65	0.0056	0.0053	0.0044	-0.0013	0.0013
0.0720	0.2050	75.05	0.0030	0.0027	0.0018	-0.0051	-0.0021
0.0980	0.2620	73.55	0.0041	0.0039	0.0031	-0.0043	-0.0011
0.1240	0.3125	72.05	0.0053	0.0052	0.0047	-0.0027	0.0003
0.1475	0.3550	70.75	0.0050	0.0050	0.0047	-0.0025	0.0003
0.1690	0.3930	69.90	0.0021	0.0021	0.0019	-0.0045	-0.0020
0.2100	0.4500	68.05	0.0051	0.0051	0.0053	0.0003	0.0021
0.2700	0.5290	65.60	0.0006	0.0007	0.0010	-0.0018	-0.0011
0.2790	0.5380	65.20	0.0017	0.0018	0.0021	-0.0004	0.0001
0.2900	0.5520	64.85	-0.0004	-0.0003	0.0000	-0.0020	-0.0017
0.3090	0.5680	64.10	0.0035	0.0036	0.0040	0.0025	0.0024
0.3390	0.5980	63.10	0.0030	0.0030	0.0033	0.0027	0.0022
0.3750	0.6330	61.90	0.0007	0.0007	0.0009	0.0012	0.0002
0.4120	0.6640	60.90	0.0006	0.0006	0.0006	0.0017	0.0003
0.4410	0.6870	60.15	0.0003	0.0002	0.0002	0.0016	-0.0000
0.4730	0.7120	59.25	-0.0009	-0.0010	-0.0011	0.0004	-0.0014
0.5075	0.7350	58.50	0.0002	0.0001	-0.0002	0.0013	-0.0005
0.5650	0.7730	57.00	0.0000	-0.0001	-0.0005	0.0002	-0.0016
0.6150	0.8020	56.20	0.0013	0.0012	0.0007	0.0006	-0.0008
0.6525	0.8230	55.50	0.0021	0.0020	0.0015	0.0004	-0.0006
0.7200	0.8550	54.30	0.0073	0.0072	0.0069	0.0037	0.0036
0.7570	0.8760	53.60	0.0058	0.0058	0.0056	0.0012	0.0018
0.7900	0.8960	53.00	0.0028	0.0028	0.0028	-0.0026	-0.0015
0.8180	0.9100	52.30	0.0030	0.0031	0.0031	-0.0030	-0.0014
0.8580	0.9280	51.70	0.0048	0.0048	0.0050	-0.0016	0.0006
0.8880	0.9440	51.10	0.0033	0.0034	0.0037	-0.0029	-0.0005
0.9180	0.9580	50.60	0.0037	0.0038	0.0040	-0.0019	0.0005
0.9450	0.9725	50.20	0.0019	0.0020	0.0023	-0.0025	-0.0005
0.9780	0.9900	49.60	-0.0002	-0.0001	0.0000	-0.0024	-0.0013
		MEAN	0.0026	0.0026	0.0025	0.0021	0.0012

PENTANE(1)-ACETONE(2)

LO T.CH.,BIEBER H.H.,KARR A.E.:J.CHEM.ENG.DATA 7(3),327(1962).

ANTOINE VAPOR PRESSURE CONSTANTS
```
          A          B          C
(1)  6.85221    1064.630    232.000
(2)  7.23967    1279.870    237.500
```

P =   760.00

EQUATION	ORDER	A12	A21	D	A122	A211
VAN LAAR	3	0.6184	0.6630	-	-	-
MARGULES	3	0.6177	0.6620	-	-	-
MARGULES	4	0.6243	0.6693	0.0240	-	-
ALPHA	2	7.7123	2.4914	-	-	-
ALPHA	3	2.8535	0.7027	-	3.7499	0.8148

			DEVIATION IN VAPOR PHASE COMPOSITION				
X	Y	TEMP.	LAAR 3	MARG 3	MARG 4	ALPHA2	ALPHA3
0.0210	0.1080	49.15	0.0333	0.0331	0.0347	0.0404	0.0270
0.0610	0.3070	45.76	0.0058	0.0056	0.0074	0.0103	-0.0044
0.1340	0.4750	39.58	0.0036	0.0036	0.0041	-0.0039	-0.0078
0.2105	0.5500	36.67	0.0154	0.0155	0.0152	0.0035	0.0073
0.2920	0.6145	34.35	0.0042	0.0042	0.0036	-0.0078	-0.0013
0.4050	0.6640	32.85	-0.0024	-0.0023	-0.0027	-0.0096	-0.0051
0.5030	0.6780	32.35	0.0086	0.0086	0.0087	0.0066	0.0066
0.6110	0.7110	31.97	0.0008	0.0007	0.0012	0.0025	-0.0029
0.7280	0.7390	31.93	0.0076	0.0075	0.0080	0.0076	-0.0002
0.8690	0.8100	32.27	0.0141	0.0141	0.0137	-0.0018	0.0008
0.9530	0.9065	33.89	0.0090	0.0091	0.0083	-0.0153	-0.0007
		MEAN	0.0095	0.0095	0.0098	0.0100	0.0058

286

PENTANE(1)-ETHYL ALCOHOL(2)

ISII N.:J.SOC.CHEM.IND.JAPAN 38,705(1935).

ANTOINE VAPOR PRESSURE CONSTANTS
	A	B	C
(1)	6.85221	1064.630	232.000
(2)	8.16290	1623.220	228.980

T = -10.00

EQUATION	ORDER	A12	A21	D	A122	A211
VAN LAAR	3	0.8479	1.1300	-	-	-
MARGULES	3	0.8468	1.0863	-	-	-
MARGULES	4	0.9614	1.2527	0.5743	-	-
ALPHA	2	218.1443	6.7571	-	-	-
ALPHA	3	11.3944	4.8885	-	194.0977	-5.0747

			DEVIATION IN VAPOR PHASE COMPOSITION				
X	Y	PRESS.	LAAR 3	MARG 3	MARG 4	ALPHA2	ALPHA3
0.0500	0.8840	50.70	-0.0179	-0.0176	-0.0027	0.0072	0.0048
0.1000	0.9290	81.80	-0.0078	-0.0074	-0.0045	-0.0000	-0.0003
0.1500	0.9440	101.30	-0.0035	-0.0032	-0.0045	-0.0017	-0.0013
0.2000	0.9510	114.40	-0.0012	-0.0009	-0.0040	-0.0019	-0.0012
0.2500	0.9550	123.40	-0.0000	0.0001	-0.0034	-0.0017	-0.0009
0.3000	0.9580	129.00	-0.0001	-0.0002	-0.0034	-0.0019	-0.0011
0.3500	0.9600	132.70	-0.0006	-0.0008	-0.0033	-0.0019	-0.0012
0.4000	0.9608	135.30	-0.0007	-0.0012	-0.0026	-0.0012	-0.0005
0.4500	0.9613	137.10	-0.0012	-0.0019	-0.0020	-0.0005	0.0000
0.5000	0.9616	138.00	-0.0020	-0.0029	-0.0017	0.0001	0.0005
0.5500	0.9617	138.40	-0.0030	-0.0040	-0.0016	0.0008	0.0010
0.6000	0.9617	138.40	-0.0042	-0.0052	-0.0017	0.0015	0.0014
0.6500	0.9617	138.50	-0.0054	-0.0065	-0.0022	0.0021	0.0016
0.7000	0.9619	139.00	-0.0067	-0.0077	-0.0032	0.0025	0.0016
0.7500	0.9626	139.00	-0.0081	-0.0088	-0.0048	0.0023	0.0010
0.8000	0.9640	139.00	-0.0091	-0.0093	-0.0069	0.0014	-0.0003
0.8500	0.9655	139.00	-0.0083	-0.0078	-0.0080	0.0004	-0.0013
0.9000	0.9676	138.90	-0.0044	-0.0032	-0.0067	-0.0009	-0.0015
0.9500	0.9733	138.70	0.0023	0.0038	-0.0017	-0.0051	0.0002
		MEAN	0.0046	0.0049	0.0036	0.0019	0.0011

PENTANE(1)-ETHYL ALCOHOL(2)

ISII N.:J.SOC.CHEM.IND.JAPAN 38,705(1935).

ANTOINE VAPOR PRESSURE CONSTANTS

	A	B	C
(1)	6.85221	1064.630	232.000
(2)	8.16290	1623.220	228.980

T =   0.00

EQUATION	ORDER	A12	A21	D	A122	A211
VAN LAAR	3	0.8459	1.2150	-	-	-
MARGULES	3	0.8403	1.1504	-	-	-
MARGULES	4	0.9683	1.3218	0.6030	-	-
ALPHA	2	192.2031	7.6168	-	-	-
ALPHA	3	.8966/ 10	.3327/ 09	-	-.5266/ 10	-.6094/ 06

DEVIATION IN VAPOR PHASE COMPOSITION

X	Y	PRESS.	LAAR 3	MARG 3	MARG 4	ALPHA2	ALPHA3
0.0500	0.8490	79.50	-0.0166	-0.0168	0.0039	0.0260	0.0736
0.1000	0.9210	134.10	-0.0203	-0.0198	-0.0152	-0.0045	0.0060
0.1500	0.9340	168.10	-0.0088	-0.0082	-0.0094	-0.0028	-0.0030
0.2000	0.9430	188.00	-0.0058	-0.0052	-0.0089	-0.0042	-0.0084
0.2500	0.9470	199.30	-0.0031	-0.0027	-0.0072	-0.0036	-0.0092
0.3000	0.9490	207.30	-0.0012	-0.0011	-0.0053	-0.0025	-0.0082
0.3500	0.9510	213.10	-0.0011	-0.0013	-0.0047	-0.0023	-0.0076
0.4000	0.9516	217.20	-0.0008	-0.0014	-0.0035	-0.0012	-0.0058
0.4500	0.9523	220.20	-0.0016	-0.0025	-0.0030	-0.0006	-0.0042
0.5000	0.9527	222.10	-0.0027	-0.0040	-0.0028	0.0001	-0.0026
0.5500	0.9530	223.80	-0.0043	-0.0060	-0.0031	0.0007	-0.0010
0.6000	0.9534	225.40	-0.0066	-0.0085	-0.0041	0.0011	0.0004
0.6500	0.9536	226.50	-0.0089	-0.0109	-0.0055	0.0016	0.0018
0.7000	0.9538	227.30	-0.0113	-0.0132	-0.0074	0.0020	0.0032
0.7500	0.9538	227.50	-0.0132	-0.0147	-0.0095	0.0026	0.0046
0.8000	0.9543	227.60	-0.0146	-0.0152	-0.0120	0.0027	0.0054
0.8500	0.9551	227.50	-0.0139	-0.0132	-0.0135	0.0026	0.0059
0.9000	0.9569	227.30	-0.0095	-0.0072	-0.0121	0.0017	0.0052
0.9500	0.9634	226.80	0.0001	0.0032	-0.0050	-0.0027	-0.0002
		MEAN	0.0076	0.0082	0.0072	0.0034	0.0082

ISOPENTANE(1)-CARBON DISULPHIDE(2)

HIRSCHBERG J.:BULL.SOC.CHIM.BELG.41,163(1932).

ANTOINE VAPOR PRESSURE CONSTANTS

	A	B	C
(1)	6.78967	1020.012	233.097
(2)	6.85145	1122.500	236.460

T = 17.00

EQUATION	ORDER	A12	A21	D	A122	A211
VAN LAAR	3	0.4741	0.2374	-	-	-
MARGULES	3	0.4048	0.1900	-	-	-
MARGULES	4	0.3810	0.1641	-0.0665	-	-
ALPHA	2	2.3529	0.4484	-	-	-
ALPHA	3	5.6769	3.4772	-	-1.7519	-3.1256

			DEVIATION IN VAPOR PHASE COMPOSITION				
X	Y	PRESS.	LAAR 3	MARG 3	MARG 4	ALPHA2	ALPHA3
0.1680	0.3800	374.80	0.0150	0.0115	0.0093	-0.0229	-0.0016
0.2850	0.4850	412.40	0.0089	0.0110	0.0114	0.0017	0.0036
0.3600	0.5450	431.30	-0.0006	0.0014	0.0021	0.0033	-0.0011
0.3750	0.5550	436.50	-0.0010	0.0008	0.0015	0.0043	-0.0007
0.4600	0.6100	451.30	-0.0025	-0.0024	-0.0023	0.0059	-0.0003
0.5200	0.6500	464.00	-0.0049	-0.0058	-0.0062	0.0017	-0.0030
0.5700	0.6750	472.00	0.0020	0.0006	-0.0002	0.0049	0.0025
0.6900	0.7500	484.60	0.0071	0.0066	0.0059	-0.0038	0.0011
0.7800	0.8100	490.00	0.0115	0.0133	0.0137	-0.0105	-0.0009
		MEAN	0.0059	0.0059	0.0058	0.0065	0.0016

CHLOROBENZENE(1)-1-NITROPROPANE(2)

LACHER J.R.,BUCK W.B.,PARRY W.H.:J.AM.CHEM.SOC.63,2422(1941).

ANTOINE VAPOR PRESSURE CONSTANTS
	A	B	C
(1)	6.97429	1431.323	217.975
(2)	7.11380	1466.963	215.184

T = 75.00

EQUATION	ORDER	A12	A21	D	A122	A211
VAN LAAR	3	0.1506	0.1510	-	-	-
MARGULES	3	0.1506	0.1510	-	-	-
MARGULES	4	0.1590	0.1601	0.0266	-	-
ALPHA	2	0.5092	0.3413	-	-	-
ALPHA	3	3.3544	2.8007	-	-2.7409	-2.1903

			DEVIATION IN VAPOR PHASE COMPOSITION				
X	Y	PRESS.	LAAR 3	MARG 3	MARG 4	ALPHA2	ALPHA3
0.1190	0.1590	119.40	-0.0003	-0.0003	0.0004	-0.0007	0.0004
0.1320	0.1740	119.60	-0.0001	-0.0001	0.0005	-0.0006	0.0002
0.1870	0.2330	121.60	0.0016	0.0016	0.0017	0.0011	0.0004
0.2890	0.3350	125.20	0.0004	0.0004	-0.0000	0.0001	-0.0018
0.4600	0.4840	128.90	0.0004	0.0004	0.0003	0.0002	0.0003
0.4720	0.4930	129.80	0.0013	0.0013	0.0013	0.0012	0.0015
0.5830	0.5870	129.50	-0.0010	-0.0010	-0.0005	-0.0013	0.0003
0.6910	0.6800	127.10	-0.0026	-0.0026	-0.0020	-0.0032	-0.0025
0.7710	0.7450	129.50	0.0044	0.0044	0.0046	0.0035	0.0015
		MEAN	0.0013	0.0013	0.0013	0.0013	0.0010

CHLOROBENZENE(1)-1-NITROPROPANE(2)

LACHER J.R.,BUCK W.B.,PARRY W.H.:J.AM.CHEM.SOC.63,2422(1941).

ANTOINE VAPOR PRESSURE CONSTANTS
            A   B   C
    (1) 6.97429   1431.323   217.975
    (2) 7.11380   1466.963   215.184

T =   120.00

EQUATION	ORDER	A12	A21	D	A122	A211
VAN LAAR	3	0.1346	0.1818	-	-	-
MARGULES	3	0.1301	0.1768	-	-	-
MARGULES	4	0.1303	0.1770	0.0004	-	-
ALPHA	2	0.4484	0.4147	-	-	-
ALPHA	3	1.3738	0.8598	-	-1.0797	-0.2272

			DEVIATION	IN VAPOR	PHASE	COMPOSITION	
X	Y	PRESS.	LAAR 3	MARG 3	MARG 4	ALPHA2	ALPHA3
0.0960	0.1180	565.20	0.0040	0.0035	0.0036	0.0075	-0.0001
0.2820	0.3140	590.20	0.0036	0.0037	0.0037	0.0033	0.0004
0.4540	0.4710	597.20	-0.0014	-0.0012	-0.0012	-0.0055	-0.0018
0.5070	0.5120	597.60	0.0016	0.0017	0.0017	-0.0028	0.0017
0.6750	0.6520	596.60	0.0005	0.0003	0.0003	-0.0018	0.0000
0.7650	0.7300	589.40	0.0016	0.0015	0.0015	0.0020	-0.0005
0.8440	0.8040	579.90	0.0040	0.0043	0.0042	0.0069	0.0003
		MEAN	0.0024	0.0023	0.0023	0.0042	0.0007

BENZENE(1)-ANILINE(2)

KORTUEM G.,FREIER H.J.:CHEM.-ING.TECHN.26,670(1954).

ANTOINE VAPOR PRESSURE CONSTANTS
	A	B	C
(1)	6.90565	1211.033	220.790
(2)	7.63851	1675.300	200.000

T = 119.30

EQUATION	ORDER	A12	A21	D	A122	A211
VAN LAAR	3	0.4957	24.5361	-	-	-
MARGULES	3	0.3128	1.3874	-	-	-
MARGULES	4	0.3789	1.4929	0.2639	-	-
ALPHA	2	31.4494	-0.5179	-	-	-
ALPHA	3	40.1438	0.7135	-	-1.1341	-1.4802

			DEVIATION IN VAPOR PHASE COMPOSITION				
X	Y	PRESS.	LAAR 3	MARG 3	MARG 4	ALPHA2	ALPHA3
0.1720	0.8650	610.00	-0.0126	-0.0160	-0.0148	-0.0045	0.0006
0.2090	0.8880	715.00	-0.0081	-0.0080	-0.0082	-0.0034	-0.0015
0.2880	0.9160	875.00	0.0015	0.0011	0.0003	0.0015	-0.0005
0.3150	0.9200	944.00	0.0064	0.0046	0.0038	0.0053	0.0027
0.4550	0.9500	1209.00	0.0070	-0.0077	-0.0070	0.0020	-0.0013
0.4720	0.9520	1250.00	0.0075	-0.0091	-0.0082	0.0022	-0.0010
0.7140	0.9750	1670.00	0.0072	-0.0474	-0.0453	0.0004	0.0012
0.8010	0.9830	1817.00	0.0028	-0.0697	-0.0700	-0.0026	0.0000
0.8790	0.9890	1957.00	-0.0050	-0.0847	-0.0908	-0.0044	-0.0003
0.8930	0.9900	1980.00	-0.0084	-0.0854	-0.0928	-0.0045	-0.0003
		MEAN	0.0067	0.0334	0.0341	0.0031	0.0009

BENZENE(1)-ANILINE(2)

KORTUEM G.,FREIER H.J.,WOERNER F.:CHEM.-ING.TECHN.25,125(1953).

ANTOINE VAPOR PRESSURE CONSTANTS

	A	B	C
(1)	6.90565	1211.033	220.790
(2)	7.63851	1675.300	200.000

P = 350.00

EQUATION	ORDER	A12	A21	D	A122	A211
VAN LAAR	3	0.5189	4.6089	-	-	-
MARGULES	3	0.4483	1.2427	-	-	-
MARGULES	4	0.5509	1.7886	0.9893	-	-
ALPHA	2	28.5876	-1.9676	-	-	-
ALPHA	3	-675.9325	-20.5985	-	722.8069	30.0065

			DEVIATION IN VAPOR PHASE COMPOSITION				
X	Y	TEMP.	LAAR 3	MARG 3	MARG 4	ALPHA2	ALPHA3
0.1175	0.8115	109.80	0.0016	-0.0005	0.0021	0.0081	0.0003
0.1285	0.8335	107.30	-0.0007	-0.0017	-0.0006	0.0029	0.0000
0.1430	0.8550	104.10	-0.0003	-0.0003	-0.0004	0.0005	0.0009
0.1480	0.8590	102.90	0.0026	0.0028	0.0025	0.0024	0.0034
0.1470	0.8670	102.10	-0.0049	-0.0046	-0.0049	-0.0068	-0.0058
0.1540	0.8695	101.70	-0.0006	-0.0001	-0.0007	-0.0015	0.0001
0.1710	0.8860	98.00	0.0017	0.0025	0.0015	-0.0012	0.0010
0.1820	0.8960	96.70	0.0005	0.0014	0.0004	-0.0018	0.0004
0.1860	0.8990	96.00	0.0008	0.0017	0.0007	-0.0016	0.0006
0.2280	0.9270	89.50	-0.0002	-0.0000	-0.0001	-0.0019	-0.0014
0.2580	0.9395	86.80	-0.0006	-0.0014	-0.0005	0.0006	-0.0005
0.2810	0.9465	84.60	0.0000	-0.0015	0.0001	0.0031	0.0008
		MEAN	0.0012	0.0015	0.0012	0.0027	0.0013

BENZENE(1)-ANILINE(2)

KORTUEM G.,FREIER H.J.,WOERNER F.:CHEM.-ING.TECHN.25,125(1953).

ANTOINE VAPOR PRESSURE CONSTANTS

	A	B	C
(1)	6.90565	1211.033	220.790
(2)	7.63851	1675.300	200.000

P = 760.00

EQUATION	ORDER	A12	A21	D	A122	A211
VAN LAAR	3	0.4866	5.4488	-	-	-
MARGULES	3	0.3848	1.2423	-	-	-
MARGULES	4	0.4893	1.7325	0.9210	-	-
ALPHA	2	19.3069	-1.4929	-	-	-
ALPHA	3	204.8067	3.2580	-	-199.1695	-3.0102

			DEVIATION IN VAPOR PHASE COMPOSITION				
X	Y	TEMP.	LAAR 3	MARG 3	MARG 4	ALPHA2	ALPHA3
0.0830	0.6200	150.90	0.0031	-0.0148	-0.0004	0.0390	-0.0130
0.0950	0.6370	150.00	0.0218	0.0084	0.0185	0.0562	0.0183
0.1025	0.6790	146.40	0.0069	-0.0039	0.0038	0.0329	0.0023
0.1240	0.7450	140.80	-0.0032	-0.0082	-0.0055	0.0118	-0.0031
0.1440	0.7900	134.90	-0.0053	-0.0066	-0.0067	-0.0003	-0.0059
0.1500	0.7980	134.00	-0.0038	-0.0043	-0.0049	0.0003	-0.0032
0.1660	0.8150	131.60	0.0020	0.0029	0.0014	0.0040	0.0046
0.1650	0.8190	130.20	-0.0004	0.0005	-0.0010	-0.0012	-0.0008
0.1730	0.8340	128.50	-0.0044	-0.0030	-0.0047	-0.0069	-0.0050
0.1820	0.8380	128.00	0.0006	0.0025	0.0006	-0.0011	0.0021
0.1890	0.8490	126.70	-0.0023	-0.0003	-0.0021	-0.0051	-0.0010
0.1905	0.8490	126.20	-0.0003	0.0018	-0.0000	-0.0036	0.0006
0.1980	0.8550	125.30	0.0009	0.0031	0.0014	-0.0025	0.0023
0.1960	0.8575	125.10	-0.0028	-0.0006	-0.0023	-0.0069	-0.0022
0.2020	0.8600	124.30	0.0004	0.0026	0.0009	-0.0040	0.0011
0.2170	0.8690	123.10	0.0031	0.0053	0.0040	-0.0005	0.0052
0.2100	0.8710	122.60	-0.0026	-0.0004	-0.0018	-0.0081	-0.0027
0.2230	0.8730	121.70	0.0047	0.0068	0.0057	0.0002	0.0059
0.2240	0.8780	121.20	0.0010	0.0031	0.0020	-0.0041	0.0017
0.2410	0.8940	118.50	-0.0023	-0.0007	-0.0011	-0.0081	-0.0024
0.2515	0.9000	116.40	-0.0007	0.0005	0.0006	-0.0074	-0.0019
0.2690	0.9120	114.80	-0.0035	-0.0031	-0.0022	-0.0091	-0.0044
0.2820	0.9130	114.20	0.0009	0.0007	0.0022	-0.0032	0.0009
0.3450	0.9380	107.20	0.0004	-0.0035	0.0009	-0.0012	-0.0012
0.4270	0.9550	101.80	0.0016	-0.0076	-0.0003	0.0062	0.0004
0.5030	0.9660	97.30	0.0014	-0.0127	-0.0037	0.0113	0.0005
		MEAN	0.0031	0.0041	0.0030	0.0090	0.0036

BENZENE(1)-BUTYL ALCOHOL(2)

ALLEN B.B.,LINGO S.P.,FELSING W.A.:J.PHYS.CHEM.43,425(1939).

ANTOINE VAPOR PRESSURE CONSTANTS
	A	B	C
(1)	6.90565	1211.033	220.790
(2)	7.54472	1405.873	183.908

T = 25.00

EQUATION	ORDER	A12	A21	D	A122	A211
VAN LAAR	3	0.7148	0.5388	-	-	-
MARGULES	3	0.6977	0.5210	-	-	-
MARGULES	4	0.4454	0.2424	-0.7617	-	-
ALPHA	2	50.9224	1.7838	-	-	-
ALPHA	3	-.3744/ 09	-.1618/ 08	-	.3078/ 09	.4573/ 04

DEVIATION IN VAPOR PHASE COMPOSITION

X	Y	PRESS.	LAAR 3	MARG 3	MARG 4	ALPHA2	ALPHA3
0.2020	0.8870	51.60	0.0086	0.0085	0.0060	-0.0013	0.0014
0.3500	0.9190	67.40	-0.0018	-0.0017	0.0017	-0.0003	-0.0039
0.5000	0.9300	77.90	-0.0005	-0.0006	-0.0016	0.0033	0.0016
0.6870	0.9410	`85.80	0.0061	0.0059	0.0027	0.0025	0.0040
0.8030	0.9550	89.30	0.0073	0.0075	0.0087	-0.0063	-0.0040
		MEAN	0.0049	0.0048	0.0041	0.0027	0.0030

BENZENE(1)-BUTYL ALCOHOL(2)

BROWN I.,SMITH F.:AUSTR.J.CHEM.12,407(1959).

ANTOINE VAPOR PRESSURE CONSTANTS

	A	B	C
(1)	6.90565	1211.033	220.790
(2)	7.54472	1405.873	183.908

T =   45.00

EQUATION	ORDER	A12	A21	D	A122	A211
VAN LAAR	3	0.4597	0.7677	-	-	-
MARGULES	3	0.4290	0.7079	-	-	-
MARGULES	4	0.5585	0.8287	0.3991	-	-
ALPHA	2	24.8776	0.5941	-	11.6836	-0.2590
ALPHA	3	14.1979	0.2910	-	11.6836	-0.2590

			DEVIATION IN VAPOR PHASE COMPOSITION				
X	Y	PRESS.	LAAR 3	MARG 3	MARG 4	ALPHA2	ALPHA3
0.1794	0.8107	114.04	0.0012	0.0009	0.0027	-0.0019	0.0005
0.2928	0.8689	148.84	0.0018	0.0029	-0.0017	-0.0012	-0.0009
0.3997	0.8946	171.29	0.0029	0.0035	0.0002	0.0009	0.0002
0.5085	0.9115	187.62	0.0020	0.0015	0.0013	0.0016	0.0003
0.5996	0.9221	197.33	0.0008	-0.0006	0.0013	0.0016	0.0001
0.6993	0.9316	205.68	-0.0002	-0.0017	0.0006	0.0014	0.0002
0.8014	0.9421	212.80	-0.0005	-0.0010	-0.0009	-0.0001	-0.0003
0.9026	0.9568	218.90	0.0025	0.0040	0.0002	-0.0030	-0.0003
0.9462	0.9677	221.29	0.0052	0.0071	0.0027	-0.0043	0.0003
		MEAN	0.0019	0.0026	0.0013	0.0018	0.0003

BENZENE(1)-BUTYL ALCOHOL(2)

MANN R.S.,SHEMILT L.W.,WALDICHUCK M.:J.CHEM.ENG.DATA 8(4),502(1963).

ANTOINE VAPOR PRESSURE CONSTANTS
```
          A          B          C
(1) 6.90565   1211.033   220.790
(2) 7.54472   1405.873   183.908
```

P = 760.00

EQUATION	ORDER	A12	A21	D	A122	A211
VAN LAAR	3	0.3568	0.6105	-	-	-
MARGULES	3	0.3312	0.5631	-	-	-
MARGULES	4	0.3562	0.5958	0.1020	-	-
ALPHA	2	6.8139	-0.1158	-	-	-
ALPHA	3	2.6183	-1.2709	-	3.1660	1.0864

			DEVIATION IN VAPOR PHASE COMPOSITION				
X	Y	TEMP.	LAAR 3	MARG 3	MARG 4	ALPHA2	ALPHA3
0.0040	0.0250	116.90	-0.0001	-0.0014	-0.0001	0.0053	0.0015
0.0400	0.2170	112.00	-0.0040	-0.0107	-0.0045	0.0229	0.0029
0.0850	0.3800	107.10	-0.0008	-0.0062	-0.0015	0.0243	0.0043
0.1340	0.5100	102.22	-0.0039	-0.0061	-0.0046	0.0103	-0.0031
0.1610	0.5600	100.22	-0.0017	-0.0023	-0.0022	0.0077	-0.0017
0.1800	0.5900	98.70	0.0005	0.0008	0.0001	0.0063	-0.0005
0.2340	0.6600	95.60	0.0015	0.0034	0.0015	0.0013	0.0010
0.3080	0.7240	92.25	0.0052	0.0077	0.0056	0.0011	0.0062
0.3690	0.7790	89.69	-0.0091	-0.0072	-0.0086	-0.0150	-0.0078
0.3970	0.7870	88.28	-0.0010	0.0005	-0.0004	-0.0080	-0.0005
0.4750	0.8190	86.39	0.0001	0.0002	0.0005	-0.0048	0.0018
0.5600	0.8470	84.54	-0.0002	-0.0015	-0.0002	-0.0022	0.0014
0.6310	0.8670	83.19	-0.0011	-0.0032	-0.0014	-0.0006	-0.0003
0.7140	0.8850	81.98	0.0009	-0.0013	0.0002	0.0042	0.0008
0.7900	0.9040	81.36	0.0001	-0.0011	-0.0006	0.0056	0.0001
0.8480	0.9200	80.87	0.0007	0.0007	0.0003	0.0064	0.0011
0.8970	0.9470	80.67	-0.0089	-0.0076	-0.0088	-0.0040	-0.0077
0.9280	0.9520	80.39	0.0001	0.0020	0.0005	0.0036	0.0016
0.9420	0.9600	80.28	-0.0005	0.0015	-0.0001	0.0022	0.0010
0.9480	0.9630	80.21	-0.0001	0.0019	0.0003	0.0022	0.0013
0.9540	0.9660	80.16	0.0004	0.0024	0.0009	0.0024	0.0019
		MEAN	0.0019	0.0033	0.0020	0.0067	0.0023

T

BENZENE(1)-BUTYL ALCOHOL(2)

YERAZUNIS S.,PLOWRIGHT J.D.,SMOLA F.M.:A.I.CH.E.J.10(5),660(1964).

ANTOINE VAPOR PRESSURE CONSTANTS

	A	B	C
(1)	6.90565	1211.033	220.790
(2)	7.54472	1405.873	183.908

P = 760.00

EQUATION	ORDER	A12	A21	D	A122	A211
VAN LAAR	3	0.3433	0.5606	-	-	-
MARGULES	3	0.3318	0.5157	-	-	-
MARGULES	4	0.4153	0.6211	0.3184	-	-
ALPHA	2	6.5558	-0.0362	-	0.5639	1.0175
ALPHA	3	4.6760	-0.9919	-	0.5639	1.0175

				DEVIATION IN VAPOR PHASE COMPOSITION				
X	Y	TEMP.	LAAR 3	MARG 3	MARG 4	ALPHA2	ALPHA3
0.0060	0.0395	116.56	-0.0035	-0.0044	0.0023	0.0039	-0.0032
0.0117	0.0693	115.74	-0.0009	-0.0024	0.0092	0.0121	-0.0005
0.0215	0.1152	114.57	0.0051	0.0029	0.0203	0.0250	0.0055
0.0647	0.2968	109.04	0.0087	0.0064	0.0271	0.0341	0.0047
0.1994	0.6026	97.82	0.0078	0.0091	0.0064	0.0079	-0.0012
0.2031	0.6076	97.67	0.0077	0.0090	0.0060	0.0075	-0.0010
0.2683	0.6823	93.87	0.0078	0.0091	0.0031	-0.0001	-0.0002
0.3030	0.7129	92.45	0.0068	0.0078	0.0018	-0.0029	0.0002
0.4230	0.7867	87.89	0.0069	0.0063	0.0047	-0.0060	0.0022
0.4329	0.7955	87.37	0.0033	0.0025	0.0013	-0.0102	-0.0020
0.5965	0.8501	84.18	0.0052	0.0027	0.0073	-0.0038	0.0010
0.7279	0.8839	82.37	0.0057	0.0037	0.0074	0.0005	-0.0005
0.8255	0.9085	81.32	0.0080	0.0078	0.0073	0.0042	-0.0001
0.8854	0.9283	80.80	0.0088	0.0100	0.0062	0.0049	0.0001
0.9075	0.9389	80.64	0.0072	0.0088	0.0042	0.0033	-0.0013
0.9265	0.9462	80.49	0.0085	0.0104	0.0054	0.0046	0.0006
0.9535	0.9617	80.35	0.0071	0.0090	0.0041	0.0035	0.0006
0.9805	0.9822	80.23	0.0034	0.0046	0.0016	0.0011	-0.0002
		MEAN	0.0063	0.0065	0.0070	0.0075	0.0014

BENZENE(1)-TERT-BUTYL ALCOHOL(2)

ALLEN B.B.,LINGO S.P.,FELSING W.A.:J.PHYS.CHEM.43,425(1939).

ANTOINE VAPOR PRESSURE CONSTANTS

	A	B	C
(1)	6.90565	1211.033	220.790
(2)	7.44427	1220.499	184.586

T = 25.00

EQUATION	ORDER	A12	A21	D	A122	A211
VAN LAAR	3	0.5025	0.6239	-	-	-
MARGULES	3	0.4958	0.6142	-	-	-
MARGULES	4	0.5598	0.6723	0.1881	-	-
ALPHA	2	6.5467	1.5083	-	-	-
ALPHA	3	4.2893	0.8842	-	1.9944	0.2344

			DEVIATION IN VAPOR PHASE COMPOSITION				
X	Y	PRESS.	LAAR 3	MARG 3	MARG 4	ALPHA2	ALPHA3
0.1980	0.5450	80.50	0.0024	0.0024	0.0029	-0.0020	-0.0010
0.3480	0.6480	92.80	0.0049	0.0053	0.0008	0.0004	0.0018
0.5010	0.7090	100.20	0.0007	0.0005	-0.0001	0.0003	-0.0004
0.6490	0.7550	104.30	-0.0027	-0.0034	-0.0007	-0.0000	-0.0024
0.8020	0.8040	106.60	0.0051	0.0051	0.0048	0.0040	0.0036
0.9000	0.8700	104.70	0.0031	0.0039	-0.0014	-0.0067	-0.0021
		MEAN	0.0032	0.0034	0.0018	0.0022	0.0019

BENZENE(1)-M-CRESOL(2)

SAVITT S.A.,OTHMER D.F.:IND.ENG.CHEM.,44,2428(1952).

ANTOINE VAPOR PRESSURE CONSTANTS
          A   B   C
(1) 6.90565   1211.033   220.790
(2) 7.53185   1875.300   201.000

P = 760.00

EQUATION	ORDER	A12	A21	D	A122	A211
VAN LAAR	3	-0.0229	0.0422	-	-	-
MARGULES	3	0.2809	-0.6739	-	-	-
MARGULES	4	-0.0398	-1.6294	-2.1933	-	-
ALPHA	2	20.5877	-0.8132	-	-	-
ALPHA	3	.1346/ 07	.6058/ 05	-	-.1352/ 07	-.5150/ 05

			DEVIATION IN VAPOR PHASE COMPOSITION				
X	Y	TEMP.	LAAR 3	MARG 3	MARG 4	ALPHA2	ALPHA3
0.0410	0.4930	190.80	-0.0925	0.0261	-0.0795	-0.0145	-0.0337
0.0870	0.6500	179.00	-0.0190	0.0394	0.0022	0.0200	0.0160
0.1530	0.7800	165.70	0.0127	0.0121	0.0205	0.0118	0.0124
0.1800	0.8280	159.20	0.0070	-0.0064	0.0075	-0.0066	-0.0056
0.3140	0.9150	137.50	0.0188	-0.0094	-0.0053	-0.0121	-0.0109
0.4080	0.9410	127.30	0.0182	-0.0010	-0.0101	-0.0095	-0.0086
0.5790	0.9500	116.00	-0.0221	0.0277	0.0140	0.0117	0.0119
0.7130	0.9760	100.10	-0.0141	0.0177	0.0144	0.0001	-0.0003
0.8440	0.9900	88.10	0.0073	0.0088	0.0090	-0.0036	-0.0049
0.9070	0.9930	83.50	0.0058	0.0066	0.0068	-0.0021	-0.0042
		MEAN	0.0217	0.0155	0.0169	0.0092	0.0108

BENZENE(1)-P-DICHLOROBENZENE(2)

MARTIN A.R.,GEORGE C.M.:J.CHEM.SOC.,1413(1933).

ANTOINE VAPOR PRESSURE CONSTANTS

	A	B	C
(1)	6.90565	1211.033	220.790
(2)	6.89797	1507.300	201.000

T = 70.00

EQUATION	ORDER	A12	A21	D	A122	A211
VAN LAAR	3	-0.1250	-2.2162	-	-	-
MARGULES	3	-0.0968	-0.3403	-	-	-
MARGULES	4	0.4386	-0.7752	-1.1998	-	-
ALPHA	2	17.9071	-0.9952	-	-	-
ALPHA	3	.7514/ 05	.3340/ 04	-	-.7832/ 05	-.3294/ 04

			DEVIATION IN VAPOR PHASE COMPOSITION				
X	Y	PRESS.	LAAR 3	MARG 3	MARG 4	ALPHA2	ALPHA3
0.1957	0.8180	100.90	0.0058	0.0042	0.0029	0.0051	0.0002
0.3838	0.9295	202.00	-0.0054	-0.0035	-0.0000	-0.0057	-0.0013
0.4911	0.9528	259.80	-0.0023	0.0020	0.0000	-0.0026	0.0006
0.6392	0.9728	339.10	0.0009	0.0062	0.0029	0.0004	0.0015
0.7694	0.9864	415.60	0.0010	0.0045	0.0040	0.0001	-0.0003
0.7877	0.9875	425.40	0.0015	0.0046	0.0045	0.0005	-0.0001
0.8881	0.9940	481.40	0.0025	0.0030	0.0038	0.0012	-0.0002
0.9517	0.9976	515.10	0.0019	0.0014	0.0019	0.0010	-0.0005
0.9829	0.9990	533.40	0.0010	0.0007	0.0009	0.0007	-0.0004
		MEAN	0.0025	0.0033	0.0023	0.0019	0.0006

BENZENE(1)-1,2-DICHLOROETHANE(2)

KIREEV V.A.,SKVORTSOVA A.A.:ZH.FIZ.KHIM.,6,63(1936).

ANTOINE VAPOR PRESSURE CONSTANTS
	A	B	C
(1)	6.90565	1211.033	220.790
(2)	7.18431	1358.500	232.000

T =   40.00

EQUATION	ORDER	A12	A21	D	A122	A211
VAN LAAR	3	-0.0006	0.0004	-	-	-
MARGULES	3	0.0504	-0.0179	-	-	-
MARGULES	4	-0.0026	-0.0653	-0.1591	-	-
ALPHA	2	0.1719	-0.0937	-	-	-
ALPHA	3	.7389/ 06	.6528/ 06	-	-.7358/ 06	-.6523/ 06

			DEVIATION IN VAPOR PHASE COMPOSITION				
X	Y	PRESS.	LAAR 3	MARG 3	MARG 4	ALPHA2	ALPHA3
0.0750	0.0850	155.00	0.0022	0.0088	0.0039	0.0015	0.0029
0.1300	0.1400	156.20	0.0097	0.0168	0.0134	0.0081	0.0080
0.1860	0.2100	158.00	0.0019	0.0076	0.0070	-0.0005	-0.0016
0.2570	0.2990	162.20	-0.0105	-0.0076	-0.0050	-0.0134	-0.0153
0.3450	0.3750	164.00	-0.0009	0.0046	0.0088	0.0022	0.0002
0.4620	0.4950	168.50	-0.0020	-0.0005	0.0016	0.0001	-0.0011
0.5600	0.5900	171.00	0.0089	0.0003	-0.0009	0.0010	0.0008
0.6400	0.6600	173.50	0.0166	0.0084	0.0056	0.0076	0.0082
0.6950	0.7150	174.50	0.0136	0.0069	0.0041	0.0045	0.0056
0.7450	0.7650	176.50	0.0099	0.0051	0.0032	0.0013	0.0026
0.7920	0.8150	177.20	0.0028	-0.0001	-0.0006	-0.0051	-0.0036
0.8480	0.8650	178.20	0.0030	0.0022	0.0038	-0.0035	-0.0022
0.8750	0.8950	179.00	-0.0031	-0.0030	-0.0007	-0.0088	-0.0076
0.9250	0.9300	179.80	0.0057	0.0066	0.0096	0.0019	0.0026
0.9450	0.9500	180.70	0.0030	0.0039	0.0067	0.0001	0.0005
		MEAN	0.0062	0.0055	0.0050	0.0040	0.0042

BENZENE(1)-1,2-DICHLOROETHANE(2)

ROSANOFF M.A.,EASELEY C.W.:J.AM.CHEM.SOC.36,979(1914).

ANTOINE VAPOR PRESSURE CONSTANTS
	A	B	C
(1)	6.90565	1211.033	220.790
(2)	7.18431	1358.500	232.000

P = 760.00

EQUATION	ORDER	A12	A21	D	A122	A211
VAN LAAR	3	-0.0002	0.0002	-	-	-
MARGULES	3	0.0532	-0.0537	-	-	-
MARGULES	4	0.0402	-0.0675	-0.0397	-	-
ALPHA	2	0.0695	-0.0457	-	-	-
ALPHA	3	.2552/ 05	.2454/ 05	-	-.2533/ 05	-.2468/ 05

DEVIATION IN VAPOR PHASE COMPOSITION

X	Y	TEMP.	LAAR 3	MARG 3	MARG 4	ALPHA2	ALPHA3
0.1298	0.1380	83.00	0.0046	0.0096	0.0087	-0.0007	0.0022
0.2021	0.2150	82.70	0.0053	0.0061	0.0059	-0.0025	-0.0014
0.3195	0.3360	82.30	0.0076	-0.0004	0.0002	-0.0031	-0.0038
0.3623	0.3740	82.20	0.0136	0.0030	0.0036	0.0023	0.0012
0.4720	0.4830	81.90	-0.0016	0.0015	0.0016	0.0034	0.0021
0.5264	0.5400	81.60	-0.0016	-0.0010	-0.0013	0.0005	-0.0006
0.6337	0.6455	81.40	0.0131	0.0028	0.0020	0.0008	0.0004
0.7761	0.7855	80.90	0.0094	0.0087	0.0086	-0.0006	0.0010
0.8803	0.8897	80.60	0.0019	0.0063	0.0072	-0.0043	-0.0010
		MEAN	0.0065	0.0044	0.0044	0.0020	0.0015

BENZENE(1)-1,2-DICHLOROETHANE(2)

SMITH F.R.,MATHESON H.B.:J.RES.NATL.BUR.STAND.20,641(1938).

ANTOINE VAPOR PRESSURE CONSTANTS
	A	B	C
(1)	6.90565	1211.033	220.790
(2)	7.18431	1358.500	232.000

P = 760.00

EQUATION	ORDER	A12	A21	D	A122	A211
VAN LAAR	3	-0.0002	0.0002	-	-	-
MARGULES	3	0.0048	-0.0046	-	-	-
MARGULES	4	0.0048	-0.0047	-0.0002	-	-
ALPHA	2	0.1094	-0.1009	-	-	-
ALPHA	3	-11.1149	-10.2065	-	11.2258	10.1068

			DEVIATION IN VAPOR PHASE COMPOSITION				
X	Y	TEMP.	LAAR 3	MARG 3	MARG 4	ALPHA2	ALPHA3
0.0500	0.0550	83.32	0.0004	0.0009	0.0009	0.0002	0.0002
0.1000	0.1100	83.14	0.0002	0.0008	0.0008	-0.0002	0.0000
0.2000	0.2170	82.79	0.0010	0.0013	0.0013	0.0002	0.0003
0.3000	0.3220	82.45	0.0013	0.0011	0.0011	0.0004	0.0005
0.4000	0.4260	82.10	-0.0009	-0.0003	-0.0003	-0.0006	-0.0006
0.5000	0.5260	81.77	-0.0854	0.0004	0.0004	0.0002	0.0002
0.6000	0.6250	81.43	0.0004	0.0004	0.0003	-0.0000	-0.0001
0.7000	0.7220	81.09	0.0007	0.0004	0.0004	-0.0003	-0.0004
0.8000	0.8160	80.76	0.0013	0.0015	0.0015	0.0004	0.0003
0.9000	0.9090	80.42	0.0007	0.0011	0.0011	0.0001	-0.0000
0.9500	0.9550	80.27	0.0001	0.0004	0.0004	-0.0002	-0.0002
		MEAN	0.0084	0.0008	0.0008	0.0003	0.0003

BENZENE(1)-1,2-DICHLOROETHANE(2)

ZAWIDZKI J.:Z.PHYS.CHEM.,35,129(1900).

ANTOINE VAPOR PRESSURE CONSTANTS
```
          A         B         C
(1) 6.90565   1211.033   220.790
(2) 7.18431   1358.500   232.000
```

T =   49.99

EQUATION	ORDER	A12	A21	D	A122	A211
VAN LAAR	3	-0.0004	0.0006	-	-	-
MARGULES	3	0.0218	-0.0471	-	-	-
MARGULES	4	-0.0357	-0.1062	-0.1718	-	-
ALPHA	2	0.1049	-0.1507	-	-	-
ALPHA	3	-.3971/ 05	-.3631/ 05	-	.3946/ 05	.3706/ 05

			DEVIATION IN VAPOR PHASE COMPOSITION				
X	Y	PRESS.	LAAR 3	MARG 3	MARG 4	ALPHA2	ALPHA3
0.0811	0.0928	238.30	0.0004	0.0021	-0.0031	-0.0036	0.0012
0.0794	0.0900	238.70	0.0013	0.0030	-0.0022	-0.0026	0.0022
0.2458	0.2698	244.00	0.0052	0.0008	0.0020	-0.0023	-0.0019
0.3434	0.3736	247.30	0.0047	-0.0036	-0.0005	-0.0027	-0.0043
0.4766	0.5058	252.00	0.0078	-0.0011	-0.0006	0.0028	-0.0003
0.4785	0.5100	251.30	0.0054	-0.0034	-0.0030	0.0005	-0.0026
0.5840	0.6119	254.80	-0.0010	0.0013	-0.0012	0.0044	0.0018
0.7021	0.7278	259.00	-0.0007	0.0037	0.0011	0.0032	0.0032
0.7073	0.7294	259.30	0.0035	0.0072	0.0047	0.0066	0.0067
0.7073	0.7362	258.80	-0.0033	0.0004	-0.0021	-0.0002	-0.0001
0.8500	0.8728	263.80	-0.0047	0.0001	0.0027	-0.0044	0.0027
0.8500	0.8842	263.30	-0.0161	-0.0113	-0.0087	-0.0158	-0.0087
		MEAN	0.0045	0.0032	0.0026	0.0041	0.0030

BENZENE(1)-DIMETHYL FORMAMIDE(2)

DELZENNE A.:CHEM.ENG.SCI.2,220(1953).

ANTOINE VAPOR PRESSURE CONSTANTS

	A	B	C
(1)	6.90565	1211.033	220.790
(2)	6.99608	1437.840	199.830

P = 760.00

EQUATION	ORDER	A12	A21	D	A122	A211
VAN LAAR	3	0.1322	0.2888	-	-	-
MARGULES	3	0.1369	0.2126	-	-	-
MARGULES	4	0.2123	0.3970	0.4479	-	-
ALPHA	2	7.9458	-0.9291	-	-	-
ALPHA	3	-25.0521	-4.5569	-	33.0337	3.7059

			DEVIATION IN VAPOR PHASE COMPOSITION				
X	Y	TEMP.	LAAR 3	MARG 3	MARG 4	ALPHA2	ALPHA3
0.0550	0.3550	140.00	-0.0342	-0.0321	-0.0134	-0.0121	-0.0119
0.1000	0.4900	131.50	-0.0036	-0.0017	0.0066	0.0096	0.0090
0.2000	0.6900	119.00	0.0045	0.0044	-0.0006	0.0031	-0.0009
0.3500	0.8350	106.00	0.0019	-0.0005	-0.0013	-0.0039	0.0002
0.5600	0.9250	95.20	-0.0036	-0.0064	-0.0003	-0.0023	-0.0028
0.6550	0.9450	91.00	-0.0014	-0.0034	0.0016	0.0028	0.0019
0.8000	0.9750	85.00	-0.0060	-0.0061	-0.0057	0.0008	0.0002
		MEAN	0.0079	0.0078	0.0042	0.0050	0.0038

BENZENE(1)-HEPTANE(2)

BROWN I.:AUSTR.J.SCI.RES.A5,530(1952).

ANTOINE VAPOR PRESSURE CONSTANTS

	A	B	C
(1)	6.90565	1211.033	220.790
(2)	6.90240	1268.115	216.900

T =   80.00

EQUATION	ORDER	A12	A21	D	A122	A211
VAN LAAR	3	0.1349	0.1569	-	-	-
MARGULES	3	0.1338	0.1559	-	-	-
MARGULES	4	0.1293	0.1507	-0.0152	-	-
ALPHA	2	1.3493	-0.1209	-	-	-
ALPHA	3	1.7870	0.0310	-	-0.4778	-0.1007

DEVIATION IN VAPOR PHASE COMPOSITION

X	Y	PRESS.	LAAR 3	MARG 3	MARG 4	ALPHA2	ALPHA3
0.0464	0.0988	454.62	0.0041	0.0039	0.0033	0.0018	0.0005
0.0861	0.1729	476.25	0.0056	0.0054	0.0048	0.0024	0.0007
0.2004	0.3473	534.38	0.0040	0.0040	0.0040	0.0008	-0.0005
0.2792	0.4412	569.49	0.0029	0.0030	0.0033	0.0004	-0.0000
0.3842	0.5464	613.53	0.0003	0.0004	0.0006	-0.0014	-0.0007
0.4857	0.6304	650.16	0.0006	0.0006	0.0006	-0.0008	0.0003
0.5824	0.7009	679.74	0.0020	0.0020	0.0017	0.0002	0.0012
0.6904	0.7759	708.78	0.0021	0.0021	0.0018	-0.0006	-0.0004
0.7842	0.8384	729.77	0.0036	0.0035	0.0035	-0.0001	-0.0010
0.8972	0.9149	748.46	0.0065	0.0065	0.0068	0.0028	0.0008
		MEAN	0.0032	0.0031	0.0030	0.0011	0.0006

BENZENE(1)-HEPTANE(2)

BRZOSTOWSKI W.:BULL.ACAD.POLON.SCI.,SER.CHIM.8,291(1960).

ANTOINE VAPOR PRESSURE CONSTANTS

	A	B	C
(1)	6.90565	1211.033	220.790
(2)	6.90240	1268.115	216.900

P = 760.00

EQUATION	ORDER	A12	A21	D	A122	A211
VAN LAAR	3	0.1145	0.1615	-	-	-
MARGULES	3	0.1090	0.1575	-	-	-
MARGULES	4	0.0946	0.1425	-0.0485	-	-
ALPHA	2	1.2260	-0.1703	-	-	-
ALPHA	3	1.3288	-0.4043	-	-0.2957	0.3526

			DEVIATION IN VAPOR PHASE COMPOSITION				
X	Y	TEMP.	LAAR 3	MARG 3	MARG 4	ALPHA2	ALPHA3
0.0150	0.0320	97.63	0.0005	0.0002	-0.0007	0.0006	-0.0020
0.0540	0.1060	96.11	0.0047	0.0039	0.0020	0.0047	-0.0019
0.0850	0.1600	95.11	0.0070	0.0062	0.0044	0.0067	-0.0011
0.1310	0.2340	93.64	0.0086	0.0080	0.0069	0.0075	0.0000
0.1500	0.2630	93.02	0.0083	0.0079	0.0072	0.0070	0.0000
0.2260	0.3670	90.75	0.0075	0.0077	0.0083	0.0048	0.0014
0.3480	0.5060	87.80	0.0033	0.0039	0.0052	0.0014	0.0014
0.4150	0.5700	86.44	0.0017	0.0024	0.0031	-0.0027	0.0011
0.4800	0.6280	85.21	-0.0013	-0.0009	-0.0008	-0.0059	-0.0017
0.5100	0.6520	84.71	-0.0015	-0.0012	-0.0013	-0.0060	-0.0020
0.5470	0.6790	84.15	-0.0002	-0.0000	-0.0005	-0.0046	-0.0011
0.5940	0.7110	83.52	0.0023	0.0023	0.0015	-0.0019	0.0005
0.6330	0.7370	82.99	0.0039	0.0038	0.0029	0.0001	0.0013
0.6930	0.7770	82.29	0.0052	0.0050	0.0041	0.0019	0.0010
0.7480	0.8140	81.70	0.0054	0.0052	0.0046	0.0026	-0.0004
0.8100	0.8540	81.12	0.0072	0.0071	0.0071	0.0051	0.0002
0.8710	0.8960	80.66	0.0072	0.0073	0.0079	0.0058	-0.0001
0.9350	0.9440	80.31	0.0053	0.0055	0.0064	0.0046	-0.0002
		MEAN	0.0045	0.0044	0.0042	0.0041	0.0010

BENZENE(1)-CYCLOHEXANE(2)

NAGATA I.:J.CHEM.ENG.DATA 7(4),461(1962).

ANTOINE VAPOR PRESSURE CONSTANTS
	A	B	C
(1)	6.90565	1211.033	220.790
(2)	6.84498	1203.526	222.863

P = 760.00

EQUATION	ORDER	A12	A21	D	A122	A211
VAN LAAR	3	0.1455	0.1606	-	-	-
MARGULES	3	0.1451	0.1601	-	-	-
MARGULES	4	0.1464	0.1614	0.0041	-	-
ALPHA	2	0.4563	0.3965	-	-	-
ALPHA	3	0.5122	0.4145	-	-0.0684	-0.0022

			DEVIATION IN VAPOR PHASE COMPOSITION				
X	Y	TEMP.	LAAR 3	MARG 3	MARG 4	ALPHA2	ALPHA3
0.1010	0.1310	79.50	-0.0003	-0.0003	-0.0002	0.0012	0.0006
0.1710	0.2110	78.90	-0.0021	-0.0021	-0.0020	-0.0007	-0.0013
0.2560	0.2930	78.40	0.0013	0.0013	0.0012	0.0020	0.0017
0.3430	0.3760	77.80	-0.0022	-0.0021	-0.0022	-0.0020	-0.0020
0.4280	0.4450	77.50	0.0018	0.0018	0.0017	0.0015	0.0018
0.5250	0.5290	77.40	-0.0018	-0.0019	-0.0018	-0.0022	-0.0018
0.5710	0.5640	77.40	0.0011	0.0011	0.0011	0.0008	0.0012
0.6650	0.6450	77.60	-0.0009	-0.0009	-0.0008	-0.0007	-0.0006
0.7590	0.7280	77.90	0.0001	0.0001	0.0001	0.0008	0.0006
0.8100	0.7770	78.20	0.0001	0.0001	0.0001	0.0011	0.0007
0.8630	0.8340	78.60	-0.0022	-0.0022	-0.0023	-0.0010	-0.0016
0.9450	0.9260	79.30	0.0007	0.0008	0.0006	0.0016	0.0011
		MEAN	0.0012	0.0012	0.0012	0.0013	0.0012

BENZENE(1)-CYCLOHEXANE(2)

RICHARDS A.R.,HARGREAVES E.,IND.ENG.CHEM.,36,805(1944).

ANTOINE VAPOR PRESSURE CONSTANTS

	A	B	C
(1)	6.90565	1211.033	220.790
(2)	6.88617	1229.973	224.104

P = 759.00

EQUATION	ORDER	A12	A21	D	A122	A211
VAN LAAR	3	0.2605	0.1505	-	-	-
MARGULES	3	0.2427	0.1330	-	-	-
MARGULES	4	0.1854	0.0880	-0.1745	-	-
ALPHA	2	0.4313	0.4376	-	-	-
ALPHA	3	-1.0851	-1.1816	-	1.3840	1.5543

DEVIATION IN VAPOR PHASE COMPOSITION

X	Y	TEMP.	LAAR 3	MARG 3	MARG 4	ALPHA2	ALPHA3
0.0450	0.0590	80.20	0.0199	0.0182	0.0123	0.0022	-0.0017
0.0660	0.0840	80.00	0.0260	0.0244	0.0180	0.0039	-0.0007
0.1050	0.1290	79.50	0.0319	0.0309	0.0257	0.0055	0.0010
0.1180	0.1450	79.30	0.0314	0.0306	0.0261	0.0044	0.0001
0.2540	0.2900	78.20	0.0199	0.0211	0.0252	-0.0017	0.0002
0.3830	0.4070	77.50	0.0079	0.0084	0.0140	-0.0047	-0.0000
0.4490	0.4610	77.40	0.0057	0.0053	0.0091	-0.0036	0.0002
0.5020	0.5020	77.40	0.0066	0.0055	0.0071	-0.0010	0.0013
0.5540	0.5450	77.40	0.0055	0.0038	0.0033	-0.0012	-0.0005
0.5970	0.5800	77.50	0.0060	0.0041	0.0021	-0.0004	-0.0011
0.6280	0.6050	77.60	0.0073	0.0054	0.0025	0.0008	-0.0007
0.6450	0.6210	77.60	0.0060	0.0041	0.0008	-0.0007	-0.0025
0.7200	0.6820	77.70	0.0123	0.0111	0.0075	0.0047	0.0022
0.7380	0.6980	77.80	0.0132	0.0122	0.0090	0.0052	0.0029
0.7530	0.7170	78.00	0.0085	0.0077	0.0048	0.0003	-0.0019
0.7980	0.7580	78.20	0.0117	0.0118	0.0104	0.0029	0.0016
0.8350	0.7980	78.30	0.0098	0.0105	0.0109	0.0008	0.0004
0.8790	0.8470	78.70	0.0082	0.0097	0.0120	-0.0004	0.0005
0.8930	0.8650	78.90	0.0058	0.0074	0.0103	-0.0024	-0.0012
0.9140	0.8890	79.00	0.0058	0.0075	0.0110	-0.0017	-0.0001
0.9280	0.9070	79.20	0.0041	0.0058	0.0095	-0.0027	-0.0010
0.9370	0.9180	79.30	0.0037	0.0055	0.0091	-0.0025	-0.0008
0.9600	0.9480	79.50	0.0016	0.0029	0.0061	-0.0030	-0.0014
		MEAN	0.0113	0.0110	0.0107	0.0025	0.0010

BENZENE(1)-CYCLOHEXANE(2)

SCATCHARD G.,WOOD S.E.,MOCHEL J.M.:J.PHYS.CHEM.,43,119(1939).

ANTOINE VAPOR PRESSURE CONSTANTS

	A	B	C
(1)	6.90565	1211.033	220.790
(2)	6.84498	1203.526	222.863

T = 39.99

EQUATION	ORDER	A12	A21	D	A122	A211
VAN LAAR	3	0.1800	0.1937	-	-	-
MARGULES	3	0.1798	0.1933	-	-	-
MARGULES	4	0.1956	0.2091	0.0479	-	-
ALPHA	2	0.5315	0.5545	-	-	-
ALPHA	3	1.2990	1.1665	-	-0.7731	-0.4910

DEVIATION IN VAPOR PHASE COMPOSITION

X	Y	PRESS.	LAAR 3	MARG 3	MARG 4	ALPHA2	ALPHA3
0.1282	0.1657	194.94	0.0005	0.0005	0.0017	0.0016	0.0003
0.2354	0.2766	200.65	0.0003	0.0003	-0.0001	0.0003	-0.0010
0.3685	0.3912	204.75	0.0029	0.0029	0.0020	0.0017	0.0017
0.4932	0.4950	206.12	-0.0010	-0.0010	-0.0011	-0.0026	-0.0011
0.6143	0.5909	205.18	-0.0010	-0.0010	-0.0002	-0.0022	-0.0005
0.7428	0.6979	201.73	0.0014	0.0014	0.0020	0.0014	0.0011
0.8656	0.8205	195.04	0.0020	0.0020	0.0009	0.0029	-0.0006
		MEAN	0.0013	0.0013	0.0011	0.0018	0.0009

311

BENZENE(1)-CYCLOHEXANE(2)

SCATCHARD G.,WOOD S.E.,MOCHEL J.M.:J.PHYS.CHEM.,43,119(1939).

ANTOINE VAPOR PRESSURE CONSTANTS

	A	B	C
(1)	6.90565	1211.033	220.790
(2)	6.84498	1203.526	222.863

T =   69.98

EQUATION	ORDER	A12	A21	D	A122	A211
VAN LAAR	3	0.1512	0.1552	-	-	-
MARGULES	3	0.1512	0.1551	-	-	-
MARGULES	4	0.1628	0.1667	0.0348	-	-
ALPHA	2	0.4407	0.4119	-	-	-
ALPHA	3	.2573/ 05	.2486/ 05	-	-.2447/ 05	-.2315/ 0

DEVIATION IN VAPOR PHASE COMPOSITION

X	Y	PRESS.	LAAR 3	MARG 3	MARG 4	ALPHA2	ALPHA3
0.1186	0.1486	567.60	0.0025	0.0025	0.0035	0.0026	0.0099
0.2409	0.2805	584.90	-0.0020	-0.0020	-0.0023	-0.0024	-0.0097
0.3759	0.3982	596.16	0.0020	0.0020	0.0013	0.0012	-0.0051
0.4945	0.4975	600.27	0.0017	0.0017	0.0017	0.0009	0.0017
0.6180	0.6027	599.32	-0.0010	-0.0010	-0.0004	-0.0016	0.0056
0.7248	0.6962	593.48	-0.0012	-0.0012	-0.0007	-0.0016	0.0055
0.8659	0.8311	577.79	0.0035	0.0035	0.0027	0.0034	-0.0076
		MEAN	0.0020	0.0020	0.0018	0.0020	0.0065

BENZENE(1)-CYCLOHEXANE(2)

WECK H.I.,HUNT H.:IND.ENG.CHEM.46,2521(1954).

ANTOINE VAPOR PRESSURE CONSTANTS
	A	B	C
(1)	6.90565	1211.033	220.790
(2)	6.88617	1229.973	224.104

P = 760.00

EQUATION	ORDER	A12	A21	D	A122	A211
VAN LAAR	3	0.2071	0.1159	-	-	-
MARGULES	3	0.1805	0.1122	-	-	-
MARGULES	4	0.2542	0.1686	0.1953	-	-
ALPHA	2	0.4000	0.3227	-	-	-
ALPHA	3	0.9985	0.5362	-	-0.7373	-0.0543

			DEVIATION IN VAPOR PHASE COMPOSITION				
X	Y	TEMP.	LAAR 3	MARG 3	MARG 4	ALPHA2	ALPHA3
0.1600	0.1860	87.85	0.0246	0.0239	0.0279	0.0088	0.0020
0.2740	0.3090	87.05	0.0099	0.0121	0.0084	0.0001	-0.0028
0.3950	0.4200	85.80	0.0022	0.0052	-0.0002	-0.0017	-0.0001
0.5090	0.5170	85.50	0.0006	0.0027	0.0006	-0.0012	0.0024
0.6140	0.6090	86.55	-0.0010	-0.0001	0.0013	-0.0038	-0.0011
0.6130	0.6080	85.93	-0.0009	-0.0000	0.0014	-0.0037	-0.0010
0.6700	0.6520	86.10	0.0061	0.0064	0.0087	0.0020	0.0032
0.7060	0.6880	86.10	0.0032	0.0033	0.0055	-0.0019	-0.0020
0.8330	0.8040	87.24	0.0113	0.0112	0.0092	0.0035	-0.0017
0.9290	0.9050	87.06	0.0130	0.0132	0.0082	0.0069	0.0008
0.9310	0.9070	88.20	0.0132	0.0135	0.0084	0.0073	0.0012
		MEAN	0.0078	0.0083	0.0073	0.0037	0.0017

U

BENZENE(1)-CYCLOHEXENE(2)

HARRISON J.M.,BERG L.:IND.ENG.CHEM.,38,117(1946).

ANTOINE VAPOR PRESSURE CONSTANTS
	A	B	C
(1)	6.90565	1211.033	220.790
(2)	6.84498	1203.526	222.863

P =   740.00

EQUATION	ORDER	A12	A21	D	A122	A211
VAN LAAR	3	0.0512	0.1965	-	-	-
MARGULES	3	0.0141	0.1189	-	-	-
MARGULES	4	0.0054	0.1115	-0.0244	-	-
ALPHA	2	0.3350	0.1556	-	-	-
ALPHA	3	0.4819	0.2094	-	-0.1937	-0.0235

			DEVIATION IN VAPOR PHASE COMPOSITION				
X	Y	TEMP.	LAAR 3	MARG 3	MARG 4	ALPHA2	ALPHA3
0.0680	0.0850	81.50	-0.0080	-0.0119	-0.0126	0.0015	-0.0006
0.3300	0.3630	79.80	-0.0092	-0.0081	-0.0075	0.0015	0.0004
0.4660	0.4870	79.40	-0.0007	0.0011	0.0015	0.0026	0.0029
0.4960	0.5180	79.20	-0.0034	-0.0020	-0.0017	-0.0016	-0.0011
0.5530	0.5690	79.10	-0.0014	-0.0011	-0.0011	-0.0020	-0.0013
0.5830	0.5950	79.10	0.0000	-0.0003	-0.0004	-0.0014	-0.0007
0.5970	0.6110	79.10	-0.0032	-0.0038	-0.0041	-0.0049	-0.0043
0.6570	0.6570	78.90	0.0047	0.0029	0.0025	0.0026	0.0030
0.6830	0.6800	78.90	0.0048	0.0027	0.0023	0.0030	0.0032
0.7300	0.7270	78.90	-0.0006	-0.0029	-0.0032	-0.0012	-0.0013
0.8420	0.8300	78.95	-0.0037	-0.0033	-0.0030	0.0022	0.0012
0.8510	0.8410	78.95	-0.0064	-0.0056	-0.0052	0.0001	-0.0010
0.8850	0.8750	78.95	-0.0084	-0.0058	-0.0052	0.0004	-0.0009
		MEAN	0.0042	0.0040	0.0039	0.0019	0.0017

BENZENE(1)-METHYLCYCLOHEXANE(2)

MYERS H.S.:IND.ENG.CHEM.48,1104(1956).

ANTOINE VAPOR PRESSURE CONSTANTS
	A	B	C
(1)	6.90565	1211.033	220.790
(2)	6.82689	1272.864	221.630

P = 760.00

EQUATION	ORDER	A12	A21	D	A122	A211
VAN LAAR	3	0.1294	0.1329	-	-	-
MARGULES	3	0.1292	0.1329	-	-	-
MARGULES	4	0.0981	0.0967	-0.1036	-	-
ALPHA	2	1.3479	-0.1810	-	-	-
ALPHA	3	-0.3930	-1.2234	-	1.5704	1.0214

DEVIATION IN VAPOR PHASE COMPOSITION

X	Y	TEMP.	LAAR 3	MARG 3	MARG 4	ALPHA2	ALPHA3
0.0150	0.0260	100.40	0.0098	0.0097	0.0076	0.0083	0.0061
0.0350	0.0720	99.50	0.0084	0.0084	0.0046	0.0055	0.0011
0.0520	0.1095	98.65	0.0063	0.0063	0.0019	0.0025	-0.0031
0.0830	0.1635	97.60	0.0118	0.0118	0.0073	0.0070	0.0003
0.1095	0.2075	96.50	0.0142	0.0141	0.0104	0.0088	0.0022
0.1670	0.2970	94.50	0.0134	0.0133	0.0120	0.0080	0.0031
0.2035	0.3520	93.20	0.0079	0.0079	0.0079	0.0029	-0.0005
0.2310	0.3855	92.40	0.0088	0.0088	0.0097	0.0042	0.0021
0.2690	0.4360	91.30	0.0025	0.0025	0.0041	-0.0016	-0.0020
0.3070	0.4830	90.25	-0.0037	-0.0037	-0.0018	-0.0073	-0.0063
0.3370	0.5115	89.50	-0.0021	-0.0020	-0.0002	-0.0053	-0.0034
0.3610	0.5320	88.90	0.0004	0.0005	0.0021	-0.0026	-0.0002
0.3885	0.5550	88.35	0.0026	0.0027	0.0040	-0.0001	0.0026
0.4200	0.5820	87.70	0.0033	0.0033	0.0041	0.0007	0.0036
0.4480	0.6085	87.15	0.0003	0.0004	0.0007	-0.0021	0.0007
0.4830	0.6370	86.50	0.0001	0.0001	-0.0002	-0.0023	0.0001
0.5100	0.6550	85.95	0.0032	0.0032	0.0024	0.0007	0.0027
0.5315	0.6750	85.50	-0.0004	-0.0004	-0.0016	-0.0031	-0.0015
0.5730	0.7025	84.90	0.0028	0.0028	0.0011	-0.0002	0.0004
0.6170	0.7330	82.25	0.0044	0.0044	0.0023	0.0003	-0.0001
0.6650	0.7670	83.50	0.0033	0.0033	0.0013	-0.0009	-0.0023
0.7050	0.7895	83.05	0.0082	0.0082	0.0065	0.0034	0.0013
0.7390	0.8150	82.55	0.0057	0.0057	0.0046	0.0005	-0.0019
0.7770	0.8380	82.10	0.0083	0.0083	0.0079	0.0026	0.0002
0.8210	0.8670	81.65	0.0089	0.0089	0.0095	0.0029	0.0009
0.8605	0.8960	81.20	0.0066	0.0066	0.0080	0.0007	-0.0007
0.9000	0.9230	80.90	0.0066	0.0066	0.0085	0.0012	0.0007
0.9335	0.9480	80.60	0.0048	0.0048	0.0068	0.0005	0.0005
0.9650	0.9730	80.35	0.0019	0.0019	0.0034	-0.0008	-0.0005
		MEAN	0.0055	0.0055	0.0049	0.0030	0.0018

BENZENE(1)-NITROMETHANE(2)

WECK H.J.,HUNT H.:IND.ENG.CHEM.46,2521(1954).

ANTOINE VAPOR PRESSURE CONSTANTS
	A	B	C
(1)	6.90565	1211.033	220.790
(2)	7.15541	1366.483	218.611

P =   760.00

EQUATION	ORDER	A12	A21	D	A122	A211
VAN LAAR	3	0.4357	0.5202	-	-	-
MARGULES	3	0.4336	0.5129	-	-	-
MARGULES	4	0.4895	0.5794	0.1953	-	-
ALPHA	2	4.0658	0.9691	-	-	-
ALPHA	3	3.0636	1.6405	-	1.9446	-1.1763

			DEVIATION IN VAPOR PHASE COMPOSITION				
X	Y	TEMP.	LAAR 3	MARG 3	MARG 4	ALPHA2	ALPHA3
0.0650	0.2510	108.30	-0.0191	-0.0195	-0.0073	-0.0120	0.0089
0.1460	0.4170	101.20	-0.0191	-0.0192	-0.0150	-0.0159	-0.0022
0.2380	0.5190	99.30	-0.0126	-0.0125	-0.0152	-0.0092	-0.0061
0.3300	0.5730	98.70	0.0027	0.0027	-0.0012	0.0086	0.0047
0.4320	0.6320	96.27	0.0002	-0.0002	-0.0013	0.0075	0.0003
0.5320	0.6790	95.30	-0.0036	-0.0042	-0.0017	0.0063	-0.0004
0.6410	0.7270	94.40	-0.0078	-0.0085	-0.0038	0.0034	0.0010
0.7580	0.7900	90.70	-0.0152	-0.0155	-0.0131	-0.0082	-0.0024
0.8760	0.8650	90.45	-0.0136	-0.0130	-0.0177	-0.0132	0.0015
		MEAN	0.0104	0.0106	0.0085	0.0094	0.0031

BENZENE(1)-PHENOL(2)

MARTIN A.R.,GEORGE C.M.:J.CHEM.SOC.,1413(1933).

ANTOINE VAPOR PRESSURE CONSTANTS
```
            A          B          C
(1)  6.90565    1211.033    220.790
(2)  7.86819    2011.400    222.000
```

T =   70.00

EQUATION	ORDER	A12	A21	D	A122	A211
VAN LAAR	3	0.2603	0.5870	-	-	-
MARGULES	3	0.2121	0.4821	-	-	-
MARGULES	4	0.3908	0.6419	0.4839	-	-
ALPHA	2	96.0854	-0.5697	-	-	-
ALPHA	3	4.8887	-1.4119	-	96.1660	0.4680

			DEVIATION IN VAPOR PHASE COMPOSITION				
X	Y	PRESS.	LAAR 3	MARG 3	MARG 4	ALPHA2	ALPHA3
0.2700	0.9694	236.00	0.0006	0.0009	0.0001	-0.0006	-0.0001
0.4115	0.9810	319.50	0.0005	0.0007	-0.0000	0.0003	0.0002
0.5769	0.9880	392.20	-0.0001	-0.0003	-0.0001	0.0003	-0.0002
0.5967	0.9884	405.80	0.0001	-0.0002	0.0000	0.0005	-0.0000
0.7785	0.9923	459.80	0.0003	0.0001	0.0001	0.0006	0.0000
0.8575	0.9937	482.20	0.0007	0.0008	0.0003	0.0005	0.0003
0.8709	0.9948	487.70	-0.0001	0.0001	-0.0004	-0.0003	-0.0005
0.8864	0.9943	495.90	0.0009	0.0011	0.0005	0.0004	0.0004
0.9188	0.9960	508.40	0.0002	0.0005	-0.0002	-0.0008	-0.0002
0.9330	0.9965	511.90	0.0002	0.0005	-0.0002	-0.0010	-0.0002
0.9678	0.9979	530.10	0.0003	0.0005	0.0000	-0.0015	0.0002
0.9702	0.9983	530.00	-0.0000	0.0002	-0.0002	-0.0018	-0.0001
		MEAN	0.0003	0.0005	0.0002	0.0007	0.0002

BENZENE(1)-PROPYL ALCOHOL(2)

LEE SIANG CHIEN:J.PHYS.CHEM.35,3558(1931).

ANTOINE VAPOR PRESSURE CONSTANTS
```
          A          B          C
(1) 6.90565    1211.033    220.790
(2) 7.85418    1497.910    204.112
```

T =   40.00

EQUATION	ORDER	A12	A21	D	A122	A211
VAN LAAR	3	0.6270	0.9703	-	-	-
MARGULES	3	0.6151	0.8981	-	-	-
MARGULES	4	0.7556	1.1130	0.6024	-	-
ALPHA	2	18.3928	2.8054	-	-	-
ALPHA	3	23.1074	3.1227	-	-4.7117	0.4259

			DEVIATION IN VAPOR PHASE COMPOSITION				
X	Y	PRESS.	LAAR 3	MARG 3	MARG 4	ALPHA2	ALPHA3
0.0990	0.5840	102.00	-0.0113	-0.0120	0.0098	0.0178	0.0169
0.1300	0.6770	114.00	-0.0464	-0.0461	-0.0362	-0.0265	-0.0273
0.2090	0.7060	134.20	0.0105	0.0116	0.0056	0.0154	0.0148
0.2910	0.7600	156.00	0.0023	0.0027	-0.0061	0.0003	0.0001
0.3600	0.7850	168.40	-0.0000	-0.0009	-0.0063	-0.0036	-0.0035
0.4160	0.7950	175.00	0.0021	-0.0001	-0.0009	-0.0008	-0.0004
0.5080	0.8130	183.50	-0.0036	-0.0078	-0.0004	-0.0024	-0.0017
0.7000	0.8370	193.00	-0.0162	-0.0213	-0.0079	-0.0001	0.0007
0.8200	0.8540	196.00	-0.0196	-0.0197	-0.0210	0.0021	0.0017
0.9610	0.9160	191.70	0.0070	0.0147	-0.0113	0.0037	-0.0021
		MEAN	0.0119	0.0137	0.0106	0.0073	0.0069

BENZENE(1)-PROPYL ALCOHOL(2)

PRABHU P.S.;VAN WINKLE M.:J.CHEM.ENG.DATA 8(2),210(1963).

ANTOINE VAPOR PRESSURE CONSTANTS
	A	B	C
(1)	6.90565	1211.033	220.790
(2)	7.85416	1497.910	204.112

P =   760.00

EQUATION	ORDER	A12	A21	D	A122	A211
VAN LAAR	3	0.4540	0.5734	-	-	-
MARGULES	3	0.4446	0.5665	-	-	-
MARGULES	4	0.4349	0.5566	-0.0340	-	-
ALPHA	2	4.1247	1.1061	-	-	-
ALPHA	3	3.9966	-0.3979	-	-1.2680	1.8941

			DEVIATION IN VAPOR PHASE COMPOSITION				
X	Y	TEMP.	LAAR 3	MARG 3	MARG 4	ALPHA2	ALPHA3
0.0490	0.1420	92.80	0.0461	0.0439	0.0419	0.0519	0.0185
0.1040	0.2960	88.40	0.0363	0.0347	0.0333	0.0323	0.0010
0.1800	0.4360	84.75	0.0248	0.0246	0.0247	0.0091	-0.0043
0.2540	0.5300	82.00	0.0132	0.0139	0.0148	-0.0099	-0.0080
0.3980	0.6220	79.00	0.0186	0.0196	0.0202	-0.0068	0.0085
0.5040	0.6800	77.40	0.0080	0.0085	0.0085	-0.0148	-0.0006
0.6400	0.7260	76.51	0.0092	0.0090	0.0082	-0.0068	-0.0014
0.7640	0.7740	76.00	0.0131	0.0127	0.0122	0.0019	-0.0031
0.7920	0.7760	76.05	0.0246	0.0244	0.0240	0.0144	0.0071
0.8340	0.8120	76.25	0.0120	0.0120	0.0120	0.0029	-0.0073
0.9160	0.8640	76.88	0.0232	0.0237	0.0246	0.0154	0.0027
0.9560	0.9160	78.25	0.0154	0.0160	0.0169	0.0098	-0.0004
		MEAN	0.0204	0.0203	0.0201	0.0147	0.0052

BENZENE(1)-ISO-PROPYL ALCOHOL(2)

NAGATA I.:J.CHEM.ENG.DATA 10,106(1965).

ANTOINE VAPOR PRESSURE CONSTANTS
	A	B	C
(1)	6.90565	1211.033	220.790
(2)	7.75634	1366.142	197.970

P = 500.00

EQUATION	ORDER	A12	A21	D	A122	A211
VAN LAAR	3	0.5090	0.7225	-	-	-
MARGULES	3	0.4956	0.6909	-	-	-
MARGULES	4	0.5880	0.7910	0.3126	-	-
ALPHA	2	4.2926	2.3459	-	-6.3771	1.9198
ALPHA	3	9.8206	2.7880	-		

DEVIATION IN VAPOR PHASE COMPOSITION

X	Y	TEMP.	LAAR 3	MARG 3	MARG 4	ALPHA2	ALPHA3
0.0390	0.1480	69.50	-0.0181	-0.0205	-0.0031	0.0121	-0.0056
0.0890	0.2620	67.10	-0.0093	-0.0115	0.0041	0.0221	0.0027
0.1420	0.3500	65.40	-0.0037	-0.0047	0.0023	0.0176	0.0031
0.1970	0.4240	63.90	-0.0059	-0.0058	-0.0068	0.0033	-0.0050
0.2550	0.4690	62.90	0.0050	0.0057	-0.0005	0.0044	0.0021
0.3350	0.5250	61.80	0.0054	0.0059	-0.0015	-0.0041	-0.0005
0.4140	0.5630	61.00	0.0074	0.0071	0.0027	-0.0055	0.0020
0.4950	0.6000	60.90	0.0001	-0.0014	-0.0007	-0.0104	-0.0016
0.5660	0.6260	60.50	-0.0028	-0.0053	-0.0003	-0.0100	-0.0021
0.6400	0.6470	60.20	-0.0028	-0.0058	0.0020	-0.0030	0.0017
0.7160	0.6740	60.10	-0.0061	-0.0087	-0.0015	0.0021	0.0007
0.7970	0.7070	60.30	-0.0057	-0.0064	-0.0050	0.0121	0.0001
0.9420	0.8280	63.00	0.0118	0.0173	-0.0009	0.0354	-0.0029
0.9760	0.8960	64.70	0.0210	0.0254	0.0108	0.0357	0.0043
		MEAN	0.0075	0.0094	0.0030	0.0127	0.0024

BENZENE(1)-ISOPROPYL ALCOHOL(2)

OLSEN A.L.,WASHBURN E.R.:J.PHYS.CHEM.41,457(1937).

ANTOINE VAPOR PRESSURE CONSTANTS
```
       A          B          C
(1) 6.90565    1211.033    220.790
(2) 7.75634    1366.142    197.970
```

T =   25.00

EQUATION	ORDER	A12	A21	D	A122	A211
VAN LAAR	3	0.5527	0.7952	-	-	-
MARGULES	3	0.5387	0.7637	-	-	-
MARGULES	4	0.6201	0.8491	0.3097	-	-
ALPHA	2	8.6338	2.4148	-	-	-
ALPHA	3	9.9590	1.3659	-	-2.9648	1.7155

			DEVIATION IN VAPOR PHASE COMPOSITION				
X	Y	PRESS.	LAAR 3	MARG 3	MARG 4	ALPHA2	ALPHA3
0.0760	0.3650	66.40	-0.0016	-0.0043	0.0132	0.0192	-0.0034
0.1640	0.5300	84.00	0.0036	0.0042	0.0034	0.0059	-0.0016
0.3000	0.6350	99.80	0.0143	0.0161	0.0066	0.0014	0.0061
0.4790	0.7120	105.80	0.0021	0.0014	0.0005	-0.0111	-0.0047
0.6380	0.7450	108.40	0.0001	-0.0029	0.0046	-0.0039	-0.0019
0.8540	0.7950	109.00	0.0135	0.0140	0.0123	0.0170	0.0086
0.9410	0.8770	104.50	0.0081	0.0118	0.0012	0.0033	-0.0084
		MEAN	0.0062	0.0078	0.0060	0.0088	0.0049

BENZENE(1)-TOLUENE(2)

GRISWOLD J.,ANDRES D.,KLEIN V.A.,TRANS.AM.INST.CHEM.ENGRS.,
        39,223(1943).

ANTOINE VAPOR PRESSURE CONSTANTS
          A   B   C
(1) 6.90565   1211.033   220.033
(2) 6.95334   1343.943   219.377

T =   120.00

EQUATION	ORDER	A12	A21	D	A122	A211
VAN LAAR	3	-0.0098	-0.2312	-	-	-
MARGULES	3	-0.0186	-0.0206	-	-	-
MARGULES	4	-0.0676	-0.0891	-0.1692	-	-
ALPHA	2	1.1766	-0.5611	-	-	-
ALPHA	3	-1.7369	-2.0537	-	2.8080	1.5820

			DEVIATION IN VAPOR PHASE COMPOSITION				
X	Y	PRESS.	LAAR 3	MARG 3	MARG 4	ALPHA2	ALPHA3
0.1170	0.2200	1132.00	0.0052	0.0032	-0.0025	0.0043	-0.0008
0.2560	0.4220	1324.00	0.0083	0.0081	0.0089	0.0077	0.0070
0.2580	0.4250	1330.00	0.0078	0.0077	0.0085	0.0073	0.0067
0.2640	0.4530	1324.00	-0.0125	-0.0125	-0.0116	-0.0130	-0.0133
0.4400	0.6390	1515.00	-0.0058	-0.0025	-0.0034	-0.0052	0.0002
0.6820	0.8240	1868.00	0.0017	0.0061	0.0034	0.0028	-0.0000
		MEAN	0.0069	0.0067	0.0064	0.0067	0.0047

BENZENE(1)-TOLUENE(2)

HEERTJES P.M.:CHEM.PROC.ENG.41,385(1960).

ANTOINE VAPOR PRESSURE CONSTANTS

	A	B	C
(1)	6.90565	1211.033	220.790
(2)	6.95334	1343.943	219.377

P = 760.00

EQUATION	ORDER	A12	A21	D	A122	A211
VAN LAAR	3	-0.0021	0.0051	-	-	-
MARGULES	3	0.0242	-0.0584	-	-	-
MARGULES	4	-0.0142	-0.1040	-0.1366	-	-
ALPHA	2	1.2903	-0.5986	-	-	-
ALPHA	3	12.7593	4.2173	-	-11.5754	-4.8208

			DEVIATION IN VAPOR PHASE COMPOSITION				
X	Y	TEMP.	LAAR 3	MARG 3	MARG 4	ALPHA2	ALPHA3
0.0420	0.0890	108.75	0.0044	0.0079	0.0029	0.0024	-0.0001
0.0470	0.1020	108.42	0.0020	0.0056	0.0003	-0.0004	-0.0030
0.1320	0.2570	104.87	0.0085	0.0097	0.0067	0.0023	0.0002
0.1830	0.3340	103.00	0.0147	0.0126	0.0124	0.0067	0.0057
0.2190	0.3950	101.52	0.0073	0.0029	0.0043	-0.0018	-0.0022
0.3250	0.5300	97.76	0.0087	-0.0002	0.0028	-0.0021	-0.0012
0.4070	0.6190	95.01	0.0072	-0.0020	-0.0020	-0.0034	-0.0023
0.4830	0.6880	92.76	0.0076	-0.0007	0.0003	-0.0014	-0.0005
0.5510	0.7420	90.76	0.0068	0.0046	0.0027	0.0011	0.0016
0.6280	0.8000	88.63	-0.0066	0.0054	0.0029	0.0001	0.0002
0.7120	0.8530	86.41	-0.8530	0.0091	0.0072	0.0023	0.0019
0.8100	0.9110	84.10	-0.0007	0.0076	0.0076	0.0005	-0.0001
0.9000	0.9580	81.99	-0.0002	0.0036	0.0049	-0.0014	-0.0018
0.9040	0.9600	81.92	-0.0003	0.0033	0.0047	-0.0016	-0.0019
0.9410	0.9730	81.18	0.0030	0.0055	0.0068	0.0022	0.0020
		MEAN	0.0621	0.0054	0.0045	0.0020	0.0016

BENZENE(1)-TOLUENE(2)

KIRSCHBAUM E.,GERTSNER H.:VERFAHRENSTECHNIK 1,10(1910).

ANTOINE VAPOR PRESSURE CONSTANTS
```
          A           B          C
(1) 6.90565    1211.033    220.790
(2) 6.95334    1343.943    219.377
```

P = 760.00

EQUATION	ORDER	A12	A21	D	A122	A211
VAN LAAR	3	-0.0001	0.0001	-	-	-
MARGULES	3	0.0052	-0.0044	-	-	-
MARGULES	4	0.1064	0.1164	0.3418	-	-
ALPHA	2	1.3995	-0.6299	-	-	-
ALPHA	3	-.3401/ 05	-.1394/ 05	-	.3390/ 05	.1404/ 05

DEVIATION IN VAPOR PHASE COMPOSITION

X	Y	TEMP.	LAAR 3	MARG 3	MARG 4	ALPHA2	ALPHA3
0.0880	0.2120	106.10	-0.0252	-0.0242	-0.0093	-0.0233	-0.0163
0.2000	0.3700	102.20	0.0059	0.0063	0.0066	0.0074	0.0121
0.3000	0.5000	98.60	0.0108	0.0104	0.0044	0.0113	0.0136
0.3970	0.6180	95.20	-0.0019	-0.0002	-0.0038	0.0001	0.0004
0.4890	0.7100	92.10	-0.0066	-0.0066	-0.0049	-0.0066	-0.0079
0.5920	0.7890	89.40	-0.0042	-0.0049	0.0006	-0.0048	-0.0073
0.7000	0.8530	86.80	0.0025	0.0022	0.0065	0.0026	-0.0003
0.8030	0.9140	84.40	-0.0015	-0.0014	-0.0019	-0.0008	-0.0031
0.9030	0.9570	82.30	0.0030	0.0032	-0.0010	0.0038	0.0033
0.9500	0.9790	81.20	0.0011	0.0012	-0.0026	0.0016	0.0025
		MEAN	0.0063	0.0061	0.0042	0.0062	0.0067

BENZENE(1)-TOLUENE(2)

ROSANOFF M.A.,BACON C.W.,SCHULZE J.F.W.:J.AM.CHEM.SOC.,
    36,1999(1914).

ANTOINE VAPOR PRESSURE CONSTANTS
        A   B   C
(1) 6.90565   1211.033   220.790
(2) 6.95334   1343.943   219.377

P =   760.00

EQUATION	ORDER	A12	A21	D	A122	A211
VAN LAAR	3	-0.0108	-0.0177	-	-	-
MARGULES	3	-0.0090	-0.0172	-	-	-
MARGULES	4	0.0048	-0.0011	0.0457	-	-
ALPHA	2	1.3475	-0.6438	-	-	-
ALPHA	3	-1.5249	-1.8505	-	2.8637	1.2387

			DEVIATION IN VAPOR PHASE COMPOSITION				
X	Y	TEMP.	LAAR 3	MARG 3	MARG 4	ALPHA2	ALPHA3
0.1000	0.2080	105.31	-0.0020	-0.0016	0.0002	0.0001	-0.0007
0.2000	0.3720	101.46	0.0003	0.0004	0.0005	0.0015	0.0004
0.3000	0.5070	98.00	0.0011	0.0009	0.0001	0.0009	0.0001
0.4000	0.6190	95.05	0.0008	0.0005	0.0000	-0.0001	-0.0000
0.5000	0.7130	92.30	-0.0003	-0.0005	-0.0002	-0.0013	-0.0003
0.6000	0.7910	89.74	-0.0003	-0.0004	0.0003	-0.0010	0.0003
0.7000	0.8570	87.29	-0.0004	-0.0005	0.0001	-0.0006	-0.0000
0.8000	0.9120	84.99	0.0004	0.0004	0.0003	0.0008	0.0000
0.9000	0.9590	82.68	0.0008	0.0008	0.0002	0.0014	-0.0001
0.9500	0.9800	81.43	0.0007	0.0007	0.0003	0.0012	0.0001
		MEAN	0.0007	0.0007	0.0002	0.0009	0.0002

BENZENE(1)-1,1,2-TRICHLOROETHYLENE(2)

PODDER C.:Z.PHYS.CHEM.(FRANKFURT)32(4-6),267(1962).

ANTOINE VAPOR PRESSURE CONSTANTS
         A   B   C
(1) 6.90565   1211.033   220.790
(2) 7.02808   1315.000   230.000

T =   70.00

EQUATION	ORDER	A12	A21	D	A122	A211
VAN LAAR	3	-0.0001	0.0003	-	-	-
MARGULES	3	0.0016	-0.0089	-	-	-
MARGULES	4	0.0037	-0.0066	0.0069	-	-
ALPHA	2	0.2315	-0.2008	-	-	-
ALPHA	3	-.3680/ 05	-.2968/ 05	-	.3683/ 05	.2971/ 05

			DEVIATION IN VAPOR PHASE COMPOSITION				
X	Y	PRESS.	LAAR 3	MARG 3	MARG 4	ALPHA2	ALPHA3
0.1372	0.1642	456.25	0.0014	0.0011	0.0012	-0.0002	-0.0002
0.3285	0.3768	476.45	0.0023	0.0007	0.0006	0.0002	0.0005
0.5296	0.5829	498.31	0.0013	0.0000	0.0001	-0.0001	-0.0001
0.7326	0.7734	521.23	-0.0002	0.0006	0.0007	-0.0001	-0.0003
0.8781	0.8998	537.48	-0.0002	0.0011	0.0009	0.0001	0.0002
		MEAN	0.0011	0.0007	0.0007	0.0001	0.0003

BENZENE(1)-1,1,2-TRICHLOROETHYLENE(2)

POUDER C.:Z.PHYS.CHEM.(FRANKFURT)35(4-6),267(1962).

ANTOINE VAPOR PRESSURE CONSTANTS
          A   B   C
(1) 6.90565   1211.033   220.790
(2) 7.02808   1315.000   230.000

P =   760.00

EQUATION	ORDER	A12	A21	D	A122	A211
VAN LAAR	3	-0.0002	0.0004	-	-	-
MARGULES	3	0.0042	-0.0092	-	-	-
MARGULES	4	0.0039	-0.0095	-0.0008	-	-
ALPHA	2	0.2182	-0.1918	-	-	-
ALPHA	3	-0.7418	-0.9738	-	0.9615	0.7828

			DEVIATION IN VAPOR PHASE COMPOSITION				
X	Y	TEMP.	LAAR 3	MARG 3	MARG 4	ALPHA2	ALPHA3
0.1370	0.1623	85.84	0.0015	0.0016	0.0016	-0.0000	0.0000
0.3270	0.3729	84.51	0.0020	0.0007	0.0007	-0.0000	-0.0001
0.5267	0.5772	83.13	-0.0002	0.0002	0.0002	0.0001	0.0001
0.7364	0.7749	81.74	-0.0003	0.0009	0.0009	-0.0001	-0.0001
0.8777	0.8985	80.84	0.0003	0.0013	0.0013	0.0001	0.0001
		MEAN	0.0008	0.0009	0.0009	0.0001	0.0001

327

CYCLOHEXANE(1)-ANILINE(2)

KORTUEM G.,FREIER H.J.,WOERNER F.:CHEM.-ING.TECHN.25,125(1953).

ANTOINE VAPOR PRESSURE CONSTANTS
	A	B	C
(1)	6.84498	1203.526	222.863
(2)	7.24179	1675.300	200.000

P = 760.00

EQUATION	ORDER	A12	A21	D	A122	A211
VAN LAAR	3	0.5952	0.6239	-	-	-
MARGULES	3	0.5952	0.6227	-	-	-
MARGULES	4	0.6609	1.1789	0.9195	-	-
ALPHA	2	68.6378	-1.0119	-	-	-
ALPHA	3	.2207/ 10	.2285/ 08	-	-.2258/ 10	.4024/ 06

			DEVIATION IN VAPOR PHASE COMPOSITION				
X	Y	TEMP.	LAAR 3	MARG 3	MARG 4	ALPHA2	ALPHA3
0.0970	0.8810	117.60	-0.0027	-0.0027	-0.0016	0.0014	-0.0010
0.1090	0.8945	112.50	0.0011	0.0011	0.0014	0.0007	0.0004
0.1290	0.9120	108.20	0.0011	0.0011	0.0006	-0.0001	0.0009
0.1440	0.9210	105.40	0.0019	0.0019	0.0013	0.0007	0.0019
0.1560	0.9320	102.40	-0.0016	-0.0016	-0.0022	-0.0038	-0.0027
0.2340	0.9540	94.30	-0.0006	-0.0006	0.0003	0.0014	0.0001
		MEAN	0.0015	0.0015	0.0012	0.0013	0.0012

CYCLOHEXANE(1)-ANILINE(2)

KORTUEM G.,FREIER H.J.:CHEM.-ING.TECHN.26,670(1954).

ANTOINE VAPOR PRESSURE CONSTANTS
	A	B	C
(1)	6.84498	1203.526	222.863
(2)	7.24179	1675.300	200.000

T = 119.30

EQUATION	ORDER	A12	A21	D	A122	A211
VAN LAAR	3	0.7024	0.6356	-	-	-
MARGULES	3	0.6993	0.6345	-	-	-
MARGULES	4	0.8470	0.7795	0.4468	-	-
ALPHA	2	96.4422	2.4959	-	-	-
ALPHA	3	.5233/ 05	.2330/ 04	-	-.3009/ 05	-.1062/ 04

			DEVIATION IN VAPOR PHASE COMPOSITION				
X	Y	PRESS.	LAAR 3	MARG 3	MARG 4	ALPHA2	ALPHA3
0.1760	0.9290	1180.00	-0.0028	-0.0028	-0.0011	-0.0063	-0.0014
0.2100	0.9280	1245.00	0.0050	0.0050	0.0049	0.0029	0.0031
0.2810	0.9400	1394.00	0.0018	0.0019	-0.0001	0.0017	-0.0022
0.4600	0.9530	1624.00	-0.0007	-0.0007	-0.0014	0.0016	-0.0016
0.5080	0.9550	1653.00	-0.0008	-0.0007	-0.0007	0.0016	-0.0006
0.5180	0.9540	1657.00	0.0007	0.0007	0.0009	0.0030	0.0010
0.5820	0.9590	1701.00	-0.0016	-0.0016	-0.0005	0.0001	-0.0003
0.6430	0.9590	1775.00	0.0014	0.0014	0.0028	0.0018	0.0029
0.8070	0.9700	1861.00	0.0024	0.0024	0.0021	-0.0055	-0.0007
0.8270	0.9710	1916.00	0.0034	0.0034	0.0027	-0.0060	-0.0009
		MEAN	0.0021	0.0021	0.0017	0.0030	0.0015

v

CYCLOHEXANE(1)-1,2-DICHLOROETHANE(2)

FORDYCE C.T.,SIMONSON D.R.:IND.ENG.CHEM.,41,104(1949).

ANTOINE VAPOR PRESSURE CONSTANTS
	A	B	C
(1)	6.84498	1203.526	222.863
(2)	7.18431	1358.500	232.000

P = 760.00

EQUATION	ORDER	A12	A21	D	A122	A211
VAN LAAR	3	0.4502	0.4107	-	-	-
MARGULES	3	0.4492	0.4094	-	-	-
MARGULES	4	0.4374	0.3978	-0.0376	-	-
ALPHA	2	2.0317	1.6511	-	-	-
ALPHA	3	0.7326	1.0056	-	1.3270	0.2302

			DEVIATION IN VAPOR PHASE COMPOSITION				
X	Y	TEMP.	LAAR 3	MARG 3	MARG 4	ALPHA2	ALPHA3
0.1190	0.2400	77.50	0.0038	0.0037	0.0025	-0.0005	0.0035
0.2420	0.3760	75.40	-0.0062	-0.0062	-0.0056	-0.0091	-0.0054
0.3800	0.4560	74.50	0.0025	0.0026	0.0034	0.0038	0.0036
0.5150	0.5290	74.10	-0.0006	-0.0007	-0.0008	0.0036	-0.0004
0.6560	0.6040	74.40	0.0011	0.0010	0.0001	0.0046	0.0004
0.7890	0.7030	75.20	-0.0021	-0.0022	-0.0024	-0.0042	-0.0021
0.9080	0.8330	77.00	-0.0003	-0.0002	0.0012	-0.0088	0.0020
		MEAN	0.0024	0.0023	0.0023	0.0049	0.0025

CYCLOHEXANE(1)-ETHYL ALCOHOL(2)

NAGAI J.,ISII N.:J.SOC.CHEM.IND.JAPAN 38,86(1935).

ANTOINE VAPOR PRESSURE CONSTANTS

	A	B	C
(1)	6.84498	1203.526	222.863
(2)	8.16290	1623.220	228.980

T =   10.00

EQUATION	ORDER	A12	A21	D	A122	A211
VAN LAAR	3	0.8173	1.0813	-	-	-
MARGULES	3	0.8102	1.0456	-	-	-
MARGULES	4	1.0177	1.2646	0.7052	-	-
ALPHA	2	33.8080	13.9900	-	-	-
ALPHA	3	83.7834	6.8357	-	-65.8686	28.4790

			DEVIATION IN VAPOR PHASE COMPOSITION				
X	Y	PRESS.	LAAR 3	MARG 3	MARG 4	ALPHA2	ALPHA3
0.0500	0.4383	40.15	-0.0614	-0.0633	0.0076	0.0680	0.0023
0.1000	0.5706	50.65	-0.0471	-0.0477	-0.0097	0.0222	-0.0052
0.1500	0.6194	55.70	-0.0206	-0.0203	-0.0081	0.0094	0.0004
0.2000	0.6468	58.60	-0.0042	-0.0036	-0.0076	0.0018	0.0017
0.2500	0.6617	60.45	0.0078	0.0083	-0.0045	-0.0003	0.0038
0.3000	0.6739	61.80	0.0122	0.0123	-0.0036	-0.0035	0.0024
0.3500	0.6842	62.70	0.0117	0.0112	-0.0037	-0.0070	-0.0004
0.4000	0.6911	63.30	0.0098	0.0086	-0.0025	-0.0084	-0.0021
0.4500	0.6949	63.75	0.0075	0.0055	0.0002	-0.0076	-0.0019
0.5000	0.6983	64.15	0.0030	0.0002	0.0016	-0.0070	-0.0022
0.5500	0.7007	64.50	-0.0023	-0.0058	0.0021	-0.0057	-0.0021
0.6000	0.7028	64.75	-0.0085	-0.0124	0.0010	-0.0041	-0.0020
0.6500	0.7036	64.95	-0.0136	-0.0176	-0.0008	-0.0011	-0.0008
0.7000	0.7036	65.10	-0.0170	-0.0206	-0.0036	0.0030	0.0013
0.7500	0.7043	65.10	-0.0185	-0.0210	-0.0081	0.0073	0.0029
0.8000	0.7061	65.00	-0.0159	-0.0165	-0.0130	0.0119	0.0040
0.8500	0.7099	64.65	-0.0051	-0.0031	-0.0151	0.0175	0.0044
0.9000	0.7221	63.70	0.0168	0.0221	-0.0106	0.0216	-0.0001
0.9500	0.7502	61.65	0.0645	0.0721	0.0218	0.0313	-0.0082
		MEAN	0.0183	0.0196	0.0066	0.0126	0.0025

CYCLOHEXANE(1)-ETHYL ALCOHOL(2)

NAGAI J.,ISII N.:J.SOC.CHEM.IND.JAPAN 38,86(1935).

ANTOINE VAPOR PRESSURE CONSTANTS

	A	B	C
(1)	6.84498	1203.526	222.863
(2)	8.16290	1623.220	228.980

T = 20.00

EQUATION	ORDER	A12	A21	D	A122	A211
VAN LAAR	3	0.8062	1.0988	-	-	-
MARGULES	3	0.7945	1.0577	-	-	-
MARGULES	4	0.9819	1.2553	0.6336	-	-
ALPHA	2	30.1805	14.4153	-	-	-
ALPHA	3	75.2512	5.1899	-	-62.3648	31.4001

DEVIATION IN VAPOR PHASE COMPOSITION

X	Y	PRESS.	LAAR 3	MARG 3	MARG 4	ALPHA2	ALPHA3
0.0500	0.4017	70.95	-0.0611	-0.0644	-0.0027	0.0741	-0.0124
0.1000	0.5248	85.75	-0.0393	-0.0409	-0.0061	0.0369	-0.0001
0.1500	0.5804	94.25	-0.0173	-0.0174	-0.0059	0.0176	0.0053
0.2000	0.6129	99.85	-0.0035	-0.0029	-0.0064	0.0053	0.0048
0.2500	0.6325	103.40	0.0059	0.0067	-0.0050	-0.0013	0.0038
0.3000	0.6490	105.70	0.0077	0.0082	-0.0065	-0.0086	-0.0011
0.3500	0.6576	107.20	0.0102	0.0101	-0.0038	-0.0110	-0.0019
0.4000	0.6641	108.10	0.0096	0.0087	-0.0016	-0.0110	-0.0030
0.4500	0.6688	108.70	0.0070	0.0051	0.0002	-0.0109	-0.0038
0.5000	0.6724	109.00	0.0026	-0.0002	0.0010	-0.0102	-0.0043
0.5500	0.6731	109.20	-0.0010	-0.0047	0.0027	-0.0069	-0.0026
0.6000	0.6736	109.40	-0.0059	-0.0101	0.0024	-0.0034	-0.0009
0.6500	0.6739	109.50	-0.0110	-0.0155	0.0002	0.0005	0.0008
0.7000	0.6742	109.60	-0.0154	-0.0196	-0.0037	0.0048	0.0026
0.7500	0.6763	109.70	-0.0193	-0.0223	-0.0103	0.0083	0.0029
0.8000	0.6770	109.60	-0.0166	-0.0175	-0.0144	0.0149	0.0053
0.8500	0.6806	109.30	-0.0065	-0.0042	-0.0159	0.0221	0.0064
0.9000	0.6932	107.90	0.0152	0.0215	-0.0100	0.0281	0.0025
0.9500	0.7224	104.50	0.0662	0.0758	0.0264	0.0420	-0.0038
		MEAN	0.0169	0.0187	0.0066	0.0167	0.0036

CYCLOHEXANE(1)-ETHYL ALCOHOL(2)

NAGAI J.,ISII N.:J.SOC.CHEM.IND.JAPAN 38,86(1935).

ANTOINE VAPOR PRESSURE CONSTANTS

	A	B	C
(1)	6.84498	1203.526	222.863
(2)	8.16290	1623.220	228.980

T = 30.00

EQUATION	ORDER	A12	A21	D	A122	A211
VAN LAAR	3	0.7431	1.1480	-	-	-
MARGULES	3	0.7188	1.0694	-	-	-
MARGULES	4	0.9322	1.3024	0.7279	-	-
ALPHA	2	22.3681	11.3268	-	-	-
ALPHA	3	194.9616	14.9074	-	-183.1015	77.4113

			DEVIATION IN VAPOR PHASE COMPOSITION				
X	Y	PRESS.	LAAR 3	MARG 3	MARG 4	ALPHA2	ALPHA3
0.0500	0.3670	116.90	-0.0798	-0.0866	-0.0216	0.0608	-0.0043
0.1000	0.4882	139.50	-0.0582	-0.0621	-0.0228	0.0358	0.0011
0.1500	0.5485	153.50	-0.0351	-0.0361	-0.0223	0.0183	0.0011
0.2000	0.5837	162.40	-0.0171	-0.0164	-0.0199	0.0076	0.0005
0.2500	0.6051	167.90	-0.0031	-0.0018	-0.0149	0.0022	0.0014
0.3000	0.6209	171.20	0.0050	0.0061	-0.0105	-0.0021	0.0011
0.3500	0.6327	174.00	0.0091	0.0093	-0.0061	-0.0051	0.0006
0.4000	0.6425	175.70	0.0094	0.0082	-0.0029	-0.0077	-0.0004
0.4500	0.6514	177.00	0.0060	0.0031	-0.0016	-0.0105	-0.0024
0.5000	0.6578	178.00	0.0016	-0.0031	-0.0005	-0.0113	-0.0032
0.5500	0.6618	178.60	-0.0034	-0.0096	-0.0000	-0.0101	-0.0024
0.6000	0.6651	178.90	-0.0098	-0.0173	-0.0019	-0.0083	-0.0017
0.6500	0.6661	179.10	-0.0154	-0.0235	-0.0047	-0.0039	0.0009
0.7000	0.6685	178.90	-0.0228	-0.0305	-0.0118	-0.0002	0.0018
0.7500	0.6702	178.60	-0.0285	-0.0343	-0.0205	0.0054	0.0032
0.8000	0.6725	178.00	-0.0311	-0.0331	-0.0302	0.0126	0.0041
0.8500	0.6777	176.90	-0.0277	-0.0237	-0.0388	0.0213	0.0025
0.9000	0.6851	175.00	-0.0067	0.0054	-0.0343	0.0373	0.0003
0.9500	0.7077	169.70	0.0484	0.0680	0.0044	0.0662	-0.0108
		MEAN	0.0220	0.0252	0.0142	0.0172	0.0023

CYCLOHEXANE(1)-ETHYL ALCOHOL(2)

WASHBURN E.R.,HANDORF B.H.:J.AM.CHEM.SOC.57,441(1935).

ANTOINE VAPOR PRESSURE CONSTANTS

	A	B	C
(1)	6.84498	1203.526	222.863
(2)	8.16290	1623.220	228.980

T = 25.00

EQUATION	ORDER	A12	A21	D	A122	A211
VAN LAAR	3	0.8195	1.1152	-	-	-
MARGULES	3	0.8050	1.0746	-	-	-
MARGULES	4	0.9656	1.2444	0.5352	-	-
ALPHA	2	34.3442	17.1043	-	-	-
ALPHA	3	59.4210	3.1287	-	-47.7889	28.2764

			DEVIATION IN VAPOR PHASE COMPOSITION				
X	Y	PRESS.	LAAR 3	MARG 3	MARG 4	ALPHA2	ALPHA3
0.1008	0.5204	108.40	-0.0433	-0.0457	-0.0156	0.0471	0.0044
0.2052	0.6304	125.60	-0.0282	-0.0278	-0.0309	-0.0121	-0.0086
0.2902	0.6468	132.60	-0.0045	-0.0039	-0.0157	-0.0118	-0.0013
0.4059	0.6490	137.90	0.0124	0.0116	0.0036	-0.0010	0.0081
0.5017	0.6576	139.40	0.0032	0.0007	0.0021	-0.0020	0.0032
0.5984	0.6632	138.90	-0.0111	-0.0150	-0.0045	-0.0008	-0.0003
0.7013	0.6687	139.10	-0.0278	-0.0315	-0.0184	0.0017	-0.0033
0.7950	0.6732	140.50	-0.0329	-0.0338	-0.0310	0.0079	-0.0034
0.8970	0.6895	136.90	-0.0040	0.0024	-0.0251	0.0179	-0.0051
		MEAN	0.0186	0.0192	0.0163	0.0114	0.0042

CYCLOHEXANE(1)-METHYLCYCLOHEXANE(2)

RICHARDS A.R.,HARGREAVES E.:IND.ENG.CHEM.36,805(1944).

ANTOINE VAPOR PRESSURE CONSTANTS
```
         A         B         C
(1) 6.84498   1203.526   222.863
(2) 6.82689   1272.864   221.630
```

P = 760.00

EQUATION	ORDER	A12	A21	D	A122	A211
VAN LAAR	3	0.1881	0.2695	-	-	-
MARGULES	3	0.1575	0.2853	-	-	-
MARGULES	4	-0.3891	-0.2596	-1.7196	-	-
ALPHA	2	1.7632	-0.0450	-	-	-
ALPHA	3	.3587/ 06	.1661/ 06	-	-.3518/ 06	-.1465/ 06

			DEVIATION IN VAPOR PHASE COMPOSITION				
X	Y	TEMP.	LAAR 3	MARG 3	MARG 4	ALPHA2	ALPHA3
0.0910	0.1850	98.30	0.0206	0.0153	-0.0524	0.0224	0.0179
0.1850	0.2780	96.00	0.0778	0.0769	0.0679	0.0801	0.0631
0.2780	0.5640	94.10	-0.0969	-0.0931	-0.0561	-0.0941	-0.1090
0.4470	0.6370	91.60	-0.0217	-0.0149	0.0026	-0.0173	-0.0171
0.5640	0.6750	90.50	0.0204	0.0256	0.0070	0.0263	0.0358
0.6370	0.7090	87.40	0.0329	0.0363	0.0054	0.0387	0.0513
0.8070	0.9020	84.20	-0.0546	-0.0554	-0.0536	-0.0487	-0.0435
0.9520	0.9650	83.60	-0.0102	-0.0115	0.0137	-0.0076	-0.0374
		MEAN	0.0419	0.0411	0.0323	0.0419	0.0469

CYCLOHEXANE(1)-NITROBENZENE(2)

BROWN I.:AUSTR.J.SCI.RES.A5,530(1952).

ANTOINE VAPOR PRESSURE CONSTANTS
	A	B	C
(1)	6.84498	1203.526	222.863
(2)	7.10779	1740.510	201.108

T =   80.00

EQUATION	ORDER	A12	A21	D	A122	A211
VAN LAAR	3	0.8522	0.6508	-	-	-
MARGULES	3	0.8545	0.6047	-	-	-
MARGULES	4	-0.0852	-0.2273	-3.2721	-	-
ALPHA	2	426.0617	4.8526	-	-	-
ALPHA	3	.4795/ 09	.5496/ 07	-	-.3214/ 09	-.3766/ 05

			DEVIATION IN VAPOR PHASE COMPOSITION				
X	Y	PRESS.	LAAR 3	MARG 3	MARG 4	ALPHA2	ALPHA3
0.1450	0.9730	322.60	0.0089	0.0089	0.0060	0.0002	0.0009
0.4400	0.9850	556.20	0.0023	0.0019	0.0049	-0.0014	-0.0030
0.4900	0.9850	573.10	0.0026	0.0023	0.0034	-0.0008	-0.0021
0.5730	0.9850	600.30	0.0034	0.0031	0.0007	-0.0002	-0.0007
0.6860	0.9850	630.30	0.0050	0.0048	-0.0003	0.0005	0.0008
0.7730	0.9830	654.10	0.0088	0.0088	0.0057	0.0029	0.0037
0.8910	0.9910	689.90	0.0043	0.0045	0.0060	-0.0045	-0.0031
		MEAN	0.0050	0.0049	0.0039	0.0015	0.0020

CYCLOHEXANE(1)-ISO-PROPYL ALCOHOL(2)

NAGATA I.:MEM.FAC.TECHNOL.,KANAZAWA UNIV.3(1),1(1963).

ANTOINE VAPOR PRESSURE CONSTANTS
	A	B	C
(1)	6.84498	1203.526	222.863
(2)	7.75634	1366.142	197.970

P =   760.00

EQUATION	ORDER	A12	A21	D	A122	A211
VAN LAAR	3	0.6090	0.7804	-	-	-
MARGULES	3	0.6076	0.7576	-	-	-
MARGULES	4	0.7276	0.9086	0.4677	-	-
ALPHA	2	5.4387	3.8773	-	-	-
ALPHA	3	15.9128	5.2163	-	-11.4214	3.9344

			DEVIATION IN VAPOR PHASE COMPOSITION				
X	Y	TEMP.	LAAR 3	MARG 3	MARG 4	ALPHA2	ALPHA3
0.0210	0.0930	80.29	-0.0114	-0.0115	0.0070	0.0185	0.0051
0.0470	0.1800	78.10	-0.0153	-0.0153	0.0093	0.0250	0.0055
0.0910	0.2790	78.85	-0.0148	-0.0145	0.0037	0.0265	0.0069
0.1470	0.3730	73.80	-0.0093	-0.0087	-0.0038	0.0093	-0.0048
0.2170	0.4420	72.13	-0.0014	-0.0009	-0.0082	-0.0002	-0.0062
0.2790	0.4810	70.88	0.0062	0.0062	-0.0053	-0.0033	-0.0030
0.3730	0.5190	70.13	0.0136	0.0124	0.0039	-0.0015	0.0059
0.4780	0.5610	69.99	0.0023	-0.0001	0.0018	-0.0090	0.0026
0.6020	0.5970	69.56	-0.0067	-0.0100	0.0038	-0.0076	0.0032
0.7480	0.6490	69.79	-0.0221	-0.0239	-0.0120	-0.0056	-0.0074
0.8820	0.7040	70.99	0.0067	0.0095	-0.0073	0.0312	-0.0029
0.9720	0.8150	74.61	0.0662	0.0704	0.0402	0.0785	0.0209
		MEAN	0.0147	0.0153	0.0089	0.0180	0.0062

CYCLOHEXANE(1)-ISO-PROPYL ALCOHOL(2)

NAGATA I.:J.CHEM.ENG.DATA 10,106(1965).

ANTOINE VAPOR PRESSURE CONSTANTS

	A	B	C
(1)	6.84498	1203.526	222.863
(2)	7.75634	1366.142	197.970

P = 500.00

EQUATION	ORDER	A12	A21	D	A122	A211
VAN LAAR	3	0.6203	0.8236	-	-	-
MARGULES	3	0.6117	0.7986	-	-	-
MARGULES	4	0.7114	0.9088	0.3452	-	-
ALPHA	2	6.9877	4.1472	-	-	-
ALPHA	3	15.4356	3.0568	-	-10.9744	5.4668

DEVIATION IN VAPOR PHASE COMPOSITION

X	Y	TEMP.	LAAR 3	MARG 3	MARG 4	ALPHA2	ALPHA3
0.0290	0.1320	67.30	-0.0060	-0.0076	0.0128	0.0399	0.0015
0.0680	0.2530	67.00	-0.0082	-0.0099	0.0118	0.0465	0.0042
0.1380	0.3940	63.10	-0.0062	-0.0066	0.0016	0.0230	-0.0018
0.2130	0.4760	61.20	0.0009	0.0014	-0.0021	0.0069	-0.0009
0.2660	0.5170	60.10	0.0021	0.0027	-0.0048	-0.0034	-0.0031
0.3130	0.5380	59.10	0.0098	0.0102	0.0018	-0.0031	0.0021
0.4080	0.5780	58.30	0.0075	0.0070	0.0019	-0.0100	0.0007
0.4750	0.5980	58.00	0.0049	0.0035	0.0031	-0.0108	0.0008
0.5560	0.6190	57.80	-0.0006	-0.0029	0.0025	-0.0105	-0.0002
0.6370	0.6320	57.80	-0.0004	-0.0031	0.0060	-0.0018	0.0046
0.7340	0.6640	57.90	-0.0123	-0.0144	-0.0067	-0.0029	-0.0053
0.8180	0.6870	58.50	-0.0047	-0.0046	-0.0056	0.0122	-0.0031
0.8840	0.7130	59.10	0.0174	0.0202	0.0073	0.0343	0.0034
0.9630	0.8160	61.90	0.0446	0.0494	0.0269	0.0517	0.0020
		MEAN	0.0090	0.0103	0.0068	0.0184	0.0024

CYCLOHEXANE(1)-TOLUENE(2)

MYERS H.S.:IND.ENG.CHEM.48,1104(1956).

ANTOINE VAPOR PRESSURE CONSTANTS
	A	B	C
(1)	6.84498	1203.526	222.863
(2)	6.95334	1343.943	219.377

P = 760.00

EQUATION	ORDER	A12	A21	D	A122	A211
VAN LAAR	3	0.1452	0.0810	-	-	-
MARGULES	3	0.1317	0.0693	-	-	-
MARGULES	4	0.1140	0.0479	-0.0574	-	-
ALPHA	2	1.7921	-0.3912	-	-	-
ALPHA	3	6.9684	1.6553	-	-5.1009	-1.8834

DEVIATION IN VAPOR PHASE COMPOSITION

X	Y	TEMP.	LAAR 3	MARG 3	MARG 4	ALPHA2	ALPHA3
0.0410	0.1020	108.25	0.0132	0.0112	0.0084	0.0036	0.0052
0.0910	0.2120	105.45	0.0144	0.0125	0.0096	0.0024	0.0038
0.1180	0.2640	103.85	0.0131	0.0118	0.0094	0.0018	0.0027
0.1430	0.3080	102.85	0.0114	0.0105	0.0088	0.0015	0.0020
0.1640	0.3480	101.75	0.0044	0.0039	0.0028	-0.0043	-0.0042
0.1920	0.3860	100.55	0.0070	0.0069	0.0064	0.0001	-0.0002
0.2170	0.4220	99.50	0.0045	0.0047	0.0047	-0.0008	-0.0013
0.2450	0.4570	98.35	0.0045	0.0049	0.0053	0.0009	0.0001
0.2430	0.4600	98.35	-0.0008	-0.0004	-0.0001	-0.0047	-0.0054
0.2730	0.4920	97.40	0.0021	0.0026	0.0032	-0.0001	-0.0009
0.2830	0.5040	96.95	0.0013	0.0018	0.0025	-0.0005	-0.0013
0.3040	0.5230	96.35	0.0048	0.0053	0.0061	0.0040	0.0032
0.3230	0.5470	95.50	0.0007	0.0011	0.0019	0.0002	-0.0005
0.3360	0.5600	95.25	0.0005	0.0010	0.0017	0.0006	-0.0001
0.3680	0.5960	94.20	-0.0047	-0.0044	-0.0038	-0.0041	-0.0045
0.3790	0.5990	93.80	0.0025	0.0028	0.0034	0.0032	0.0028
0.4160	0.6330	92.75	0.0014	0.0014	0.0017	0.0022	0.0021
0.4520	0.6620	91.85	0.0025	0.0023	0.0023	0.0031	0.0034
0.5040	0.7020	90.55	0.0035	0.0030	0.0026	0.0031	0.0038
0.5330	0.7240	89.75	0.0032	0.0027	0.0020	0.0019	0.0029
0.5590	0.7490	88.85	-0.0028	-0.0034	-0.0042	-0.0051	-0.0040
0.5990	0.7740	87.95	-0.0000	-0.0006	-0.0016	-0.0037	-0.0025
0.6020	0.7770	88.00	-0.0011	-0.0017	-0.0027	-0.0048	-0.0036
0.6340	0.7940	87.35	0.0032	0.0026	0.0016	-0.0017	-0.0005
0.6720	0.8110	86.55	0.0105	0.0101	0.0092	0.0043	0.0053
0.7270	0.8520	85.25	0.0033	0.0031	0.0026	-0.0049	-0.0044
0.7630	0.8640	84.80	0.0123	0.0123	0.0121	0.0034	0.0031
0.7800	0.8770	84.45	0.0090	0.0092	0.0091	-0.0002	-0.0008
0.8140	0.8950	83.75	0.0101	0.0104	0.0107	0.0004	-0.0010
0.8740	0.9260	82.70	0.0113	0.0118	0.0126	0.0022	-0.0008
0.9640	0.9730	81.10	0.0097	0.0101	0.0107	0.0056	0.0023
		MEAN	0.0056	0.0055	0.0053	0.0026	0.0025

CYCLOHEXANE(1)-2,2,3-TRIMETHYLBUTANE(2)

HARRISON J.M.,BERG L.:IND.ENG.CHEM.38,117(1946).

ANTOINE VAPOR PRESSURE CONSTANTS
```
          A          B          C
(1) 6.84498    1203.526    222.863
(2) 6.79230    1200.563    226.050
```

P =   744.00

EQUATION	ORDER	A12	A21	D	A122	A211
VAN LAAR	3	0.0908	0.0995	-	-	-
MARGULES	3	0.0909	0.0990	-	-	-
MARGULES	4	0.1124	0.1207	0.0716	-	-
ALPHA	2	0.2784	0.2160	-	-	-
ALPHA	3	9.7699	10.1922	-	-8.8164	-9.8983

			DEVIATION IN VAPOR PHASE COMPOSITION				
X	Y	TEMP.	LAAR 3	MARG 3	MARG 4	ALPHA2	ALPHA3
0.0420	0.0660	80.00	-0.0153	-0.0153	-0.0137	-0.0138	-0.0001
0.1180	0.1490	79.90	-0.0125	-0.0125	-0.0111	-0.0092	0.0026
0.2250	0.2550	79.75	-0.0076	-0.0076	-0.0084	-0.0032	-0.0009
0.2790	0.3060	79.70	-0.0060	-0.0060	-0.0075	-0.0013	-0.0025
0.3790	0.3960	79.55	-0.0027	-0.0027	-0.0043	0.0022	-0.0019
0.4350	0.4430	79.50	0.0010	0.0009	-0.0000	0.0059	0.0018
0.4840	0.4910	79.45	-0.0032	-0.0033	-0.0035	0.0017	-0.0016
0.5170	0.5180	79.45	-0.0008	-0.0009	-0.0006	0.0041	0.0016
0.5220	0.5220	79.45	-0.0003	-0.0004	-0.0000	0.0045	0.0022
0.6650	0.6540	79.45	-0.0037	-0.0038	-0.0020	0.0009	0.0032
0.7570	0.7480	79.50	-0.0114	-0.0114	-0.0103	-0.0072	-0.0030
0.8340	0.8230	79.60	-0.0099	-0.0099	-0.0104	-0.0064	-0.0025
0.8710	0.8600	79.70	-0.0083	-0.0083	-0.0095	-0.0052	-0.0023
0.9120	0.9000	79.80	-0.0037	-0.0037	-0.0055	-0.0013	0.0000
0.9400	0.9270	79.90	0.0010	0.0010	-0.0009	0.0028	0.0029
0.9630	0.9520	80.00	0.0029	0.0029	0.0014	0.0041	0.0035
		MEAN	0.0056	0.0057	0.0056	0.0046	0.0021

METHYLCYCLOPENTANE(1)-BENZENE(2)

GRISWOLD J.,LUDWIG E.E.:IND.ENG.CHEM.,35,117(1943).

ANTOINE VAPOR PRESSURE CONSTANTS

	A	B	C
(1)	6.86283	1186.059	226.042
(2)	6.90565	1211.033	220.790

P = 760.00

EQUATION	ORDER	A12	A21	D	A122	A211
VAN LAAR	3	0.1441	0.1365	-	-	-
MARGULES	3	0.1439	0.1364	-	-	-
MARGULES	4	0.1494	0.1416	0.0182	-	-
ALPHA	2	0.7214	0.0930	-	-	-
ALPHA	3	0.9546	0.3923	-	-0.1519	-0.3257

			DEVIATION IN VAPOR PHASE COMPOSITION				
X	Y	TEMP.	LAAR 3	MARG 3	MARG 4	ALPHA2	ALPHA3
0.0297	0.0526	79.64	-0.0017	-0.0018	-0.0013	-0.0033	-0.0014
0.1080	0.1668	77.62	0.0009	0.0009	0.0014	-0.0022	0.0011
0.1751	0.2533	76.62	-0.0007	-0.0007	-0.0007	-0.0034	-0.0009
0.3017	0.3870	74.85	0.0008	0.0009	0.0003	0.0002	0.0002
0.3806	0.4598	74.00	0.0015	0.0016	0.0011	0.0022	0.0010
0.4450	0.5179	73.43	-0.0004	-0.0004	-0.0007	0.0009	-0.0007
0.5737	0.6255	72.84	-0.0016	-0.0016	-0.0014	0.0000	-0.0014
0.6434	0.6795	72.06	0.0011	0.0011	0.0015	0.0021	0.0013
0.7206	0.7442	71.97	-0.0005	-0.0005	-0.0001	-0.0003	-0.0002
0.8224	0.8299	71.54	0.0005	0.0005	0.0005	-0.0008	0.0004
0.9030	0.9034	71.47	0.0002	0.0002	-0.0001	-0.0015	-0.0002
0.9180	0.9174	71.53	0.0004	0.0004	0.0001	-0.0013	0.0000
0.9373	0.9360	71.65	0.0005	0.0005	0.0001	-0.0011	0.0001
0.9450	0.9442	71.68	-0.0002	-0.0002	-0.0005	-0.0016	-0.0006
0.9518	0.9503	71.80	0.0004	0.0004	0.0001	-0.0009	0.0001
		MEAN	0.0008	0.0008	0.0007	0.0014	0.0006

METHYLCYCLOPENTANE(1)-BENZENE(2)

MYERS H.S.:IND.ENG.CHEM.48,1104(1956).

ANTOINE VAPOR PRESSURE CONSTANTS

	A	B	C
(1)	6.86283	1186.059	226.042
(2)	6.90565	1211.033	220.790

P = 760.00

EQUATION	ORDER	A12	A21	D	A122	A211
VAN LAAR	3	0.1794	0.1294	-	-	-
MARGULES	3	0.1743	0.1241	-	-	-
MARGULES	4	0.1641	0.1129	-0.0333	-	-
ALPHA	2	0.7358	0.1459	-	-	-
ALPHA	3	0.1853	-0.0477	-	0.6500	0.0810

			DEVIATION IN VAPOR PHASE COMPOSITION				
X	Y	TEMP.	LAAR 3	MARG 3	MARG 4	ALPHA2	ALPHA3
0.0125	0.0220	79.85	0.0016	0.0014	0.0009	-0.0007	0.0005
0.0250	0.0450	79.45	0.0011	0.0007	-0.0001	-0.0030	-0.0009
0.0450	0.0760	79.00	0.0040	0.0035	0.0023	-0.0022	0.0011
0.0695	0.1130	78.35	0.0055	0.0049	0.0037	-0.0022	0.0020
0.0960	0.1550	77.80	0.0019	0.0014	0.0003	-0.0065	-0.0017
0.1350	0.2060	77.15	0.0024	0.0021	0.0014	-0.0057	-0.0007
0.1630	0.2370	76.65	0.0054	0.0053	0.0049	-0.0019	0.0028
0.1950	0.2780	76.10	0.0008	0.0009	0.0008	-0.0052	-0.0012
0.2300	0.3160	75.60	0.0002	0.0004	0.0006	-0.0044	-0.0012
0.2725	0.3580	75.10	0.0007	0.0009	0.0015	-0.0019	0.0001
0.3120	0.3950	74.60	0.0011	0.0013	0.0020	0.0002	0.0009
0.3410	0.4250	74.20	-0.0025	-0.0022	-0.0016	-0.0024	-0.0025
0.3920	0.4670	73.75	0.0003	0.0004	0.0009	0.0018	0.0005
0.4080	0.4815	73.60	-0.0005	-0.0004	0.0000	0.0014	-0.0003
0.4300	0.5000	73.40	-0.0003	-0.0003	0.0000	0.0019	-0.0003
0.4480	0.5140	73.25	0.0008	0.0007	0.0009	0.0032	0.0007
0.4580	0.5230	73.15	0.0001	0.0001	0.0002	0.0026	0.0000
0.4890	0.5480	72.95	0.0008	0.0006	0.0006	0.0034	0.0005
0.5140	0.5680	72.80	0.0014	0.0011	0.0009	0.0039	0.0008
0.5300	0.5800	72.70	0.0025	0.0022	0.0019	0.0048	0.0018
0.5600	0.6070	72.50	0.0001	-0.0003	-0.0008	0.0020	-0.0009
0.6010	0.6380	72.30	0.0027	0.0023	0.0016	0.0038	0.0013
0.6310	0.6640	72.20	0.0014	0.0009	0.0002	0.0017	-0.0002
0.6770	0.7040	72.00	-0.0004	-0.0008	-0.0015	-0.0015	-0.0023
0.6890	0.7130	71.95	0.0006	0.0003	-0.0004	-0.0009	-0.0013
0.7490	0.7620	71.80	0.0026	0.0025	0.0021	-0.0008	0.0007
0.7900	0.7990	71.75	0.0013	0.0014	0.0013	-0.0033	-0.0004
0.8080	0.8150	71.70	0.0013	0.0014	0.0015	-0.0038	-0.0003
0.8300	0.8360	71.65	0.0000	0.0003	0.0005	-0.0055	-0.0014
0.8450	0.8460	71.65	0.0037	0.0039	0.0044	-0.0021	0.0024
0.8750	0.8750	71.65	0.0023	0.0027	0.0034	-0.0036	0.0014
0.9210	0.9210	71.75	-0.0000	0.0004	0.0013	-0.0052	-0.0004
0.9640	0.9640	71.80	-0.0007	-0.0004	0.0003	-0.0038	-0.0007
		MEAN	0.0016	0.0015	0.0014	0.0029	0.0010

METHYLCYCLOPENTANE(1)-TOLUENE(2)

MYERS H.S.:IND.ENG.CHEM.48,1104(1956).

ANTOINE VAPOR PRESSURE CONSTANTS

	A	B	C
(1)	6.86283	1186.059	226.042
(2)	6.95334	1343.943	219.377

P = 760.00

EQUATION	ORDER	A12	A21	D	A122	A211
VAN LAAR	3	0.1264	0.0654	-	-	-
MARGULES	3	0.1162	0.0499	-	-	-
MARGULES	4	0.0840	0.0063	-0.1255	-	-
ALPHA	2	2.5102	-0.5548	-	-	-
ALPHA	3	2.2148	-0.6561	-	0.2805	0.1010

			DEVIATION IN VAPOR PHASE COMPOSITION				
X	Y	TEMP.	LAAR 3	MARG 3	MARG 4	ALPHA2	ALPHA3
0.0230	0.0680	108.80	0.0138	0.0125	0.0082	0.0081	0.0078
0.0400	0.1270	107.20	0.0080	0.0064	0.0007	-0.0001	-0.0005
0.0635	0.1930	105.35	0.0071	0.0057	-0.0003	-0.0024	-0.0028
0.0920	0.2575	103.40	0.0112	0.0102	0.0053	0.0018	0.0014
0.1060	0.2900	102.30	0.0093	0.0085	0.0043	0.0001	-0.0003
0.1100	0.2960	102.00	0.0116	0.0109	0.0070	0.0026	0.0022
0.1510	0.3725	99.30	0.0127	0.0127	0.0112	0.0055	0.0052
0.1910	0.4460	96.80	0.0038	0.0042	0.0047	-0.0014	-0.0016
0.2200	0.4880	95.30	0.0031	0.0035	0.0051	-0.0007	-0.0008
0.2670	0.5490	92.90	0.0016	0.0019	0.0043	-0.0006	-0.0006
0.3125	0.6020	90.75	-0.0009	-0.0008	0.0016	-0.0022	-0.0021
0.3335	0.6250	89.85	-0.0025	-0.0027	-0.0005	-0.0036	-0.0035
0.3710	0.6570	88.50	0.0007	0.0003	0.0019	-0.0000	0.0002
0.4125	0.6950	87.00	-0.0014	-0.0022	-0.0015	-0.0024	-0.0022
0.4550	0.7250	85.35	0.0023	0.0013	0.0010	0.0006	0.0008
0.4980	0.7480	84.10	0.0104	0.0092	0.0081	0.0080	0.0081
0.5310	0.7770	83.05	0.0038	0.0025	0.0005	0.0005	0.0006
0.5525	0.7920	82.30	0.0027	0.0015	-0.0005	-0.0012	-0.0011
0.5680	0.8020	81.80	0.0025	0.0012	-0.0009	-0.0019	-0.0019
0.6220	0.8310	80.30	0.0052	0.0042	0.0018	-0.0006	-0.0006
0.6750	0.8570	78.95	0.0079	0.0071	0.0050	0.0007	0.0006
0.7270	0.8825	77.70	0.0083	0.0079	0.0065	0.0002	-0.0000
0.7960	0.9150	76.05	0.0073	0.0075	0.0072	-0.0015	-0.0016
0.8460	0.9360	74.80	0.0074	0.0078	0.0084	-0.0011	-0.0013
0.8920	0.9540	73.70	0.0075	0.0080	0.0091	0.0001	-0.0000
0.9290	0.9690	72.90	0.0063	0.0068	0.0080	0.0005	0.0005
0.9660	0.9825	72.20	0.0059	0.0063	0.0071	0.0027	0.0027
		MEAN	0.0061	0.0057	0.0045	0.0019	0.0019

HEXANE(1)-BENZENE(2)

PRABHU P.S.,VAN WINKLE M.:J.CHEM.ENG.DATA 8(2),210(1963).

ANTOINE VAPOR PRESSURE CONSTANTS

	A	B	C
(1)	6.87776	1171.530	224.366
(2)	6.90565	1211.033	220.790

P = 760.00

EQUATION	ORDER	A12	A21	D	A122	A211
VAN LAAR	3	0.2704	0.1290	-	-	-
MARGULES	3	0.2294	0.1092	-	-	-
MARGULES	4	0.2597	0.1400	0.0986	-	-
ALPHA	2	0.8898	0.1133	-	-	-
ALPHA	3	2.3407	1.7147	-	-1.0383	-1.6897

			DEVIATION IN VAPOR PHASE COMPOSITION				
X	Y	TEMP.	LAAR 3	MARG 3	MARG 4	ALPHA2	ALPHA3
0.0730	0.1400	77.60	0.0092	0.0045	0.0086	-0.0153	-0.0016
0.1720	0.2680	75.10	0.0120	0.0122	0.0129	-0.0066	0.0029
0.2680	0.3760	73.40	0.0000	0.0031	0.0011	-0.0062	-0.0049
0.3720	0.4600	72.00	0.0067	0.0096	0.0074	0.0098	0.0050
0.4620	0.5400	70.90	0.0009	0.0021	0.0014	0.0068	0.0002
0.5850	0.6440	70.00	-0.0033	-0.0041	-0.0024	0.0002	-0.0046
0.6920	0.7250	69.40	0.0039	0.0028	0.0049	0.0014	0.0008
0.7920	0.8070	69.10	0.0065	0.0067	0.0071	-0.0015	0.0018
0.8280	0.8380	69.00	0.0067	0.0074	0.0069	-0.0026	0.0016
0.8830	0.8880	68.90	0.0052	0.0066	0.0049	-0.0046	0.0002
0.9470	0.9500	68.80	0.0010	0.0023	0.0003	-0.0059	-0.0024
0.9620	0.9640	68.80	0.0007	0.0019	0.0002	-0.0047	-0.0019
		MEAN	0.0047	0.0053	0.0048	0.0055	0.0023

HEXANE(1)-BENZENE(2)

TONBERG C.O.,JOHNSTON F.:IND.ENG.CHEM.,25,733(1933).

ANTOINE VAPOR PRESSURE CONSTANTS

	A	B	C
(1)	6.87776	1171.530	224.366
(2)	6.90565	1211.033	220.790

P = 735.00

EQUATION	ORDER	A12	A21	D	A122	A211
VAN LAAR	3	0.1733	0.2103	-	-	-
MARGULES	3	0.1729	0.2063	-	-	-
MARGULES	4	0.1934	0.2337	0.0822	-	-
ALPHA	2	1.1582	0.1138	-	-	-
ALPHA	3	1.3852	0.3982	-	-0.1384	-0.3125

			DEVIATION IN VAPOR PHASE COMPOSITION				
X	Y	TEMP.	LAAR 3	MARG 3	MARG 4	ALPHA2	ALPHA3
0.0040	0.0090	78.40	-0.0006	-0.0006	-0.0003	-0.0004	-0.0001
0.0250	0.0600	77.70	-0.0095	-0.0095	-0.0077	-0.0084	-0.0067
0.0430	0.0830	77.30	0.0014	0.0014	0.0037	0.0032	0.0056
0.0750	0.1530	76.00	-0.0127	-0.0127	-0.0103	-0.0102	-0.0073
0.2150	0.3330	72.80	0.0011	0.0011	0.0000	0.0049	0.0056
0.4300	0.5450	69.80	-0.0050	-0.0053	-0.0057	-0.0008	-0.0033
0.7130	0.7530	67.80	-0.0028	-0.0031	-0.0010	0.0008	0.0007
0.7630	0.7900	67.70	-0.0026	-0.0028	-0.0015	0.0006	0.0012
0.8890	0.8930	67.50	-0.0032	-0.0030	-0.0046	-0.0016	-0.0001
0.9650	0.9660	67.40	-0.0037	-0.0035	-0.0051	-0.0032	-0.0024
0.9900	0.9900	67.30	-0.0011	-0.0010	-0.0017	-0.0010	-0.0007
		MEAN	0.0040	0.0040	0.0038	0.0032	0.0031

W

HEXANE(1)-CHLOROBENZENE(2)

BROWN I.:AUSTR.J.SCI.RES.A5,530(1952).

ANTOINE VAPOR PRESSURE CONSTANTS
	A	B	C
(1)	6.87776	1171.530	224.366
(2)	6.97429	1431.323	217.975

T = 65.00

EQUATION	ORDER	A12	A21	D	A122	A211
VAN LAAR	3	0.2948	0.2142	-	-	-
MARGULES	3	0.2892	0.2035	-	-	-
MARGULES	4	0.2552	0.1547	-0.1397	-	-
ALPHA	2	12.1908	-0.2404	-	-	-
ALPHA	3	6.7755	0.0261	-	7.4440	-0.8265

			DEVIATION IN VAPOR PHASE COMPOSITION				
X	Y	PRESS.	LAAR 3	MARG 3	MARG 4	ALPHA2	ALPHA3
0.0830	0.5440	166.40	0.0134	0.0124	0.0051	-0.0146	0.0063
0.1440	0.6790	222.30	0.0001	0.0000	-0.0015	-0.0132	-0.0027
0.2010	0.7440	264.10	-0.0005	-0.0004	0.0007	-0.0045	-0.0011
0.2840	0.8030	319.70	-0.0008	-0.0007	0.0011	0.0025	-0.0000
0.3940	0.8520	382.30	0.0000	-0.0001	0.0005	0.0056	0.0005
0.4380	0.8660	403.90	0.0015	0.0012	0.0011	0.0064	0.0014
0.4850	0.8820	428.30	0.0002	-0.0002	-0.0008	0.0038	-0.0005
0.5400	0.8960	453.80	0.0019	0.0014	0.0002	0.0031	0.0003
0.5910	0.9100	477.90	0.0012	0.0007	-0.0008	-0.0002	-0.0011
0.6790	0.9290	516.30	0.0033	0.0030	0.0016	-0.0035	0.0001
0.8060	0.9570	578.00	0.0032	0.0033	0.0032	-0.0115	0.0000
0.9270	0.9840	638.40	0.0013	0.0015	0.0022	-0.0154	0.0001
		MEAN	0.0023	0.0021	0.0016	0.0070	0.0012

346

HEXANE(1)-CHLOROBENZENE(2)

BROWN I.:AUSTR.J.SCI.RES.A5,530(1952).

ANTOINE VAPOR PRESSURE CONSTANTS

	A	B	C
(1)	6.87776	1171.350	224.366
(2)	6.97429	1431.323	217.975

P = 759.80

EQUATION	ORDER	A12	A21	D	A122	A211
VAN LAAR	3	0.2373	0.1530	-	-	-
MARGULES	3	0.2363	0.1229	-	-	-
MARGULES	4	0.1301	-0.0282	-0.4548	-	-
ALPHA	2	7.2541	-0.5932	-	-	-
ALPHA	3	-3.9785	-2.0644	-	11.1383	1.2228

DEVIATION IN VAPOR PHASE COMPOSITION

X	Y	TEMP.	LAAR 3	MARG 3	MARG 4	ALPHA2	ALPHA3
0.0180	0.1180	127.56	0.0201	0.0200	-0.0025	0.0128	0.0117
0.0490	0.2820	121.06	0.0262	0.0263	-0.0031	0.0134	0.0120
0.0810	0.4060	115.66	0.0227	0.0230	0.0013	0.0091	0.0084
0.1090	0.4910	111.53	0.0156	0.0158	0.0030	0.0030	0.0029
0.1460	0.5770	106.62	0.0077	0.0076	0.0047	-0.0031	-0.0024
0.2000	0.6660	101.04	0.0000	-0.0006	0.0050	-0.0073	-0.0057
0.3090	0.7690	92.70	0.0006	-0.0012	0.0064	-0.0020	0.0002
0.4190	0.8350	86.84	-0.0002	-0.0025	-0.0010	-0.0015	-0.0003
0.5160	0.8720	82.66	0.0046	0.0025	-0.0012	0.0019	0.0013
0.5910	0.8960	80.19	0.0066	0.0051	-0.0005	0.0022	0.0001
0.5930	0.8960	80.13	0.0073	0.0057	0.0001	0.0028	0.0007
0.5940	0.8960	80.17	0.0075	0.0060	0.0004	0.0031	0.0009
0.6440	0.9120	78.31	0.0070	0.0060	0.0003	0.0009	-0.0018
0.7370	0.9340	75.70	0.0102	0.0100	0.0067	0.0014	0.0003
0.7900	0.9500	74.17	0.0071	0.0073	0.0061	-0.0030	-0.0009
0.7930	0.9500	74.14	0.0077	0.0080	0.0069	-0.0024	-0.0000
0.8470	0.9650	72.72	0.0049	0.0054	0.0062	-0.0059	0.0006
		MEAN	0.0092	0.0090	0.0033	0.0045	0.0030

HEXANE(1)-ETHYL ALCOHOL(2)

HO J.C.K.,LU B.C.Y.:J.CHEM.ENG.DATA 8(4),549(1963).

ANTOINE VAPOR PRESSURE CONSTANTS

	A	B	C
(1)	6.87776	1171.530	224.366
(2)	8.16290	1623.220	228.980

T =   55.00

EQUATION	ORDER	A12	A21	D	A122	A211
VAN LAAR	3	0.8911	0.8303	-	-	-
MARGULES	3	0.8861	0.8323	-	-	-
MARGULES	4	1.0924	1.0524	0.7571	-	-
ALPHA	2	17.8105	10.1089	-	-	-
ALPHA	3	20.4765	11.0939	-	-1.8758	0.0977

			DEVIATION IN VAPOR PHASE COMPOSITION				
X	Y	PRESS.	LAAR 3	MARG 3	MARG 4	ALPHA2	ALPHA3
0.0110	0.1780	345.30	-0.0533	-0.0545	-0.0002	-0.0209	-0.0166
0.0700	0.4310	491.20	-0.0041	-0.0057	0.0481	0.0054	0.0073
0.1000	0.4870	538.50	0.0043	0.0031	0.0367	-0.0022	-0.0017
0.1410	0.5200	588.40	0.0248	0.0241	0.0352	0.0045	0.0039
0.2060	0.5680	624.00	0.0215	0.0214	0.0100	-0.0077	-0.0089
0.3030	0.5830	654.40	0.0333	0.0339	0.0118	0.0064	0.0052
0.3320	0.5950	660.30	0.0248	0.0255	0.0041	0.0004	-0.0006
0.3870	0.6060	668.40	0.0173	0.0182	0.0018	-0.0010	-0.0016
0.4090	0.6090	673.20	0.0150	0.0159	0.0025	-0.0006	-0.0010
0.4980	0.6210	674.50	0.0041	0.0050	0.0066	-0.0006	-0.0002
0.5370	0.6220	676.50	0.0037	0.0046	0.0129	0.0032	0.0039
0.6030	0.6350	676.70	-0.0063	-0.0056	0.0124	-0.0016	-0.0004
0.7240	0.6500	674.60	-0.0019	-0.0016	0.0198	0.0010	0.0027
0.8100	0.6820	673.50	0.0026	0.0025	0.0083	-0.0116	-0.0103
0.9000	0.7040	654.90	0.0630	0.0626	0.0349	0.0085	0.0073
0.9430	0.7460	640.50	0.0922	0.0917	0.0501	0.0139	0.0099
0.9610	0.8020	602.00	0.0765	0.0761	0.0342	-0.0062	-0.0118
		MEAN	0.0264	0.0266	0.0194	0.0056	0.0055

HEXANE(1)-ETHYL ALCOHOL(2)

ISII N.:J.SOC.CHEM.IND.JAPAN 38,659(1935).

ANTOINE VAPOR PRESSURE CONSTANTS
	A	B	C
(1)	6.87776	1171.530	224.366
(2)	8.16290	1623.220	228.980

T = -10.00

EQUATION	ORDER	A12	A21	D	A122	A211
VAN LAAR	3	0.9132	0.9539	-	-	-
MARGULES	3	0.9144	0.9517	-	-	-
MARGULES	4	1.0791	1.1195	0.5673	-	-
ALPHA	2	65.4571	12.5104	-	-	-
ALPHA	3	12.9808	8.7515	-	43.1159	-5.8055

			DEVIATION IN VAPOR PHASE COMPOSITION				
X	Y	PRESS.	LAAR 3	MARG 3	MARG 4	ALPHA2	ALPHA3
0.0500	0.6474	17.30	-0.0293	-0.0288	0.0239	0.0243	0.0116
0.1000	0.7441	23.25	-0.0081	-0.0078	0.0142	0.0032	0.0002
0.1500	0.7793	26.28	0.0038	0.0039	0.0096	-0.0028	-0.0019
0.2000	0.7978	28.04	0.0085	0.0085	0.0052	-0.0057	-0.0031
0.2500	0.8067	29.13	0.0118	0.0118	0.0038	-0.0049	-0.0017
0.3000	0.8124	29.85	0.0124	0.0123	0.0026	-0.0039	-0.0007
0.3500	0.8151	30.30	0.0123	0.0122	0.0031	-0.0017	0.0012
0.4000	0.8184	30.67	0.0094	0.0091	0.0023	-0.0012	0.0010
0.4500	0.8207	30.95	0.0059	0.0057	0.0021	-0.0003	0.0009
0.5000	0.8221	31.20	0.0026	0.0023	0.0026	0.0009	0.0011
0.5500	0.8233	31.42	-0.0007	-0.0010	0.0031	0.0020	0.0010
0.6000	0.8248	31.56	-0.0040	-0.0043	0.0030	0.0026	0.0004
0.6500	0.8262	31.65	-0.0061	-0.0063	0.0030	0.0032	-0.0002
0.7000	0.8275	31.66	-0.0061	-0.0063	0.0033	0.0040	-0.0004
0.7500	0.8291	31.60	-0.0034	-0.0035	0.0042	0.0047	-0.0002
0.8000	0.8323	31.48	0.0026	0.0026	0.0055	0.0043	0.0001
0.8500	0.8402	31.18	0.0111	0.0112	0.0067	0.0003	-0.0007
0.9000	0.8566	30.68	0.0223	0.0225	0.0092	-0.0095	-0.0018
0.9500	0.8897	29.93	0.0346	0.0348	0.0167	-0.0267	0.0015
		MEAN	0.0103	0.0103	0.0065	0.0056	0.0016

HEXANE(1)-ETHYL ALCOHOL(2)

ISII N.:J.SOC.CHEM.IND.JAPAN 38,659(1935).

ANTOINE VAPOR PRESSURE CONSTANTS

	A	B	C
(1)	6.87776	1171.530	224.366
(2)	8.16290	1623.220	228.980

T = 0.00

EQUATION	ORDER	A12	A21	D	A122	A211
VAN LAAR	3	0.8898	1.0982	-	-	-
MARGULES	3	0.8858	1.0783	-	-	-
MARGULES	4	0.9843	1.2018	0.3905	-	-
ALPHA	2	63.6159	13.3177	-	-	-
ALPHA	3	12.0012	4.3015	-	25.5021	-0.2948

			DEVIATION IN VAPOR PHASE COMPOSITION				
X	Y	PRESS.	LAAR 3	MARG 3	MARG 4	ALPHA2	ALPHA3
0.0500	0.5880	28.40	-0.0192	-0.0202	0.0120	0.0720	0.0175
0.1000	0.7240	41.30	-0.0233	-0.0233	-0.0105	0.0112	-0.0075
0.1500	0.7636	46.75	-0.0069	-0.0065	-0.0044	0.0007	-0.0022
0.2000	0.7840	50.00	0.0017	0.0022	-0.0012	-0.0042	0.0005
0.2500	0.7965	52.10	0.0054	0.0058	-0.0002	-0.0070	0.0013
0.3000	0.8060	53.60	0.0047	0.0049	-0.0016	-0.0098	-0.0004
0.3500	0.8095	54.60	0.0053	0.0052	-0.0002	-0.0084	0.0007
0.4000	0.8123	55.40	0.0035	0.0030	-0.0004	-0.0073	0.0002
0.4500	0.8154	55.80	-0.0010	-0.0018	-0.0025	-0.0073	-0.0021
0.5000	0.8157	56.15	-0.0043	-0.0055	-0.0033	-0.0050	-0.0027
0.6000	0.8040	56.43	-0.0014	-0.0032	0.0044	0.0112	0.0061
0.7000	0.8038	56.36	-0.0094	-0.0111	-0.0020	0.0157	0.0024
0.8000	0.8100	56.17	-0.0136	-0.0140	-0.0111	0.0149	-0.0039
0.9000	0.8305	55.27	0.0032	0.0053	-0.0072	0.0056	-0.0038
0.9500	0.8658	53.82	0.0215	0.0243	0.0061	-0.0125	0.0019
		MEAN	0.0083	0.0091	0.0045	0.0128	0.0036

HEXANE(1)-ETHYL ALCOHOL(2)

ISII N.:J.SOC.CHEM.IND.JAPAN 38,659(1935).

ANTOINE VAPOR PRESSURE CONSTANTS
	A	B	C
(1)	6.87776	1171.530	224.366
(2)	8.16290	1623.220	228.980

T = 10.00

EQUATION	ORDER	A12	A21	D	A122	A211
VAN LAAR	3	0.8487	1.0706	-	-	-
MARGULES	3	0.8442	1.0452	-	-	-
MARGULES	4	1.0128	1.2257	0.5878	-	-
ALPHA	2	53.3105	13.3593	-	-	-
ALPHA	3	18.8753	6.4666	-	17.6871	-0.6390

			DEVIATION IN VAPOR PHASE COMPOSITION				
X	Y	PRESS.	LAAR 3	MARG 3	MARG 4	ALPHA2	ALPHA3
0.0500	0.5778	54.00	-0.0720	-0.0732	-0.0144	0.0419	0.0037
0.1000	0.6888	70.70	-0.0413	-0.0415	-0.0145	0.0109	-0.0024
0.1500	0.7293	79.80	-0.0177	-0.0175	-0.0100	0.0019	-0.0002
0.2000	0.7506	85.00	-0.0044	-0.0041	-0.0077	-0.0024	0.0008
0.2500	0.7650	88.50	0.0012	0.0015	-0.0078	-0.0061	-0.0004
0.3000	0.7725	91.00	0.0054	0.0054	-0.0056	-0.0062	0.0003
0.3500	0.7774	93.00	0.0069	0.0065	-0.0037	-0.0056	0.0007
0.4000	0.7810	94.50	0.0060	0.0052	-0.0022	-0.0049	0.0005
0.4500	0.7830	95.40	0.0041	0.0028	-0.0006	-0.0034	0.0006
0.5000	0.7852	95.90	0.0001	-0.0017	-0.0005	-0.0025	-0.0004
0.5500	0.7861	96.30	-0.0039	-0.0060	-0.0004	-0.0007	-0.0006
0.6000	0.7867	96.60	-0.0083	-0.0106	-0.0011	0.0012	-0.0008
0.6500	0.7876	96.50	-0.0129	-0.0153	-0.0033	0.0028	-0.0013
0.7000	0.7876	96.50	-0.0155	-0.0176	-0.0055	0.0054	-0.0006
0.7500	0.7884	96.40	-0.0165	-0.0180	-0.0085	0.0075	0.0001
0.8000	0.7902	96.30	-0.0139	-0.0142	-0.0111	0.0095	0.0017
0.8500	0.7973	95.70	-0.0080	-0.0067	-0.0140	0.0076	0.0015
0.9000	0.8143	94.80	0.0034	0.0064	-0.0140	-0.0003	-0.0003
0.9500	0.8529	91.80	0.0231	0.0271	-0.0027	-0.0172	-0.0010
		MEAN	0.0139	0.0148	0.0067	0.0073	0.0009

HEXANE(1)-ETHYL ALCOHOL(2)

ISII N.:J.SOC.CHEM.IND.JAPAN 38,659(1935).

ANTOINE VAPOR PRESSURE CONSTANTS

	A	B	C
(1)	6.87776	1171.530	224.366
(2)	8.16290	1623.220	228.980

T =   20.00

EQUATION	ORDER	A12	A21	D	A122	A211
VAN LAAR	3	0.7625	1.0974	-	-	-
MARGULES	3	0.7388	1.0485	-	-	-
MARGULES	4	0.8057	1.1244	0.2362	-	-
ALPHA	2	32.1031	8.0573	-	-	-
ALPHA	3	4.7735	3.3914	-	18.9568	-1.3013

			DEVIATION IN VAPOR PHASE COMPOSITION				
X	Y	PRESS.	LAAR 3	MARG 3	MARG 4	ALPHA2	ALPHA3
0.0500	0.4821	80.90	-0.0565	-0.0648	-0.0418	0.0595	0.0233
0.1000	0.6331	109.30	-0.0546	-0.0590	-0.0472	0.0147	0.0006
0.1500	0.6959	127.60	-0.0407	-0.0422	-0.0386	-0.0026	-0.0048
0.2000	0.7260	138.30	-0.0262	-0.0259	-0.0272	-0.0073	-0.0033
0.2500	0.7443	145.10	-0.0166	-0.0155	-0.0192	-0.0093	-0.0023
0.3000	0.7553	150.00	-0.0096	-0.0084	-0.0128	-0.0089	-0.0008
0.3500	0.7635	153.10	-0.0064	-0.0055	-0.0094	-0.0085	-0.0007
0.4000	0.7679	154.70	-0.0039	-0.0036	-0.0062	-0.0061	0.0004
0.4500	0.7702	155.80	-0.0026	-0.0032	-0.0042	-0.0028	0.0017
0.5000	0.7717	156.40	-0.0032	-0.0046	-0.0037	0.0006	0.0025
0.5500	0.7736	156.80	-0.0060	-0.0084	-0.0056	0.0030	0.0019
0.6000	0.7754	157.20	-0.0101	-0.0133	-0.0090	0.0052	0.0011
0.6500	0.7767	157.20	-0.0144	-0.0180	-0.0128	0.0079	0.0007
0.7000	0.7792	157.60	-0.0197	-0.0233	-0.0181	0.0096	-0.0000
0.7500	0.7830	158.10	-0.0248	-0.0276	-0.0237	0.0106	-0.0002
0.8000	0.7910	158.40	-0.0304	-0.0313	-0.0303	0.0085	-0.0007
0.8500	0.8094	160.00	-0.0385	-0.0366	-0.0403	-0.0018	-0.0045
0.9000	0.8352	156.00	-0.0385	-0.0329	-0.0423	-0.0140	-0.0018
0.9500	0.8854	151.80	-0.0293	-0.0211	-0.0344	-0.0342	0.0037
		MEAN	0.0227	0.0234	0.0225	0.0113	0.0029

HEXANE(1)-ETHYL ALCOHOL(2)

ISII N.:J.SOC.CHEM.IND.JAPAN 38,659(1935).

ANTOINE VAPOR PRESSURE CONSTANTS
```
          A          B          C
(1) 6.87776    1171.530    224.366
(2) 8.16290    1623.220    228.980
```

T =   30.00

EQUATION	ORDER	A12	A21	D	A122	A211
VAN LAAR	3	0.7046	1.1962	-	-	-
MARGULES	3	0.6589	1.0899	-	-	-
MARGULES	4	0.7810	1.2294	0.4295	-	-
ALPHA	2	24.9691	6.9747	-	-	-
ALPHA	3	21.1045	1.9835	-	-6.0597	5.0240

			DEVIATION IN VAPOR PHASE COMPOSITION				
X	Y	PRESS.	LAAR 3	MARG 3	MARG 4	ALPHA2	ALPHA3
0.0500	0.4273	128.00	-0.0621	-0.0774	-0.0373	0.0637	-0.0028
0.1000	0.5778	165.80	-0.0555	-0.0645	-0.0423	0.0279	-0.0047
0.1500	0.6461	191.30	-0.0378	-0.0412	-0.0341	0.0110	-0.0005
0.2000	0.6840	207.90	-0.0228	-0.0226	-0.0248	0.0025	0.0027
0.2500	0.7088	219.80	-0.0127	-0.0108	-0.0176	-0.0032	0.0031
0.3000	0.7257	228.60	-0.0060	-0.0036	-0.0118	-0.0065	0.0026
0.3500	0.7378	235.30	-0.0020	-0.0001	-0.0074	-0.0084	0.0014
0.4000	0.7461	240.30	0.0004	0.0010	-0.0039	-0.0085	0.0005
0.4500	0.7544	243.10	-0.0014	-0.0024	-0.0041	-0.0100	-0.0026
0.5000	0.7593	245.50	-0.0031	-0.0061	-0.0042	-0.0090	-0.0038
0.5500	0.7613	246.80	-0.0046	-0.0096	-0.0043	-0.0057	-0.0030
0.6000	0.7633	247.20	-0.0085	-0.0151	-0.0070	-0.0026	-0.0026
0.6500	0.7652	247.00	-0.0140	-0.0217	-0.0120	0.0005	-0.0023
0.7000	0.7655	246.90	-0.0190	-0.0269	-0.0172	0.0056	0.0000
0.7500	0.7663	246.90	-0.0245	-0.0310	-0.0238	0.0108	0.0027
0.8000	0.7693	246.20	-0.0303	-0.0331	-0.0315	0.0154	0.0051
0.8500	0.7773	244.30	-0.0350	-0.0315	-0.0389	0.0178	0.0060
0.9000	0.7994	238.30	-0.0387	-0.0263	-0.0457	0.0128	0.0008
0.9500	0.8458	227.00	-0.0279	-0.0075	-0.0366	0.0027	-0.0064
		MEAN	0.0214	0.0227	0.0213	0.0118	0.0028

HEXANE(1)-PROPYL ALCOHOL(2)

PRABHU P.S.,VAN WINKLE M.:J.CHEM.ENG.DATA 8(2),210(1963).

ANTOINE VAPOR PRESSURE CONSTANTS

	A	B	C
(1)	6.87776	1171.530	224.366
(2)	7.85418	1497.910	204.112

P = 760.00

EQUATION	ORDER	A12	A21	D	A122	A211
VAN LAAR	3	0.6724	0.7893	-	-	-
MARGULES	3	0.6680	0.7824	-	-	-
MARGULES	4	0.6575	0.7683	-0.0404	-	-
ALPHA	2	14.3698	2.1658	-	-	-
ALPHA	3	0.6277	1.5130	-	15.0470	-1.6055

			DEVIATION IN VAPOR PHASE COMPOSITION				
X	Y	TEMP.	LAAR 3	MARG 3	MARG 4	ALPHA2	ALPHA3
0.0240	0.0256	89.60	0.1856	0.1842	0.1810	0.2343	0.2491
0.0600	0.4900	82.00	-0.0795	-0.0809	-0.0840	-0.0396	-0.0231
0.1440	0.6620	74.60	-0.0449	-0.0452	-0.0459	-0.0316	-0.0212
0.2360	0.7280	71.90	-0.0255	-0.0253	-0.0248	-0.0180	-0.0137
0.2620	0.7160	71.20	0.0015	0.0017	0.0023	0.0084	0.0112
0.3700	0.7600	70.00	-0.0053	-0.0051	-0.0047	0.0062	0.0035
0.4760	0.7860	68.40	-0.0097	-0.0098	-0.0100	0.0063	-0.0007
0.6200	0.8000	67.70	-0.0058	-0.0062	-0.0071	0.0182	0.0082
0.7520	0.8360	67.00	-0.0183	-0.0186	-0.0192	0.0044	0.0009
0.7840	0.8560	66.40	-0.0285	-0.0286	-0.0289	-0.0094	-0.0084
0.9040	0.9160	66.20	-0.0298	-0.0293	-0.0281	-0.0326	-0.0015
0.9750	0.9700	67.20	-0.0105	-0.0100	-0.0090	-0.0254	0.0068
		MEAN	0.0371	0.0371	0.0371	0.0362	0.0290

TRIEHYLAMINE(1)-ETHYL ALCOHOL(2)

COPP J.L.,EVERETT D.H.:DISC.FARAD.SOC.15,174(1953).

ANTOINE VAPOR PRESSURE CONSTANTS

	A	B	C
(1)	7.18658	1341.300	222.000
(2)	8.16290	1623.220	228.980

T =   64.85

EQUATION	ORDER	A12	A21	D	A122	A211
VAN LAAR	3	0.1871	0.4930	-	-	-
MARGULES	3	0.1104	0.3811	-	-	-
MARGULES	4	0.0612	0.3342	-0.1394	-	-
ALPHA	2	0.6500	1.1588	-	-	-
ALPHA	3	0.8615	0.5079	-	-0.5536	0.9614

			DEVIATION IN VAPOR PHASE COMPOSITION				
X	Y	PRESS.	LAAR 3	MARG 3	MARG 4	ALPHA2	ALPHA3
0.1040	0.1270	447.25	-0.0134	-0.0215	-0.0253	0.0138	-0.0004
0.2405	0.2640	459.15	-0.0169	-0.0182	-0.0175	0.0060	0.0001
0.3040	0.3185	462.10	-0.0146	-0.0121	-0.0099	0.0008	0.0002
0.3705	0.3705	463.30	-0.0103	-0.0055	-0.0029	-0.0033	0.0005
0.4415	0.4230	462.70	-0.0060	-0.0009	0.0008	-0.0068	-0.0001
0.6055	0.5370	451.15	0.0009	0.0003	-0.0017	-0.0058	-0.0001
0.6540	0.5740	445.55	-0.0020	-0.0046	-0.0070	-0.0056	-0.0022
0.7130	0.6170	437.60	-0.0035	-0.0075	-0.0094	0.0005	0.0003
0.7730	0.6655	428.35	-0.0081	-0.0111	-0.0114	0.0078	0.0034
0.8485	0.7480	409.70	-0.0274	-0.0243	-0.0206	0.0082	-0.0013
0.9280	0.8570	377.80	-0.0407	-0.0263	-0.0185	0.0097	-0.0012
		MEAN	0.0131	0.0120	0.0114	0.0062	0.0009

TRIEHYLAMINE(1)-ETHYL ALCOHOL(2)

COPP J.L.,EVERETT D.H.:DISC.FARAD.SOC.15,174(1953).

ANTOINE VAPOR PRESSURE CONSTANTS

	A	B	C
(1)	7.18658	1341.300	222.000
(2)	8.16290	1623.220	228.980

T =   49.60

EQUATION	ORDER	A12	A21	D	A122	A211
VAN LAAR	3	0.1672	0.5581	-	-	-
MARGULES	3	0.0778	0.3797	-	-	-
MARGULES	4	0.0018	0.3019	-0.2403	-	-
ALPHA	2	0.6940	0.8370	-	-	-
ALPHA	3	0.5898	-0.0643	-	-0.2659	1.1514

DEVIATION IN VAPOR PHASE COMPOSITION

X	Y	PRESS.	LAAR 3	MARG 3	MARG 4	ALPHA2	ALPHA3
0.1305	0.1655	225.15	-0.0173	-0.0262	-0.0306	0.0128	-0.0025
0.2105	0.2480	230.80	-0.0168	-0.0202	-0.0195	0.0117	0.0024
0.3260	0.3585	236.25	-0.0169	-0.0122	-0.0071	-0.0005	0.0007
0.4185	0.4345	238.55	-0.0118	-0.0055	-0.0021	-0.0066	-0.0002
0.4575	0.4645	238.95	-0.0094	-0.0040	-0.0022	-0.0082	-0.0009
0.5510	0.5315	238.50	-0.0028	-0.0024	-0.0051	-0.0073	-0.0008
0.6740	0.6190	234.45	-0.0007	-0.0081	-0.0138	-0.0006	0.0002
0.7930	0.7170	227.25	-0.0150	-0.0214	-0.0220	0.0078	0.0008
0.8690	0.7955	219.10	-0.0335	-0.0287	-0.0219	0.0118	0.0014
0.9455	0.9040	205.10	-0.0512	-0.0295	-0.0181	0.0055	-0.0031
		MEAN	0.0175	0.0158	0.0142	0.0073	0.0013

TRIEHYLAMINE(1)-ETHYL ALCOHOL(2)

COPP J.L.,EVERETT D.H.:DISC.FARAD.SOC.15,174(1953).

ANTOINE VAPOR PRESSURE CONSTANTS

	A	B	C
(1)	7.18658	1341.300	222.000
(2)	8.16290	1623.220	228.980

T =   34.85

EQUATION	ORDER	A12	A21	D	A122	A211
VAN LAAR	3	0.1715	0.6885	-	-	-
MARGULES	3	0.0417	0.4266	-	-	-
MARGULES	4	-0.0572	0.3236	-0.3089	-	-
ALPHA	2	0.7881	0.5375	-	-	-
ALPHA	3	0.2299	-0.6862	-	0.1519	1.4407

DEVIATION IN VAPOR PHASE COMPOSITION

X	Y	PRESS.	LAAR 3	MARG 3	MARG 4	ALPHA2	ALPHA3
0.0810	0.1090	104.80	-0.0053	-0.0202	-0.0282	0.0181	0.0011
0.1875	0.2435	109.15	-0.0162	-0.0248	-0.0264	0.0124	0.0002
0.2865	0.3535	113.10	-0.0228	-0.0189	-0.0136	-0.0013	-0.0025
0.3450	0.4050	114.90	-0.0179	-0.0090	-0.0027	-0.0025	0.0016
0.4010	0.4550	116.25	-0.0170	-0.0060	-0.0012	-0.0073	-0.0001
0.5265	0.5490	118.20	-0.0071	-0.0013	-0.0035	-0.0056	0.0014
0.5825	0.5905	118.80	-0.0065	-0.0059	-0.0110	-0.0051	-0.0007
0.7250	0.6930	117.60	-0.0120	-0.0230	-0.0284	0.0047	-0.0008
0.8905	0.8440	112.75	-0.0544	-0.0442	-0.0324	0.0126	0.0011
0.9415	0.9100	109.15	-0.0680	-0.0401	-0.0251	0.0079	-0.0009
		MEAN	0.0227	0.0193	0.0172	0.0078	0.0010

357

TRIETHYLAMINE(1)-WATER(2)

KOHLER F.:MONATSH.82,913(1951).

ANTOINE VAPOR PRESSURE CONSTANTS

	A	B	C
(1)	7.18658	1341.300	222.000
(2)	7.96681	1668.210	228.000

T =   0.00

EQUATION	ORDER	A12	A21	D	A122	A211
VAN LAAR	3	0.8289	0.9780	-	-	-
MARGULES	3	0.8287	0.9623	-	-	-
MARGULES	4	1.0162	1.1816	0.6645	-	-
ALPHA	2	46.6266	10.9613	-	-	-
ALPHA	3	5.8841	24.0333	-	84.5734	-23.9019

			DEVIATION IN VAPOR PHASE COMPOSITION				
X	Y	PRESS.	LAAR 3	MARG 3	MARG 4	ALPHA2	ALPHA3
0.1000	0.7310	16.70	-0.1041	-0.1037	-0.0722	-0.0361	-0.0067
0.2000	0.7490	17.50	-0.0212	-0.0212	-0.0240	0.0010	0.0061
0.3000	0.7650	18.30	-0.0039	-0.0043	-0.0155	0.0057	0.0021
0.4000	0.7750	18.90	-0.0031	-0.0041	-0.0107	0.0070	-0.0005
0.5000	0.7840	19.40	-0.0112	-0.0127	-0.0095	0.0055	-0.0032
0.6000	0.7900	19.75	-0.0199	-0.0215	-0.0094	0.0055	-0.0020
0.7000	0.7980	20.05	-0.0280	-0.0292	-0.0154	0.0033	0.0005
0.8000	0.8160	20.15	-0.0333	-0.0333	-0.0313	-0.0073	0.0026
0.9000	0.8710	20.10	-0.0388	-0.0368	-0.0618	-0.0467	-0.0009
		MEAN	0.0293	0.0297	0.0278	0.0131	0.0027

TRIETHYLAMINE(1)-WATER(2)

KOHLER F.:MONATSH.82,913(1951).

ANTOINE VAPOR PRESSURE CONSTANTS
	A	B	C
(1)	7.18658	1341.300	222.000
(2)	7.96681	1668.210	228.000

T = 10.00

EQUATION	ORDER	A12	A21	D	A122	A211
VAN LAAR	3	0.8937	0.9257	-	-	-
MARGULES	3	0.8948	0.9237	-	-	-
MARGULES	4	1.1212	1.1769	0.7851	-	-
ALPHA	2	51.4786	15.0368	-	-	-
ALPHA	3	19.1154	74.3853	-	192.7164	-73.7119

			DEVIATION IN VAPOR PHASE COMPOSITION				
X	Y	PRESS.	LAAR 3	MARG 3	MARG 4	ALPHA2	ALPHA3
0.1000	0.7210	33.55	-0.1005	-0.1002	-0.0617	-0.0435	-0.0055
0.2000	0.7260	34.00	-0.0152	-0.0152	-0.0187	-0.0014	0.0037
0.3000	0.7330	34.50	0.0040	0.0038	-0.0110	0.0093	0.0035
0.4000	0.7420	35.15	0.0007	0.0004	-0.0092	0.0099	0.0001
0.5000	0.7510	35.60	-0.0104	-0.0108	-0.0078	0.0074	-0.0028
0.6000	0.7600	36.00	-0.0224	-0.0227	-0.0081	0.0037	-0.0038
0.7000	0.7670	36.50	-0.0262	-0.0264	-0.0093	0.0020	0.0012
0.8000	0.7870	36.60	-0.0258	-0.0258	-0.0231	-0.0108	0.0029
0.9000	0.8420	36.05	-0.0192	-0.0189	-0.0492	-0.0500	-0.0011
		MEAN	0.0249	0.0249	0.0220	0.0153	0.0027

TRIETHYLAMINE(1)-WATER(2)

KOHLER F.:MONATSH.82,913(1951).

ANTOINE VAPOR PRESSURE CONSTANTS
	A	B	C
(1)	7.18658	1341.300	222.000
(2)	7.96681	1668.210	228.000

T =   18.00

EQUATION	ORDER	A12	A21	D	A122	A211
VAN LAAR	3	0.9566	0.9115	-	-	-
MARGULES	3	0.9542	0.9129	-	-	-
MARGULES	4	1.1636	1.1401	0.7147	-	-
ALPHA	2	65.2797	22.0669	-	-	-
ALPHA	3	.1762/ 04	.2188/ 05	-	.5239/ 05	-.2266/ 05

			DEVIATION IN VAPOR PHASE COMPOSITION				
X	Y	PRESS.	LAAR 3	MARG 3	MARG 4	ALPHA2	ALPHA3
0.1000	0.7160	54.90	-0.0945	-0.0950	-0.0591	-0.0417	-0.0024
0.2000	0.7170	55.10	-0.0154	-0.0154	-0.0186	-0.0062	-0.0014
0.3000	0.7190	55.35	0.0017	0.0020	-0.0123	0.0052	-0.0009
0.4000	0.7220	55.80	-0.0007	-0.0003	-0.0101	0.0095	-0.0006
0.5000	0.7250	56.10	-0.0092	-0.0088	-0.0067	0.0115	0.0009
0.6000	0.7300	56.35	-0.0185	-0.0182	-0.0048	0.0107	0.0027
0.7000	0.7430	56.55	-0.0272	-0.0270	-0.0109	0.0019	0.0007
0.8000	0.7680	56.50	-0.0275	-0.0276	-0.0250	-0.0171	-0.0028
0.9000	0.8240	55.20	-0.0139	-0.0142	-0.0425	-0.0594	0.0006
		MEAN	0.0232	0.0232	0.0211	0.0181	0.0014

360

TOLUENE(1)-BUTYL ALCOHOL(2)

MANN R.S.,SHEMILT L.W.:J.CHEM.ENG.DATA 8,189(1963).

ANTOINE VAPOR PRESSURE CONSTANTS

	A	B	C
(1)	6.95334	1343.943	219.377
(2)	7.54472	1405.873	183.908

P = 760.00

EQUATION	ORDER	A12	A21	D	A122	A211
VAN LAAR	3	0.3713	0.5671	-	-	-
MARGULES	3	0.3462	0.5405	-	-	-
MARGULES	4	0.3584	0.5513	0.0362	-	-
ALPHA	2	2.6667	1.2563	-	-	-
ALPHA	3	3.2704	0.0926	-	-1.6346	1.7688

			DEVIATION IN VAPOR PHASE COMPOSITION				
X	Y	TEMP.	LAAR 3	MARG 3	MARG 4	ALPHA2	ALPHA3
0.0280	0.0750	116.05	-0.0005	-0.0036	-0.0022	0.0159	-0.0047
0.0960	0.2210	112.90	-0.0016	-0.0057	-0.0039	0.0233	-0.0075
0.1650	0.3210	110.50	0.0074	0.0056	0.0062	0.0246	0.0040
0.2270	0.3990	109.00	0.0043	0.0048	0.0044	0.0126	0.0035
0.3180	0.4870	107.60	-0.0004	0.0018	0.0007	-0.0027	0.0010
0.4150	0.5540	106.40	-0.0007	0.0014	0.0006	-0.0098	0.0002
0.4870	0.5950	106.00	-0.0023	-0.0011	-0.0014	-0.0126	-0.0020
0.5320	0.6170	105.80	-0.0023	-0.0019	-0.0018	-0.0120	-0.0026
0.5580	0.6270	105.70	-0.0002	-0.0002	0.0000	-0.0091	-0.0008
0.6140	0.6530	105.60	-0.0012	-0.0020	-0.0015	-0.0073	-0.0023
0.6680	0.6750	105.50	0.0009	-0.0005	0.0001	-0.0015	-0.0006
0.6750	0.6760	105.50	0.0030	0.0016	0.0023	0.0012	0.0015
0.7010	0.6870	105.50	0.0040	0.0025	0.0031	0.0044	0.0025
0.7660	0.7200	105.60	0.0037	0.0025	0.0029	0.0104	0.0023
0.8590	0.7840	106.30	0.0014	0.0025	0.0018	0.0173	0.0009
0.8710	0.7940	106.50	0.0017	0.0031	0.0023	0.0185	0.0014
0.9480	0.8940	108.10	-0.0060	-0.0026	-0.0041	0.0105	-0.0059
		MEAN	0.0024	0.0026	0.0023	0.0114	0.0026

361

TOLUENE(1)-FURFURALDEHYDE(2)

THORNTON J.D.,GARNER F.H.:J.APPL.CHEM.1,SUPPL.NO1,S74(1951).

ANTOINE VAPOR PRESSURE CONSTANTS
	A	B	C
(1)	6.95334	1343.943	219.377
(2)	6.48323	1112.311	147.256

P =   760.00

EQUATION	ORDER	A12	A21	D	A122	A211
VAN LAAR	3	0.4520	0.4845	-	-	-
MARGULES	3	0.4547	0.4777	-	-	-
MARGULES	4	0.5780	0.7234	0.7073	-	-
ALPHA	2	8.8796	-0.0945	-	-	-
ALPHA	3	19.3209	3.6245	-	-6.0539	-4.1620

			DEVIATION IN VAPOR PHASE COMPOSITION				
X	Y	TEMP.	LAAR 3	MARG 3	MARG 4	ALPHA2	ALPHA3
0.0270	0.2507	153.30	-0.0359	-0.0350	0.0011	-0.0392	0.0121
0.0520	0.4002	147.00	-0.0513	-0.0504	-0.0185	-0.0585	-0.0075
0.0586	0.4374	145.20	-0.0595	-0.0586	-0.0295	-0.0680	-0.0194
0.0734	0.4500	145.00	-0.0208	-0.0201	0.0021	-0.0260	0.0161
0.1712	0.6372	135.40	-0.0039	-0.0040	-0.0146	-0.0002	0.0020
0.3365	0.7682	127.00	-0.0046	-0.0054	-0.0147	0.0149	-0.0043
0.5276	0.8426	122.70	-0.0115	-0.0123	0.0002	0.0167	0.0056
0.6689	0.8998	117.60	-0.0275	-0.0278	-0.0123	-0.0051	-0.0028
0.7789	0.9331	115.10	-0.0282	-0.0281	-0.0214	-0.0146	-0.0020
0.8968	0.9640	112.50	-0.0157	-0.0153	-0.0237	-0.0161	0.0017
0.9536	0.9830	111.30	-0.0084	-0.0081	-0.0185	-0.0134	-0.0001
		MEAN	0.0243	0.0241	0.0143	0.0248	0.0067

362

TOLUENE(1)-OCTANE(2)

BROMILEY E.C.,QUIGGLE D.:IND.ENG.CHEM.25,1136(1933).

ANTOINE VAPOR PRESSURE CONSTANTS
	A	B	C
(1)	6.95334	1343.943	219.377
(2)	6.92377	1355.126	209.517

P = 760.00

EQUATION	ORDER	A12	A21	D	A122	A211
VAN LAAR	3	0.0836	0.0858	-	-	-
MARGULES	3	0.0832	0.0861	-	-	-
MARGULES	4	-0.0276	-0.0365	-0.3381	-	-
ALPHA	2	0.8336	-0.1253	-	-	-
ALPHA	3	1.1616	0.0051	-	-0.3709	-0.0902

			DEVIATION IN VAPOR PHASE COMPOSITION				
X	Y	TEMP.	LAAR 3	MARG 3	MARG 4	ALPHA2	ALPHA3
0.0970	0.1640	23.00	0.0411	0.0410	0.0252	-0.0039	-0.0060
0.2030	0.2970	120.00	0.0060	0.0060	0.0040	0.0061	0.0045
0.3000	0.4100	119.00	0.0039	0.0040	0.0088	0.0035	0.0031
0.3460	0.4605	118.00	0.0014	0.0014	0.0066	0.0003	0.0004
0.4075	0.5265	117.30	-0.0047	-0.0047	-0.0012	-0.0066	-0.0060
0.4800	0.5855	115.00	0.0030	0.0031	0.0026	-0.0008	0.0003
0.5270	0.6235	115.80	0.0050	0.0051	0.0020	0.0010	0.0022
0.6145	0.6975	114.00	0.0038	0.0038	-0.0022	-0.0022	-0.0014
0.7215	0.7785	112.00	0.0074	0.0075	0.0037	-0.0003	-0.0004
0.7570	0.8030	112.30	0.0101	0.0101	0.0085	0.0024	0.0018
0.7945	0.8325	112.00	0.0093	0.0093	0.0105	0.0016	0.0006
0.9075	0.9225	110.90	0.0057	0.0057	0.0133	0.0001	-0.0016
		MEAN	0.0085	0.0085	0.0074	0.0024	0.0023

TOLUENE(1)-OCTANE(2)

YORK R.,HOLMES R.C.:IND.ENG.CHEM.34,345(1942).

ANTOINE VAPOR PRESSURE CONSTANTS
```
        A          B         C
(1) 6.95334    1343.943   219.377
(2) 6.92377    1355.126   209.517
```

P =   50.00

EQUATION	ORDER	A12	A21	D	A122	A211
VAN LAAR	3	0.1695	0.2154	-	-	-
MARGULES	3	0.1715	0.2076	-	-	-
MARGULES	4	0.2464	0.2990	0.2808	-	-
ALPHA	2	1.6947	-0.1615	-		-
ALPHA	3	16.2803	7.6085	-	-13.3892	-7.4891

DEVIATION IN VAPOR PHASE COMPOSITION

X	Y	TEMP.	LAAR 3	MARG 3	MARG 4	ALPHA2	ALPHA3
0.0010	0.0130	50.40	-0.0103	-0.0103	-0.0098	-0.0103	-0.0091
0.0110	0.0530	49.80	-0.0240	-0.0239	-0.0191	-0.0240	-0.0135
0.0380	0.1170	48.80	-0.0222	-0.0219	-0.0112	-0.0224	-0.0004
0.0530	0.1600	48.30	-0.0317	-0.0312	-0.0196	-0.0318	-0.0085
0.0890	0.2120	47.50	-0.0108	-0.0103	-0.0002	-0.0106	0.0098
0.1380	0.2850	46.70	0.0011	0.0016	0.0061	0.0022	0.0141
0.1710	0.3400	45.80	-0.0037	-0.0033	-0.0027	-0.0022	0.0041
0.2050	0.4000	45.00	-0.0170	-0.0168	-0.0196	-0.0150	-0.0137
0.2650	0.4600	44.20	-0.0053	-0.0054	-0.0117	-0.0018	-0.0065
0.3250	0.5200	43.40	-0.0035	-0.0040	-0.0106	0.0014	-0.0061
0.3600	0.5500	42.80	-0.0009	-0.0016	-0.0072	0.0046	-0.0033
0.4400	0.6100	41.70	0.0057	0.0047	0.0032	0.0124	0.0061
0.4530	0.6250	41.20	0.0010	-0.0001	-0.0007	0.0076	0.0017
0.5350	0.6700	40.50	0.0149	0.0137	0.0177	0.0225	0.0205
0.6050	0.7400	40.20	-0.0089	-0.0100	-0.0036	-0.0008	0.0008
0.6700	0.8050	39.30	-0.0322	-0.0331	-0.0264	-0.0248	-0.0202
0.7400	0.8350	38.90	-0.0184	-0.0189	-0.0144	-0.0118	-0.0054
0.8450	0.8900	38.20	-0.0057	-0.0055	-0.0055	-0.0015	0.0034
0.8800	0.9100	37.90	-0.0018	-0.0014	-0.0055	0.0014	0.0044
0.9100	0.9300	37.60	-0.0004	0.0000	-0.0053	0.0018	0.0026
0.9350	0.9500	37.40	-0.0020	-0.0015	-0.0071	-0.0005	-0.0016
0.9550	0.9630	37.00	0.0004	0.0008	-0.0043	0.0013	-0.0012
		MEAN	0.0101	0.0100	0.0097	0.0097	0.0071

TOLUENE(1)-OCTANE(2)

YORK R.,HOLMES R.C.:IND.ENG.CHEM.34,345(1942).

ANTOINE VAPOR PRESSURE CONSTANTS

	A	B	C
(1)	6.95334	1343.943	219.377
(2)	6.92377	1355.126	209.517

P = 100.00

EQUATION	ORDER	A12	A21	D	A122	A211
VAN LAAR	3	0.1432	0.4535	-	-	-
MARGULES	3	0.0978	0.3209	-	-	-
MARGULES	4	0.0921	0.3140	-0.0219	-	-
ALPHA	2	1.7888	-0.1570	-	-	-
ALPHA	3	0.5011	-0.9773	-	1.0559	0.8382

			DEVIATION IN VAPOR PHASE COMPOSITION				
X	Y	TEMP.	LAAR 3	MARG 3	MARG 4	ALPHA2	ALPHA3
0.0010	0.0140	65.10	-0.0116	-0.0118	-0.0119	-0.0112	-0.0114
0.0280	0.0700	64.60	-0.0052	-0.0101	-0.0106	0.0034	-0.0015
0.0300	0.0770	64.20	-0.0077	-0.0128	-0.0134	0.0013	-0.0038
0.0350	0.1080	64.00	-0.0277	-0.0334	-0.0340	-0.0176	-0.0233
0.0980	0.2150	63.40	-0.0099	-0.0164	-0.0170	0.0088	0.0008
0.1300	0.2550	62.30	0.0059	0.0013	0.0009	0.0255	0.0185
0.1700	0.3300	61.20	-0.0056	-0.0074	-0.0074	0.0133	0.0084
0.2350	0.4350	59.60	-0.0202	-0.0178	-0.0174	-0.0047	-0.0055
0.3250	0.5520	58.40	-0.0336	-0.0287	-0.0281	-0.0237	-0.0201
0.4200	0.6150	57.00	-0.0071	-0.0039	-0.0037	-0.0027	0.0028
0.5050	0.6650	55.60	0.0099	0.0092	0.0090	0.0113	0.0159
0.6120	0.7450	55.20	0.0001	-0.0058	-0.0063	0.0022	0.0039
0.7000	0.8230	54.40	-0.0278	-0.0356	-0.0361	-0.0219	-0.0230
0.8000	0.8450	53.80	0.0026	-0.0025	-0.0026	0.0163	0.0133
0.8500	0.8920	53.30	-0.0177	-0.0189	-0.0187	0.0004	-0.0027
0.8910	0.9150	53.00	-0.0168	-0.0139	-0.0135	0.0041	0.0015
0.9250	0.9420	52.60	-0.0208	-0.0146	-0.0141	0.0004	-0.0014
0.9430	0.9560	52.20	-0.0205	-0.0133	-0.0128	-0.0006	-0.0020
		MEAN	0.0139	0.0143	0.0143	0.0094	0.0089

TOLUENE(1)-OCTANE(2)

YORK R.,HOLMES R.C.:IND.ENG.CHEM.34,345(1942).

ANTOINE VAPOR PRESSURE CONSTANTS
	A	B	C
(1)	6.95334	1343.943	219.377
(2)	6.92377	1355.126	209.517

P = 200.00

EQUATION	ORDER	A12	A21	D	A122	A211
VAN LAAR	3	0.1127	1.4834	-	-	-
MARGULES	3	0.0416	0.3096	-	-	-
MARGULES	4	0.1410	0.4266	0.3556	-	-
ALPHA	2	1.5533	-0.2526	-	-	-
ALPHA	3	16.5298	6.8245	-	-14.5405	-6.5978

			DEVIATION IN VAPOR PHASE COMPOSITION				
X	Y	TEMP.	LAAR 3	MARG 3	MARG 4	ALPHA2	ALPHA3
0.0120	0.0820	82.00	-0.0566	-0.0601	-0.0552	-0.0520	-0.0479
0.0600	0.1300	81.00	-0.0095	-0.0203	-0.0075	0.0075	0.0153
0.1030	0.2350	79.40	-0.0370	-0.0481	-0.0374	-0.0149	-0.0100
0.1900	0.3450	78.00	-0.0101	-0.0150	-0.0158	0.0125	0.0109
0.2650	0.4550	76.20	-0.0189	-0.0193	-0.0265	-0.0022	-0.0063
0.3650	0.5600	74.80	-0.0097	-0.0103	-0.0171	-0.0029	-0.0065
0.4750	0.6450	73.80	0.0096	0.0006	0.0010	0.0061	0.0058
0.5500	0.7000	72.80	0.0156	-0.0017	0.0036	0.0069	0.0089
0.6650	0.8020	71.00	-0.0067	-0.0352	-0.0272	-0.0181	-0.0140
0.7500	0.8420	70.00	0.0023	-0.0286	-0.0238	-0.0049	-0.0016
0.8350	0.8800	69.70	0.0041	-0.0176	-0.0199	0.0096	0.0093
0.8900	0.9200	69.60	-0.0168	-0.0210	-0.0282	0.0044	0.0004
0.9400	0.9560	69.50	-0.0442	-0.0172	-0.0265	0.0014	-0.0056
		MEAN	0.0185	0.0227	0.0223	0.0110	0.0110

TOLUENE(1)-OCTANE(2)

YORK R.,HOLMES R.C.:IND.ENG.CHEM.34,345(1942).

ANTOINE VAPOR PRESSURE CONSTANTS
          A   B   C
(1) 6.95334   1343.943   219.377
(2) 6.92377   1355.126   209.517

P =   300.00

EQUATION	ORDER	A12	A21	D	A122	A211
VAN LAAR	3	0.1113	1.7623	-	-	-
MARGULES	3	0.0613	0.3048	-	-	-
MARGULES	4	0.1641	0.4185	0.3598	-	-
ALPHA	2	1.5015	-0.2351	-	-	-
ALPHA	3	15.4628	6.9511	-	-13.2920	-6.8658

			DEVIATION IN VAPOR PHASE COMPOSITION				
X	Y	TEMP.	LAAR 3	MARG 3	MARG 4	ALPHA2	ALPHA3
0.0200	0.0760	93.10	-0.0353	-0.0390	-0.0313	-0.0278	-0.0186
0.0380	0.1210	92.30	-0.0450	-0.0508	-0.0392	-0.0322	-0.0198
0.0910	0.2000	91.40	-0.0276	-0.0345	-0.0224	-0.0052	0.0056
0.1400	0.2550	90.70	-0.0021	-0.0068	-0.0007	0.0233	0.0284
0.2020	0.3750	88.80	-0.0298	-0.0310	-0.0335	-0.0062	-0.0075
0.2800	0.4750	87.30	-0.0279	-0.0274	-0.0359	-0.0108	-0.0165
0.4000	0.5800	85.70	-0.0012	-0.0065	-0.0124	0.0031	-0.0028
0.5300	0.6700	83.10	0.0256	0.0058	0.0093	0.0172	0.0157
0.6050	0.7360	82.00	0.0164	-0.0122	-0.0051	0.0039	0.0052
0.6900	0.8170	81.30	-0.0084	-0.0441	-0.0365	-0.0214	-0.0179
0.7500	0.8400	80.80	0.0034	-0.0331	-0.0281	-0.0064	-0.0026
0.8450	0.8830	80.60	0.0058	-0.0186	-0.0215	0.0105	0.0119
0.9050	0.9300	80.40	-0.0220	-0.0224	-0.0304	0.0026	0.0007
0.9320	0.9560	80.20	-0.0450	-0.0259	-0.0350	-0.0052	-0.0086
		MEAN	0.0211	0.0256	0.0244	0.0126	0.0116

TOLUENE(1)-OCTANE(2)

YORK R.,HOLMES R.C.:IND.ENG.CHEM.34,345(1942).

ANTOINE VAPOR PRESSURE CONSTANTS

	A	B	C
(1)	6.95334	1343.943	219.377
(2)	6.92377	1355.126	209.517

P = 400.00

EQUATION	ORDER	A12	A21	D	A122	A211
VAN LAAR	3	0.1719	1.0494	-	-	-
MARGULES	3	0.1145	0.4363	-	-	-
MARGULES	4	0.1013	0.4214	-0.0444	-	-
ALPHA	2	2.0391	-0.0621	-	-	-
ALPHA	3	0.9246	0.0922	-	1.6638	-0.4335

			DEVIATION IN VAPOR PHASE COMPOSITION				
X	Y	TEMP.	LAAR 3	MARG 3	MARG 4	ALPHA2	ALPHA3
0.0180	0.0670	101.60	-0.0259	-0.0301	-0.0310	-0.0148	-0.0066
0.0520	0.1500	100.00	-0.0364	-0.0443	-0.0461	-0.0110	0.0056
0.0950	0.2330	98.50	-0.0366	-0.0440	-0.0457	-0.0020	0.0175
0.1650	0.3650	97.00	-0.0519	-0.0537	-0.0542	-0.0145	0.0011
0.2200	0.4400	95.40	-0.0479	-0.0457	-0.0453	-0.0145	-0.0045
0.2880	0.5100	94.50	-0.0334	-0.0296	-0.0287	-0.0076	-0.0047
0.3050	0.5400	94.40	-0.0443	-0.0407	-0.0398	-0.0205	-0.0192
0.4030	0.5950	93.70	-0.0015	-0.0038	-0.0032	0.0106	0.0046
0.4750	0.6330	93.50	0.0206	0.0104	0.0104	0.0257	0.0173
0.5050	0.6700	92.50	0.0066	-0.0071	-0.0074	0.0091	0.0003
0.5450	0.6950	92.00	0.0099	-0.0088	-0.0093	0.0100	0.0015
0.5800	0.7050	91.50	0.0228	0.0002	-0.0006	0.0217	0.0140
0.6660	0.8020	91.20	-0.0249	-0.0548	-0.0557	-0.0244	-0.0278
0.7550	0.8400	90.70	-0.0209	-0.0501	-0.0506	-0.0110	-0.0075
0.8200	0.8620	90.30	-0.0181	-0.0370	-0.0368	0.0058	0.0144
0.8400	0.8950	89.60	-0.0442	-0.0575	-0.0570	-0.0148	-0.0049
0.8900	0.9220	89.20	-0.0562	-0.0489	-0.0478	-0.0091	0.0027
0.9300	0.9560	88.90	-0.0776	-0.0470	-0.0456	-0.0144	-0.0035
0.9630	0.9700	88.70	-0.0720	-0.0238	-0.0226	-0.0025	0.0052
		MEAN	0.0343	0.0336	0.0336	0.0128	0.0086

TOLUENE(1)-PHENOL(2)

DRICKAMER H.G.,BROWN G.G.,WHITE R.R.:TRANS.AM.INST.CHEM.ENGRS.,
41,555(1945).

ANTOINE VAPOR PRESSURE CONSTANTS

	A	B	C
(1)	6.95334	1343.943	219.377
(2)	7.57893	1817.000	205.000

P = 760.00

EQUATION	ORDER	A12	A21	D	A122	A211
VAN LAAR	3	0.3561	0.2954	-	-	-
MARGULES	3	0.3583	0.2832	-	-	-
MARGULES	4	0.2684	0.1305	-0.4342	-	-
ALPHA	2	11.8396	-0.4648	-	-	-
ALPHA	3	-2.1323	-1.6652	-	13.8654	0.8505

			DEVIATION IN VAPOR PHASE COMPOSITION				
X	Y	TEMP.	LAAR 3	MARG 3	MARG 4	ALPHA2	ALPHA3
0.0435	0.3410	172.70	0.0107	0.0116	-0.0163	0.0229	0.0221
0.0872	0.5120	159.40	0.0316	0.0323	0.0170	0.0284	0.0285
0.1186	0.6210	153.80	0.0051	0.0055	-0.0007	-0.0014	-0.0006
0.1248	0.6250	149.40	0.0225	0.0229	0.0181	0.0073	0.0083
0.2190	0.7850	142.20	-0.0175	-0.0178	-0.0119	-0.0232	-0.0213
0.2750	0.8070	133.80	0.0100	0.0094	0.0153	-0.0005	0.0013
0.4080	0.8725	128.30	0.0023	0.0015	0.0020	-0.0005	0.0002
0.4800	0.8901	126.70	0.0052	0.0044	0.0018	0.0047	0.0043
0.5898	0.9159	122.20	0.0073	0.0067	0.0016	0.0047	0.0026
0.6348	0.9280	120.20	0.0053	0.0049	-0.0002	0.0012	-0.0015
0.6512	0.9260	120.00	0.0105	0.0101	0.0052	0.0061	0.0032
0.7400	0.9463	119.70	0.0062	0.0061	0.0032	0.0002	-0.0023
0.7730	0.9536	119.40	0.0049	0.0049	0.0032	-0.0021	-0.0037
0.8012	0.9545	115.60	0.0107	0.0107	0.0101	0.0011	0.0008
0.8840	0.9750	112.70	0.0051	0.0053	0.0069	-0.0065	-0.0012
0.9108	0.9796	112.20	0.0051	0.0053	0.0072	-0.0063	0.0007
0.9394	0.9861	113.30	0.0033	0.0034	0.0053	-0.0068	0.0010
0.9770	0.9948	111.10	0.0012	0.0013	0.0023	-0.0048	0.0007
0.9910	0.9980	111.10	0.0004	0.0005	0.0009	-0.0024	0.0003
0.9939	0.9986	110.50	0.0004	0.0004	0.0007	-0.0017	0.0003
0.9973	0.9993	110.50	0.0002	0.0002	0.0004	-0.0007	0.0002
		MEAN	0.0079	0.0079	0.0062	0.0064	0.0050

TOLUENE(1)-ISOPROPYL ALCOHOL(2)

KIREEV V.A.,SHEINKER YU.N.,PERESLENI E.M.:ZH.FIZ.KHIM.
26,352(1952).

ANTOINE VAPOR PRESSURE CONSTANTS
	A	B	C
(1)	6.95334	1343.943	219.377
(2)	7.75634	1366.142	197.970

P =   760.00

EQUATION	ORDER	A12	A21	D	A122	A211
VAN LAAR	3	0.4939	0.5680	-	-	-
MARGULES	3	0.4892	0.5661	-	-	-
MARGULES	4	0.4472	0.5303	-0.1261	-	-
ALPHA	2	1.5774	6.7202	-	-1.4378	6.0677
ALPHA	3	1.7760	2.0105	-	-1.4378	6.0677

			DEVIATION IN VAPOR PHASE COMPOSITION				
X	Y	TEMP.	LAAR 3	MARG 3	MARG 4	ALPHA2	ALPHA3
0.0670	0.0820	81.60	-0.0068	-0.0072	-0.0107	0.0270	-0.0021
0.1420	0.1460	81.20	-0.0074	-0.0077	-0.0095	0.0202	0.0009
0.1850	0.1760	81.20	-0.0082	-0.0083	-0.0086	0.0118	0.0006
0.2200	0.1930	81.40	-0.0046	-0.0046	-0.0037	0.0094	0.0041
0.2580	0.2190	81.50	-0.0106	-0.0104	-0.0085	-0.0027	-0.0027
0.2960	0.2350	81.50	-0.0086	-0.0083	-0.0057	-0.0061	-0.0019
0.3240	0.2450	81.80	-0.0067	-0.0063	-0.0035	-0.0074	-0.0007
0.4260	0.2790	82.20	-0.0025	-0.0020	0.0002	-0.0110	0.0005
0.5310	0.3100	83.20	0.0006	0.0009	0.0009	-0.0087	0.0025
0.6030	0.3370	84.00	-0.0025	-0.0023	-0.0039	-0.0086	-0.0000
0.6880	0.3780	85.40	-0.0103	-0.0103	-0.0129	-0.0089	-0.0056
0.7440	0.4020	86.60	-0.0059	-0.0060	-0.0082	0.0028	0.0011
0.7970	0.4380	88.50	-0.0065	-0.0066	-0.0073	0.0112	0.0037
0.8510	0.4980	91.00	-0.0143	-0.0143	-0.0119	0.0142	-0.0003
0.8970	0.5660	94.40	-0.0144	-0.0141	-0.0079	0.0242	0.0035
0.9220	0.6230	96.60	-0.0174	-0.0170	-0.0084	0.0255	0.0020
0.9700	0.8210	104.60	-0.0431	-0.0425	-0.0320	-0.0027	-0.0239
		MEAN	0.0100	0.0099	0.0084	0.0119	0.0033

METHYLCYCLOHEXANE(1)-PHENOL(2)

DRICKAMER H.G.,BROWN G.G.,WHITE R.R.:TRANS.AM.INST.CHEM.ENGRS.
           41,555(1945).

ANTOINE VAPOR PRESSURE CONSTANTS
         A   B   C
(1) 6.82689   1272.864   221.630
(2) 7.57893   1817.000   205.000

P =   760.00

EQUATION	ORDER	A12	A21	D	A122	A211
VAN LAAR	3	0.5083	0.7481	-	-	-
MARGULES	3	0.5242	0.6949	-	-	-
MARGULES	4	0.7918	0.9245	0.9393	-	-
ALPHA	2	22.8658	0.7070	-	-	-
ALPHA	3	-4.0598	-1.9051	-	21.4909	1.5171

			DEVIATION IN VAPOR PHASE COMPOSITION				
X	Y	TEMP.	LAAR 3	MARG 3	MARG 4	ALPHA2	ALPHA3
0.1140	0.6775	150.00	0.0719	0.0770	0.0989	0.0393	0.0184
0.2620	0.8710	130.00	0.0181	0.0186	0.0048	-0.0284	-0.0203
0.3800	0.8840	120.00	0.0407	0.0395	0.0302	-0.0040	0.0090
0.6520	0.9088	112.20	0.0434	0.0415	0.0464	0.0111	0.0018
0.9026	0.9384	105.60	0.0386	0.0394	0.0363	0.0097	-0.0029
0.9518	0.9683	102.50	0.0187	0.0196	0.0154	-0.0070	-0.0005
0.9611	0.9741	102.20	0.0151	0.0159	0.0119	-0.0088	0.0002
0.9874	0.9914	101.70	0.0047	0.0051	0.0031	-0.0079	0.0007
0.9952	0.9963	101.10	0.0022	0.0024	0.0015	-0.0036	0.0007
		MEAN	0.0281	0.0288	0.0276	0.0133	0.0061

METHYLCYCLOHEXANE(1)-TOLUENE(2)

QUIGGLE D.,FENSKE M.R.:J.AM.CHEM.SOC.59,1829(1937).

ANTOINE VAPOR PRESSURE CONSTANTS
	A	B	C
(1)	6.82689	1272.864	221.630
(2)	6.95334	1343.943	219.377

P = 760.00

EQUATION	ORDER	A12	A21	D	A122	A211
VAN LAAR	3	0.1289	0.0776	-	-	-
MARGULES	3	0.1194	0.0705	-	-	-
MARGULES	4	0.1227	0.0740	0.0105	-	-
ALPHA	2	0.5663	-0.0122	-	-	-
ALPHA	3	1.1386	0.5247	-	-0.5031	-0.5535

			DEVIATION IN VAPOR PHASE COMPOSITION				
X	Y	TEMP.	LAAR 3	MARG 3	MARG 4	ALPHA2	ALPHA3
0.0500	0.0750	109.55	0.0061	0.0051	0.0055	-0.0001	0.0021
0.1000	0.1430	108.55	0.0087	0.0079	0.0083	0.0008	0.0035
0.1500	0.2100	107.65	0.0049	0.0045	0.0047	-0.0025	-0.0002
0.2000	0.2700	106.90	0.0026	0.0027	0.0027	-0.0031	-0.0016
0.2500	0.3260	106.20	0.0003	0.0007	0.0006	-0.0033	-0.0028
0.3000	0.3780	105.60	-0.0010	-0.0005	-0.0006	-0.0027	-0.0030
0.3500	0.4240	105.00	0.0015	0.0020	0.0018	0.0012	0.0003
0.4000	0.4700	104.50	0.0022	0.0026	0.0024	0.0030	0.0017
0.4500	0.5150	104.00	0.0028	0.0029	0.0028	0.0040	0.0026
0.5000	0.5600	103.55	0.0024	0.0023	0.0023	0.0035	0.0022
0.5500	0.6040	103.15	0.0024	0.0021	0.0021	0.0029	0.0019
0.6000	0.6500	102.75	-0.0000	-0.0004	-0.0003	-0.0005	-0.0010
0.6500	0.6940	102.45	-0.0007	-0.0011	-0.0009	-0.0024	-0.0024
0.7000	0.7370	102.15	-0.0004	-0.0008	-0.0006	-0.0034	-0.0030
0.7500	0.7780	101.90	0.0018	0.0017	0.0018	-0.0024	-0.0015
0.8000	0.8180	101.65	0.0053	0.0054	0.0054	0.0001	0.0013
0.8500	0.8600	101.40	0.0069	0.0073	0.0072	0.0014	0.0027
0.9000	0.9060	101.20	0.0049	0.0054	0.0052	-0.0002	0.0009
0.9500	0.9540	101.00	0.0012	0.0017	0.0015	-0.0022	-0.0014
		MEAN	0.0030	0.0030	0.0030	0.0021	0.0019

METHYLCYCLOHEXANE(1)-TOLUENE(2)

SCHNEIDER G.:Z.PHYS.CHEM.(FRANKFURT)27(3-4),171(1961).

ANTOINE VAPOR PRESSURE CONSTANTS

	A	B	C
(1)	6.82689	1272.864	221.630
(2)	6.95334	1343.943	219.377

T = 100.02

EQUATION	ORDER	A12	A21	D	A122	A211
VAN LAAR	3	0.1276	0.0694	-	-	-
MARGULES	3	0.1224	0.0852	-	-	-
MARGULES	4	0.1263	0.0693	0.0115	-	-
ALPHA	2	0.6331	0.0005	-	-	-
ALPHA	3	1.0855	0.5232	-	-0.3374	-0.5709

			DEVIATION IN VAPOR PHASE COMPOSITION				
X	Y	PRESS.	LAAR 3	MARG 3	MARG 4	ALPHA2	ALPHA3
0.1002	0.1523	593.84	0.0029	0.0023	0.0028	-0.0035	0.0011
0.2000	0.2775	624.22	0.0003	0.0003	0.0003	-0.0039	-0.0012
0.2000	0.2772	624.59	0.0006	0.0006	0.0006	-0.0036	-0.0009
0.2990	0.3800	648.85	0.0017	0.0019	0.0017	0.0011	0.0010
0.3005	0.3815	649.59	0.0016	0.0019	0.0017	0.0011	0.0010
0.4003	0.4775	671.54	-0.0001	0.0001	-0.0001	0.0019	-0.0002
0.4003	0.4785	670.58	-0.0011	-0.0009	-0.0011	0.0009	-0.0012
0.4995	0.5655	689.38	-0.0003	-0.0003	-0.0003	0.0023	-0.0001
0.5995	0.6505	705.29	0.0005	0.0003	0.0005	0.0018	0.0004
0.6000	0.6500	705.59	0.0014	0.0012	0.0014	0.0027	0.0013
0.6995	0.7360	717.89	-0.0001	-0.0003	-0.0001	-0.0013	-0.0008
0.7002	0.7365	719.13	-0.0000	-0.0002	-0.0000	-0.0012	-0.0008
0.7995	0.8205	727.49	0.0010	0.0011	0.0010	-0.0026	-0.0004
0.7995	0.8195	727.55	0.0020	0.0021	0.0020	-0.0016	0.0006
0.9005	0.9080	735.61	0.0019	0.0022	0.0019	-0.0022	0.0004
		MEAN	0.0010	0.0010	0.0010	0.0021	0.0007

METHYLCYCLOHEXANE(1)-TOLUENE(2)

WEBER J.H.:IND.ENG.CHEM.47,454(1955).

ANTOINE VAPOR PRESSURE CONSTANTS

	A	B	C
(1)	6.82689	1272.864	221.630
(2)	6.95334	1343.943	219.377

P = 200.00

EQUATION	ORDER	A12	A21	D	A122	A211
VAN LAAR	3	0.1294	0.1229	-	-	-
MARGULES	3	0.1288	0.1234	-	-	-
MARGULES	4	0.1873	0.1837	0.1910	-	-
ALPHA	2	0.8848	-0.0637	-	-	-
ALPHA	3	6.9319	4.5231	-	-5.5146	-4.5651

DEVIATION IN VAPOR PHASE COMPOSITION

X	Y	TEMP.	LAAR 3	MARG 3	MARG 4	ALPHA2	ALPHA3
0.0020	0.0070	69.30	-0.0032	-0.0032	-0.0026	-0.0032	-0.0022
0.0170	0.0390	68.70	-0.0071	-0.0072	-0.0034	-0.0076	-0.0006
0.0740	0.1430	67.30	-0.0146	-0.0147	-0.0074	-0.0155	-0.0020
0.1085	0.1870	66.60	-0.0067	-0.0068	-0.0009	-0.0072	0.0045
0.1530	0.2570	65.70	-0.0155	-0.0156	-0.0127	-0.0151	-0.0076
0.2100	0.3125	64.80	-0.0005	-0.0005	-0.0016	0.0015	0.0032
0.2820	0.3875	63.80	0.0040	0.0040	-0.0001	0.0079	0.0041
0.3295	0.4385	63.40	0.0007	0.0008	-0.0038	0.0059	-0.0001
0.4805	0.5775	61.60	-0.0021	-0.0020	-0.0025	0.0046	-0.0015
0.5100	0.6045	61.40	-0.0045	-0.0044	-0.0037	0.0021	-0.0031
0.5950	0.6730	61.10	-0.0044	-0.0043	-0.0009	0.0017	-0.0003
0.6865	0.7440	60.70	-0.0032	-0.0031	0.0007	0.0012	0.0028
0.6925	0.7525	60.70	-0.0070	-0.0069	-0.0032	-0.0027	-0.0009
0.7730	0.8170	60.30	-0.0079	-0.0079	-0.0064	-0.0056	-0.0019
0.8300	0.8530	60.15	0.0019	0.0019	0.0008	0.0027	0.0067
0.8975	0.9175	60.10	-0.0069	-0.0070	-0.0107	-0.0074	-0.0047
0.8845	0.9025	60.00	-0.0028	-0.0028	-0.0062	-0.0031	-0.0000
0.9875	0.9950	59.70	-0.0063	-0.0063	-0.0078	-0.0066	-0.0067
		MEAN	0.0055	0.0055	0.0042	0.0056	0.0029

374

METHYLCYCLOHEXANE(1)-TOLUENE(2)

WEBER J.H.:IND.ENG.CHEM.47,454(1955).

ANTOINE VAPOR PRESSURE CONSTANTS
	A	B	C
(1)	6.82689	1272.864	221.630
(2)	6.95334	1343.943	219.377

P = 400.00

EQUATION	ORDER	A12	A21	D	A122	A211
VAN LAAR	3	0.1067	0.1278	-	-	-
MARGULES	3	0.1055	0.1266	-	-	-
MARGULES	4	0.0938	0.1137	-0.0375	-	-
ALPHA	2	0.7508	-0.0500	-	-	-
ALPHA	3	-1.0237	-1.1441	-	1.7900	0.9700

			DEVIATION IN VAPOR PHASE COMPOSITION				
X	Y	TEMP.	LAAR 3	MARG 3	MARG 4	ALPHA2	ALPHA3
0.0050	0.0013	89.40	0.0073	0.0073	0.0071	0.0074	0.0075
0.0540	0.0870	88.60	0.0016	0.0015	0.0002	0.0022	0.0031
0.1145	0.1820	87.60	-0.0050	-0.0051	-0.0062	-0.0040	-0.0020
0.1930	0.2805	86.45	-0.0024	-0.0024	-0.0026	-0.0011	0.0016
0.2570	0.3580	85.70	-0.0066	-0.0065	-0.0060	-0.0049	-0.0026
0.3570	0.4550	84.40	-0.0010	-0.0009	-0.0002	0.0010	0.0015
0.4130	0.5080	83.70	-0.0015	-0.0015	-0.0010	0.0006	-0.0003
0.4820	0.5680	83.00	-0.0006	-0.0005	-0.0006	0.0018	-0.0006
0.5485	0.6220	82.40	0.0013	0.0013	0.0008	0.0038	0.0007
0.6370	0.6940	81.60	0.0010	0.0010	0.0002	0.0036	0.0012
0.6960	0.7445	81.25	-0.0025	-0.0026	-0.0033	-0.0000	-0.0007
0.7980	0.8320	80.50	-0.0083	-0.0083	-0.0083	-0.0063	-0.0028
0.8830	0.8990	80.20	-0.0048	-0.0047	-0.0039	-0.0033	0.0025
0.9920	0.9920	79.70	0.0003	0.0003	0.0005	0.0004	0.0014
		MEAN	0.0032	0.0031	0.0029	0.0029	0.0020

HEPTANE(1)-1-BROMOBUTANE(2)

SMITH C.P.,ENGEL E.W.:J.AM.CHEM.SOC.51,2646(1929).

ANTOINE VAPOR PRESSURE CONSTANTS
	A	B	C
(1)	6.90240	1268.115	216.900
(2)	6.92254	1298.608	219.700

T = 50.00

EQUATION	ORDER	A12	A21	D	A122	A211
VAN LAAR	3	0.1792	0.1570	-	-	-
MARGULES	3	0.1783	0.1559	-	-	-
MARGULES	4	0.1773	0.1547	-0.0034	-	-
ALPHA	2	0.6009	0.3607	-	-	-
ALPHA	3	0.3143	0.3332	-	0.3745	-0.0810

			DEVIATION IN VAPOR PHASE COMPOSITION				
X	Y	PRESS.	LAAR 3	MARG 3	MARG 4	ALPHA2	ALPHA3
0.0479	0.0733	130.80	0.0011	0.0010	0.0009	-0.0012	0.0018
0.1195	0.1719	135.50	-0.0031	-0.0031	-0.0032	-0.0060	-0.0019
0.2065	0.2661	139.80	-0.0010	-0.0010	-0.0010	-0.0026	0.0001
0.2877	0.3415	143.50	0.0017	0.0018	0.0018	0.0017	0.0025
0.3412	0.3957	144.80	-0.0050	-0.0050	-0.0049	-0.0041	-0.0046
0.3667	0.4096	145.50	0.0029	0.0029	0.0030	0.0042	0.0032
0.4164	0.4536	146.40	0.0004	0.0004	0.0004	0.0023	0.0004
0.4818	0.5075	147.40	-0.0003	-0.0003	-0.0003	0.0020	-0.0004
0.5677	0.5782	148.90	-0.0016	-0.0017	-0.0017	0.0004	-0.0017
0.6671	0.6582	149.50	0.0008	0.0007	0.0006	0.0015	0.0012
0.7638	0.7453	148.40	-0.0009	-0.0009	-0.0009	-0.0020	0.0005
0.8829	0.8652	145.40	-0.0029	-0.0028	-0.0027	-0.0056	-0.0007
		MEAN	0.0018	0.0018	0.0018	0.0028	0.0016

HEPTANE(1)-BUTYL ALCOHOL(2)

SMITH C.P.,ENGEL E.W.:J.AM.CHEM.SOC.51,2660(1929).

ANTOINE VAPOR PRESSURE CONSTANTS
```
         A          B          C
(1) 6.90240    1268.115    216.900
(2) 7.54472    1405.873    183.908
```

T =   50.00

EQUATION	ORDER	A12	A21	D	A122	A211
VAN LAAR	3	0.5226	0.9114	-	-	-
MARGULES	3	0.4736	0.8277	-	-	-
MARGULES	4	0.6138	0.9751	0.4319	-	-
ALPHA	2	17.2586	2.1195	-	-	-
ALPHA	3	15.7846	-1.7656	-	-10.2520	4.5062

DEVIATION IN VAPOR PHASE COMPOSITION

X	Y	PRESS.	LAAR 3	MARG 3	MARG 4	ALPHA2	ALPHA3
0.2166	0.6828	113.50	0.0466	0.0465	0.0458	0.0506	0.0195
0.2612	0.7627	123.50	-0.0027	-0.0013	-0.0057	-0.0049	-0.0157
0.3195	0.8017	131.60	-0.0131	-0.0110	-0.0167	-0.0206	-0.0160
0.3600	0.8021	135.90	0.0013	0.0033	-0.0019	-0.0086	0.0016
0.3631	0.8121	133.60	-0.0077	-0.0058	-0.0108	-0.0177	-0.0072
0.4571	0.8166	140.70	0.0116	0.0120	0.0106	-0.0006	0.0127
0.4690	0.8168	141.90	0.0137	0.0139	0.0130	0.0015	0.0145
0.6359	0.8520	148.60	0.0003	-0.0027	0.0022	-0.0078	-0.0050
0.7788	0.8599	151.30	0.0068	0.0051	0.0059	0.0049	-0.0036
0.8020	0.8598	151.20	0.0104	0.0095	0.0084	0.0091	-0.0011
0.8953	0.8746	151.50	0.0213	0.0260	0.0149	0.0177	0.0029
0.9312	0.9076	149.40	0.0088	0.0153	0.0014	0.0011	-0.0134
		MEAN	0.0120	0.0127	0.0114	0.0121	0.0095

377

Y

HEPTANE(1)-METHYLCYCLOHEXANE(2)

BROMILEY E.C.,QUIGGLE D.:IND.ENG.CHEM.25,1136(1933).

ANTOINE VAPOR PRESSURE CONSTANTS

	A	B	C
(1)	6.90240	1268.115	216.900
(2)	6.82689	1272.864	221.630

P = 760.00

EQUATION	ORDER	A12	A21	D	A122	A211
VAN LAAR	3	0.0001	-0.0000	-	-	-
MARGULES	3	-0.0076	0.0026	-	-	-
MARGULES	4	-0.0055	0.0051	0.0070	-	-
ALPHA	2	0.0752	-0.0883	-	-	-
ALPHA	3	-.4552/ 05	-.4231/ 05	-	.4547/ 05	.4251/ 0

DEVIATION IN VAPOR PHASE COMPOSITION

X	Y	TEMP.	LAAR 3	MARG 3	MARG 4	ALPHA2	ALPHA3
0.0310	0.0350	100.70	-0.0017	-0.0022	-0.0021	-0.0017	-0.0005
0.0580	0.0620	100.60	0.0001	-0.0006	-0.0005	0.0002	0.0013
0.0950	0.1030	100.50	-0.0015	-0.0025	-0.0023	-0.0014	-0.0005
0.1330	0.1430	100.40	-0.0013	-0.0023	-0.0022	-0.0011	-0.0003
0.1800	0.1920	100.30	-0.0008	-0.0018	-0.0018	-0.0005	0.0000
0.2160	0.2290	100.20	0.0006	-0.0010	-0.0010	0.0002	0.0006
0.2715	0.2890	100.00	0.0001	-0.0030	-0.0031	-0.0019	-0.0018
0.3170	0.3330	100.00	0.0002	0.0002	0.0001	0.0012	0.0011
0.3630	0.3810	99.90	-0.0010	-0.0004	-0.0005	0.0005	0.0002
0.4010	0.4200	99.80	-0.0015	-0.0005	-0.0006	0.0004	-0.0001
0.4560	0.4750	99.60	-0.0010	0.0002	0.0002	0.0012	0.0005
0.5010	0.5210	99.30	-0.0020	-0.0006	-0.0005	0.0005	-0.0002
0.5990	0.6180	99.00	-0.0019	-0.0004	-0.0003	0.0010	0.0003
0.6470	0.6660	98.90	-0.0028	-0.0015	-0.0013	0.0002	-0.0004
0.7090	0.7280	98.80	-0.0045	-0.0034	-0.0033	-0.0015	-0.0018
0.7960	0.8100	98.60	-0.0027	-0.0022	-0.0022	-0.0001	0.0005
0.8430	0.8535	98.55	-0.0013	-0.0011	-0.0012	0.0009	0.0021
0.9310	0.9400	98.50	-0.0046	-0.0047	-0.0049	-0.0034	-0.0006
0.9800	0.9860	98.42	-0.0047	-0.0047	-0.0048	-0.0043	-0.0004
		MEAN	0.0018	0.0018	0.0017	0.0012	0.0007

HEPTANE(1)-TOLUENE(2)

BROMILEY E.C.,QUIGGLE D.:ING.ENG.CHEM.25,1136(1933).

ANTOINE VAPOR PRESSURE CONSTANTS

	A	B	C
(1)	6.90240	1268.115	216.900
(2)	6.95334	1343.943	219.377

P = 760.00

EQUATION	ORDER	A12	A21	D	A122	A211
VAN LAAR	3	0.1743	0.0748	-	-	-
MARGULES	3	0.1443	0.0492	-	-	-
MARGULES	4	0.1278	0.0292	-0.0541	-	-
ALPHA	2	0.7529	0.0100	-	-	-
ALPHA	3	5.0927	3.3475	-	-4.0646	-3.2569

			DEVIATION IN VAPOR PHASE COMPOSITION				
X	Y	TEMP.	LAAR 3	MARG 3	MARG 4	ALPHA2	ALPHA3
0.0795	0.1335	108.80	0.0049	0.0020	-0.0001	-0.0060	0.0011
0.1985	0.2870	106.40	0.0005	0.0007	0.0005	-0.0032	-0.0021
0.3510	0.4390	103.80	0.0022	0.0025	0.0032	0.0062	0.0026
0.5005	0.5785	101.80	-0.0018	-0.0034	-0.0039	-0.0001	-0.0017
0.6780	0.7245	100.20	0.0056	0.0044	0.0033	-0.0024	0.0004
		MEAN	0.0030	0.0026	0.0022	0.0036	0.0016

HEPTANE(1)-TOLUENE(2)

HIPKIN H.,MYERS H.S.:IND.ENG.CHEM.46,2524(1954).

ANTOINE VAPOR PRESSURE CONSTANTS
	A	B	C
(1)	6.90240	1268.115	216.900
(2)	6.95334	1343.943	219.377

P = 760.00

EQUATION	ORDER	A12	A21	D	A122	A211
VAN LAAR	3	0.1511	0.0992	-	-	-
MARGULES	3	0.1451	0.0913	-	-	-
MARGULES	4	0.1203	0.0611	-0.0862	-	-
ALPHA	2	0.7964	0.0116	-	-	-
ALPHA	3	-0.8584	-1.0498	-	1.6389	0.9496

			DEVIATION IN VAPOR PHASE COMPOSITION				
X	Y	TEMP.	LAAR 3	MARG 3	MARG 4	ALPHA2	ALPHA3
0.0300	0.0530	109.70	0.0036	0.0031	0.0008	-0.0010	-0.0013
0.0740	0.1240	108.50	0.0055	0.0049	0.0019	-0.0022	-0.0022
0.1220	0.1910	107.30	0.0076	0.0073	0.0052	-0.0002	0.0004
0.1230	0.1940	107.30	0.0060	0.0057	0.0036	-0.0018	-0.0012
0.1840	0.2710	106.00	0.0056	0.0057	0.0054	-0.0003	0.0011
0.1930	0.2820	105.80	0.0051	0.0052	0.0052	-0.0004	0.0010
0.2280	0.3230	105.10	0.0034	0.0036	0.0043	-0.0007	0.0009
0.2400	0.3370	105.00	0.0023	0.0025	0.0035	-0.0012	0.0004
0.2940	0.3950	104.00	-0.0002	0.0000	0.0016	-0.0017	-0.0004
0.3290	0.4300	103.60	-0.0013	-0.0011	0.0004	-0.0016	-0.0007
0.3450	0.4450	103.30	-0.0011	-0.0010	0.0005	-0.0011	-0.0005
0.3990	0.4920	102.70	0.0013	0.0012	0.0021	0.0022	0.0018
0.4110	0.5050	102.50	-0.0010	-0.0012	-0.0004	0.0000	-0.0007
0.4700	0.5550	102.00	0.0005	0.0000	-0.0002	0.0014	-0.0005
0.5270	0.6020	101.20	0.0019	0.0013	0.0001	0.0019	-0.0010
0.5880	0.6500	100.80	0.0048	0.0040	0.0020	0.0031	-0.0002
0.6550	0.7030	100.20	0.0071	0.0064	0.0042	0.0030	0.0004
0.7420	0.7740	99.60	0.0077	0.0074	0.0062	0.0006	0.0004
0.8130	0.8330	99.20	0.0075	0.0077	0.0080	-0.0011	0.0013
0.8680	0.8820	99.00	0.0046	0.0051	0.0066	-0.0040	0.0001
0.9060	0.9170	98.90	0.0018	0.0024	0.0044	-0.0059	-0.0013
0.9520	0.9580	98.80	0.0002	0.0007	0.0025	-0.0048	-0.0012
		MEAN	0.0036	0.0035	0.0031	0.0018	0.0009

380

HEPTANE(1)-TOLUENE(2)

ROSE A.,WILLIAMS E.T.:IND.ENG.CHEM.47,1528(1955).

ANTOINE VAPOR PRESSURE CONSTANTS
```
            A          B          C
(1)  6.90240    1268.115    216.900
(2)  6.95334    1343.943    219.377
```

P =   760.00

EQUATION	ORDER	A12	A21	D	A122	A211
VAN LAAR	3	0.1491	0.1055	-	-	-
MARGULES	3	0.1440	0.1007	-	-	-
MARGULES	4	0.1379	0.0943	-0.0188	-	-
ALPHA	2	0.7977	-0.0141	-	-	-
ALPHA	3	0.4559	-0.0550	-	0.4416	-0.0446

			DEVIATION IN VAPOR PHASE COMPOSITION				
X	Y	TEMP.	LAAR 3	MARG 3	MARG 4	ALPHA2	ALPHA3
0.1000	0.1660	107.73	0.0024	0.0019	0.0012	-0.0055	-0.0010
0.2000	0.2940	105.62	0.0024	0.0024	0.0024	-0.0029	0.0004
0.3000	0.4005	103.88	0.0019	0.0021	0.0025	0.0010	0.0014
0.4000	0.4970	102.59	-0.0012	-0.0011	-0.0008	0.0008	-0.0011
0.5000	0.5825	101.52	-0.0003	-0.0005	-0.0005	0.0024	-0.0005
0.6000	0.6640	100.60	0.0010	0.0007	0.0004	0.0022	0.0000
0.7000	0.7440	99.82	0.0027	0.0025	0.0021	0.0009	0.0010
0.8000	0.8275	99.26	0.0014	0.0014	0.0015	-0.0032	-0.0004
0.9000	0.9120	98.43	0.0010	0.0013	0.0017	-0.0042	-0.0002
		MEAN	0.0016	0.0016	0.0015	0.0026	0.0007

HEPTANE(1)-TOLUENE(2)

STEINHAUSER H.H.,WHITE R.H.:IND.ENG.CHEM.41,2912(1949).

ANTOINE VAPOR PRESSURE CONSTANTS

	A	B	C
(1)	6.90240	1268.115	216.900
(2)	6.95334	1343.943	219.377

P = 760.00

EQUATION	ORDER	A12	A21	D	A122	A211
VAN LAAR	3	0.1473	0.1018	-	-	-
MARGULES	3	0.1420	0.0959	-	-	-
MARGULES	4	0.1328	0.0855	-0.0297	-	-
ALPHA	2	0.8078	-0.0004	-	-	-
ALPHA	3	0.1367	-0.2450	-	0.7629	0.1147

			DEVIATION IN VAPOR PHASE COMPOSITION				
X	Y	TEMP.	LAAR 3	MARG 3	MARG 4	ALPHA2	ALPHA3
0.0250	0.0480	110.75	-0.0007	-0.0012	-0.0019	-0.0042	-0.0022
0.0620	0.1070	108.60	0.0032	0.0026	0.0014	-0.0029	0.0007
0.1290	0.2050	106.80	0.0029	0.0025	0.0017	-0.0035	0.0008
0.1850	0.2750	105.65	0.0031	0.0030	0.0028	-0.0015	0.0020
0.2350	0.3330	104.80	0.0015	0.0016	0.0018	-0.0010	0.0013
0.2500	0.3490	104.50	0.0015	0.0017	0.0020	-0.0003	0.0015
0.2860	0.3960	103.83	-0.0084	-0.0082	-0.0077	-0.0089	-0.0080
0.3540	0.4540	102.95	-0.0008	-0.0006	-0.0001	0.0008	-0.0002
0.4120	0.5040	102.25	0.0019	0.0019	0.0022	0.0043	0.0023
0.4480	0.5410	101.78	-0.0035	-0.0036	-0.0035	-0.0010	-0.0035
0.4550	0.5400	101.72	0.0036	0.0034	0.0035	0.0060	0.0035
0.4970	0.5770	101.35	0.0024	0.0022	0.0020	0.0046	0.0020
0.5680	0.6370	100.70	0.0018	0.0014	0.0009	0.0025	0.0006
0.5800	0.6470	100.60	0.0017	0.0013	0.0008	0.0021	0.0004
0.6920	0.7420	99.73	-0.0013	-0.0016	-0.0022	-0.0047	-0.0032
0.8430	0.8640	98.90	0.0014	0.0017	0.0020	-0.0058	0.0006
0.9400	0.9480	98.50	-0.0003	0.0001	0.0008	-0.0054	-0.0001
0.9750	0.9760	98.40	0.0021	0.0023	0.0027	-0.0005	0.0023
0.9940	0.9930	98.35	0.0017	0.0018	0.0019	0.0010	0.0018
		MEAN	0.0023	0.0023	0.0022	0.0032	0.0020

HEPTANE(1)-TOLUENE(2)

YERAZUNIS S.,PLOWRIGHT J.D.,SMOLA F.M.:A.I.CH.E.J.10(5),660(1964).

ANTOINE VAPOR PRESSURE CONSTANTS

	A	B	C
(1)	6.90240	1268.115	216.900
(2)	6.95334	1343.943	219.377

P = 760.00

EQUATION	ORDER	A12	A21	D	A122	A211
VAN LAAR	3	0.1512	0.0975	-	-	-
MARGULES	3	0.1443	0.0913	-	-	-
MARGULES	4	0.1460	0.0933	0.0065	-	-
ALPHA	2	0.7675	-0.0223	-	-	-
ALPHA	3	1.0270	0.3958	-	-0.1154	-0.4878

DEVIATION IN VAPOR PHASE COMPOSITION

X	Y	TEMP.	LAAR 3	MARG 3	MARG 4	ALPHA2	ALPHA3
0.0256	0.0470	109.80	0.0017	0.0012	0.0013	-0.0031	-0.0001
0.0445	0.0794	109.24	0.0024	0.0017	0.0019	-0.0046	-0.0003
0.0676	0.1162	108.60	0.0032	0.0026	0.0027	-0.0054	-0.0000
0.1087	0.1762	107.57	0.0040	0.0036	0.0037	-0.0055	0.0004
0.1622	0.2464	106.39	0.0037	0.0038	0.0038	-0.0044	0.0004
0.2681	0.3672	104.52	0.0010	0.0017	0.0015	-0.0019	-0.0009
0.3492	0.4459	103.33	0.0015	0.0020	0.0018	0.0019	0.0003
0.4474	0.5345	102.10	0.0012	0.0012	0.0012	0.0035	0.0002
0.4762	0.5594	101.85	0.0012	0.0010	0.0010	0.0036	0.0002
0.5096	0.5880	101.47	0.0011	0.0008	0.0008	0.0033	-0.0001
0.6525	0.7057	100.25	0.0024	0.0017	0.0019	0.0014	0.0002
0.7479	0.7845	99.58	0.0023	0.0019	0.0020	-0.0018	-0.0005
0.8380	0.8585	99.07	0.0032	0.0033	0.0033	-0.0029	0.0003
0.8771	0.8924	98.87	0.0022	0.0025	0.0024	-0.0039	-0.0004
0.9154	0.9245	98.70	0.0025	0.0030	0.0028	-0.0028	0.0004
0.9503	0.9556	98.58	0.0013	0.0017	0.0016	-0.0026	-0.0001
0.9700	0.9728	98.51	0.0011	0.0014	0.0013	-0.0015	0.0002
		MEAN	0.0021	0.0021	0.0021	0.0032	0.0003

383

2,2,3-TRIMETHYLBUTANE(1)-BENZENE(2)

HARRISON J.M.,BERG L.:IND.ENG.CHEM.,38,117(1946).

ANTOINE VAPOR PRESSURE CONSTANTS
```
        A          B          C
(1)  6.79230   1200.563   226.050
(2)  6.90565   1211.033   220.790
```

P =   736.00

EQUATION	ORDER	A12	A21	D	A122	A211
VAN LAAR	3	0.2789	0.1790	-	-	-
MARGULES	3	0.2645	0.1658	-	-	-
MARGULES	4	0.2673	0.1690	0.0094	-	-
ALPHA	2	0.6041	0.7317	-	-	-
ALPHA	3	0.2206	1.0595	-	0.6544	-0.6391

			DEVIATION IN VAPOR PHASE COMPOSITION				
X	Y	TEMP.	LAAR 3	MARG 3	MARG 4	ALPHA2	ALPHA3
0.0280	0.0500	78.30	-0.0017	-0.0028	-0.0026	-0.0071	-0.0012
0.0550	0.0870	77.70	0.0019	0.0005	0.0008	-0.0062	0.0029
0.0950	0.1430	77.10	-0.0020	-0.0032	-0.0029	-0.0112	-0.0003
0.1420	0.2000	76.60	-0.0067	-0.0072	-0.0070	-0.0146	-0.0044
0.2050	0.2550	76.20	-0.0015	-0.0012	-0.0012	-0.0058	0.0011
0.2890	0.3190	75.80	0.0043	0.0051	0.0049	0.0052	0.0065
0.3530	0.3780	75.70	-0.0057	-0.0050	-0.0052	-0.0018	-0.0041
0.4100	0.4130	75.60	0.0017	0.0021	0.0019	0.0073	0.0026
0.4330	0.4330	75.60	-0.0012	-0.0011	-0.0012	0.0048	-0.0006
0.4680	0.4610	75.60	-0.0031	-0.0033	-0.0033	0.0031	-0.0030
0.5200	0.4990	75.70	-0.0017	-0.0024	-0.0023	0.0041	-0.0021
0.5700	0.5390	75.80	-0.0026	-0.0036	-0.0034	0.0020	-0.0032
0.6400	0.5860	76.20	0.0079	0.0068	0.0070	0.0098	0.0077
0.7170	0.6620	76.60	0.0006	-0.0002	0.0001	-0.0015	0.0017
0.7810	0.7300	77.10	-0.0046	-0.0047	-0.0046	-0.0101	-0.0019
0.8590	0.8220	78.00	-0.0112	-0.0102	-0.0103	-0.0196	-0.0064
0.9330	0.9020	78.90	0.0015	0.0030	0.0027	-0.0060	0.0062
		MEAN	0.0035	0.0037	0.0036	0.0071	0.0033

2,4-DIMETHYLPENTANE(1)-BENZENE(2)

RICHARDS A.R.,HARGREAVES E.:IND.ENG.CHEM.,36,805(1944).

ANTOINE VAPOR PRESSURE CONSTANTS
	A	B	C
(1)	6.82621	1192.041	221.634
(2)	6.90565	1211.033	220.790

P =   757.00

EQUATION	ORDER	A12	A21	D	A122	A211
VAN LAAR	3	0.2897	0.1774	-	-	-
MARGULES	3	0.2676	0.1639	-	-	-
MARGULES	4	0.3209	0.2214	0.1640	-	-
ALPHA	2	0.6413	0.7553	-	-	-
ALPHA	3	3.5498	4.6366	-	-2.3018	-3.8949

			DEVIATION IN VAPOR PHASE COMPOSITION				
X	Y	TEMP.	LAAR 3	MARG 3	MARG 4	ALPHA2	ALPHA3
0.0780	0.1250	77.30	-0.0027	-0.0051	0.0013	-0.0122	0.0027
0.1060	0.1640	77.00	-0.0075	-0.0095	-0.0039	-0.0167	-0.0035
0.1400	0.1950	76.50	-0.0014	-0.0026	0.0014	-0.0090	0.0007
0.1620	0.2190	76.30	-0.0033	-0.0040	-0.0013	-0.0096	-0.0024
0.1920	0.2420	76.10	0.0018	0.0018	0.0029	-0.0024	0.0015
0.2240	0.2680	75.90	0.0038	0.0044	0.0040	0.0019	0.0026
0.2510	0.2920	75.70	0.0022	0.0032	0.0018	0.0023	0.0006
0.2810	0.3180	75.50	0.0002	0.0015	-0.0007	0.0022	-0.0017
0.3350	0.3570	75.40	0.0026	0.0040	0.0013	0.0075	0.0010
0.3780	0.3920	75.40	-0.0003	0.0010	-0.0014	0.0061	-0.0013
0.4320	0.4320	75.20	-0.0003	0.0006	-0.0008	0.0069	-0.0003
0.4800	0.4690	75.40	-0.0016	-0.0011	-0.0011	0.0056	-0.0005
0.5250	0.5040	75.30	-0.0023	-0.0022	-0.0008	0.0040	-0.0004
0.5720	0.5410	75.50	-0.0024	-0.0027	-0.0001	0.0023	0.0001
0.6160	0.5760	75.50	-0.0014	-0.0020	0.0014	0.0011	0.0012
0.6620	0.6140	75.90	-0.0001	-0.0007	0.0028	-0.0003	0.0022
0.7000	0.6520	76.10	-0.0040	-0.0045	-0.0014	-0.0066	-0.0023
0.7400	0.6860	76.30	-0.0002	-0.0004	0.0016	-0.0055	0.0004
0.7850	0.7310	76.60	-0.0000	0.0002	0.0002	-0.0081	-0.0012
0.8390	0.7850	77.20	0.0043	0.0053	0.0022	-0.0063	0.0007
		MEAN	0.0021	0.0028	0.0016	0.0058	0.0014

ETHYLBENZENE(1)-STYRENE(2)

CHAIYAVECH P.,VAN WINKLE M.:J.CHEM.ENG.DATA 4,54(1959).

ANTOINE VAPOR PRESSURE CONSTANTS

	A	B	C
(1)	6.95719	1424.255	213.206
(2)	6.92409	1420.000	206.000

P = 20.00

EQUATION	ORDER	A12	A21	D	A122	A211
VAN LAAR	3	-0.0005	0.0005	-	-	-
MARGULES	3	0.0638	-0.0820	-	-	-
MARGULES	4	0.0933	-0.0591	0.1217	-	-
ALPHA	2	0.4586	-0.2797	-	-	-
ALPHA	3	-2.1315	-1.8147	-	2.7067	1.4062

			DEVIATION IN VAPOR PHASE COMPOSITION				
X	Y	TEMP.	LAAR 3	MARG 3	MARG 4	ALPHA2	ALPHA3
0.0555	0.0900	45.10	-0.0094	-0.0018	0.0012	-0.0112	-0.0065
0.0782	0.1200	44.90	-0.0076	0.0005	0.0035	-0.0102	-0.0044
0.1150	0.1620	44.60	0.0005	0.0076	0.0096	-0.0032	0.0032
0.1390	0.2010	44.40	-0.0066	-0.0013	-0.0001	-0.0112	-0.0048
0.2405	0.3180	43.62	0.0034	-0.0018	-0.0036	-0.0040	-0.0016
0.2715	0.3550	43.37	0.0029	-0.0056	-0.0078	-0.0052	-0.0047
0.3350	0.4055	42.95	0.0238	0.0099	0.0078	0.0149	0.0111
0.4600	0.5350	41.98	-0.0015	0.0064	0.0071	0.0145	0.0052
0.5450	0.6300	41.30	-0.0012	-0.0046	-0.0017	0.0007	-0.0073
0.7550	0.8250	39.75	-0.0017	-0.0024	-0.0006	-0.0121	-0.0044
0.8740	0.9100	39.10	0.0032	0.0089	0.0073	-0.0034	0.0067
0.9650	0.9750	38.70	0.0017	0.0050	0.0035	-0.0004	0.0037
		MEAN	0.0053	0.0046	0.0045	0.0076	0.0053

ETHYLBENZENE(1)-STYRENE(2)

CHAIYAVECH P.,VAN WINKLE M.:J.CHEM.ENG.DATA 4,54(1959).

ANTOINE VAPOR PRESSURE CONSTANTS

	A	B	C
(1)	6.95719	1424.255	213.206
(2)	6.92409	1420.000	206.000

P = 50.00

EQUATION	ORDER	A12	A21	D	A122	A211
VAN LAAR	3	-0.0033	0.0065	-	-	-
MARGULES	3	0.0587	-0.0067	-	-	-
MARGULES	4	0.0345	-0.0344	-0.0792	-	-
ALPHA	2	0.4651	-0.2156	-	-	-
ALPHA	3	.1275/ 06	.9181/ 05	-	-.1268/ 06	-.9180/ 05

DEVIATION IN VAPOR PHASE COMPOSITION

X	Y	TEMP.	LAAR 3	MARG 3	MARG 4	ALPHA2	ALPHA3
0.0550	0.0750	65.10	-0.0017	0.0093	0.0067	0.0032	0.0057
0.0885	0.1222	64.76	-0.1222	0.0090	0.0005	0.0013	0.0019
0.1750	0.2410	63.95	-0.0101	0.0002	-0.0005	-0.0073	-0.0096
0.2950	0.3650	63.00	0.0089	0.0108	0.0121	0.0074	0.0045
0.3600	0.4380	62.45	0.0078	0.0052	0.0065	0.0036	0.0016
0.3900	0.4780	62.20	-0.0002	-0.0047	-0.0036	-0.0054	-0.0071
0.5000	0.5850	61.30	0.0042	-0.0047	-0.0052	-0.0049	-0.0045
0.5710	0.6440	60.80	0.0125	0.0028	0.0014	0.0015	0.0032
0.6850	0.7460	59.90	0.0118	0.0038	0.0022	-0.0008	0.0022
0.7260	0.7740	59.60	0.0183	0.0116	0.0104	0.0058	0.0088
0.8740	0.9040	58.40	0.0052	0.0037	0.0050	-0.0035	-0.0031
0.9280	0.9450	58.03	0.0041	0.0036	0.0054	-0.0010	-0.0035
		MEAN	0.0173	0.0058	0.0055	0.0039	0.0046

387

ETHYLBENZENE(1)-STYRENE(2)

FRIED V.,PICK J.,HALA E.,VILIM O.:COLL.CZECH.CHEM.
COMMUN.22,1535(1956).

ANTOINE VAPOR PRESSURE CONSTANTS
	A	B	C
(1)	6.95719	1424.255	213.206
(2)	6.92409	1420.000	206.000

P =   50.00

EQUATION	ORDER	A12	A21	D	A122	A211
VAN LAAR	3	0.0332	0.0312	-	-	-
MARGULES	3	0.0334	0.0309	-	-	-
MARGULES	4	0.0007	-0.0035	-0.1137	-	-
ALPHA	2	0.5144	-0.2453	-	-	-
ALPHA	3	.2438/ 05	.1690/ 05	-	-.2429/ 05	-.1676/ 05

			DEVIATION IN VAPOR PHASE COMPOSITION				
X	Y	TEMP.	LAAR 3	MARG 3	MARG 4	ALPHA2	ALPHA3
0.1050	0.1450	64.60	0.0057	0.0057	0.0030	0.0046	0.0033
0.3130	0.4050	62.70	-0.0038	-0.0039	-0.0008	-0.0046	-0.0074
0.4500	0.5380	61.60	0.0035	0.0035	0.0045	0.0033	0.0028
0.5050	0.6000	61.20	-0.0061	-0.0062	-0.0066	-0.0064	-0.0057
0.6100	0.6800	60.40	0.0086	0.0086	0.0062	0.0083	0.0108
0.8350	0.8800	59.00	-0.0057	-0.0057	-0.0053	-0.0065	-0.0047
0.9120	0.9320	58.50	0.0018	0.0018	0.0036	0.0011	0.0005
0.9650	0.9700	58.20	0.0039	0.0039	0.0053	0.0035	0.0002
		MEAN	0.0049	0.0049	0.0044	0.0048	0.0044

ETHYLBENZENE(1)-STYRENE(2)

MALYUSOV V.A.,MALAFEEV N.A.,ZHAVORONKOV N.M.:ZH.FIZ.KHIM.
31,699(1957).

ANTOINE VAPOR PRESSURE CONSTANTS
	A	B	C
(1)	6.95719	1424.255	213.206
(2)	6.92409	1420.000	206.000

P = 50.00

EQUATION	ORDER	A12	A21	D	A122	A211
VAN LAAR	3	0.0397	0.0302	-	-	-
MARGULES	3	0.0412	0.0267	-	-	-
MARGULES	4	-0.0343	-0.0550	-0.2439	-	-
ALPHA	2	0.5194	-0.2366	-	-	-
ALPHA	3	-8.3797	-6.4215	-	8.8281	6.1055

			DEVIATION IN VAPOR PHASE COMPOSITION				
X	Y	TEMP.	LAAR 3	MARG 3	MARG 4	ALPHA2	ALPHA3
0.0240	0.0360	66.00	0.0007	0.0008	-0.0039	-0.0001	-0.0016
0.0540	0.0840	65.00	-0.0032	-0.0030	-0.0105	-0.0046	-0.0075
0.1410	0.1980	64.00	0.0011	0.0013	-0.0035	-0.0008	-0.0044
0.2190	0.2840	62.80	0.0114	0.0113	0.0126	0.0096	-0.0009
0.3080	0.3665	62.20	0.0293	0.0289	0.0338	0.0284	0.0230
0.3740	0.5240	61.80	-0.0588	-0.0594	-0.0549	-0.0592	-0.0623
0.4690	0.5500	61.60	0.0088	0.0081	0.0087	0.0090	0.0092
0.5390	0.6100	61.20	0.0141	0.0134	0.0108	0.0142	0.0167
0.6110	0.6740	61.00	0.0143	0.0137	0.0089	0.0142	0.0186
0.6930	0.7650	60.60	-0.0065	-0.0068	-0.0113	-0.0071	-0.0015
0.8200	0.8640	59.40	-0.0019	-0.0017	-0.0005	-0.0033	-0.0006
0.8330	0.8720	59.20	0.0004	0.0006	0.0025	-0.0010	-0.0003
0.8930	0.9240	58.80	-0.0048	-0.0046	-0.0002	-0.0062	0.0029
		MEAN	0.0119	0.0118	0.0125	0.0121	0.0115

ETHYLBENZENE(1)-STYRENE(2)

WHITE W.S.,VAN WINKLE M.:IND.ENG.CHEM.46,1284(1954).

ANTOINE VAPOR PRESSURE CONSTANTS
	A	B	C
(1)	6.95719	1424.255	213.206
(2)	6.92409	1420.000	206.000

P = 100.00

EQUATION	ORDER	A12	A21	D	A122	A211
VAN LAAR	3	0.0522	0.1772	-	-	-
MARGULES	3	0.0427	0.1107	-	-	-
MARGULES	4	0.0810	0.1554	0.1297	-	-
ALPHA	2	0.6387	-0.1972	-	-	-
ALPHA	3	3.7993	2.1252	-	-3.0133	-2.3382

DEVIATION IN VAPOR PHASE COMPOSITION

X	Y	TEMP.	LAAR 3	MARG 3	MARG 4	ALPHA2	ALPHA3
0.0910	0.1440	80.72	-0.0100	-0.0109	-0.0069	-0.0052	-0.0006
0.1410	0.2110	80.15	-0.0092	-0.0097	-0.0073	-0.0037	-0.0001
0.2350	0.3240	79.33	-0.0047	-0.0044	-0.0056	0.0001	0.0005
0.3190	0.4150	78.64	-0.0005	-0.0003	-0.0027	0.0027	0.0010
0.4120	0.5110	77.86	-0.0003	-0.0013	-0.0027	0.0010	-0.0015
0.5220	0.6110	76.98	0.0025	-0.0007	0.0007	0.0028	0.0011
0.6190	0.6990	76.19	-0.0030	-0.0076	-0.0047	-0.0019	-0.0020
0.7640	0.8140	75.03	-0.0052	-0.0085	-0.0074	0.0003	0.0024
0.8870	0.9140	74.25	-0.0131	-0.0113	-0.0141	-0.0033	-0.0014
		MEAN	0.0054	0.0061	0.0058	0.0023	0.0012

M-XYLENE(1)-ANILINE(2)

CHU J.C.,KHARBANDA O.P.,BROOKS R.F.,WANG S.L.:IND.ENG.CHEM.
46,754(1954).

ANTOINE VAPOR PRESSURE CONSTANTS
```
          A          B          C
(1) 7.00908    1462.266    215.105
(2) 7.24179    1675.300    200.000
```

P = 745.00

EQUATION	ORDER	A12	A21	D	A122	A211
VAN LAAR	3	0.2754	0.7164	-	-	-
MARGULES	3	0.2831	0.4797	-	-	-
MARGULES	4	0.5673	0.9177	1.1681	-	-
ALPHA	2	5.5950	-0.2830	-	-	-
ALPHA	3	-.1592/ 06	-.3424/ 05	-	.1482/ 06	.2891/ 05

			DEVIATION IN VAPOR PHASE COMPOSITION				
X	Y	TEMP.	LAAR 3	MARG 3	MARG 4	ALPHA2	ALPHA3
0.1000	0.4550	167.00	-0.0664	-0.0621	-0.0145	-0.0467	-0.0031
0.1950	0.5800	160.00	-0.0045	-0.0036	-0.0080	0.0053	0.0026
0.3400	0.7150	153.00	0.0118	0.0044	-0.0077	0.0129	0.0006
0.5300	0.8300	146.00	-0.0027	-0.0172	0.0001	-0.0020	-0.0044
0.7150	0.8900	143.00	-0.0097	-0.0200	-0.0036	0.0008	0.0061
0.8200	0.9300	141.00	-0.0251	-0.0254	-0.0292	-0.0075	-0.0036
		MEAN	0.0200	0.0221	0.0105	0.0125	0.0034

P-XYLENE(1)-ANILINE(2)

CHU J.C.,KHARBANDA O.P.,BROOKS R.F.,WANG S.L.:IND.ENG.CHEM.
46,754(1954).

ANTOINE VAPOR PRESSURE CONSTANTS
	A	B	C
(1)	6.99052	1453.430	215.307
(2)	7.24179	1675.300	200.000

P = 745.00

EQUATION	ORDER	A12	A21	D	A122	A211
VAN LAAR	3	0.3200	0.5862	-	-	-
MARGULES	3	0.3153	0.4962	-	-	-
MARGULES	4	0.4151	0.6687	0.4634	-	-
ALPHA	2	6.0435	-0.1437	-	-	-
ALPHA	3	6.5576	0.2357	-	-0.0671	-0.4529

			DEVIATION IN VAPOR PHASE COMPOSITION				
X	Y	TEMP.	LAAR 3	MARG 3	MARG 4	ALPHA2	ALPHA3
0.0750	0.3700	171.00	-0.0345	-0.0346	-0.0138	-0.0193	-0.0111
0.1400	0.5000	165.00	-0.0032	-0.0026	0.0018	0.0073	0.0116
0.2650	0.6750	156.00	-0.0064	-0.0072	-0.0148	-0.0040	-0.0058
0.4850	0.8000	148.00	0.0023	-0.0032	0.0024	0.0063	0.0029
0.7250	0.8900	142.00	-0.0137	-0.0174	-0.0101	-0.0032	-0.0022
0.8300	0.9200	140.00	-0.0132	-0.0124	-0.0141	-0.0017	0.0011
		MEAN	0.0122	0.0129	0.0095	0.0070	0.0058

1-OCTENE(1)-ETHYLBENZENE(2)

WEBER J.H.:IND.ENG.CHEM.48,134(1956).

ANTOINE VAPOR PRESSURE CONSTANTS

	A	B	C
(1)	6.93263	1353.500	212.764
(2)	6.95719	1424.255	213.206

P = 760.00

EQUATION	ORDER	A12	A21	D	A122	A211
VAN LAAR	3	0.0761	0.0555	-	-	-
MARGULES	3	0.0733	0.0544	-	-	-
MARGULES	4	0.0864	0.0682	0.0433	-	-
ALPHA	2	0.7025	-0.2254	-	-	-
ALPHA	3	5.5017	3.2711	-	-4.5271	-3.5428

			DEVIATION IN VAPOR PHASE COMPOSITION				
X	Y	TEMP.	LAAR 3	MARG 3	MARG 4	ALPHA2	ALPHA3
0.0410	0.0720	135.30	-0.0027	-0.0029	-0.0016	-0.0046	0.0021
0.0820	0.1490	134.20	-0.0163	-0.0166	-0.0151	-0.0188	-0.0107
0.1360	0.2050	133.10	0.0038	0.0037	0.0045	0.0019	0.0082
0.1630	0.2460	132.50	-0.0017	-0.0018	-0.0014	-0.0030	0.0017
0.2380	0.3350	131.20	0.0003	0.0004	-0.0001	0.0013	0.0016
0.3160	0.4200	129.70	0.0009	0.0012	0.0002	0.0041	0.0011
0.3250	0.4360	129.50	-0.0056	-0.0054	-0.0064	-0.0023	-0.0056
0.4670	0.5700	127.40	-0.0015	-0.0013	-0.0015	0.0037	-0.0008
0.5560	0.6520	126.10	-0.0042	-0.0042	-0.0036	0.0006	-0.0025
0.6380	0.7170	125.20	0.0002	0.0002	0.0011	0.0038	0.0029
0.6580	0.7330	124.80	0.0008	0.0007	0.0017	0.0040	0.0036
0.7630	0.8230	123.80	-0.0047	-0.0047	-0.0043	-0.0037	-0.0015
0.8350	0.8760	123.00	-0.0014	-0.0014	-0.0016	-0.0017	0.0015
0.8630	0.9030	122.70	-0.0068	-0.0067	-0.0073	-0.0075	-0.0043
0.9090	0.9320	122.30	-0.0006	-0.0006	-0.0014	-0.0016	0.0012
0.9470	0.9610	121.80	-0.0008	-0.0008	-0.0015	-0.0017	0.0003
		MEAN	0.0033	0.0033	0.0033	0.0040	0.0031

z

OCTANE(1)-ETHYLBENZENE(2)

YANG C.P.,VAN WINKLE M.:IND.ENG.CHEM.47,293(1955).

ANTOINE VAPOR PRESSURE CONSTANTS
	A	B	C
(1)	6.92377	1355.126	209.517
(2)	6.95719	1424.255	213.206

P = 760.00

EQUATION	ORDER	A12	A21	D	A122	A211
VAN LAAR	3	0.0825	0.1030	-	-	-
MARGULES	3	0.0812	0.1014	-	-	-
MARGULES	4	0.0849	0.1052	0.0115	-	-
ALPHA	2	0.6141	-0.0570	-	-	-
ALPHA	3	0.8017	0.0211	-	-0.2141	-0.0511

			DEVIATION IN VAPOR PHASE COMPOSITION				
X	Y	TEMP.	LAAR 3	MARG 3	MARG 4	ALPHA2	ALPHA3
0.0490	0.0750	135.10	0.0004	0.0003	0.0006	0.0007	-0.0003
0.0960	0.1380	134.10	0.0041	0.0040	0.0044	0.0044	0.0031
0.1500	0.2140	133.20	-0.0010	-0.0011	-0.0008	-0.0008	-0.0021
0.2010	0.2750	132.40	0.0001	0.0001	0.0001	-0.0001	-0.0010
0.2510	0.3320	131.70	0.0000	0.0001	-0.0001	-0.0003	-0.0009
0.3040	0.3880	131.00	0.0007	0.0008	0.0006	0.0002	0.0000
0.3590	0.4420	130.20	0.0022	0.0023	0.0021	0.0015	0.0018
0.4180	0.5010	129.70	-0.0004	-0.0004	-0.0005	-0.0013	-0.0007
0.4660	0.5410	129.10	0.0034	0.0035	0.0034	0.0025	0.0033
0.5160	0.5900	128.80	-0.0015	-0.0015	-0.0014	-0.0024	-0.0015
0.5700	0.6360	128.20	-0.0012	-0.0012	-0.0011	-0.0020	-0.0012
0.6240	0.6810	127.80	-0.0009	-0.0010	-0.0008	-0.0017	-0.0011
0.6710	0.7190	127.30	-0.0001	-0.0002	0.0000	-0.0008	-0.0005
0.7200	0.7580	126.90	0.0011	0.0010	0.0012	0.0006	0.0005
0.7720	0.8010	126.60	0.0008	0.0008	0.0008	0.0005	-0.0000
0.8220	0.8410	126.30	0.0022	0.0023	0.0022	0.0021	0.0012
0.8690	0.8810	126.00	0.0019	0.0020	0.0018	0.0019	0.0007
0.9090	0.9150	125.90	0.0024	0.0025	0.0022	0.0025	0.0013
0.9570	0.9640	125.80	-0.0039	-0.0038	-0.0040	-0.0037	-0.0046
		MEAN	0.0015	0.0015	0.0015	0.0016	0.0014

OCTANE(1)-ETHYLBENZENE(2)

YANG C.P.,VAN WINKLE M.:IND.ENG.CHEM.47,293(1955).

ANTOINE VAPOR PRESSURE CONSTANTS
	A	B	C
(1)	6.92377	1355.126	209.517
(2)	6.95719	1424.255	213.206

P = 500.00

EQUATION	ORDER	A12	A21	D	A122	A211
VAN LAAR	3	0.0871	0.1087	-	-	-
MARGULES	3	0.0869	0.1060	-	-	-
MARGULES	4	0.1221	0.1426	0.1100	-	-
ALPHA	2	0.6594	-0.0649	-	-	-
ALPHA	3	6.2180	4.0971	-	-5.2998	-4.0488

DEVIATION IN VAPOR PHASE COMPOSITION

X	Y	TEMP.	LAAR 3	MARG 3	MARG 4	ALPHA2	ALPHA3
0.0950	0.1510	119.00	-0.0077	-0.0077	-0.0039	-0.0067	-0.0003
0.2060	0.2860	117.50	-0.0013	-0.0013	-0.0013	0.0000	0.0011
0.3140	0.4030	116.10	-0.0000	-0.0001	-0.0023	0.0013	-0.0010
0.4180	0.5030	115.00	0.0011	0.0009	-0.0006	0.0023	-0.0002
0.5200	0.5960	114.00	-0.0011	-0.0014	-0.0008	0.0001	-0.0007
0.6240	0.6820	113.10	0.0001	-0.0002	0.0019	0.0014	0.0025
0.7240	0.7680	112.50	-0.0044	-0.0046	-0.0031	-0.0031	-0.0011
0.8210	0.8460	112.00	-0.0031	-0.0030	-0.0038	-0.0018	-0.0010
0.9090	0.9160	111.50	0.0015	0.0017	-0.0009	0.0024	0.0008
		MEAN	0.0023	0.0023	0.0021	0.0021	0.0010

395

OCTANE(1)-ETHYLBENZENE(2)

YANG C.P.,VAN WINKLE M.:IND.ENG.CHEM.47,293(1955).

ANTOINE VAPOR PRESSURE CONSTANTS
	A	B	C
(1)	6.92377	1355.126	209.517
(2)	6.95719	1424.255	213.206

P = 200.00

EQUATION	ORDER	A12	A21	D	A122	A211
VAN LAAR	3	0.1063	0.1302	-	-	-
MARGULES	3	0.1069	0.1267	-	-	-
MARGULES	4	0.1747	0.1981	0.2143	-	-
ALPHA	2	0.7792	-0.0440	-	-	-
ALPHA	3	10.5602	7.2752	-	-9.1303	-7.1533

			DEVIATION IN VAPOR PHASE COMPOSITION				
X	Y	TEMP.	LAAR 3	MARG 3	MARG 4	ALPHA2	ALPHA3
0.0930	0.1630	90.40	-0.0150	-0.0148	-0.0070	-0.0135	0.0000
0.1980	0.2940	89.00	-0.0081	-0.0081	-0.0077	-0.0059	-0.0027
0.3180	0.4100	87.80	0.0072	0.0070	0.0026	0.0100	0.0059
0.4140	0.5090	87.00	-0.0003	-0.0007	-0.0036	0.0028	-0.0021
0.5190	0.6030	86.10	-0.0033	-0.0038	-0.0026	-0.0001	-0.0026
0.6170	0.6840	85.50	-0.0046	-0.0051	-0.0010	-0.0014	-0.0006
0.7160	0.7620	84.90	-0.0042	-0.0044	-0.0011	-0.0013	0.0019
0.8190	0.8440	84.30	-0.0036	-0.0035	-0.0048	-0.0013	0.0015
0.9090	0.9190	84.00	-0.0027	-0.0024	-0.0076	-0.0014	-0.0019
		MEAN	0.0054	0.0056	0.0042	0.0042	0.0021

OCTANE(1)-ETHYLBENZENE(2)

YANG C.P.,VAN WINKLE M.:IND.ENG.CHEM.47,293(1955).

ANTOINE VAPOR PRESSURE CONSTANTS
	A	B	C
(1)	6.92377	1355.126	209.517
(2)	6.95719	1424.255	213.206

P = 50.00

EQUATION	ORDER	A12	A21	D	A122	A211
VAN LAAR	3	0.1007	0.1812	-	-	-
MARGULES	3	0.0960	0.1585	-	-	-
MARGULES	4	0.1634	0.2300	0.2167	-	-
ALPHA	2	0.8988	-0.0676	-	-	-
ALPHA	3	8.9481	5.5082	-	-7.5670	-5.3837

DEVIATION IN VAPOR PHASE COMPOSITION

X	Y	TEMP.	LAAR 3	MARG 3	MARG 4	ALPHA2	ALPHA3
0.0880	0.1600	55.80	-0.0153	-0.0158	-0.0078	-0.0099	0.0009
0.1900	0.2960	54.50	-0.0109	-0.0109	-0.0101	-0.0049	-0.0023
0.3030	0.4140	53.60	0.0011	0.0012	-0.0034	0.0052	0.0018
0.4120	0.5190	52.80	0.0030	0.0022	-0.0008	0.0052	0.0012
0.5160	0.6130	51.90	-0.0012	-0.0029	-0.0016	0.0002	-0.0014
0.5990	0.6810	51.40	-0.0035	-0.0056	-0.0016	-0.0017	-0.0008
0.7200	0.7750	50.60	-0.0068	-0.0084	-0.0050	-0.0031	-0.0001
0.8190	0.8480	50.20	-0.0059	-0.0059	-0.0070	-0.0002	0.0016
0.9100	0.9200	50.00	-0.0049	-0.0035	-0.0087	0.0009	-0.0010
		MEAN	0.0058	0.0062	0.0051	0.0035	0.0012

OCTANE(1)-PROPIONIC ACID(2)

JOHNSON A.I.,WARD D.M.,FURTER W.F.:CAN.J.TECH.34,514(1957).

ANTOINE VAPOR PRESSURE CONSTANTS

	A	B	C
(1)	6.92377	1355.126	209.517
(2)	7.35027	1497.775	194.120

P = 750.00

EQUATION	ORDER	A12	A21	D	A122	A211
VAN LAAR	3	0.5808	0.4236	-	-	-
MARGULES	3	0.5668	0.4092	-	-	-
MARGULES	4	0.5419	0.3888	-0.0814	-	-
ALPHA	2	2.9252	1.3944	-	-	-
ALPHA	3	4.3018	2.1974	-	-1.2280	-0.5448

			DEVIATION IN VAPOR PHASE COMPOSITION				
X	Y	TEMP.	LAAR 3	MARG 3	MARG 4	ALPHA2	ALPHA3
0.0510	0.1200	136.80	0.0878	0.0849	0.0794	0.0393	0.0414
0.0820	0.2340	133.20	0.0518	0.0496	0.0449	-0.0060	-0.0049
0.1350	0.3370	129.20	0.0387	0.0381	0.0363	-0.0202	-0.0214
0.3610	0.5120	122.60	0.0328	0.0337	0.0366	0.0068	0.0038
0.4610	0.5660	121.90	0.0228	0.0226	0.0240	0.0069	0.0058
0.5950	0.6480	121.40	0.0012	-0.0002	-0.0013	-0.0111	-0.0096
0.6490	0.6600	119.80	0.0190	0.0175	0.0157	0.0030	0.0050
0.7010	0.6900	121.80	0.0161	0.0148	0.0129	-0.0003	0.0019
0.7660	0.7310	122.00	0.0180	0.0173	0.0159	-0.0038	-0.0023
0.8880	0.8270	122.90	0.0277	0.0289	0.0301	-0.0023	-0.0052
0.9380	0.8780	123.70	0.0339	0.0354	0.0372	0.0076	0.0029
0.9670	0.9180	124.80	0.0322	0.0334	0.0349	0.0139	0.0094
		MEAN	0.0318	0.0314	0.0308	0.0101	0.0095

2,2,4-TRIMETHYLPENTANE(1)-FURFURALDEHYDE(2)

THORNTON J.D.,GARNER F.H.:J.APPL.CHEM.1,SUPPL.NO1,S74(1951).

ANTOINE VAPOR PRESSURE CONSTANTS

	A	B	C
(1)	6.81189	1257.840	220.735
(2)	6.48323	1112.311	147.256

P = 760.00

EQUATION	ORDER	A12	A21	D	A122	A211
VAN LAAR	3	0.8715	1.0021	-	-	-
MARGULES	3	0.8702	0.9932	-	-	-
MARGULES	4	0.9263	1.0602	0.2169	-	-
ALPHA	2	71.0196	5.9139	-	-	-
ALPHA	3	2.3322	3.9907	-	64.4199	-3.9816

			DEVIATION IN VAPOR PHASE COMPOSITION				
X	Y	TEMP.	LAAR 3	MARG 3	MARG 4	ALPHA2	ALPHA3
0.0440	0.6900	126.80	-0.0578	-0.0581	-0.0393	0.0256	0.0199
0.1635	0.8591	106.40	-0.0028	-0.0026	-0.0025	-0.0019	-0.0005
0.2795	0.8904	103.50	-0.0028	-0.0028	-0.0052	-0.0063	-0.0045
0.2468	0.8851	102.20	-0.0000	0.0001	-0.0021	-0.0062	-0.0043
0.3720	0.8989	102.20	-0.0043	-0.0044	-0.0061	-0.0048	-0.0037
0.5118	0.8965	101.70	-0.0024	-0.0028	-0.0020	0.0063	0.0052
0.6406	0.8971	101.90	-0.0069	-0.0074	-0.0045	0.0109	0.0069
0.7798	0.9055	99.80	-0.0091	-0.0093	-0.0077	0.0075	0.0012
0.9036	0.9364	99.10	-0.0103	-0.0098	-0.0133	-0.0158	-0.0099
0.9657	0.9613	99.00	0.0035	0.0040	0.0000	-0.0262	0.0048
		MEAN	0.0100	0.0101	0.0083	0.0112	0.0061

2,2,4-TRIMETHYLPENTANE(1)-METHYLCYCLOHEXANE(2)

HARRISON J.M.,BERG L.:IND.ENG.CHEM.38,117(1946).

ANTOINE VAPOR PRESSURE CONSTANTS
```
           A          B          C
(1) 6.81189    1257.840   220.735
(2) 6.82689    1272.864   221.630
```

P =   741.00

EQUATION	ORDER	A12	A21	D	A122	A211
VAN LAAR	3	0.7557	-0.0081	-	-	-
MARGULES	3	0.0358	-0.0003	-	-	-
MARGULES	4	0.0945	0.0819	0.2263	-	-
ALPHA	2	0.0943	-0.0379	-	-	-
ALPHA	3	9.6638	9.8964	-	-9.2095	-10.2285

			DEVIATION IN VAPOR PHASE COMPOSITION				
X	Y	TEMP.	LAAR 3	MARG 3	MARG 4	ALPHA2	ALPHA3
0.0400	0.0480	99.80	0.0048	-0.0032	0.0008	-0.0045	0.0033
0.0880	0.1070	99.65	-0.0108	-0.0104	-0.0058	-0.0119	-0.0031
0.1400	0.1630	99.50	-0.0133	-0.0119	-0.0095	-0.0127	-0.0060
0.1900	0.2040	99.40	-0.0026	-0.0019	-0.0021	-0.0012	0.0025
0.2450	0.2570	99.30	0.0012	0.0001	-0.0023	0.0027	0.0032
0.3400	0.3500	99.10	0.0055	0.0006	-0.0025	0.0067	0.0027
0.4070	0.4160	98.95	0.0075	-0.0000	-0.0011	0.0080	0.0023
0.4760	0.4900	98.85	0.0028	-0.0067	-0.0045	0.0026	-0.0038
0.6950	0.7070	98.55	0.0020	-0.0076	-0.0013	-0.0005	-0.0013
0.7940	0.8090	98.40	-0.0043	-0.0110	-0.0103	-0.0070	-0.0028
0.8790	0.8880	98.30	-0.0021	-0.0056	-0.0108	-0.0043	0.0036
		MEAN	0.0052	0.0054	0.0046	0.0056	0.0031

2,2,4-TRIMETHYLPENANE(1)-OCTANE(2)

BROMILEY E.C.,QUIGGLE D.:IND.ENG.CHEM.25,1136(1933).

ANTOINE VAPOR PRESSURE CONSTANTS

	A	B	C
(1)	6.81189	1257.840	220.735
(2)	6.92377	1355.126	209.517

P = 200.00

EQUATION	ORDER	A12	A21	D	A122	A211
VAN LAAR	3	0.0116	0.0914	-	-	-
MARGULES	3	0.0043	0.0295	-	-	-
MARGULES	4	0.0370	0.0720	0.1151	-	-
ALPHA	2	1.0885	-0.5465	-	-	-
ALPHA	3	-5.9295	-3.8396	-	7.0551	3.3063

			DEVIATION IN VAPOR PHASE COMPOSITION				
X	Y	TEMP.	LAAR 3	MARG 3	MARG 4	ALPHA2	ALPHA3
0.0510	0.1050	124.00	-0.0053	-0.0064	-0.0022	-0.0040	-0.0025
0.1070	0.1970	123.00	0.0015	0.0003	0.0040	0.0036	0.0058
0.1600	0.2900	121.00	-0.0061	-0.0070	-0.0053	-0.0045	-0.0023
0.2080	0.3560	119.80	-0.0016	-0.0022	-0.0023	-0.0004	0.0013
0.2480	0.4090	118.90	-0.0003	-0.0006	-0.0017	0.0006	0.0017
0.2810	0.4500	118.30	0.0007	0.0006	-0.0010	0.0015	0.0016
0.3210	0.5000	117.00	-0.0007	-0.0008	-0.0025	-0.0007	-0.0026
0.3600	0.5350	116.00	0.0084	0.0083	0.0069	0.0080	0.0006
0.4640	0.6500	113.40	-0.0015	-0.0021	-0.0015	-0.0026	0.0027
0.5600	0.7350	110.70	-0.0030	-0.0043	-0.0021	-0.0044	-0.0030
0.6580	0.8100	108.20	-0.0044	-0.0061	-0.0039	-0.0052	-0.0057
0.7390	0.8540	106.00	0.0051	0.0034	0.0046	0.0051	0.0025
0.7880	0.8900	105.00	-0.0015	-0.0028	-0.0027	-0.0007	0.0004
0.8350	0.9150	103.50	0.0001	-0.0006	-0.0015	0.0015	0.0010
0.8920	0.9450	101.90	0.0002	0.0005	-0.0012	0.0024	0.0016
0.9330	0.9690	100.00	-0.0031	-0.0021	-0.0038	-0.0008	-0.0014
		MEAN	0.0027	0.0030	0.0030	0.0029	0.0023

2,2,4-TRIMETHYLPENTANE(1)-PHENOL(2)

DRICKAMER H.G.,BROWN G.G.,WHITE R.R.:TRANS.AM.INST.CHEM.ENGRS.
41,555(1945)

ANTOINE VAPOR PRESSURE CONSTANTS
	A	B	C
(1)	6.81189	1257.840	220.735
(2)	7.57893	1817.000	205.000

P = 760.00

EQUATION	ORDER	A12	A21	D	A122	A211
VAN LAAR	3	1.1582	0.9338	-	-	-
MARGULES	3	1.0592	0.9559	-	-	-
MARGULES	4	1.7943	1.2474	1.5975	-	-
ALPHA	2	48.0900	2.1458	-	-	-
ALPHA	3	-5.0173	-0.6255	-	42.6663	0.1565

			DEVIATION IN VAPOR PHASE COMPOSITION				
X	Y	TEMP.	LAAR 3	MARG 3	MARG 4	ALPHA2	ALPHA3
0.2380	0.8880	125.60	0.0416	0.0408	0.0320	-0.0019	0.0005
0.4270	0.9364	113.30	-0.0032	0.0004	-0.0221	-0.0190	-0.0141
0.6660	0.9140	107.80	0.0178	0.0214	0.0202	0.0194	0.0118
0.9015	0.9459	103.90	0.0175	0.0172	0.0100	0.0012	-0.0086
0.9530	0.9693	101.10	0.0114	0.0108	0.0013	-0.0133	0.0003
0.9546	0.9704	101.10	0.0109	0.0103	0.0009	-0.0140	0.0003
0.9795	0.9864	100.60	0.0046	0.0042	-0.0025	-0.0181	0.0015
0.9892	0.9914	100.60	0.0037	0.0035	-0.0007	-0.0133	0.0025
0.9959	0.9966	100.00	0.0015	0.0014	-0.0003	-0.0073	0.0012
		MEAN	0.0125	0.0122	0.0100	0.0119	0.0045

2,2,4-TRIMETHYLPENTANE(1)-TOLUENE(2)

DRICKAMER H.G.,BROWN G.G.,WHITE R.R.:TRANS.AM.INST.
           CHEM.ENGRS.41,555(1944).

ANTOINE VAPOR PRESSURE CONSTANTS
          A   B   C
(1) 6.81189   1257.840   220.735
(2) 6.95334   1343.943   219.377

P = 760.00

EQUATION	ORDER	A12	A21	D	A122	A211
VAN LAAR	3	0.1175	0.0962	-	-	-
MARGULES	3	0.1164	0.0946	-	-	-
MARGULES	4	0.1060	0.0840	-0.0351	-	-
ALPHA	2	0.6914	-0.0447	-	-	-
ALPHA	3	-0.1468	-0.5431	-	0.8675	0.4314

DEVIATION IN VAPOR PHASE COMPOSITION

X	Y	TEMP.	LAAR 3	MARG 3	MARG 4	ALPHA2	ALPHA3
0.0202	0.0345	110.30	0.0006	0.0006	-0.0001	-0.0010	-0.0005
0.0699	0.1125	109.10	0.0017	0.0016	0.0004	-0.0023	-0.0008
0.0736	0.1238	109.00	-0.0041	-0.0042	-0.0054	-0.0082	-0.0066
0.1400	0.2030	107.30	0.0086	0.0086	0.0080	0.0041	0.0062
0.2180	0.3020	107.20	0.0031	0.0032	0.0036	0.0005	0.0023
0.2864	0.3806	107.30	-0.0024	-0.0023	-0.0015	-0.0028	-0.0018
0.4097	0.5050	103.90	-0.0069	-0.0069	-0.0063	-0.0061	-0.0068
0.4712	0.5490	103.30	0.0040	0.0039	0.0041	0.0052	0.0038
0.5204	0.5960	103.10	-0.0006	-0.0007	-0.0009	0.0007	-0.0011
0.5793	0.6392	102.80	0.0060	0.0058	0.0052	0.0068	0.0051
0.6180	0.6746	101.70	0.0032	0.0031	0.0023	0.0031	0.0017
0.7220	0.7673	101.10	-0.0036	-0.0037	-0.0044	-0.0054	-0.0054
0.8020	0.8290	100.20	0.0011	0.0011	0.0010	-0.0021	-0.0004
0.8870	0.9010	100.00	0.0006	0.0007	0.0012	-0.0029	0.0000
0.9181	0.9260	99.60	0.0022	0.0024	0.0030	-0.0009	0.0019
0.9661	0.9685	99.40	0.0015	0.0016	0.0021	-0.0003	0.0015
		MEAN	0.0031	0.0031	0.0031	0.0033	0.0029

2,2,4-TRIMETHYLPENTANE(1)-TOLUENE(2)

PRENGLE H.W.,PALM G.F.:IND.ENG.CHEM.49,1769(1957).

ANTOINE VAPOR PRESSURE CONSTANTS

	A	B	C
(1)	6.81189	1257.840	220.735
(2)	6.95334	1343.943	219.377

T = 100.00

EQUATION	ORDER	A12	A21	D	A122	A211
VAN LAAR	3	0.1716	0.1185	-	-	-
MARGULES	3	0.1650	0.1123	-	-	-
MARGULES	4	0.1634	0.1106	-0.0049	-	-
ALPHA	2	0.8361	0.0522	-	0.3148	-0.4372
ALPHA	3	0.7308	0.3511	-	0.3148	-0.4372

			DEVIATION IN VAPOR PHASE COMPOSITION				
X	Y	PRESS.	LAAR 3	MARG 3	MARG 4	ALPHA2	ALPHA3
0.1000	0.1703	604.30	0.0011	0.0004	0.0002	-0.0080	0.0006
0.2000	0.2980	643.60	-0.0005	-0.0004	-0.0004	-0.0058	-0.0002
0.3000	0.4022	676.00	-0.0017	-0.0014	-0.0013	-0.0014	-0.0011
0.4000	0.4905	702.00	0.0006	0.0008	0.0009	0.0046	0.0010
0.5000	0.5753	721.30	0.0001	-0.0001	-0.0001	0.0049	-0.0000
0.6000	0.6559	736.20	0.0012	0.0007	0.0007	0.0041	0.0007
0.7000	0.7395	748.40	-0.0009	-0.0012	-0.0013	-0.0016	-0.0011
0.8000	0.8223	758.50	-0.0003	-0.0002	-0.0002	-0.0047	0.0001
0.9000	0.9092	769.50	-0.0005	-0.0001	0.0001	-0.0062	0.0005
		MEAN	0.0008	0.0006	0.0006	0.0046	0.0006

2,2,4-TRIMETHYLPENTANE(1)-TOLUENE(2)

PRENGLE H.W.,PALM G.F.:IND.ENG.CHEM.49,1769(1957).

ANTOINE VAPOR PRESSURE CONSTANTS
	A	B	C
(1)	6.81189	1257.840	220.735
(2)	6.95334	1343.943	219.377

P =   760.00

EQUATION	ORDER	A12	A21	D	A122	A211
VAN LAAR	3	0.1644	0.1173	-	-	-
MARGULES	3	0.1594	0.1118	-	-	-
MARGULES	4	0.1487	0.1006	-0.0333	-	-
ALPHA	2	0.7986	0.0363	-	-	-
ALPHA	3	0.0962	-0.2048	-	0.8188	0.1144

			DEVIATION IN VAPOR PHASE COMPOSITION				
X	Y	TEMP.	LAAR 3	MARG 3	MARG 4	ALPHA2	ALPHA3
0.1000	0.1647	107.90	0.0030	0.0025	0.0013	-0.0048	0.0008
0.2000	0.2938	105.77	0.0003	0.0003	0.0002	-0.0047	-0.0003
0.3000	0.3996	104.10	-0.0014	-0.0012	-0.0005	-0.0017	-0.0008
0.4000	0.4899	102.72	0.0002	0.0002	0.0007	0.0030	0.0006
0.5000	0.5750	101.72	0.0003	0.0000	-0.0001	0.0038	-0.0001
0.6000	0.6554	101.03	0.0021	0.0016	0.0011	0.0041	0.0010
0.7000	0.7394	100.41	-0.0001	-0.0004	-0.0010	-0.0011	-0.0013
0.8000	0.8223	99.90	0.0003	0.0004	0.0005	-0.0039	-0.0000
0.9000	0.9092	99.50	-0.0002	0.0003	0.0010	-0.0053	0.0006
		MEAN	0.0009	0.0008	0.0007	0.0036	0.0006

2,2,4-TRIMETHYLPENTANE(1)-TOLUENE(2)

THORNTON J.D.,GARNER F.H.:J.APPL.CHEM.1,SUPPL.NO1,S74(1951).

ANTOINE VAPOR PRESSURE CONSTANTS

	A	B	C
(1)	6.81189	1257.840	220.735
(2)	6.95334	1343.943	219.377

P =   760.00

EQUATION	ORDER	A12	A21	D	A122	A211
VAN LAAR	3	0.0005	-0.0002	-	-	-
MARGULES	3	0.2098	0.0179	-	-	-
MARGULES	4	0.2190	0.0282	0.0313	-	-
ALPHA	2	0.6658	0.0539	-	-	-
ALPHA	3	-0.4674	-0.0760	-	1.4741	-0.1896

X	Y	TEMP.	DEVIATION IN VAPOR PHASE COMPOSITION				
			LAAR 3	MARG 3	MARG 4	ALPHA2	ALPHA3
0.0400	0.0700	109.70	-0.0160	0.0092	0.0102	-0.0062	0.0039
0.0800	0.1307	108.90	-0.0241	0.0131	0.0142	-0.0082	0.0067
0.1700	0.2540	107.30	-0.0334	0.0032	0.0034	-0.0144	-0.0005
0.2680	0.3694	105.70	0.0038	-0.0153	-0.0159	-0.0201	-0.0151
0.3620	0.4560	104.50	-0.0131	-0.0196	-0.0202	-0.0137	-0.0180
0.4650	0.4550	103.40	0.0920	0.0688	0.0687	0.0799	0.0694
0.5070	0.5860	102.90	0.0023	-0.0260	-0.0258	-0.0153	-0.0264
0.6243	0.6797	101.80	0.0184	-0.0155	-0.0147	-0.0120	-0.0184
0.7480	0.7825	100.90	0.0229	-0.0034	-0.0030	-0.0133	-0.0067
0.8650	0.8780	100.00	0.0215	0.0095	0.0091	-0.0083	0.0083
0.9430	0.9490	99.50	0.0096	0.0061	0.0055	-0.0067	0.0062
		MEAN	0.0234	0.0172	0.0173	0.0180	0.0163

ALPHA-METHYLSTYRENE(1)-PHENOL(2)

SHCHERBAK L.I.,BYK S.SH.,AEROV M.E.:ZH.PRIKL.KHIM.28,1120(1955).

ANTOINE VAPOR PRESSURE CONSTANTS

	A	B	C
(1)	6.92366	1486.880	202.400
(2)	7.57893	1817.000	205.000

P = 760.00

EQUATION	ORDER	A12	A21	D	A122	A211
VAN LAAR	3	0.1984	0.3003	-	-	-
MARGULES	3	0.1908	0.2806	-	-	-
MARGULES	4	0.2396	0.3371	0.1500	-	-
ALPHA	2	1.5340	0.0531	-	-	-
ALPHA	3	0.5490	-0.1557	-	1.1501	0.0096

			DEVIATION IN VAPOR PHASE COMPOSITION				
X	Y	TEMP.	LAAR 3	MARG 3	MARG 4	ALPHA2	ALPHA3
0.0750	0.1650	176.50	-0.0080	-0.0091	-0.0015	-0.0016	0.0047
0.1800	0.3650	173.40	-0.0423	-0.0428	-0.0403	-0.0357	-0.0308
0.2300	0.3750	170.40	0.0129	0.0129	0.0128	0.0166	0.0196
0.2400	0.3900	170.70	0.0090	0.0090	0.0086	0.0131	0.0156
0.2700	0.4250	168.90	0.0088	0.0089	0.0076	0.0110	0.0122
0.3200	0.5000	167.70	-0.0152	-0.0152	-0.0171	-0.0140	-0.0148
0.3950	0.5650	166.50	-0.0136	-0.0140	-0.0153	-0.0128	-0.0159
0.4250	0.5750	166.90	-0.0006	-0.0011	-0.0019	0.0013	-0.0024
0.5100	0.6200	165.70	0.0169	0.0158	0.0171	0.0196	0.0153
0.5950	0.6950	164.60	-0.0015	-0.0029	-0.0003	0.0028	-0.0002
0.6880	0.7550	164.30	-0.0034	-0.0047	-0.0023	0.0038	0.0042
0.7300	0.8000	163.10	-0.0211	-0.0220	-0.0206	-0.0136	-0.0111
0.8450	0.8700	162.40	-0.0134	-0.0127	-0.0156	-0.0041	0.0039
		MEAN	0.0128	0.0132	0.0124	0.0115	0.0116

2,2,5-TRIMETHYLHEXANE(1)-ETHYLBENZENE(2)

WEBER J.H.:IND.ENG.CHEM.48,134(1956).

ANTOINE VAPOR PRESSURE CONSTANTS

	A	B	C
(1)	6.83531	1324.049	210.737
(2)	6.95719	1424.255	213.206

P = 760.00

EQUATION	ORDER	A12	A21	D	A122	A211
VAN LAAR	3	0.1186	0.1508	-	-	-
MARGULES	3	0.1181	0.1462	-	-	-
MARGULES	4	0.1513	0.1836	0.1093	-	-
ALPHA	2	0.8507	-0.0179	-	-	-
ALPHA	3	1.9081	0.9246	-	-0.8958	-0.9730

			DEVIATION IN VAPOR PHASE COMPOSITION				
X	Y	TEMP.	LAAR 3	MARG 3	MARG 4	ALPHA2	ALPHA3
0.0450	0.0900	134.70	-0.0128	-0.0129	-0.0092	-0.0113	-0.0067
0.0580	0.1040	134.40	-0.0060	-0.0060	-0.0020	-0.0041	0.0011
0.1110	0.1880	133.30	-0.0107	-0.0107	-0.0074	-0.0079	-0.0026
0.1610	0.2480	132.00	-0.0032	-0.0032	-0.0017	0.0000	0.0038
0.2480	0.3530	130.50	-0.0044	-0.0045	-0.0058	-0.0010	-0.0008
0.2950	0.3970	129.80	0.0018	0.0017	-0.0003	0.0053	0.0039
0.3740	0.4760	128.60	0.0003	0.0001	-0.0017	0.0036	0.0006
0.4840	0.5790	127.20	-0.0056	-0.0061	-0.0059	-0.0025	-0.0058
0.5420	0.6220	126.90	-0.0011	-0.0017	-0.0003	0.0021	-0.0005
0.6810	0.7300	125.80	-0.0001	-0.0005	0.0018	0.0032	0.0033
0.6920	0.7420	125.70	-0.0036	-0.0040	-0.0018	-0.0005	0.0000
0.7340	0.7770	125.30	-0.0059	-0.0062	-0.0046	-0.0028	-0.0016
0.7650	0.7970	125.20	-0.0017	-0.0018	-0.0009	0.0014	0.0031
0.8220	0.8470	124.70	-0.0061	-0.0060	-0.0067	-0.0033	-0.0010
0.8990	0.9080	124.40	-0.0024	-0.0021	-0.0047	-0.0004	0.0018
0.9600	0.9700	124.30	-0.0091	-0.0088	-0.0111	-0.0081	-0.0070
		MEAN	0.0047	0.0048	0.0041	0.0036	0.0027

408

NAPHTHALENE(1)-1-HEXADECENE(2)

WARD S.H.,VAN WINKLE M.:IND.ENG.CHEM.46,338(1954).

ANTOINE VAPOR PRESSURE CONSTANTS
	A	B	C
(1)	6.84577	1606.529	187.227
(2)	6.93600	1755.200	148.000

P = 200.00

EQUATION	ORDER	A12	A21	D	A122	A211
VAN LAAR	3	-0.0071	0.0100	-	-	-
MARGULES	3	0.1730	-0.1995	-	-	-
MARGULES	4	0.0054	-0.4509	-0.6929	-	-
ALPHA	2	4.4530	-0.7410	-	-	-
ALPHA	3	-0.4869	-1.9361	-	4.4278	1.2540

			DEVIATION IN VAPOR PHASE COMPOSITION				
X	Y	TEMP.	LAAR 3	MARG 3	MARG 4	ALPHA2	ALPHA3
0.0630	0.2470	223.90	0.0111	0.0657	0.0255	0.0203	0.0053
0.1800	0.5300	212.80	0.0214	0.0345	0.0355	0.0109	0.0016
0.3350	0.7320	198.90	0.0237	0.0034	0.0117	-0.0056	-0.0025
0.5000	0.8440	187.20	-0.0087	0.0086	0.0024	-0.0073	-0.0004
0.6490	0.9010	179.40	-0.0088	0.0241	0.0162	0.0003	0.0024
0.7920	0.9430	174.10	0.0239	0.0266	0.0259	0.0037	-0.0005
0.8830	0.9670	171.10	0.0170	0.0202	0.0220	0.0038	-0.0012
0.9620	0.9900	168.80	0.0053	0.0068	0.0079	0.0004	-0.0018
		MEAN	0.0150	0.0237	0.0184	0.0065	0.0020

NAPHTHALENE(1)-TETRADECANE(2)

HAYNES S.,VAN WINKLE M.:IND.ENG.CHEM.46,334(1954).

ANTOINE VAPOR PRESSURE CONSTANTS

	A	B	C
(1)	7.18400	1815.300	206.100
(2)	7.31430	1930.400	183.800

P = 10.00

EQUATION	ORDER	A12	A21	D	A122	A211
VAN LAAR	3	-0.0067	0.0329	-	-	-
MARGULES	3	0.0588	-0.2054	-	-	-
MARGULES	4	-0.0932	-0.4209	-0.5801	-	-
ALPHA	2	3.3768	-0.8432	-	-	-
ALPHA	3	18.8734	2.1914	-	-16.1259	-2.9749

			DEVIATION IN VAPOR PHASE COMPOSITION				
X	Y	TEMP.	LAAR 3	MARG 3	MARG 4	ALPHA2	ALPHA3
0.0600	0.2040	117.20	0.0223	0.0346	0.0016	0.0151	-0.0003
0.1650	0.4590	111.40	0.0295	0.0186	0.0126	0.0082	0.0008
0.2670	0.6220	106.70	0.0258	-0.0004	0.0075	-0.0017	-0.0012
0.3790	0.7380	102.20	0.0248	-0.0023	0.0015	-0.0027	0.0008
0.4710	0.8070	98.90	0.0216	0.0018	-0.0017	-0.0025	0.0007
0.5720	0.8660	95.60	0.0152	0.0068	-0.0009	-0.0028	-0.0013
0.6750	0.9090	93.10	0.0097	0.0134	0.0075	0.0010	0.0004
0.7670	0.9410	90.80	-0.0148	0.0146	0.0131	0.0023	0.0002
0.8810	0.9730	88.30	-0.0149	0.0104	0.0125	0.0028	-0.0001
		MEAN	0.0199	0.0114	0.0065	0.0043	0.0006

NAPHTHALENE(1)-TETRADECANE(2)

HAYNES S.,VAN WINKLE M.:IND.ENG.CHEM.46,334(1954).

ANTOINE VAPOR PRESSURE CONSTANTS
```
          A          B          C
(1) 6.84577    1606.529    187.227
(2) 6.99570    1725.460    165.750
```

P =   100.00

EQUATION	ORDER	A12	A21	D	A122	A211
VAN LAAR	3	-0.0023	0.0031	-	-	-
MARGULES	3	0.0940	-0.0866	-	-	-
MARGULES	4	0.0376	-0.1603	-0.2090	-	-
ALPHA	2	2.0371	-0.6422	-	-	-
ALPHA	3	11.3115	2.4234	-	-9.2869	-3.0292

			DEVIATION IN VAPOR PHASE COMPOSITION				
X	Y	TEMP.	LAAR 3	MARG 3	MARG 4	ALPHA2	ALPHA3
0.0830	0.2100	173.90	0.0042	0.0262	0.0168	0.0052	0.0046
0.1410	0.3310	171.10	0.0035	0.0192	0.0150	0.0007	0.0001
0.2270	0.4740	167.30	0.0049	0.0056	0.0079	-0.0045	-0.0049
0.3100	0.5730	164.40	0.0149	0.0046	0.0084	0.0014	0.0012
0.4260	0.6890	160.30	0.0143	0.0013	0.0021	0.0001	0.0003
0.5110	0.7550	157.80	-0.0060	0.0050	0.0029	0.0014	0.0018
0.6430	0.8430	154.20	-0.0049	0.0084	0.0048	-0.0017	-0.0013
0.7380	0.8910	151.70	0.0145	0.0140	0.0122	0.0005	0.0006
0.8220	0.9310	149.70	0.0109	0.0128	0.0132	-0.0008	-0.0010
0.9060	0.9640	147.70	0.0078	0.0106	0.0121	0.0009	0.0002
0.9430	0.9780	146.90	0.0055	0.0077	0.0090	0.0011	0.0003
		MEAN	0.0083	0.0105	0.0095	0.0016	0.0015

NAPHTHALENE(1)-TETRADECANE(2)

HAYNES S.,VAN WINKLE M.:IND.ENG.CHEM.46,334(1954).

ANTOINE VAPOR PRESSURE CONSTANTS

	A	B	C
(1)	6.84577	1606.529	187.227
(2)	6.99570	1725.460	165.750

P = 200.00

EQUATION	ORDER	A12	A21	D	A122	A211
VAN LAAR	3	0.0004	-0.0001	-	-	-
MARGULES	3	0.0864	-0.0110	-	-	-
MARGULES	4	0.0806	-0.0184	-0.0214	-	-
ALPHA	2	1.8612	-0.5819	-	-	-
ALPHA	3	93.6061	32.0590	-	-91.7787	-31.8542

			DEVIATION IN VAPOR PHASE COMPOSITION				
X	Y	TEMP.	LAAR 3	MARG 3	MARG 4	ALPHA2	ALPHA3
0.0560	0.1460	198.60	-0.0103	0.0085	0.0075	-0.0014	-0.0031
0.1100	0.2540	195.60	-0.0035	0.0170	0.0163	0.0057	0.0034
0.2050	0.4190	190.80	-0.0012	0.0092	0.0093	0.0016	-0.0003
0.3050	0.5510	186.40	0.0023	0.0028	0.0032	-0.0007	-0.0015
0.4120	0.6580	182.20	0.0112	0.0032	0.0033	0.0007	0.0013
0.4980	0.7310	179.40	0.0133	0.0025	0.0023	-0.0011	0.0002
0.5940	0.7990	176.70	0.0145	0.0041	0.0037	-0.0020	-0.0003
0.6970	0.8600	173.90	0.0146	0.0073	0.0070	-0.0018	-0.0005
0.8130	0.9180	171.40	0.0125	0.0095	0.0095	-0.0005	-0.0009
0.9100	0.9570	169.40	0.0122	0.0118	0.0120	0.0047	0.0016
0.9620	0.9800	168.30	0.0076	0.0077	0.0078	0.0040	-0.0006
		MEAN	0.0094	0.0076	0.0075	0.0022	0.0012

NAPHTHALENE(1)-TETRADECANE(2)

HAYNES S.,VAN WINKLE M.:IND.ENG.CHEM.46,334(1954).

ANTOINE VAPOR PRESSURE CONSTANTS
	A	B	C
(1)	6.84577	1606.529	187.227
(2)	6.99570	1725.460	165.750

P = 400.00

EQUATION	ORDER	A12	A21	D	A122	A211
VAN LAAR	3	0.0750	0.0582	-	-	-
MARGULES	3	0.0736	0.0571	-	-	-
MARGULES	4	0.0706	0.0541	-0.0103	-	-
ALPHA	2	1.6480	-0.5442	-	-	-
ALPHA	3	5.2093	0.8181	-	-3.5811	-1.3228

DEVIATION IN VAPOR PHASE COMPOSITION

X	Y	TEMP.	LAAR 3	MARG 3	MARG 4	ALPHA2	ALPHA3
0.0540	0.1300	223.30	0.0049	0.0047	0.0043	0.0008	-0.0000
0.1030	0.2310	220.60	0.0063	0.0061	0.0058	0.0006	-0.0005
0.2120	0.4100	215.60	0.0067	0.0067	0.0069	0.0014	0.0003
0.4220	0.6490	208.00	0.0024	0.0024	0.0026	0.0002	-0.0003
0.5280	0.7370	204.40	0.0013	0.0013	0.0012	-0.0008	-0.0002
0.6220	0.8030	201.70	0.0007	0.0007	0.0005	-0.0015	-0.0008
0.6820	0.8390	200.30	0.0020	0.0020	0.0018	-0.0004	0.0002
0.7240	0.8630	199.20	0.0025	0.0025	0.0024	-0.0000	0.0003
0.8150	0.9110	196.90	0.0035	0.0035	0.0035	0.0007	0.0004
0.8650	0.9360	195.60	0.0033	0.0034	0.0034	0.0007	-0.0000
0.9150	0.9600	195.00	0.0027	0.0027	0.0028	0.0007	-0.0003
0.9380	0.9710	194.40	0.0021	0.0022	0.0022	0.0005	-0.0006
0.9640	0.9820	193.90	0.0026	0.0026	0.0027	0.0015	0.0006
		MEAN	0.0032	0.0031	0.0031	0.0008	0.0004

413

NAPHTHALENE(1)-TETRADECANE(2)

HAYNES S.,VAN WINKLE M.:IND.ENG.CHEM.46,334(1954).

ANTOINE VAPOR PRESSURE CONSTANTS
	A	B	C
(1)	6.84577	1606.529	187.227
(2)	6.99570	1725.460	165.750

P = 760.00

EQUATION	ORDER	A12	A21	D	A122	A211
VAN LAAR	3	0.0651	0.1009	-	-	-
MARGULES	3	0.0614	0.0955	-	-	-
MARGULES	4	0.0684	0.1026	0.0235	-	-
ALPHA	2	1.4895	-0.4797	-	-	-
ALPHA	3	3.0661	0.0723	-	-1.6688	-0.4933

DEVIATION IN VAPOR PHASE COMPOSITION

X	Y	TEMP.	LAAR 3	MARG 3	MARG 4	ALPHA2	ALPHA3
0.0520	0.1160	249.40	0.0021	0.0015	0.0025	0.0035	0.0004
0.0950	0.2000	247.20	0.0038	0.0033	0.0042	0.0052	0.0014
0.2090	0.3890	241.90	0.0014	0.0014	0.0012	0.0011	-0.0013
0.2890	0.4920	238.60	0.0023	0.0026	0.0020	0.0008	0.0002
0.4640	0.6690	231.90	0.0009	0.0010	0.0009	-0.0021	-0.0001
0.5450	0.7330	229.40	0.0009	0.0008	0.0010	-0.0017	0.0003
0.6250	0.7900	226.90	-0.0001	-0.0004	-0.0000	-0.0023	-0.0008
0.6970	0.8330	224.70	0.0024	0.0021	0.0025	0.0008	0.0014
0.8220	0.9050	221.70	0.0015	0.0015	0.0015	0.0011	-0.0003
0.8830	0.9380	220.30	0.0010	0.0012	0.0009	0.0009	-0.0011
0.9190	0.9560	219.40	0.0019	0.0021	0.0018	0.0019	-0.0002
0.9520	0.9730	218.90	0.0020	0.0022	0.0019	0.0021	0.0003
0.9760	0.9850	218.30	0.0025	0.0026	0.0024	0.0025	0.0014
		MEAN	0.0018	0.0017	0.0018	0.0020	0.0007

NAPHTHALENE(1)-TETRADECANE(2)

WARD S.H.,VAN WINKLE M.:IND.ENG.CHEM.46,338(1954).

ANTOINE VAPOR PRESSURE CONSTANTS
	A	B	C
(1)	6.84577	1606.529	187.227
(2)	6.99570	1725.460	165.750

P = 200.00

EQUATION	ORDER	A12	A21	D	A122	A211
VAN LAAR	3	0.2283	0.1007	-	-	-
MARGULES	3	0.1960	0.0707	-	-	-
MARGULES	4	0.1465	0.0081	-0.1833	-	-
ALPHA	2	2.2925	-0.3781	-	-	-
ALPHA	3	1.8240	-0.6436	-	0.3562	0.2855

			DEVIATION IN VAPOR PHASE COMPOSITION				
X	Y	TEMP.	LAAR 3	MARG 3	MARG 4	ALPHA2	ALPHA3
0.0560	0.1520	199.70	0.0374	0.0323	0.0228	0.0089	0.0054
0.1350	0.3200	195.10	0.0350	0.0342	0.0302	0.0091	0.0059
0.2260	0.4770	190.30	0.0073	0.0093	0.0116	-0.0073	-0.0084
0.3600	0.6160	184.40	0.0066	0.0071	0.0101	-0.0015	0.0017
0.4530	0.6940	180.70	0.0054	0.0042	0.0042	-0.0015	0.0004
0.6220	0.7990	174.90	0.0156	0.0135	0.0100	0.0016	0.0023
0.7320	0.8600	172.40	0.0166	0.0160	0.0138	-0.0009	-0.0015
0.8370	0.9120	170.30	0.0172	0.0180	0.0188	-0.0003	-0.0017
0.9550	0.9720	168.30	0.0096	0.0105	0.0120	0.0014	0.0007
0.9600	0.9740	168.20	0.0097	0.0105	0.0120	0.0023	0.0016
0.9850	0.9900	167.90	0.0040	0.0043	0.0050	0.0008	0.0006
		MEAN	0.0150	0.0145	0.0137	0.0031	0.0027

ALPHA PINENE(1)-BETA PINENE(2)

TUCKER W.C.,HAWKINS J.E.:IND.ENG.CHEM.46,2387(1954).

ANTOINE VAPOR PRESSURE CONSTANTS
	A	B	C
(1)	6.91664	1482.214	211.416
(2)	6.85950	1489.740	208.225

P = 300.00

EQUATION	ORDER	A12	A21	D	A122	A211
VAN LAAR	3	-0.0052	0.0003	-	-	-
MARGULES	3	-0.0133	-0.0563	-	-	-
MARGULES	4	0.0145	-0.0276	0.0855	-	-
ALPHA	2	0.2185	-0.3113	-	-	-
ALPHA	3	-13.4709	-11.0050	-	13.7114	10.8966

			DEVIATION IN VAPOR PHASE COMPOSITION				
X	Y	TEMP.	LAAR 3	MARG 3	MARG 4	ALPHA2	ALPHA3
0.0460	0.0510	131.21	-0.0001	0.0068	0.0091	0.0048	0.0043
0.1480	0.1910	130.00	-0.0055	-0.0112	-0.0096	-0.0134	-0.0018
0.2530	0.2980	129.11	0.0106	0.0012	0.0002	0.0016	0.0075
0.3550	0.4210	128.22	-0.0003	-0.0102	-0.0118	-0.0072	-0.0050
0.4720	0.5190	127.13	0.0224	0.0155	0.0151	0.0199	0.0178
0.5290	0.6010	126.57	-0.0036	-0.0081	-0.0075	-0.0037	-0.0076
0.6590	0.7140	125.45	0.0048	0.0066	0.0082	0.0092	0.0036
0.7380	0.8110	124.74	-0.0225	-0.0174	-0.0165	-0.0165	-0.0207
0.8740	0.9150	123.89	-0.0132	-0.0065	-0.0079	-0.0077	0.0036
0.9530	0.9560	123.00	0.0081	0.0117	0.0103	0.0108	-0.0043
		MEAN	0.0091	0.0095	0.0096	0.0095	0.0076

ALPHA PINENE(1)-BETA PINENE(2)

TUCKER W.C.,HAWKINS J.E.:IND.ENG.CHEM.46,2387(1954).

ANTOINE VAPOR PRESSURE CONSTANTS
```
        A         B          C
(1) 6.91664   1482.214   211.416
(2) 6.85950   1489.740   208.225
```

P =   500.00

EQUATION	ORDER	A12	A21	D	A122	A211
VAN LAAR	3	0.0750	0.0184	-	-	-
MARGULES	3	0.0345	0.0188	-	-	-
MARGULES	4	0.1970	0.2022	0.5431	-	-
ALPHA	2	0.3809	-0.1952	-	-	-
ALPHA	3	-.1476/ 06	-.1134/ 06	-	.1453/ 06	.1126/ 06

DEVIATION IN VAPOR PHASE COMPOSITION

X	Y	TEMP.	LAAR 3	MARG 3	MARG 4	ALPHA2	ALPHA3
0.0480	0.0730	149.57	-0.0042	-0.0073	0.0090	-0.0081	0.0059
0.1290	0.1930	147.96	-0.0227	-0.0240	-0.0126	-0.0247	-0.0177
0.2450	0.2980	146.94	0.0033	0.0056	-0.0007	0.0070	0.0082
0.3570	0.4150	145.95	0.0042	0.0078	-0.0022	0.0113	0.0101
0.4430	0.5070	144.80	-0.0012	0.0021	-0.0012	0.0065	0.0049
0.5080	0.5750	144.38	-0.0059	-0.0031	0.0007	0.0014	0.0000
0.6630	0.7210	143.13	-0.0073	-0.0062	0.0060	-0.0027	-0.0033
0.7380	0.8040	142.19	-0.0233	-0.0228	-0.0153	-0.0205	-0.0210
0.8630	0.8780	141.47	0.0100	0.0100	0.0016	0.0105	0.0090
0.9510	0.9530	140.52	0.0076	0.0076	-0.0041	0.0074	0.0036
		MEAN	0.0090	0.0097	0.0054	0.0100	0.0084

ALPHA PINENE(1)-BETA PINENE(2)

TUCKER W.C.,HAWKINS J.E.:IND.ENG.CHEM.46,2387(1954).

ANTOINE VAPOR PRESSURE CONSTANTS
```
        A          B          C
(1) 6.91664    1482.214    211.416
(2) 6.85950    1489.740    208.225
```

P =   750.00

EQUATION	ORDER	A12	A21	D	A122	A211
VAN LAAR	3	0.0244	0.1366	-	-	-
MARGULES	3	0.0436	0.0470	-	-	-
MARGULES	4	0.2682	0.2932	0.6712	-	-
ALPHA	2	0.4413	-0.1647	-	-	-
ALPHA	3	.3796/ 05	.3036/ 05	-	-.3637/ 05	-.3044/ 0!

			DEVIATION IN VAPOR PHASE COMPOSITION				
X	Y	TEMP.	LAAR 3	MARG 3	MARG 4	ALPHA2	ALPHA3
0.0480	0.1170	164.63	-0.0526	-0.0504	-0.0255	-0.0497	-0.0136
0.2410	0.3030	163.16	-0.0017	-0.0007	-0.0027	0.0032	0.0072
0.3480	0.4100	161.60	0.0092	0.0065	-0.0031	0.0117	0.0089
0.4370	0.5120	160.60	-0.0012	-0.0071	-0.0110	-0.0012	-0.0058
0.5170	0.5850	160.08	0.0035	-0.0046	0.0004	0.0017	-0.0025
0.6570	0.7060	158.67	0.0090	0.0003	0.0121	0.0061	0.0050
0.7400	0.8010	157.73	-0.0161	-0.0227	-0.0177	-0.0178	-0.0165
0.8630	0.8810	156.84	0.0024	0.0023	-0.0125	0.0052	0.0088
		MEAN	0.0120	0.0118	0.0106	0.0121	0.0085

DECANE(1)-TRANS-DECALINE(2)

STRUCK R.T.,KINNEY C.R.:IND.ENG.CHEM.42,77(1950).

ANTOINE VAPOR PRESSURE CONSTANTS
```
            A          B          C
(1) 6.95367    1501.268    194.480
(2) 6.90464    1570.300    203.000
```

P =   50.00

EQUATION	ORDER	A12	A21	D	A122	A211
VAN LAAR	3	-0.0028	0.0035	-	-	-
MARGULES	3	0.1214	-0.1281	-	-	-
MARGULES	4	0.3774	0.0933	0.6343	-	-
ALPHA	2	0.2228	-0.1523	-	-	-
ALPHA	3	-.4890/ 06	-.4090/ 06	-	.4863/ 06	.4087/ 06

			DEVIATION IN VAPOR PHASE COMPOSITION				
X	Y	TEMP.	LAAR 3	MARG 3	MARG 4	ALPHA2	ALPHA3
0.2560	0.2950	95.35	0.0203	0.0136	0.0117	-0.0005	-0.0004
0.2570	0.2950	95.35	0.0214	0.0146	0.0125	0.0006	0.0007
0.4690	0.5160	94.00	-0.0030	-0.0066	-0.0142	-0.0006	-0.0011
0.6370	0.6760	93.00	-0.0001	0.0047	0.0088	0.0015	0.0015
0.6370	0.6780	93.00	-0.0021	0.0027	0.0068	-0.0005	-0.0005
0.8440	0.8660	91.95	0.0124	0.0207	0.0097	-0.0007	-0.0003
		MEAN	0.0099	0.0105	0.0106	0.0007	0.0007

DECANE(1)-TRANS-DECALINE(2)

STRUCK R.T.,KINNEY C.R.:IND.ENG.CHEM.42,77(1950).

ANTOINE VAPOR PRESSURE CONSTANTS
	A	B	C
(1)	7.33883	1719.860	213.800
(2)	7.24657	1774.400	221.000

P = 20.00

EQUATION	ORDER	A12	A21	D	A122	A211
VAN LAAR	3	-0.0095	0.0144	-	-	-
MARGULES	3	0.1644	-0.1545	-	-	-
MARGULES	4	0.6191	0.3370	1.3117	-	-
ALPHA	2	0.1476	-0.1270	-	-	-
ALPHA	3	-.1440/ 04	-.1239/ 04	-	.1448/ 04	.1233/ 0

DEVIATION IN VAPOR PHASE COMPOSITION

X	Y	TEMP.	LAAR 3	MARG 3	MARG 4	ALPHA2	ALPHA3
0.2580	0.2850	72.90	0.0261	0.0242	0.0186	0.0001	-0.0007
0.2620	0.2880	72.90	0.0275	0.0247	0.0181	0.0014	0.0006
0.4600	0.4940	72.03	-0.0011	-0.0037	-0.0067	0.0001	0.0010
0.4660	0.5020	72.00	-0.0066	-0.0060	-0.0078	-0.0018	-0.0009
0.7330	0.7590	70.90	-0.0068	0.0166	0.0253	-0.0002	-0.0001
0.8420	0.8580	70.45	0.0123	0.0268	0.0043	0.0013	0.0001
		MEAN	0.0134	0.0170	0.0135	0.0008	0.0006

DECANE(1)-TRANS-DECALINE(2)

STRUCK R.T.,KINNEY C.R.:IND.ENG.CHEM.42,77(1950).

ANTOINE VAPOR PRESSURE CONSTANTS
          A   B   C
(1) 7.33883   1719.860   213.800
(2) 7.24657   1774.400   221.000

P =   10.00

EQUATION	ORDER	A12	A21	D	A122	A211
VAN LAAR	3	-0.0131	0.0079	-	-	-
MARGULES	3	0.2191	-0.1157	-	-	-
MARGULES	4	0.5334	0.1682	0.8226	-	-
ALPHA	2	0.1125	-0.0984	-	-	-
ALPHA	3	7.2784	6.2459	-	-7.2207	-6.3088

			DEVIATION IN VAPOR PHASE COMPOSITION				
X	Y	TEMP.	LAAR 3	MARG 3	MARG 4	ALPHA2	ALPHA3
0.2620	0.2820	58.15	-0.0052	0.0380	0.0322	0.0010	0.0001
0.4970	0.5230	57.30	-0.0041	-0.0026	-0.0073	0.0003	0.0008
0.5000	0.5280	57.30	-0.0038	-0.0048	-0.0091	-0.0017	-0.0012
0.6840	0.7050	56.65	0.0295	0.0057	0.0134	0.0012	0.0014
0.7320	0.7530	56.50	0.0255	0.0094	0.0129	-0.0010	-0.0011
0.8740	0.8840	56.10	0.0162	0.0218	0.0041	0.0010	0.0001
		MEAN	0.0141	0.0137	0.0132	0.0010	0.0008

DODECANE(1)-1-HEXADECENE(2)

KEISTLER G.R.,VAN WINKLE M.:IND.ENG.CHEM.44,622(1952).

ANTOINE VAPOR PRESSURE CONSTANTS
```
          A          B          C
(1) 6.98059    1625.928    180.311
(2) 6.93600    1755.200    148.000
```

P =   760.00

EQUATION	ORDER	A12	A21	D	A122	A211
VAN LAAR	3	-0.0031	0.0059	-	-	-
MARGULES	3	0.0311	-0.0861	-	-	-
MARGULES	4	-0.0270	-0.1895	-0.2739	-	-
ALPHA	2	3.2382	-0.8815	-	-	-
ALPHA	3	-.3153/ 05	-.7078/ 04	-	.3177/ 05	.7456/ 0

			DEVIATION IN VAPOR PHASE COMPOSITION				
X	Y	TEMP.	LAAR 3	MARG 3	MARG 4	ALPHA2	ALPHA3
0.1020	0.3260	271.10	-0.0010	0.0030	-0.0041	0.0018	-0.0034
0.1400	0.4070	267.20	0.0056	0.0057	0.0033	0.0058	0.0043
0.2870	0.6430	252.30	0.0078	-0.0026	0.0017	-0.0023	-0.0002
0.4030	0.7580	243.20	0.0069	-0.0024	-0.0025	-0.0036	-0.0027
0.5610	0.8560	234.00	-0.0015	0.0063	0.0015	0.0036	0.0027
0.7770	0.9510	222.80	-0.0039	0.0012	0.0003	-0.0009	-0.0009
		MEAN	0.0044	0.0036	0.0022	0.0030	0.0024

DODECANE(1)-1-HEXADECENE(2)

KEISTLER G.R.,VAN WINKLE M.:IND.ENG.CHEM.44,622(1952).

ANTOINE VAPOR PRESSURE CONSTANTS
	A	B	C
(1)	6.98059	1625.928	180.311
(2)	6.93600	1755.200	148.000

P = 200.00

EQUATION	ORDER	A12	A21	D	A122	A211
VAN LAAR	3	0.0820	0.0320	-	-	-
MARGULES	3	0.0657	0.0202	-	-	-
MARGULES	4	0.1216	0.1267	0.2711	-	-
ALPHA	2	5.5097	-0.9768	-	-	-
ALPHA	3	-2.8829	-2.1824	-	8.5139	1.2881

			DEVIATION IN VAPOR PHASE COMPOSITION				
X	Y	TEMP.	LAAR 3	MARG 3	MARG 4	ALPHA2	ALPHA3
0.0970	0.4130	215.20	-0.0006	-0.0026	0.0059	0.0019	0.0044
0.1960	0.6270	204.00	-0.0065	-0.0059	-0.0081	-0.0062	-0.0056
0.3350	0.7720	193.90	0.0045	0.0050	0.0032	0.0054	0.0043
0.4800	0.8720	185.10	-0.0032	-0.0035	-0.0008	-0.0016	-0.0022
0.6660	0.9430	176.80	-0.0056	-0.0059	-0.0030	-0.0011	0.0005
0.8310	0.9800	170.70	-0.0054	-0.0052	-0.0057	0.0006	-0.0000
		MEAN	0.0043	0.0047	0.0044	0.0028	0.0028

DODECANE(1)-1-HEXADECENE(2)

KEISTLER G.R.,VAN WINKLE M.:IND.ENG.CHEM.44,622(1952).

ANTOINE VAPOR PRESSURE CONSTANTS
	A	B	C
(1)	6.98059	1625.928	180.311
(2)	6.93600	1755.200	148.000

P =   50.00

EQUATION	ORDER	A12	A21	D	A122	A211
VAN LAAR	3	0.0023	-0.0004	-	-	-
MARGULES	3	0.0835	-0.0423	-	-	-
MARGULES	4	0.0904	-0.0283	0.0352	-	-
ALPHA	2	8.0490	-1.0341	-	-	-
ALPHA	3	.2109/ 06	.2313/ 05	-	-.2109/ 06	-.2580/ 0

			DEVIATION IN VAPOR PHASE COMPOSITION				
X	Y	TEMP.	LAAR 3	MARG 3	MARG 4	ALPHA2	ALPHA3
0.1080	0.5280	170.60	-0.0003	-0.0016	-0.0007	-0.0008	0.0007
0.1820	0.6760	162.90	-0.0005	0.0002	0.0000	-0.0008	-0.0002
0.3180	0.8190	153.30	0.0036	-0.0007	-0.0010	-0.0005	-0.0010
0.4940	0.9080	144.60	0.0065	0.0006	0.0009	0.0021	0.0013
0.6390	0.9530	137.90	0.0016	-0.0015	-0.0012	0.0002	-0.0000
0.8120	0.9880	132.70	-0.0059	-0.0062	-0.0062	-0.0026	-0.0006
		MEAN	0.0031	0.0018	0.0017	0.0012	0.0006

DODECANE(1)-1-HEXADECENE(2)

KEISTLER G.R.,VAN WINKLE M.:IND.ENG.CHEM.44,622(1952).

ANTOINE VAPOR PRESSURE CONSTANTS
	A	B	C
(1)	6.98059	1625.928	180.311
(2)	6.93600	1755.200	148.000

P =   20.00

EQUATION	ORDER	A12	A21	D	A122	A211
VAN LAAR	3	-0.0033	0.0088	-	-	-
MARGULES	3	0.0945	-0.2423	-	-	-
MARGULES	4	0.0861	-0.2618	-0.0483	-	-
ALPHA	2	9.2751	-1.0293	-	-	-
ALPHA	3	183.8465	16.6710	-	-172.7101	-19.9805

			DEVIATION IN VAPOR PHASE COMPOSITION				
X	Y	TEMP.	LAAR 3	MARG 3	MARG 4	ALPHA2	ALPHA3
0.0800	0.4850	151.30	-0.0186	0.0018	0.0003	-0.0104	0.0007
0.1660	0.6810	142.40	0.0079	-0.0005	-0.0002	-0.0039	-0.0019
0.3080	0.8240	132.90	0.0235	0.0015	0.0019	0.0046	0.0030
0.3880	0.8750	128.90	0.0177	-0.0015	-0.0016	0.0007	-0.0009
0.4700	0.9100	125.30	0.0138	0.0004	-0.0000	0.0004	-0.0008
0.6430	0.9610	118.40	-0.0060	0.0010	0.0006	-0.0027	-0.0023
0.7630	0.9800	113.80	-0.0062	0.0026	0.0026	-0.0004	0.0017
		MEAN	0.0134	0.0013	0.0010	0.0033	0.0016

425

DODECANE(1)-1-OCTADECENE(2)

JORDAN B.T.,VAN WINKLE M.:IND.ENG.CHEM.43,2908(1951).

ANTOINE VAPOR PRESSURE CONSTANTS

	A	B	C
(1)	6.98059	1625.928	180.311
(2)	6.90100	1789.400	131.000

P = 200.00

EQUATION	ORDER	A12	A21	D	A122	A211
VAN LAAR	3	-0.0526	-0.1683	-	-	-
MARGULES	3	-0.0482	-0.1062	-	-	-
MARGULES	4	-0.0667	-0.1663	-0.1305	-	-
ALPHA	2	9.2837	-1.2642	-	-	-
ALPHA	3	29.3347	-0.4653	-	-22.7300	-0.6301

			DEVIATION IN VAPOR PHASE COMPOSITION				
X	Y	TEMP.	LAAR 3	MARG 3	MARG 4	ALPHA2	ALPHA3
0.0550	0.3400	250.40	-0.0007	0.0003	-0.0042	0.0394	-0.0049
0.0930	0.5180	242.20	-0.0251	-0.0246	-0.0269	0.0046	-0.0282
0.1120	0.5230	240.70	0.0252	0.0254	0.0242	0.0529	0.0275
0.2030	0.7210	225.40	0.0191	0.0188	0.0198	0.0211	0.0212
0.2230	0.7680	221.70	0.0014	0.0011	0.0021	-0.0015	0.0017
0.2360	0.8200	218.30	-0.0318	-0.0320	-0.0311	-0.0392	-0.0344
0.3300	0.8600	210.60	0.0071	0.0073	0.0071	-0.0008	0.0087
0.3940	0.9010	202.30	0.0043	0.0048	0.0040	-0.0054	0.0032
0.4600	0.9350	198.60	-0.0051	-0.0044	-0.0056	-0.0105	-0.0048
0.5850	0.9620	188.10	0.0008	0.0016	0.0005	0.0013	0.0002
0.7150	0.9820	179.70	0.0000	0.0005	0.0001	0.0076	0.0000
		MEAN	0.0110	0.0110	0.0114	0.0168	0.0122

DODECANE(1)-1-OCTADECENE(2)

JORDAN B.T.,VAN WINKLE M.:IND.ENG.CHEM.43,2908(1951).

ANTOINE VAPOR PRESSURE CONSTANTS
	A	B	C
(1)	6.98059	1625.928	180.311
(2)	6.90100	1789.400	131.000

P = 400.00

EQUATION	ORDER	A12	A21	D	A122	A211
VAN LAAR	3	-0.0589	-0.0569	-	-	-
MARGULES	3	-0.0591	-0.0566	-	-	-
MARGULES	4	-0.0314	0.0262	0.1880	-	-
ALPHA	2	7.7596	-1.2162	-	-	-
ALPHA	3	11.6986	-1.4994	-	-5.7271	0.5816

			DEVIATION IN VAPOR PHASE COMPOSITION				
X	Y	TEMP.	LAAR 3	MARG 3	MARG 4	ALPHA2	ALPHA3
0.0340	0.2090	283.30	-0.0143	-0.0144	-0.0078	0.0288	-0.0040
0.0560	0.3430	277.30	-0.0439	-0.0439	-0.0378	0.0034	-0.0334
0.1130	0.4860	259.10	0.0292	0.0291	0.0306	0.0519	0.0249
0.1970	0.6960	249.10	-0.0021	-0.0021	-0.0040	0.0040	-0.0027
0.2500	0.7740	241.80	-0.0053	-0.0053	-0.0070	-0.0084	-0.0065
0.3400	0.8470	232.30	0.0044	0.0044	0.0045	-0.0038	0.0049
0.4050	0.8880	226.90	0.0022	0.0022	0.0035	-0.0052	0.0042
0.4260	0.9100	221.80	-0.0067	-0.0067	-0.0052	-0.0164	-0.0074
0.6400	0.9640	208.70	0.0005	0.0005	0.0020	0.0042	0.0021
0.7410	0.9820	201.80	-0.0022	-0.0022	-0.0016	0.0067	-0.0016
		MEAN	0.0111	0.0111	0.0104	0.0133	0.0092

DODECANE(1)-1-OCTADECENE(2)

JORDAN B.T.,VAN WINKLE M.:IND.ENG.CHEM.43,2908(1951).

ANTOINE VAPOR PRESSURE CONSTANTS
	A	B	C
(1)	6.98059	1625.928	180.311
(2)	6.90100	1789.400	131.000

P = 760.00

EQUATION	ORDER	A12	A21	D	A122	A211
VAN LAAR	3	-0.0751	-0.0635	-	-	-
MARGULES	3	-0.0783	-0.0578	-	-	-
MARGULES	4	0.0300	0.2178	0.6587	-	-
ALPHA	2	5.6133	-1.2772	-	-	-
ALPHA	3	-136.4168	-21.1106	-	143.3638	25.9716

			DEVIATION IN VAPOR PHASE COMPOSITION				
X	Y	TEMP.	LAAR 3	MARG 3	MARG 4	ALPHA2	ALPHA3
0.0370	0.2610	303.80	-0.0772	-0.0781	-0.0533	-0.0557	0.0195
0.0820	0.3800	296.90	-0.0230	-0.0239	-0.0070	0.0004	-0.0259
0.1040	0.4330	294.70	-0.0086	-0.0093	0.0008	0.0136	0.0047
0.1500	0.5210	286.90	0.0241	0.0238	0.0222	0.0365	0.0382
0.2180	0.6630	277.40	0.0110	0.0112	0.0028	0.0126	0.0165
0.2830	0.7640	267.20	-0.0010	-0.0005	-0.0066	-0.0076	-0.0049
0.3350	0.8180	261.10	-0.0039	-0.0034	-0.0052	-0.0115	-0.0103
0.4270	0.8910	249.80	-0.0115	-0.0109	-0.0061	-0.0174	-0.0185
0.4480	0.8900	249.20	-0.0006	-0.0001	0.0058	-0.0040	-0.0055
0.5440	0.9290	241.50	-0.0004	-0.0001	0.0075	0.0040	0.0015
0.6920	0.9730	231.70	-0.0076	-0.0075	-0.0033	0.0084	0.0083
		MEAN	0.0153	0.0153	0.0110	0.0156	0.0140

TETRADECANE(1)-1-HEXADECENE(2)

RASMUSSEN R.R.,VAN WINKLE M.:IND.ENG.CHEM.42,2121(1950).

ANTOINE VAPOR PRESSURE CONSTANTS

	A	B	C
(1)	6.99570	1725.460	165.750
(2)	6.93600	1755.200	148.000

P =   10.00

EQUATION	ORDER	A12	A21	D	A122	A211
VAN LAAR	3	0.0101	-0.3496	-	-	-
MARGULES	3	-0.0074	0.0356	-	-	-
MARGULES	4	-0.0082	0.0343	-0.0034	-	-
ALPHA	2	2.2841	-0.7836	-	-	-
ALPHA	3	43.7323	10.9805	-	-42.0376	-11.9768

			DEVIATION IN VAPOR PHASE COMPOSITION				
X	Y	TEMP.	LAAR 3	MARG 3	MARG 4	ALPHA2	ALPHA3
0.0700	0.1910	143.80	0.0052	0.0012	0.0010	0.0082	-0.0015
0.1430	0.3500	141.10	0.0049	0.0018	0.0017	0.0072	0.0030
0.1600	0.3830	140.60	0.0034	0.0008	0.0008	0.0056	0.0026
0.2070	0.4700	138.90	-0.0041	-0.0054	-0.0054	-0.0031	-0.0035
0.2390	0.5170	137.90	-0.0033	-0.0038	-0.0038	-0.0029	-0.0021
0.3420	0.6390	134.70	0.0020	0.0028	0.0028	0.0009	0.0030
0.3580	0.6630	134.00	-0.0049	-0.0041	-0.0040	-0.0065	-0.0044
0.3990	0.6950	133.10	0.0025	0.0033	0.0033	0.0012	0.0031
0.4630	0.7520	131.50	-0.0002	0.0001	0.0001	-0.0012	-0.0001
0.5830	0.8320	127.80	0.0026	0.0013	0.0012	0.0020	0.0015
0.6950	0.8960	125.90	-0.0027	-0.0056	-0.0057	-0.0014	-0.0030
0.8150	0.9430	123.60	0.0006	-0.0034	-0.0034	0.0025	0.0012
		MEAN	0.0030	0.0028	0.0028	0.0036	0.0024

TETRADECANE(1)-1-HEXADECENE(2)

RASMUSSEN R.R.,VAN WINKLE M.:IND.ENG.CHEM.42,2121(1950).

ANTOINE VAPOR PRESSURE CONSTANTS
         A   B   C
(1) 6.99570   1725.460   165.750
(2) 6.93600   1755.200   148.000

P =   20.00

EQUATION	ORDER	A12	A21	D	A122	A211
VAN LAAR	3	-0.0001	0.0000	-	-	-
MARGULES	3	-0.0206	0.0047	-	-	-
MARGULES	4	-0.0533	-0.0442	-0.1295	-	-
ALPHA	2	1.9000	-0.7460	-	-	-
ALPHA	3	256.3776	80.0091	-	-258.6365	-81.7410

			DEVIATION IN VAPOR PHASE COMPOSITION				
X	Y	TEMP.	LAAR 3	MARG 3	MARG 4	ALPHA2	ALPHA3
0.1190	0.2720	157.90	0.0105	0.0063	0.0029	0.0119	0.0017
0.1430	0.3260	157.10	-0.0012	-0.0016	-0.0038	0.0033	-0.0032
0.1850	0.3990	155.70	-0.0006	-0.0005	-0.0007	0.0026	0.0008
0.2850	0.5430	152.40	0.0026	0.0024	0.0043	0.0012	0.0045
0.3180	0.5910	151.30	-0.0053	-0.0047	-0.0030	-0.0071	-0.0034
0.4310	0.7030	148.20	-0.0026	-0.0002	-0.0006	-0.0039	-0.0010
0.5430	0.7890	145.00	-0.0001	0.0025	0.0002	-0.0004	0.0003
0.7100	0.8870	141.30	0.0003	0.0017	0.0001	0.0028	0.0007
0.8480	0.9500	138.30	-0.0019	-0.0016	-0.0008	0.0014	-0.0004
		MEAN	0.0028	0.0024	0.0018	0.0038	0.0018

TETRADECANE(1)-1-HEXADECENE(2)

RASMUSSEN R.R.,VAN WINKLE M.:IND.ENG.CHEM.42,2121(1950).

ANTOINE VAPOR PRESSURE CONSTANTS

	A	B	C
(1)	6.99570	1725.460	165.750
(2)	6.93600	1755.200	148.000

P = 50.00

EQUATION	ORDER	A12	A21	D	A122	A211
VAN LAAR	3	-0.0007	0.0001	-	-	-
MARGULES	3	-0.0347	0.0040	-	-	-
MARGULES	4	-0.0161	0.0300	0.0729	-	-
ALPHA	2	1.5447	-0.7321	-	-	-
ALPHA	3	5.8359	0.7237	-	-4.4283	-1.4632

			DEVIATION IN VAPOR PHASE COMPOSITION				
X	Y	TEMP.	LAAR 3	MARG 3	MARG 4	ALPHA2	ALPHA3
0.0790	0.1750	183.30	0.0035	-0.0009	0.0017	0.0057	0.0015
0.0860	0.1910	182.80	-0.0015	-0.0026	-0.0001	0.0040	-0.0002
0.1350	0.2870	181.30	-0.0005	-0.0049	-0.0036	0.0010	-0.0021
0.2150	0.4140	178.70	0.0060	0.0016	0.0009	0.0043	0.0039
0.2550	0.4790	177.20	-0.0021	-0.0047	-0.0059	-0.0039	-0.0031
0.4140	0.6630	172.20	-0.0048	-0.0015	-0.0018	-0.0041	-0.0017
0.4790	0.7180	170.20	-0.0013	0.0029	0.0035	0.0007	0.0028
0.5910	0.8070	167.20	-0.0057	-0.0015	-0.0000	-0.0009	-0.0004
0.6950	0.8720	164.60	-0.0064	-0.0034	-0.0022	0.0002	-0.0006
0.7670	0.9110	162.90	-0.0073	-0.0054	-0.0049	-0.0002	-0.0015
0.8070	0.9270	162.20	-0.0041	-0.0028	-0.0028	0.0029	0.0015
		MEAN	0.0039	0.0029	0.0025	0.0026	0.0018

431

TETRADECANE(1)-1-HEXADECENE(2)

RASMUSSEN R.R.,VAN WINKLE M.:IND.ENG.CHEM.42,2121(1950).

ANTOINE VAPOR PRESSURE CONSTANTS
	A	B	C
(1)	6.99570	1725.460	165.750
(2)	6.93600	1755.200	148.000

P = 100.00

EQUATION	ORDER	A12	A21	D	A122	A211
VAN LAAR	3	-0.0222	-0.0547	-	-	-
MARGULES	3	-0.0186	-0.0420	-	-	-
MARGULES	4	-0.0270	-0.0516	-0.0264	-	-
ALPHA	2	1.2961	-0.6658	-	-	-
ALPHA	3	0.5387	-1.1014	-	0.6573	0.4996

			DEVIATION IN VAPOR PHASE COMPOSITION				
X	Y	TEMP.	LAAR 3	MARG 3	MARG 4	ALPHA2	ALPHA3
0.0780	0.1590	203.70	0.0031	0.0037	0.0024	0.0048	0.0004
0.1430	0.2790	200.80	-0.0000	0.0003	-0.0004	0.0012	-0.0034
0.2230	0.4000	198.30	0.0029	0.0029	0.0029	0.0035	0.0007
0.2710	0.4640	196.70	0.0047	0.0046	0.0049	0.0047	0.0033
0.3260	0.5350	195.00	0.0020	0.0019	0.0023	0.0015	0.0016
0.3660	0.5830	193.70	-0.0004	-0.0004	-0.0001	-0.0013	-0.0003
0.4230	0.6470	191.80	-0.0047	-0.0046	-0.0045	-0.0062	-0.0043
0.4950	0.7110	190.00	-0.0020	-0.0016	-0.0017	-0.0035	-0.0013
0.5430	0.7510	188.50	-0.0017	-0.0011	-0.0014	-0.0033	-0.0014
0.5600	0.7600	188.30	0.0026	0.0032	0.0028	0.0011	0.0028
0.5910	0.7840	187.20	0.0023	0.0029	0.0026	0.0007	0.0019
0.6870	0.8480	185.00	0.0035	0.0041	0.0038	0.0024	0.0016
0.7750	0.9030	182.90	-0.0004	-0.0001	-0.0002	-0.0011	-0.0039
0.8870	0.9510	180.60	0.0057	0.0055	0.0058	0.0055	0.0018
		MEAN	0.0026	0.0026	0.0026	0.0029	0.0021

TETRADECANE(1)-1-HEXADECENE(2)

RASMUSSEN R.R.,VAN WINKLE M.:IND.ENG.CHEM.42,2121(1950).

ANTOINE VAPOR PRESSURE CONSTANTS
```
           A          B          C
(1) 6.99570    1725.460    165.750
(2) 6.93600    1755.200    148.000
```

P =   200.00

EQUATION	ORDER	A12	A21	D	A122	A211
VAN LAAR	3	-0.0009	0.0036	-	-	-
MARGULES	3	0.0182	-0.0836	-	-	-
MARGULES	4	-0.0008	-0.1071	-0.0642	-	-
ALPHA	2	1.0995	-0.5804	-	-	-
ALPHA	3	.3407/ 05	.1563/ 05	-	-.3425/ 05	-.1575/ 05

			DEVIATION IN VAPOR PHASE COMPOSITION				
X	Y	TEMP.	LAAR 3	MARG 3	MARG 4	ALPHA2	ALPHA3
0.0700	0.1350	224.40	0.0090	0.0101	0.0074	0.0019	-0.0026
0.1430	0.2620	222.20	0.0110	0.0070	0.0055	-0.0009	-0.0013
0.2300	0.3900	219.40	0.0140	0.0031	0.0035	-0.0010	0.0005
0.3030	0.4790	217.20	0.0193	0.0048	0.0058	0.0033	0.0052
0.3740	0.5590	215.20	0.0197	0.0042	0.0050	0.0040	0.0055
0.4790	0.6710	212.40	0.0096	-0.0029	-0.0033	-0.0040	-0.0037
0.5170	0.7030	211.30	0.0103	-0.0002	-0.0009	-0.0023	-0.0025
0.6390	0.7990	208.70	0.0056	0.0039	0.0027	-0.0018	-0.0032
0.7350	0.8630	206.70	-0.0039	0.0069	0.0063	-0.0010	-0.0027
0.8400	0.9180	204.40	-0.0021	0.0132	0.0136	0.0054	0.0044
		MEAN	0.0105	0.0056	0.0054	0.0026	0.0032

TETRADECANE(1)-1-HEXADECENE(2)

RASMUSSEN R.R.,VAN WINKLE M.:IND.ENG.CHEM.42,2121(1950).

ANTOINE VAPOR PRESSURE CONSTANTS
```
        A         B          C
(1) 6.99570   1725.460   165.750
(2) 6.93600   1755.200   148.000
```

P =   400.00

EQUATION	ORDER	A12	A21	D	A122	A211
VAN LAAR	3	-0.0011	0.0024	-	-	-
MARGULES	3	0.0508	-0.0906	-	-	-
MARGULES	4	-0.0753	-0.2300	-0.4280	-	-
ALPHA	2	0.9361	-0.5014	-	-	-
ALPHA	3	-0.6547	-1.4346	-	1.5041	0.9839

			DEVIATION IN VAPOR PHASE COMPOSITION				
X	Y	TEMP.	LAAR 3	MARG 3	MARG 4	ALPHA2	ALPHA3
0.0620	0.1110	248.50	0.0103	0.0178	0.0012	0.0026	-0.0011
0.1270	0.2150	245.70	0.0193	0.0236	0.0128	0.0052	0.0006
0.1750	0.2870	245.00	0.0219	0.0209	0.0184	0.0048	0.0007
0.3350	0.4960	240.30	0.0216	0.0054	0.0151	-0.0004	-0.0001
0.4630	0.6290	238.10	0.0186	0.0017	0.0032	-0.0011	0.0015
0.5030	0.6710	235.90	0.0127	-0.0019	-0.0039	-0.0062	-0.0035
0.5430	0.7030	235.00	0.0131	0.0023	-0.0026	-0.0032	-0.0007
0.6290	0.7670	232.90	-0.0077	0.0114	0.0034	0.0024	0.0035
0.7280	0.8400	230.60	-0.0064	0.0136	0.0083	0.0010	-0.0004
0.8000	0.8860	229.40	0.0097	0.0153	0.0149	0.0020	-0.0008
0.8390	0.9090	228.60	0.0099	0.0156	0.0177	0.0029	-0.0003
		MEAN	0.0137	0.0118	0.0092	0.0029	0.0012

TETRADECANE(1)-1-HEXADECENE(2)

RASMUSSEN R.R.,VAN WINKLE M.:IND.ENG.CHEM.42,2121(1950).

ANTOINE VAPOR PRESSURE CONSTANTS
	A	B	C
(1)	6.99570	1725.460	165.750
(2)	6.93600	1755.200	148.000

P =   760.00

EQUATION	ORDER	A12	A21	D	A122	A211
VAN LAAR	3	-0.0063	0.0052	-	-	-
MARGULES	3	0.1231	-0.1506	-	-	-
MARGULES	4	0.0663	-0.2105	-0.1653	-	-
ALPHA	2	0.7809	-0.4185	-	-	-
ALPHA	3	-.2416/ 05	-.1360/ 05	-	.2416/ 05	.1349/ 05

			DEVIATION IN VAPOR PHASE COMPOSITION				
X	Y	TEMP.	LAAR 3	MARG 3	MARG 4	ALPHA2	ALPHA3
0.0950	0.1590	271.80	0.0105	0.0321	0.0239	-0.0018	-0.0017
0.2070	0.3180	268.50	0.0171	0.0216	0.0204	-0.0016	-0.0015
0.3030	0.4310	265.70	0.0152	0.0114	0.0137	0.0036	0.0040
0.3580	0.4940	264.40	-0.0139	0.0032	0.0057	0.0022	0.0026
0.5430	0.6790	260.00	-0.0131	-0.0037	-0.0051	-0.0030	-0.0026
0.6220	0.7430	258.30	0.0169	0.0057	0.0034	-0.0006	-0.0004
0.6790	0.7900	257.20	0.0169	0.0092	0.0072	-0.0030	-0.0031
0.7510	0.8390	255.80	0.0193	0.0195	0.0189	0.0011	0.0004
0.8630	0.9120	253.90	0.0156	0.0239	0.0259	0.0041	0.0021
		MEAN	0.0154	0.0145	0.0138	0.0023	0.0020

TETRADECANE(1)-1-HEXADECENE(2)

WARD S.H.,VAN WINKLE M.:IND.ENG.CHEM.46,338(1954).

ANTOINE VAPOR PRESSURE CONSTANTS

	A	B	C
(1)	6.99570	1725.460	165.750
(2)	6.93600	1755.200	148.000

P = 200.00

EQUATION	ORDER	A12	A21	D	A122	A211
VAN LAAR	3	-0.0584	-0.0384	-	-	-
MARGULES	3	-0.0566	-0.0342	-	-	-
MARGULES	4	-0.0335	-0.0077	0.0773	-	-
ALPHA	2	1.0587	-0.6803	-	-	-
ALPHA	3	0.1095	-1.1848	-	0.8945	0.5582

			DEVIATION IN VAPOR PHASE COMPOSITION				
X	Y	TEMP.	LAAR 3	MARG 3	MARG 4	ALPHA2	ALPHA3
0.0580	0.1140	229.00	-0.0055	-0.0053	-0.0023	-0.0005	-0.0028
0.1520	0.2730	226.20	-0.0034	-0.0033	-0.0020	0.0020	-0.0010
0.2840	0.4580	222.20	0.0043	0.0044	0.0030	0.0056	0.0045
0.4110	0.6160	218.30	-0.0035	-0.0032	-0.0040	-0.0047	-0.0037
0.5020	0.7010	215.50	0.0004	0.0008	0.0013	-0.0006	0.0011
0.5610	0.7550	213.60	-0.0030	-0.0026	-0.0016	-0.0033	-0.0018
0.6580	0.8240	210.30	0.0004	0.0007	0.0019	0.0019	0.0022
0.8090	0.9170	207.00	-0.0024	-0.0025	-0.0026	0.0019	-0.0005
0.9470	0.9800	204.10	-0.0011	-0.0012	-0.0020	0.0015	-0.0006
		MEAN	0.0027	0.0027	0.0023	0.0024	0.0020

CARBON DISULPHIDE(1)-ACETONE(2)

HIRSCHBERG J.:BULL.SOC.CHIM.BELG.41,163(1932).

ANTOINE VAPOR PRESSURE CONSTANTS
```
        A          B          C
(1) 6.85145   1122.500   236.460
(2) 7.23967   1279.870   237.500
```

T =   29.20

EQUATION	ORDER	A12	A21	D	A122	A211
VAN LAAR	3	0.5587	0.6220	-	-	-
MARGULES	3	0.5564	0.6197	-	-	-
MARGULES	4	0.5521	0.6151	-0.0133	-	-
ALPHA	2	5.1310	2.5742	-	-	-
ALPHA	3	3.5132	0.4826	-	0.0049	1.9839

			DEVIATION IN VAPOR PHASE COMPOSITION				
X	Y	PRESS.	LAAR 3	MARG 3	MARG 4	ALPHA2	ALPHA3
0.1350	0.3650	406.50	0.0214	0.0210	0.0205	0.0215	-0.0008
0.2950	0.5250	490.40	0.0070	0.0072	0.0074	-0.0016	0.0004
0.3400	0.5500	502.00	0.0051	0.0053	0.0055	-0.0035	0.0011
0.4600	0.6000	523.50	0.0011	0.0012	0.0013	-0.0047	0.0006
0.5000	0.6150	526.40	-0.0012	-0.0012	-0.0012	-0.0058	-0.0015
0.5680	0.6350	530.80	-0.0006	-0.0007	-0.0009	-0.0030	-0.0013
0.6700	0.6650	534.00	0.0029	0.0028	0.0026	0.0024	-0.0005
0.7700	0.7000	528.70	0.0127	0.0127	0.0126	0.0099	0.0038
0.8500	0.7550	518.00	0.0127	0.0129	0.0132	0.0038	-0.0023
		MEAN	0.0072	0.0072	0.0073	0.0062	0.0014

CARBON DISULPHIDE(1)-ACETONE(2)

ROSANOFF M.A.,EASELEY C.W.:J.AM.CHEM.SOC.31,953(1909).

ANTOINE VAPOR PRESSURE CONSTANTS
```
         A          B          C
(1) 6.85145    1122.500    236.460
(2) 7.23967    1279.870    237.500
```

P = 760.00

EQUATION	ORDER	A12	A21	D	A122	A211
VAN LAAR	3	0.5329	0.7243	-	-	-
MARGULES	3	0.5206	0.7022	-	-	-
MARGULES	4	0.5613	0.7495	0.1477	-	-
ALPHA	2	5.2778	2.6996	-	-	-
ALPHA	3	7.1173	1.2426	-	-3.5348	2.6160

			DEVIATION IN VAPOR PHASE COMPOSITION				
X	Y	TEMP.	LAAR 3	MARG 3	MARG 4	ALPHA2	ALPHA3
0.0190	0.0832	54.00	-0.0026	-0.0043	0.0016	0.0190	-0.0035
0.0476	0.1850	51.40	-0.0068	-0.0094	-0.0008	0.0257	-0.0084
0.1340	0.3510	46.60	0.0157	0.0147	0.0179	0.0366	0.0142
0.1858	0.4430	44.00	-0.0059	-0.0059	-0.0063	0.0030	-0.0083
0.2912	0.5275	41.40	0.0012	0.0022	-0.0014	-0.0059	-0.0020
0.3798	0.5740	40.30	0.0022	0.0029	0.0002	-0.0103	-0.0009
0.4477	0.5980	39.80	0.0046	0.0045	0.0038	-0.0084	0.0022
0.5360	0.6270	39.30	0.0027	0.0015	0.0035	-0.0075	0.0015
0.6530	0.6610	39.10	-0.0002	-0.0023	0.0019	-0.0025	0.0008
0.7894	0.7050	39.30	0.0039	0.0031	0.0044	0.0115	0.0035
0.8023	0.7230	39.60	-0.0078	-0.0083	-0.0076	0.0007	-0.0086
0.8799	0.7600	40.50	0.0102	0.0121	0.0076	0.0200	0.0024
0.9683	0.8860	43.50	0.0190	0.0222	0.0152	0.0220	0.0031
		MEAN	0.0064	0.0072	0.0056	0.0133	0.0046

CARBON DISULPHIDE(1)-ACETONE(2)

ZAWIDZKI J.:Z.PHYS.CHEM.35,129(1900).

ANTOINE VAPOR PRESSURE CONSTANTS

	A	B	C
(1)	6.85145	1122.500	236.460
(2)	7.23967	1279.870	237.500

T = 35.17

EQUATION	ORDER	A12	A21	D	A122	A211
VAN LAAR	3	0.5751	0.8008	-	-	-
MARGULES	3	0.5591	0.7750	-	-	-
MARGULES	4	0.6132	0.8281	0.1845	-	-
ALPHA	2	6.6437	3.2807	-	-	-
ALPHA	3	7.1875	1.4685	-	-2.4064	2.5891

			DEVIATION IN VAPOR PHASE COMPOSITION				
X	Y	PRESS.	LAAR 3	MARG 3	MARG 4	ALPHA2	ALPHA3
0.0624	0.2506	441.70	-0.0025	-0.0061	0.0062	0.0348	0.0030
0.0670	0.2674	447.50	-0.0066	-0.0101	0.0018	0.0302	-0.0011
0.0711	0.2724	451.80	-0.0008	-0.0042	0.0074	0.0356	0.0048
0.1212	0.3794	505.20	-0.0021	-0.0038	0.0021	0.0235	0.0029
0.1330	0.4012	514.80	-0.0046	-0.0059	-0.0014	0.0181	0.0001
0.1857	0.4666	553.80	-0.0011	-0.0008	-0.0016	0.0094	0.0016
0.1991	0.4834	562.50	-0.0039	-0.0033	-0.0051	0.0039	-0.0018
0.2085	0.5005	567.30	-0.0119	-0.0111	-0.0135	-0.0058	-0.0102
0.2761	0.5403	598.50	0.0010	0.0026	-0.0026	-0.0027	0.0001
0.2869	0.5452	602.90	0.0028	0.0045	-0.0009	-0.0020	0.0016
0.3502	0.5759	622.20	0.0047	0.0062	0.0010	-0.0044	0.0020
0.3551	0.5795	623.40	0.0032	0.0047	-0.0004	-0.0061	0.0005
0.4058	0.5986	634.10	0.0034	0.0044	0.0005	-0.0069	0.0003
0.4141	0.6015	635.10	0.0033	0.0041	0.0006	-0.0070	0.0001
0.4474	0.6094	640.60	0.0055	0.0058	0.0035	-0.0043	0.0026
0.4530	0.6141	641.80	0.0023	0.0026	0.0005	-0.0072	-0.0005
0.4933	0.6242	646.00	0.0026	0.0022	0.0018	-0.0054	0.0004
0.4974	0.6254	646.20	0.0023	0.0019	0.0016	-0.0054	0.0003
0.5702	0.6433	652.00	-0.0000	-0.0016	0.0010	-0.0028	0.0002
0.5730	0.6441	652.50	-0.0003	-0.0019	0.0008	-0.0028	0.0000
0.6124	0.6529	653.90	-0.0014	-0.0035	0.0004	-0.0005	0.0005
0.6146	0.6543	653.60	-0.0024	-0.0045	-0.0006	-0.0012	-0.0004
0.6161	0.6550	653.60	-0.0028	-0.0049	-0.0010	-0.0015	-0.0008
0.6713	0.6682	655.00	-0.0048	-0.0074	-0.0025	0.0018	-0.0004
0.7220	0.6827	654.60	-0.0072	-0.0097	-0.0051	0.0043	-0.0010
0.7197	0.6836	654.60	-0.0087	-0.0112	-0.0066	0.0026	-0.0026
0.8280	0.7207	645.10	-0.0041	-0.0043	-0.0046	0.0144	0.0024
0.9191	0.7989	614.10	0.0003	0.0039	-0.0041	0.0142	-0.0021
0.9245	0.8029	610.30	0.0042	0.0079	-0.0005	0.0172	0.0009
0.9350	0.8181	601.30	0.0057	0.0098	0.0008	0.0170	0.0008
0.9407	0.8261	595.50	0.0077	0.0119	0.0027	0.0179	0.0019
0.9549	0.8523	582.10	0.0098	0.0141	0.0048	0.0171	0.0020
0.9620	0.8723	574.20	0.0061	0.0103	0.0012	0.0118	-0.0024
0.9692	0.8902	564.00	0.0063	0.0103	0.0018	0.0105	-0.0025
		MEAN	0.0040	0.0059	0.0027	0.0103	0.0016

439

CARBON DISULPHIDE(1)-BENZENE(2)

HIRSCHBERG J.:BULL.SOC.CHIM.BELG.41,163(1932)

ANTOINE VAPOR PRESSURE CONSTANTS
         A   B   C
(1) 6.85145   1122.500   236.460
(2) 6.90565   1211.033   220.790

T =   19.90

EQUATION	ORDER	A12	A21	D	A122	A211
VAN LAAR	3	0.2187	0.1315	-	-	-
MARGULES	3	0.1988	0.1190	-	-	-
MARGULES	4	0.1302	0.0503	-0.1855	-	-
ALPHA	2	4.1539	-0.3962	-	-	-
ALPHA	3	5.8829	-0.0192	-	-1.6996	-0.2989

			DEVIATION IN VAPOR PHASE COMPOSITION				
X	Y	PRESS.	LAAR 3	MARG 3	MARG 4	ALPHA2	ALPHA3
0.2000	0.5400	143.00	0.0141	0.0137	0.0105	-0.0000	-0.0002
0.3100	0.6640	171.50	0.0015	0.0023	0.0044	0.0005	0.0002
0.4400	0.7600	201.30	-0.0033	-0.0028	-0.0017	-0.0001	-0.0001
0.5350	0.8100	217.00	-0.0004	-0.0003	-0.0009	0.0006	0.0008
0.5840	0.8350	226.10	-0.0009	-0.0010	-0.0022	-0.0022	-0.0019
0.6400	0.8550	238.10	0.0052	0.0051	0.0038	0.0010	0.0012
0.7880	0.9100	262.00	0.0125	0.0128	0.0134	0.0004	-0.0001
		MEAN	0.0054	0.0054	0.0053	0.0007	0.0006

CARBON DISULPHIDE(1)-BENZENE(2)

SAMESHIMA J.:J.AM.CHEM.SOC.40,1482(1918).

ANTOINE VAPOR PRESSURE CONSTANTS

	A	B	C
(1)	6.85145	1122.500	236.460
(2)	6.90565	1211.033	220.790

T =   20.00

EQUATION	ORDER	A12	A21	D	A122	A211
VAN LAAR	3	0.1384	0.1515	-	-	-
MARGULES	3	0.1366	0.1532	-	-	-
MARGULES	4	0.0184	0.0203	-0.4220	-	-
ALPHA	2	3.9528	-0.4715	-	-	-
ALPHA	3	-1.7984	-1.9743	-	5.3355	1.3885

DEVIATION IN VAPOR PHASE COMPOSITION

X	Y	PRESS.	LAAR 3	MARG 3	MARG 4	ALPHA2	ALPHA3
0.1220	0.3839	110.70	0.0296	0.0294	0.0161	0.0134	0.0025
0.3749	0.7235	181.90	-0.0012	-0.0009	0.0061	-0.0069	-0.0025
0.5376	0.8150	215.70	0.0042	0.0044	0.0010	-0.0001	0.0025
0.6253	0.8537	232.40	0.0060	0.0061	0.0002	0.0008	-0.0004
0.7518	0.8998	253.50	0.0107	0.0107	0.0075	0.0031	-0.0009
0.8865	0.9517	277.20	0.0080	0.0079	0.0109	-0.0006	0.0002
		MEAN	0.0100	0.0099	0.0070	0.0041	0.0015

CARBON DISULPHIDE(1)-BENZENE(2)

SAMESHIMA J.:J.AM.CHEM.SOC.40,1482(1918)

ANTOINE VAPOR PRESSURE CONSTANTS
	A	B	C
(1)	6.85145	1122.500	236.460
(2)	6.90565	1211.033	220.790

T =   25.00

EQUATION	ORDER	A12	A21	D	A122	A211
VAN LAAR	3	0.1594	0.1505	-	-	-
MARGULES	3	0.1593	0.1503	-	-	-
MARGULES	4	0.1566	0.1462	-0.0115	-	-
ALPHA	2	3.9204	-0.4541	-	-	-
ALPHA	3	3.2543	-0.3738	-	0.9975	-0.1740

DEVIATION IN VAPOR PHASE COMPOSITION

X	Y	PRESS.	LAAR 3	MARG 3	MARG 4	ALPHA2	ALPHA3
0.0581	0.2376	117.40	0.0068	0.0067	0.0062	-0.0084	0.0004
0.1286	0.4180	144.00	0.0050	0.0050	0.0048	-0.0090	-0.0013
0.2337	0.5765	177.80	0.0068	0.0068	0.0070	0.0009	0.0029
0.3698	0.7114	220.10	-0.0024	-0.0024	-0.0023	-0.0015	-0.0043
0.5014	0.7861	251.10	0.0054	0.0054	0.0053	0.0076	0.0036
0.6998	0.8825	298.70	0.0020	0.0020	0.0018	-0.0010	-0.0019
0.8176	0.9257	323.10	0.0058	0.0058	0.0058	-0.0013	0.0008
0.8847	0.9526	337.40	0.0044	0.0044	0.0045	-0.0036	-0.0003
0.9532	0.9798	351.20	0.0028	0.0028	0.0029	-0.0028	-0.0000
0.9789	0.9902	356.20	0.0019	0.0020	0.0020	-0.0011	0.0005
		MEAN	0.0043	0.0043	0.0043	0.0037	0.0016

CARBON DISULPHIDE(1)-CARBON TETRACHLORIDE(2)

ROSANOFF M.A.,EASELEY C.W.:J.AM.CHEM.SOC.31,953(1909).

ANTOINE VAPOR PRESSURE CONSTANTS
	A	B	C
(1)	6.85145	1122.500	236.460
(2)	6.93390	1242.430	230.000

P =   760.00

EQUATION	ORDER	A12	A21	D	A122	A211
VAN LAAR	3	0.1038	0.0905	-	-	-
MARGULES	3	0.1038	0.0891	-	-	-
MARGULES	4	0.0846	0.0610	-0.0803	-	-
ALPHA	2	1.9785	-0.4831	-	-	-
ALPHA	3	1.3800	-0.6872	-	0.6031	0.1865

			DEVIATION IN VAPOR PHASE COMPOSITION				
X	Y	TEMP.	LAAR 3	MARG 3	MARG 4	ALPHA2	ALPHA3
0.0296	0.0823	74.90	0.0028	0.0028	0.0003	0.0006	0.0007
0.0615	0.1555	73.10	0.0097	0.0097	0.0066	0.0062	0.0064
0.1106	0.2660	70.30	0.0041	0.0042	0.0022	0.0001	0.0004
0.1435	0.3325	68.60	-0.0019	-0.0019	-0.0027	-0.0059	-0.0056
0.2585	0.4950	63.80	0.0026	0.0026	0.0042	0.0007	0.0008
0.3908	0.6340	59.30	0.0019	0.0018	0.0025	0.0016	0.0014
0.5318	0.7470	55.30	0.0003	0.0002	-0.0012	-0.0005	-0.0009
0.6630	0.8290	52.30	0.0020	0.0019	0.0001	-0.0007	-0.0009
0.7574	0.8780	50.40	0.0054	0.0054	0.0046	0.0013	0.0014
0.8604	0.9320	48.50	0.0035	0.0036	0.0042	-0.0012	-0.0006
		MEAN	0.0034	0.0034	0.0028	0.0019	0.0019

443

CARBON DISULPHIDE(1)-CHLOROFORM(2)

HIRSCHBERG J.:BULL.SOC.CHIM.BELG.41,163(1932).

ANTOINE VAPOR PRESSURE CONSTANTS
	A	B	C
(1)	6.85145	1122.500	236.460
(2)	6.90328	1163.000	227.000

T =   20.00

EQUATION	ORDER	A12	A21	D	A122	A211
VAN LAAR	3	0.2537	0.1408	-	-	-
MARGULES	3	0.2223	0.1288	-	-	-
MARGULES	4	0.2736	0.1768	0.1465	-	-
ALPHA	2	1.5607	-0.0451	-	-	-
ALPHA	3	1.7201	0.2046	-	-0.0181	-0.2972

			DEVIATION IN VAPOR PHASE COMPOSITION				
X	Y	PRESS.	LAAR 3	MARG 3	MARG 4	ALPHA2	ALPHA3
0.1820	0.3400	211.20	0.0172	0.0166	0.0187	-0.0019	0.0015
0.3000	0.4780	236.00	0.0024	0.0047	0.0017	-0.0018	-0.0018
0.4800	0.6300	263.00	-0.0028	-0.0013	-0.0022	0.0009	-0.0012
0.5920	0.7040	276.00	0.0052	0.0057	0.0071	0.0053	0.0039
0.7550	0.8200	290.00	0.0050	0.0052	0.0059	-0.0048	-0.0037
0.8920	0.9100	296.40	0.0122	0.0130	0.0103	-0.0005	0.0019
		MEAN	0.0075	0.0077	0.0077	0.0025	0.0023

WATER(1)-ACETIC ACID(2)

BROWN I.,EWALD A.H.:AUSTR.J.SCI.RES.A3,306(1950).

ANTOINE VAPOR PRESSURE CONSTANTS

	A	B	C
(1)	7.96681	1668.210	228.000
(2)	7.18807	1416.700	211.000

P =   760.00

EQUATION	ORDER	A12	A21	D	A122	A211
VAN LAAR	3	-0.0010	0.0009	-	-	-
MARGULES	3	0.0668	-0.0562	-	-	-
MARGULES	4	0.1238	0.0151	0.2017	-	-
ALPHA	2	0.7774	-0.3315	-	-	-
ALPHA	3	103.9603	62.3029	-	-101.6652	-62.0202

DEVIATION IN VAPOR PHASE COMPOSITION

X	Y	TEMP.	LAAR 3	MARG 3	MARG 4	ALPHA2	ALPHA3
0.0002	0.0002	117.96	0.0002	0.0002	0.0003	0.0001	0.0005
0.0002	0.0003	117.92	0.0000	0.0001	0.0002	0.0000	0.0003
0.0002	0.0004	117.91	0.0000	0.0001	0.0002	-0.0000	0.0004
0.0034	0.0069	117.64	-0.0007	0.0003	0.0013	-0.0009	0.0032
0.0055	0.0112	117.51	-0.0013	0.0003	0.0018	-0.0016	0.0042
0.0474	0.0979	115.03	-0.0149	-0.0056	0.0017	-0.0170	-0.0058
0.0812	0.1446	113.81	-0.0064	0.0048	0.0121	-0.0099	-0.0013
0.1497	0.2382	111.51	0.0029	0.0113	0.0145	-0.0029	0.0002
0.2198	0.3273	109.84	0.0087	0.0104	0.0090	0.0007	0.0002
0.2917	0.4071	108.16	0.0163	0.0114	0.0079	0.0071	0.0048
0.3378	0.4573	107.36	0.0172	0.0098	0.0063	0.0080	0.0053
0.4198	0.5496	105.85	-0.0049	-0.0008	-0.0021	-0.0001	-0.0024
0.5359	0.6591	104.17	-0.0036	-0.0015	0.0017	-0.0026	-0.0033
0.6463	0.7524	102.86	0.0106	0.0022	0.0068	-0.0046	-0.0036
0.7388	0.8217	101.92	0.0117	0.0084	0.0109	-0.0033	-0.0017
0.8251	0.8783	101.24	0.0147	0.0160	0.0150	0.0023	0.0033
0.9210	0.9429	100.54	0.0108	0.0139	0.0106	0.0039	0.0025
0.9676	0.9761	100.24	0.0053	0.0071	0.0050	0.0022	-0.0008
0.9891	0.9921	100.07	0.0017	0.0024	0.0015	0.0006	-0.0021
		MEAN	0.0069	0.0056	0.0057	0.0036	0.0024

445

WATER(1)-ACETIC ACID(2)

RIVENC G.:MEM.SER.CHIM.ETAT.38,311(1953).

ANTOINE VAPOR PRESSURE CONSTANTS

	A	B	C
(1)	7.96681	1668.210	228.000
(2)	7.18807	1416.700	211.000

P = 760.00

EQUATION	ORDER	A12	A21	D	A122	A211
VAN LAAR	3	0.2743	0.0578	-	-	-
MARGULES	3	0.1486	0.0045	-	-	-
MARGULES	4	0.2019	0.0726	0.1912	-	-
ALPHA	2	1.0925	-0.2045	-	-	-
ALPHA	3	30.1882	18.6354	-	-27.0005	-18.9553

DEVIATION IN VAPOR PHASE COMPOSITION

X	Y	TEMP.	LAAR 3	MARG 3	MARG 4	ALPHA2	ALPHA3
0.0055	0.0320	118.00	-0.0139	-0.0180	-0.0163	-0.0206	-0.0108
0.0530	0.1330	116.00	-0.0020	-0.0145	-0.0066	-0.0298	0.0016
0.1250	0.2400	114.00	0.0015	-0.0010	0.0039	-0.0172	0.0010
0.2060	0.3380	112.00	0.0009	0.0055	0.0049	-0.0021	0.0002
0.2970	0.4370	110.00	-0.0016	0.0030	-0.0003	0.0060	-0.0015
0.3940	0.5330	108.00	-0.0035	-0.0025	-0.0044	0.0073	-0.0027
0.5100	0.6300	106.00	0.0039	0.0010	0.0034	0.0108	0.0048
0.6490	0.7510	104.00	-0.0021	-0.0044	0.0001	-0.0042	-0.0015
0.8030	0.8660	102.00	-0.0005	0.0018	0.0017	-0.0104	-0.0007
0.9594	0.9725	100.25	0.0010	0.0034	0.0006	-0.0040	0.0005
		MEAN	0.0031	0.0055	0.0042	0.0112	0.0025

WATER(1)-ANILINE(2)

HACK C.W.,VAN WINKLE M.:IND.ENG.CHEM.46,2392(1954)

ANTOINE VAPOR PRESSURE CONSTANTS
	A	B	C
(1)	7.96681	1668.210	228.000
(2)	7.24179	1675.300	200.000

P = 745.00

EQUATION	ORDER	A12	A21	D	A122	A211
VAN LAAR	3	0.6128	7.3011	-	-	-
MARGULES	3	0.5702	1.4535	-	-	-
MARGULES	4	0.5460	1.0900	-0.5629	-	-
ALPHA	2	53.6651	-1.4573	-	-	-
ALPHA	3	-104.3343	-6.2247	-	150.1705	11.0981

			DEVIATION IN VAPOR PHASE COMPOSITION				
X	Y	TEMP.	LAAR 3	MARG 3	MARG 4	ALPHA2	ALPHA3
0.0105	0.3420	168.00	-0.0070	-0.0248	-0.0343	0.0263	-0.0052
0.0170	0.4830	160.00	-0.0244	-0.0420	-0.0513	0.0050	-0.0267
0.0250	0.5730	152.00	-0.0081	-0.0228	-0.0304	0.0136	-0.0143
0.0425	0.7000	140.00	0.0032	-0.0046	-0.0081	0.0125	-0.0050
0.0590	0.7660	131.00	0.0113	0.0080	0.0069	0.0134	0.0037
0.0760	0.8150	126.00	0.0091	0.0086	0.0089	0.0089	0.0053
0.0930	0.8580	121.00	-0.0013	-0.0002	0.0007	-0.0031	-0.0023
0.1170	0.8890	115.80	-0.0012	0.0008	0.0019	-0.0036	0.0013
0.1530	0.9250	109.80	-0.0079	-0.0060	-0.0052	-0.0098	-0.0021
0.2000	0.9440	105.00	-0.0045	-0.0039	-0.0041	-0.0046	0.0009
0.2470	0.9470	101.00	0.0064	0.0052	0.0040	0.0080	-0.0001
		MEAN	0.0077	0.0115	0.0142	0.0099	0.0061

WATER(1)-DIMETHYLFORMAMIDE(2)

MICHALSKI H.,MICHALOWSKI S.,SERWINSKI M.,STRUMILLO C.:
ZESZYTY NAUK.POL.LODZ.,NR.46,73(1962).

ANTOINE VAPOR PRESSURE CONSTANTS
	A	B	C
(1)	7.96681	1668.210	228.000
(2)	6.99608	1437.840	199.830

P = 760.00

EQUATION	ORDER	A12	A21	D	A122	A211
VAN LAAR	3	0.0982	0.5939	-	-	-
MARGULES	3	0.0568	0.2582	-	-	-
MARGULES	4	0.1375	0.3623	0.3500	-	-
ALPHA	2	5.0971	-0.7015	-	-	-
ALPHA	3	9.9731	0.2236	-	-4.6860	-0.9045

			DEVIATION IN VAPOR PHASE COMPOSITION				
X	Y	TEMP.	LAAR 3	MARG 3	MARG 4	ALPHA2	ALPHA3
0.0430	0.2250	145.00	-0.0178	-0.0284	-0.0092	-0.0110	-0.0070
0.0820	0.3590	139.70	-0.0175	-0.0271	-0.0118	-0.0091	-0.0051
0.1020	0.4080	137.40	-0.0109	-0.0188	-0.0074	-0.0024	0.0011
0.1720	0.5510	130.00	-0.0057	-0.0071	-0.0079	0.0012	0.0028
0.2900	0.7020	121.50	-0.0024	0.0005	-0.0055	0.0011	0.0006
0.3260	0.7340	119.70	-0.0012	0.0015	-0.0034	0.0015	0.0007
0.3850	0.7790	117.10	-0.0006	0.0006	-0.0015	0.0010	0.0000
0.4580	0.8230	114.20	0.0002	-0.0012	0.0003	0.0011	0.0002
0.5000	0.8450	112.90	-0.0004	-0.0034	-0.0001	0.0003	-0.0004
0.5490	0.8700	111.20	-0.0035	-0.0084	-0.0036	-0.0028	-0.0032
0.6160	0.8930	109.10	-0.0010	-0.0078	-0.0023	0.0001	0.0001
0.6650	0.9100	107.60	-0.0020	-0.0095	-0.0045	-0.0003	0.0000
0.7190	0.9260	105.90	-0.0025	-0.0099	-0.0062	0.0002	0.0008
0.7720	0.9400	104.50	-0.0031	-0.0093	-0.0075	0.0011	0.0018
0.8520	0.9620	102.70	-0.0073	-0.0093	-0.0109	-0.0003	0.0004
0.8950	0.9730	101.90	-0.0093	-0.0079	-0.0110	-0.0007	-0.0002
0.9270	0.9810	101.30	-0.0102	-0.0061	-0.0097	-0.0007	-0.0005
0.9470	0.9860	100.90	-0.0101	-0.0047	-0.0081	-0.0006	-0.0006
0.9590	0.9890	100.60	-0.0095	-0.0037	-0.0068	-0.0004	-0.0005
0.9750	0.9930	100.40	-0.0077	-0.0021	-0.0045	-0.0001	-0.0003
		MEAN	0.0062	0.0084	0.0061	0.0016	0.0013

WATER(1)-HYDRAZINE(2)

BURTLE J.G.:IND.ENG.CHEM. 44,1675(1952)

ANTOINE VAPOR PRESSURE CONSTANTS

	A	B	C
(1)	7.96681	1668.210	228.000
(2)	7.77306	1620.000	218.000

P =   124.80

EQUATION	ORDER	A12	A21	D	A122	A211
VAN LAAR	3	-0.6638	-1.5120	-	-	-
MARGULES	3	-0.4971	-1.2080	-	-	-
MARGULES	4	-0.9237	-1.6627	-1.1688	-	-
ALPHA	2	-0.8433	-1.0080	-	-	-
ALPHA	3	-1.6035	-1.2771	-	1.0466	0.2294

			DEVIATION IN VAPOR PHASE COMPOSITION				
X	Y	TEMP.	LAAR 3	MARG 3	MARG 4	ALPHA2	ALPHA3
0.0105	0.0035	66.80	0.0004	0.0021	-0.0013	-0.0017	0.0012
0.1528	0.0690	69.10	0.0071	0.0154	0.0003	-0.0116	0.0114
0.3223	0.2503	71.70	-0.0153	-0.0170	-0.0084	-0.0184	-0.0160
0.4522	0.4267	73.90	0.0069	0.0048	0.0084	0.0227	0.0071
0.5013	0.5174	74.20	0.0068	0.0082	0.0039	0.0234	0.0085
0.5142	0.5498	74.00	-0.0009	0.0015	-0.0048	0.0149	0.0010
0.6797	0.8443	69.70	-0.0039	0.0033	-0.0055	-0.0132	-0.0031
0.7991	0.9660	63.80	-0.0089	-0.0089	-0.0053	-0.0216	-0.0081
0.9040	0.9921	58.90	0.0010	-0.0007	0.0025	-0.0022	0.0038
		MEAN	0.0057	0.0069	0.0045	0.0144	0.0067

WATER(1)-HYDRAZINE(2)

BURTLE J.G.:IND.ENG.CHEM. 44,1675(1952)

ANTOINE VAPOR PRESSURE CONSTANTS
	A	B	C
(1)	7.96681	1668.210	228.000
(2)	7.77306	1620.000	218.000

P = 281.80

EQUATION	ORDER	A12	A21	D	A122	A211
VAN LAAR	3	-0.6750	-1.3409	-	-	-
MARGULES	3	-0.5453	-1.1493	-	-	-
MARGULES	4	-0.3804	-0.9766	0.4523	-	-
ALPHA	2	-0.8173	-1.0256	-	-	-
ALPHA	3	-2.0827	-1.7960	-	1.6121	0.8685

			DEVIATION IN VAPOR PHASE COMPOSITION				
X	Y	TEMP.	LAAR 3	MARG 3	MARG 4	ALPHA2	ALPHA3
0.0124	0.0054	86.50	-0.0010	0.0003	0.0027	-0.0030	0.0012
0.1817	0.1069	89.60	-0.0118	-0.0066	-0.0020	-0.0240	-0.0009
0.3185	0.2274	91.90	0.0055	0.0034	-0.0002	0.0078	0.0023
0.4442	0.4267	93.20	0.0021	-0.0018	-0.0042	0.0181	-0.0053
0.5034	0.5405	93.30	-0.0010	-0.0024	-0.0010	0.0141	-0.0023
0.5405	0.6054	93.00	0.0053	0.0061	0.0096	0.0170	0.0090
0.6866	0.8679	88.40	-0.0144	-0.0110	-0.0080	-0.0215	-0.0052
0.7778	0.9406	83.80	0.0017	0.0019	0.0011	-0.0066	0.0032
0.9091	0.9921	77.60	0.0005	-0.0007	-0.0024	0.0006	-0.0023
		MEAN	0.0048	0.0038	0.0035	0.0125	0.0035

450

WATER(1)-HYDRAZINE(2)

BURTLE J.G.:IND.ENG.CHEM. 44,1675(1952)

ANTOINE VAPOR PRESSURE CONSTANTS
```
           A          B          C
(1) 7.96681    1668.210    228.000
(2) 7.77306    1620.000    218.000
```

P =   411.20

EQUATION	ORDER	A12	A21	D	A122	A211
VAN LAAR	3	-0.6288	-1.2501	-	-	-
MARGULES	3	-0.5105	-1.0723	-	-	-
MARGULES	4	-0.4407	-0.9971	0.2010	-	-
ALPHA	2	-0.7809	-1.0134	-	-	-
ALPHA	3	-1.8601	-1.5532	-	1.4051	0.5770

DEVIATION IN VAPOR PHASE COMPOSITION

X	Y	TEMP.	LAAR 3	MARG 3	MARG 4	ALPHA2	ALPHA3
0.0194	0.0140	96.80	-0.0064	-0.0044	-0.0030	-0.0093	-0.0034
0.1620	0.0970	99.40	-0.0105	-0.0050	-0.0027	-0.0230	0.0008
0.3237	0.2475	102.20	0.0011	-0.0014	-0.0033	0.0041	0.0007
0.4499	0.4408	103.60	0.0039	0.0004	-0.0004	0.0208	-0.0006
0.4848	0.5088	103.40	-0.0011	-0.0031	-0.0028	0.0163	-0.0039
0.5478	0.6216	102.80	0.0026	0.0040	0.0061	0.0163	0.0048
0.6833	0.8521	98.40	-0.0088	-0.0051	-0.0033	-0.0113	-0.0009
0.7816	0.9461	93.50	-0.0060	-0.0057	-0.0060	-0.0116	-0.0014
0.7991	0.9541	92.40	-0.0027	-0.0029	-0.0035	-0.0078	0.0011
0.9065	0.9915	86.90	-0.0006	-0.0019	-0.0027	-0.0005	0.0004
		MEAN	0.0044	0.0034	0.0034	0.0121	0.0018

WATER(1)-HYDRAZINE(2)

BURTLE J.G.:IND.ENG.CHEM. 44,1675(1952)

ANTOINE VAPOR PRESSURE CONSTANTS

	A	B	C
(1)	7.96681	1668.210	228.000
(2)	7.77306	1620.000	218.000

P = 560.40

EQUATION	ORDER	A12	A21	D	A122	A211
VAN LAAR	3	-0.6477	-1.2146	-	-	-
MARGULES	3	-0.5539	-1.0668	-	-	-
MARGULES	4	-0.5188	-1.0290	0.1005	-	-
ALPHA	2	-0.7837	-1.0041	-	-	-
ALPHA	3	.5542/ 05	.5060/ 05	-	-.5717/ 05	-.5726/ 0

			DEVIATION IN VAPOR PHASE COMPOSITION				
X	Y	TEMP.	LAAR 3	MARG 3	MARG 4	ALPHA2	ALPHA3
0.0263	0.0140	105.20	-0.0041	-0.0021	-0.0013	-0.0075	-0.0191
0.1697	0.0937	107.90	-0.0040	-0.0003	0.0007	-0.0145	0.0691
0.3118	0.2188	110.20	0.0128	0.0109	0.0101	0.0142	0.1104
0.4522	0.4523	111.30	-0.0013	-0.0032	-0.0036	0.0109	0.0415
0.5153	0.5789	110.90	-0.0110	-0.0101	-0.0094	-0.0015	-0.0112
0.6833	0.8386	106.40	0.0076	0.0114	0.0123	-0.0007	-0.0739
0.8030	0.9522	100.30	0.0019	0.0017	0.0015	-0.0054	-0.0470
0.9006	0.9898	95.50	-0.0003	-0.0014	-0.0018	-0.0011	0.0299
		MEAN	0.0054	0.0052	0.0051	0.0070	0.0503

WATER(1)-HYDRAZINE(2)

BURTLE J.G.:IND.ENG.CHEM. 44,1675(1952)

ANTOINE VAPOR PRESSURE CONSTANTS

	A	B	C
(1)	7.96681	1668.210	228.000
(2)	7.77306	1620.000	218.000

P = 700.60

EQUATION	ORDER	A12	A21	D	A122	A211
VAN LAAR	3	-0.5970	-1.1557	-	-	-
MARGULES	3	-0.4972	-1.0014	-	-	-
MARGULES	4	-0.4463	-0.9447	0.1490	-	-
ALPHA	2	-0.7462	-0.9867	-	-	-
ALPHA	3	-1.7179	-1.5331	-	1.2290	0.5966

			DEVIATION IN VAPOR PHASE COMPOSITION				
X	Y	TEMP.	LAAR 3	MARG 3	MARG 4	ALPHA2	ALPHA3
0.0213	0.0124	111.70	-0.0036	-0.0016	-0.0005	-0.0064	-0.0014
0.1604	0.0955	114.20	-0.0062	-0.0015	0.0002	-0.0174	-0.0003
0.3264	0.2558	116.90	0.0034	0.0012	-0.0002	0.0064	0.0005
0.4511	0.4477	117.60	0.0048	0.0024	0.0021	0.0188	0.0020
0.5142	0.5692	117.20	-0.0042	-0.0038	-0.0026	0.0088	-0.0030
0.6816	0.8415	112.60	-0.0044	-0.0007	0.0007	-0.0086	0.0004
0.7946	0.9437	106.80	0.0011	0.0011	0.0007	-0.0056	0.0009
0.9116	0.9960	101.50	-0.0055	-0.0068	-0.0075	-0.0063	-0.0070
		MEAN	0.0041	0.0024	0.0018	0.0098	0.0019

WATER(1)-HYDRAZINE(2)

LOBRY DE BRUYN,DITO,PROC.AKAD.WETENSCH.AMST. 5,171(1902)

ANTOINE VAPOR PRESSURE CONSTANTS
```
           A          B          C
(1) 7.96681    1668.210    228.000
(2) 7.77306    1620.000    218.000
```

P = 771.00

EQUATION	ORDER	A12	A21	D	A122	A211
VAN LAAR	3	-0.7993	-1.0034	-	-	-
MARGULES	3	-0.7836	-0.9832	-	-	-
MARGULES	4	-0.8050	-1.0068	-0.0558	-	-
ALPHA	2	-0.8176	-1.1473	-	-	-
ALPHA	3	.1514/ 22	.2361/ 22	-	-.6429/ 08	.7031/ 05

DEVIATION IN VAPOR PHASE COMPOSITION

X	Y	TEMP.	LAAR 3	MARG 3	MARG 4	ALPHA2	ALPHA3
0.2640	0.1630	119.25	0.0079	0.0080	0.0078	0.0071	0.2278
0.2730	0.1900	119.50	-0.0080	-0.0079	-0.0080	-0.0085	0.2008
0.3170	0.2450	119.90	-0.0019	-0.0020	-0.0019	-0.0014	0.1458
0.3420	0.2800	120.25	0.0024	0.0022	0.0024	0.0032	0.1108
0.4150	0.4150	120.50	-0.0015	-0.0017	-0.0015	-0.0010	-0.0242
0.4400	0.4700	120.50	-0.0077	-0.0078	-0.0077	-0.0076	-0.0792
0.4520	0.4720	120.45	0.0141	0.0141	0.0141	0.0140	-0.0812
0.4670	0.5125	120.35	0.0034	0.0035	0.0034	0.0032	-0.1217
0.4820	0.5540	120.20	-0.0082	-0.0080	-0.0082	-0.0086	-0.1632
		MEAN	0.0061	0.0061	0.0061	0.0061	0.1283

WATER(1)-PHENOL(2)

SCHREINEMAKERS F.A.H.:Z.PHYS.CHEM.,35,459(1900)

ANTOINE VAPOR PRESSURE CONSTANTS
```
          A          B          C
(1) 7.96681    1668.210    228.000
(2) 7.57893    1817.000    205.000
```

T =   56.30

EQUATION	ORDER	A12	A21	D	A122	A211
VAN LAAR	3	0.5043	1.4547	-	-	-
MARGULES	3	0.4213	1.3038	-	-	-
MARGULES	4	0.8985	1.4409	1.4639	-	-
ALPHA	2	155.9620	1.8670	-	-	-
ALPHA	3	226.1948	4.6958	-	9.4900	-1.8958

			DEVIATION IN VAPOR PHASE COMPOSITION				
X	Y	PRESS.	LAAR 3	MARG 3	MARG 4	ALPHA2	ALPHA3
0.4148	0.9746	102.00	0.0005	0.0046	-0.0035	-0.0010	-0.0010
0.5612	0.9746	118.00	0.0052	0.0052	0.0038	0.0029	0.0027
0.6136	0.9795	122.00	0.0010	-0.0007	0.0001	-0.0011	-0.0012
0.6994	0.9795	124.00	0.0012	-0.0034	0.0005	0.0002	0.0003
0.7770	0.9840	126.00	-0.0043	-0.0111	-0.0054	-0.0033	-0.0030
0.9685	0.9840	126.00	-0.0015	0.0001	-0.0024	0.0008	0.0012
0.9772	0.9849	127.00	0.0006	0.0027	0.0001	0.0008	0.0010
0.9849	0.9867	126.50	0.0023	0.0045	0.0022	0.0005	0.0003
0.9888	0.9890	127.00	0.0023	0.0042	0.0023	-0.0006	-0.0010
0.9961	0.9950	125.00	0.0015	0.0025	0.0016	-0.0019	-0.0028
		MEAN	0.0020	0.0039	0.0022	0.0013	0.0014

WATER(1)-PHENOL(2)

SCHREINEMAKERS F.A.H.:Z.PHYS.CHEM.,35,459(1900)

ANTOINE VAPOR PRESSURE CONSTANTS
```
          A          B          C
(1) 7.96681    1668.210    228.000
(2) 7.57893    1817.000    205.000
```

T =   75.00

EQUATION	ORDER	A12	A21	D	A122	A211
VAN LAAR	3	0.3323	1.3754	-	-	-
MARGULES	3	0.4252	0.9245	-	-	-
MARGULES	4	0.9505	1.2951	1.8002	-	-
ALPHA	2	64.3534	0.4360	-	-	-
ALPHA	3	283.8664	4.3296	-	-226.3917	-0.1179

			DEVIATION IN VAPOR PHASE COMPOSITION				
X	Y	PRESS.	LAAR 3	MARG 3	MARG 4	ALPHA2	ALPHA3
0.3212	0.9506	177.00	0.0016	0.0080	-0.0074	-0.0018	-0.0038
0.4148	0.9506	218.00	0.0140	0.0152	0.0060	0.0081	0.0075
0.5276	0.9731	259.00	0.0003	-0.0033	-0.0043	-0.0070	-0.0064
0.6136	0.9731	280.00	0.0045	-0.0021	0.0015	-0.0030	-0.0018
0.7321	0.9782	289.00	0.0027	-0.0063	-0.0003	-0.0040	-0.0023
0.7737	0.9782	292.00	0.0033	-0.0057	-0.0004	-0.0027	-0.0010
0.8437	0.9795	294.00	0.0020	-0.0047	-0.0028	-0.0019	-0.0003
0.8673	0.9795	294.00	0.0019	-0.0033	-0.0033	-0.0011	0.0003
0.9425	0.9806	294.00	0.0008	0.0045	-0.0026	0.0014	0.0014
0.9473	0.9806	294.00	0.0011	0.0054	-0.0019	0.0018	0.0016
0.9628	0.9812	294.00	0.0020	0.0080	0.0005	0.0029	0.0017
0.9847	0.9849	294.00	0.0044	0.0101	0.0049	0.0040	0.0002
0.9918	0.9895	293.00	0.0038	0.0077	0.0044	0.0028	-0.0023
0.9953	0.9932	293.00	0.0026	0.0052	0.0031	0.0016	-0.0036
		MEAN	0.0032	0.0064	0.0031	0.0032	0.0025

2.2. Electrolytes

```
HYDROGEN BROMIDE(1)-WATER(2)

     HAASE R.,NAAS H.,THUMM H.: Z.PHYSIK.CHEM.(FRANKFURT)
              37,220(1963).


     T =    25.00


     ORDER          A12        B12         C12

        1        -12.4033    30.3264        -
        2         51.5301   -50.7888    -46.2502


                           DEV. IN VAPOR PHASE COMPN.
     X          Y       PRESS.     ORDER1     ORDER2

  0.1050     0.0001     13.87     -0.0001     0.0003
  0.1200     0.0004     12.19     -0.0002     0.0005
  0.1340     0.0010     10.29     -0.0001     0.0012
  0.1470     0.0050      8.35     -0.0013     0.0006
  0.1510     0.0070      7.84     -0.0014     0.0005
  0.1590     0.0150      6.98     -0.0025    -0.0011
  0.1680     0.0300      6.47      0.0000    -0.0016
  0.1940     0.1960      5.60      0.0643     0.0227
  0.2070     0.4920      6.50      0.0341    -0.0087
  0.2210     0.7950     11.70     -0.0094    -0.0121
  0.2350     0.9340     28.90     -0.0133     0.0018

             MEAN      0.0115     0.0046
```

HYDROGEN BROMIDE(1)-WATER(2)

VREVSKII M.S.,ZAVARITSKII N.N.SHARLOV L.E.: ZH. RUSS. FIZ. KHIM. OBSHCH. 54,360(1924).

T = 19.93

ORDER	A12	B12	C12
1	-12.4077	29.8858	-
2	51.4376	-51.5634	-45.7890

			DEV. IN VAPOR PHASE COMPN.	
X	Y	PRESS.	ORDER1	ORDER2
0.1694	0.0253	5.10	-0.0027	-0.0003
0.1960	0.1832	4.30	0.0269	0.0072
0.2138	0.5562	4.90	-0.0038	-0.0182
0.2273	0.8000	3.10	-0.0094	0.0055
		MEAN	0.0107	0.0078

HYDROGEN CHLORIDE(1)-WATER(2)

STORONKIN A.V.,SUSAREV M.P.:VESTNIK LENINGR.UNIV.
        6,119(1952).

T =   25.00

ORDER	A12	B12	C12
1	-5.5923	17.5465	-
2	9.7138	-0.4355	-11.9442

			DEV. IN VAPOR PHASE COMPN.	
X	Y	PRESS.	ORDER1	ORDER2
0.0300	0.0000	22.08	0.0000	0.0000
0.0500	0.0002	20.50	-0.0001	0.0000
0.0800	0.0025	17.83	-0.0010	-0.0002
0.1000	0.0106	15.73	-0.0015	-0.0001
0.1100	0.0215	14.66	-0.0013	0.0000
0.1225	0.0496	13.74	0.0016	0.0010
0.1300	0.0825	13.21	0.0028	-0.0001
0.1400	0.1507	13.21	0.0075	0.0008
0.1450	0.2010	13.35	0.0077	-0.0004
0.1500	0.2590	13.81	0.0098	0.0012
0.1600	0.4124	15.44	-0.0002	-0.0051
0.1700	0.5783	19.47	-0.0120	-0.0070
0.1800	0.7137	26.90	-0.0094	0.0060
		MEAN	0.0042	0.0017

HYDROGEN CHLORIDE(1)-WATER(2)

HAASE R.,NAAS H.,THUMM H.: Z.PHYSIK.CHEM.(FRANKFURT)
37,220(1963).

T =   25.00

ORDER	A12	B12	C12
1	-6.0111	18.6746	-
2	6.8830	2.9425	-9.7662

			DEV. IN VAPOR PHASE COMPN.	
X	Y	PRESS.	ORDER1	ORDER2
0.0830	0.0030	17.95	-0.0014	-0.0003
0.0980	0.0090	16.40	-0.0024	-0.0004
0.1120	0.0240	14.55	-0.0025	0.0001
0.1260	0.0620	13.80	0.0007	0.0013
0.1390	0.1430	13.40	0.0067	0.0003
0.1530	0.2940	14.10	0.0263	0.0079
0.1650	0.4960	17.25	0.0164	-0.0064
0.1780	0.6960	24.20	0.0110	-0.0048
0.1900	0.8290	38.00	0.0054	-0.0005
0.2010	0.9040	62.20	0.0023	0.0025
0.2130	0.9580	103.00	-0.0074	-0.0045
0.2230	0.9720	160.00	-0.0009	0.0024
		MEAN	0.0070	0.0026

HYDROGEN CHLORIDE(1) - WATER(2)

SUSAREV M.P.,PROKOF'EVA R.V.: ZHUR. FIZ. KHIM. 37,2408(1'63).

T =   25.00

ORDER	A12	B12	C12
1	-5.7367	17.8810	-
2	13.7188	-4.8674	-15.2124

			DEV. IN VAPOR PHASE COMPN.	
X	Y	PRESS.	ORDER1	ORDER2
0.0400	0.0001	21.29	-0.0001	0.0000
0.1200	0.0416	14.05	-0.0014	-0.0002
0.1300	0.0798	13.35	0.0013	0.0006
0.1400	0.1499	13.21	0.0024	-0.0002
0.1450	0.2012	13.33	0.0011	-0.0017
0.1500	0.2590	13.84	0.0032	0.0016
0.1600	0.4129	15.53	-0.0067	-0.0001
		MEAN	0.0023	0.0006

HYDROGEN CHLORIDE(1)-WATER(2)

VREVSKII M.S.,ZAVARITSKII N.N.SHARLOV L.E.: ZH. RUSS. FIZ. KHIM. OBSHCH. 54,360(1924).

T =   19.95

ORDER	A12	B12	C12
1	-5.7556	18.2152	-
2	15.2494	-7.1208	-16.1103

			DEV. IN VAPOR PHASE COMPN.	
X	Y	PRESS.	ORDER1	ORDER2
0.0860	0.0057	12.30	-0.0028	-0.0008
0.0978	0.0108	12.20	-0.0024	0.0003
0.0978	0.0096	11.60	-0.0012	0.0015
0.1116	0.0321	10.30	-0.0059	-0.0043
0.1356	0.1145	9.30	0.0291	0.0115
0.1413	0.1617	9.20	0.0392	0.0125
0.1747	0.6845	16.10	0.0096	-0.0236
0.1950	0.8885	38.50	-0.0038	-0.0063
0.2240	0.9790	145.30	-0.0042	0.0018
		MEAN	0.0109	0.0069

462

HYDROGEN CHLORIDE(1)-WATER(2)

VREVSKII M.S.,ZAVARITSKII N.N.,SHARLOV L.E.: ZH. RUSS. FIZ. KHIM.
OBSHCH. 54,360(1924).

T = 55.20

ORDER	A12	B12	C12
1	-4.3784	14.8339	-
2	11.1689	-3.2337	-12.2449

			DEV. IN VAPOR PHASE COMPN.	
X	Y	PRESS.	ORDER1	ORDER2
0.0520	0.0007	114.70	-0.0004	0.0001
0.0860	0.0083	89.90	-0.0014	0.0001
0.0978	0.0197	83.90	-0.0024	-0.0005
0.1290	0.1270	74.10	0.0039	-0.0032
0.1340	0.1614	74.00	0.0094	-0.0001
0.1413	0.2340	76.30	0.0090	-0.0029
0.1514	0.3465	86.50	0.0207	0.0102
0.1676	0.5838	111.00	0.0046	0.0104
0.1676	0.5886	110.50	-0.0002	0.0056
0.1747	0.7050	145.80	-0.0289	-0.0143
		MEAN	0.0081	0.0048

HYDROCHLORIC ACID(1) - WATER(2)

DVORAK K.,BOUBLIK T.: COLL.CZECH.CHEM.COMM. 28,1249 (1963).

T =   75.90

ORDER	A12	B12	C12
1	-3.5015	12.7059	-
2	-39.4411	51.6088	29.5381

			DEV. IN VAPOR PHASE COMPN.	
X	Y	PRESS.	ORDER1	ORDER2
0.1011	0.0339	NOT	0.0005	0.0002
0.1043	0.0418	GIVEN	0.0002	0.0003
0.1063	0.0471		0.0005	0.0008
0.1088	0.0590		-0.0036	-0.0033
0.1117	0.0642		0.0014	0.0017
0.1145	0.0764		0.0006	0.0003
0.1147	0.0775		0.0004	0.0000
		MEAN	0.0010	0.0009

HYDROGEN FLUORIDE(1) - WATER(2)

BROSHEER J.C.,LENFESTY F.A.,ELMORE K.L.: IND. ENG. CHEM.
39,424(1947).

T = 25.00

ORDER	A12	B12	C12
1	-0.1947	0.8253	-
2	6.4250	-7.5773	-4.8470

			DEV. IN VAPOR PHASE COMPN.	
X	Y	PRESS.	ORDER1	ORDER2
0.0180	0.0020	23.51	-0.0017	-0.0002
0.0358	0.0038	22.97	-0.0026	0.0002
0.0545	0.0058	22.43	-0.0028	0.0006
0.0896	0.0120	21.38	-0.0030	0.0001
0.1167	0.0181	20.97	-0.0015	-0.0002
0.1334	0.0225	20.12	0.0001	0.0000
0.1538	0.0306	19.46	0.0010	-0.0012
0.1826	0.0414	18.65	0.0059	0.0012
0.2299	0.0767	16.68	0.0047	0.0001
0.2688	0.1232	15.42	-0.0056	-0.0003
		MEAN	0.0029	0.0004

465

HYDROGEN FLUORIDE(1) - WATER(2)

BROSHEER J.C.,LENFESTY F.A.,ELMORE K.L.: IND. ENG. CHEM.
39,424(1947).

T =   40.00

ORDER	A12	B12	C12
1	-0.1413	0.7407	-
2	6.1657	-7.2446	-4.6314

			DEV. IN VAPOR PHASE COMPN.	
X	Y	PRESS.	ORDER1	ORDER2
0.0180	0.0021	54.17	-0.0018	-0.0004
0.0381	0.0044	52.76	-0.0029	-0.0001
0.0552	0.0066	52.38	-0.0033	0.0001
0.0944	0.0131	49.69	-0.0023	0.0004
0.1124	0.0172	48.25	-0.0011	0.0004
0.1278	0.0224	47.67	-0.0008	-0.0006
0.1538	0.0311	45.32	0.0019	-0.0003
0.1931	0.0500	42.18	0.0059	0.0007
0.2225	0.0733	40.22	0.0043	-0.0005
0.2679	0.1247	36.56	-0.0057	0.0002
		MEAN	0.0030	0.0004

466

HYDROGEN FLUORIDE(1) - WATER(2)

BROSHEER J.C.,LENFESTY F.A.,ELMORE K.L.: IND. ENG. CHEM.
39,424(1947).

T =   60.00

ORDER	A12	B12	C12
1	-0.0533	0.5953	-
2	5.7385	-6.7600	-4.2341

			DEV. IN VAPOR PHASE COMPN.	
X	Y	PRESS.	ORDER1	ORDER2
0.0201	0.0025	146.97	-0.0021	-0.0004
0.0372	0.0046	145.27	-0.0030	-0.0001
0.0557	0.0073	141.93	-0.0035	0.0001
0.0823	0.0124	137.41	-0.0035	-0.0001
0.1093	0.0186	130.22	-0.0021	0.0000
0.1371	0.0267	126.89	0.0007	0.0004
0.1641	0.0383	121.25	0.0026	-0.0002
0.1922	0.0532	115.34	0.0051	0.0003
0.2204	0.0741	110.38	0.0050	0.0003
0.2450	0.0999	105.77	0.0003	-0.0014
0.2736	0.1341	99.55	-0.0063	0.0009
		MEAN	0.0031	0.0004

HYDROGEN FLUORIDE(1) - WATER(2)

BROSHEER J.C.,LENFESTY F.A.,ELMORE K.L.: IND. ENG. CHEM. 39,424(1947).

T =   75.00

ORDER	A12	B12	C12
1	-0.0034	0.5286	-
2	5.9281	-7.0088	-4.3358

			DEV. IN VAPOR PHASE COMPN.	
X	Y	PRESS.	ORDER1	ORDER2
0.0177	0.0023	283.06	-0.0019	-0.0003
0.0373	0.0054	277.29	-0.0037	-0.0003
0.0562	0.0079	271.94	-0.0037	0.0003
0.0818	0.0130	263.52	-0.0036	0.0002
0.1084	0.0192	257.24	-0.0019	0.0004
0.1390	0.0294	245.10	0.0005	-0.0001
0.1599	0.0383	238.02	0.0023	-0.0004
0.1922	0.0560	222.15	0.0049	-0.0004
0.2214	0.0764	216.12	0.0067	0.0015
0.2470	0.1049	204.56	0.0007	-0.0012
0.2755	0.1412	195.15	-0.0075	0.0003
		MEAN	0.0034	0.0005

HYDROGEN FLUORIDE(1)-WATER(2)

MUNTER P.A.,AEPLI O.T.,KOSSATZ R.A.:IND.ENG.CHEM.
39,3,427(1947).

P =   760.00

ORDER	A12	B12	C12
1	-1.7710	4.0462	-
2	7.3144	-8.7715	-5.2015

			DEV. IN VAPOR PHASE COMPN.	
X	Y	TEMP.	ORDER1	ORDER2
0.0495	0.0078	101.60	-0.0075	0.0037
0.0921	0.0183	102.80	-0.0159	0.0032
0.1894	0.0640	106.80	-0.0289	0.0002
0.2280	0.1059	108.40	-0.0323	-0.0080
0.2794	0.1781	110.30	-0.0132	-0.0083
0.3382	0.3053	111.70	0.0251	-0.0008
0.3440	0.3208	112.00	0.0286	0.0002
0.3518	0.3403	112.10	0.0351	0.0039
0.3578	0.3571	112.30	0.0386	0.0055
0.3583	0.3582	112.40	0.0392	0.0060
0.3662	0.3859	112.10	0.0384	0.0034
0.3967	0.4748	111.40	0.0532	0.0174
0.4439	0.6330	108.70	0.0406	0.0210
0.5028	0.8104	101.70	-0.0012	0.0078
0.5219	0.8620	98.90	-0.0204	-0.0047
0.5604	0.9218	90.90	-0.0294	-0.0059
0.5817	0.9581	86.60	-0.0446	-0.0197
0.6166	0.9889	79.00	-0.0490	-0.0249
0.6382	0.9856	74.60	-0.0335	-0.0111
0.6984	0.9867	61.60	-0.0118	0.0040
0.7976	0.9922	45.10	-0.0004	0.0063
0.8793	0.9945	33.50	0.0027	0.0052
		MEAN	0.0268	0.0078

NITRIC ACID(1)-WATER(2)

WILSON G.L.,MILES F.D.:TRANS.FARADAY SOC. 36,225,356(1940).

T =   15.00

ORDER		A12	B12	C12
1		-2.5147	5.2069	-
2		-0.2850	2.0439	-1.2550

			DEV. IN VAPOR PHASE COMPN.	
X	Y	PRESS.	ORDER1	ORDER2
0.2220	0.0570	5.81	-0.0139	-0.0098
0.3000	0.1500	4.14	0.0269	0.0254
0.4000	0.5280	4.17	0.0055	-0.0070
0.5330	0.8935	8.64	-0.0087	-0.0080
0.7200	0.9844	19.85	0.0041	0.0064
		MEAN	0.0118	0.0113

470

NITRIC ACID(1)-WATER(2)

WILSON G.L.,MILES F.D.:TRANS.FARADAY SOC. 36,225,356(1940).

T =   20.00

ORDER	A12	B12	C12
1	-2.3655	4.9552	-
2	-0.1289	1.7808	-1.2563

| | | | DEV. IN VAPOR PHASE COMPN. | |
X	Y	PRESS.	ORDER1	ORDER2
0.2220	0.0610	8.02	-0.0149	-0.0104
0.3000	0.1530	5.82	0.0278	0.0262
0.4000	0.5210	5.91	0.0068	-0.0060
0.5330	0.8860	11.84	-0.0096	-0.0091
0.7200	0.9826	26.49	0.0041	0.0067
		MEAN	0.0126	0.0117

NITRIC ACID(1)-WATER(2)

VANDONI M.R.,LANDY M.:J.CHIM.PHYS. 49,99(1952).

T =   20.00

ORDER	A12	B12	C12
1	-3.0676	6.1004	-
2	0.5784	0.9696	-2.1330

			DEV. IN VAPOR PHASE COMPN.	
X	Y	PRESS.	ORDER1	ORDER2
0.1604	0.0070	10.21	-0.0008	0.0016
0.1636	0.0080	9.90	-0.0011	0.0013
0.1956	0.0180	8.06	-0.0013	0.0019
0.2227	0.0360	7.19	-0.0033	-0.0001
0.2253	0.0360	7.29	-0.0013	0.0019
0.2574	0.0740	6.44	-0.0037	-0.0025
0.2856	0.1250	5.80	-0.0034	-0.0061
0.3060	0.1710	5.22	0.0022	-0.0043
0.4073	0.5350	5.34	0.0361	0.0225
0.4480	0.7350	6.76	-0.0126	-0.0173
0.4540	0.7150	6.80	0.0263	0.0230
0.5386	0.9296	10.10	-0.0166	-0.0087
0.7208	0.9920	27.54	0.0009	0.0038
		MEAN	0.0084	0.0073

NITRIC ACID(1) - WATER(2)

YAKIMOV M.A.,MISHIN V.YA.: RADIOKHIMIYA 6,545 (1964).

T =   25.00

ORDER	A12	B12	C12
1	-2.8778	5.8024	
2	-2.6277	5.4483	-0.1389

			DEV. IN VAPOR PHASE COMPN.	
X	Y	PRESS.	ORDER1	ORDER2
0.0980	0.0004	17.66	0.0004	0.0004
0.1107	0.0016	16.94	-0.0002	-0.0001
0.1354	0.0032	14.94	0.0002	0.0003
0.1656	0.0092	12.48	-0.0007	-0.0004
0.2107	0.0294	10.85	-0.0018	-0.0014
0.2680	0.0895	8.57	0.0043	0.0047
0.3096	0.2043	7.71	-0.0122	-0.0122
0.3634	0.3940	7.47	-0.0058	-0.0660
0.3967	0.4924	7.79	0.0353	0.0343
0.4129	0.5912	8.46	0.0019	0.0009
0.4727	0.7859	11.14	0.0023	-0.0019
0.5043	0.8749	13.58	-0.0185	-0.0185
0.5487	0.9170	18.03	0.0020	0.0022
0.6022	0.9572	23.91	0.0028	0.0031
0.6576	0.9789	31.26	0.0020	0.0022
0.7270	0.9956	41.65	-0.0031	-0.0030
		MEAN	0.0058	0.0058

NITRIC ACID(1) - WATER(2)

YAKIMOV M.A.,MISHIN V.YA.: RADIOKHIMIYA 6,545 (1964).

T =   35.00

ORDER	A12	B12	C12
1	-2.4956	5.1964	-
2	-1.8189	4.2421	-0.3919

			DEV. IN VAPOR PHASE COMPN.	
X	Y	PRESS.	ORDER1	ORDER2
0.0696	0.0001	35.30	0.0003	0.0003
0.0891	0.0008	32.77	0.0001	0.0003
0.1153	0.0022	29.87	0.0003	0.0006
0.1443	0.0063	26.45	0.0001	0.0006
0.1717	0.0133	23.57	0.0003	0.0011
0.2074	0.0333	20.01	-0.0010	-0.0001
0.2429	0.0675	17.18	0.0004	0.0012
0.2716	0.1117	15.73	0.0033	0.0034
0.3197	0.2555	12.68	-0.0157	-0.0175
0.3662	0.3993	14.38	0.0090	0.0059
0.4294	0.6484	16.02	-0.0011	-0.0031
0.5035	0.8228	23.79	0.0180	0.0187
0.5404	0.9104	28.85	-0.0134	-0.0120
		MEAN	0.0048	0.0050

NITRIC ACID(1) - WATER(2)

YAKIMOV M.A.,MISHIN V.YA.: RADIOKHIMIYA 6,545 (1964).

T =   50.00

ORDER		A12	B12	C12
1		-2.3969	4.9675	-
2		0.2274	1.2569	-1.4945

			DEV. IN VAPOR PHASE COMPN.	
X	Y	PRESS.	ORDER1	ORDER2
0.0761	0.0005	76.03	0.0000	0.0007
0.1001	0.0016	69.38	-0.0001	0.0011
0.1469	0.0090	58.28	-0.0020	0.0008
0.1615	0.0135	56.31	-0.0030	0.0004
0.2011	0.0282	47.26	-0.0005	0.0040
0.2630	0.1000	38.20	-0.0049	-0.0023
0.3287	0.2579	34.60	-0.0029	0.0094
0.3910	0.4809	34.07	-0.0007	-0.0131
0.4493	0.6598	39.93	0.0244	0.0172
0.4849	0.7647	43.80	0.0154	0.0135
0.5358	0.8844	57.49	-0.0096	-0.0058
0.5810	0.9351	70.26	-0.0092	-0.0034
		MEAN	0.0061	0.0060

475

NITRIC ACID(1) - WATER(2)

BOUBLIK T.,KUCHYNKA K.:COLL.CZECH.CHEM.COMM. 25,579(1960).

P =   50.00

ORDER	A12	B12	C12
1	-1.0593	2.7272	-
2	-3.5374	6.2673	1.3004

			DEV. IN VAPOR PHASE COMPN.	
X	Y	TEMP.	ORDER1	ORDER2
0.0930	0.0060	43.00	-0.0009	-0.0036
0.1530	0.0100	48.20	0.0138	0.0055
0.2400	0.0900	55.00	0.0136	0.0000
0.2830	0.1650	56.90	0.0116	0.0016
0.3530	0.3450	57.80	-0.0037	0.0002
0.3570	0.3580	57.90	-0.0061	-0.0014
0.3850	0.4390	57.60	-0.0113	-0.0014
0.3920	0.4580	57.10	-0.0112	-0.0002
0.5460	0.7950	50.40	-0.0017	0.0050
0.5510	0.8040	50.20	-0.0031	0.0031
0.5760	0.8480	47.40	-0.0123	-0.0089
0.6640	0.9230	38.50	-0.0032	-0.0066
0.6810	0.9190	36.60	0.0117	0.0075
0.7130	0.9410	32.60	0.0067	0.0016
0.7270	0.9500	32.50	0.0039	-0.0015
0.8870	0.9900	24.60	0.0017	-0.0017
		MEAN	0.0073	0.0031

NITRIC ACID(1) - WATER(2)

BOUBLIK T.,KUCHYNKA K.:COLL.CZECH.CHEM.COMM. 25,579(1960).

P = 100.00

ORDER	A12	B12	C12
1	-1.0892	2.7594	-
2	-4.1497	7.1267	1.6338

			DEV. IN VAPOR PHASE COMPN.	
X	Y	TEMP.	ORDER1	ORDER2
0.1490	0.0120	61.60	0.0090	0.0006
0.2120	0.0540	65.60	0.0127	-0.0013
0.2500	0.1060	68.10	0.0092	-0.0042
0.3100	0.2240	71.60	0.0054	0.0021
0.3540	0.3430	72.60	-0.0046	0.0031
0.3890	0.4430	71.70	-0.0099	0.0048
0.3980	0.4630	71.50	-0.0053	0.0107
0.4330	0.5760	70.10	-0.0245	-0.0063
0.5520	0.8170	61.90	-0.0168	-0.0116
0.6290	0.8810	55.60	0.0106	0.0065
0.6540	0.9040	54.00	0.0081	0.0020
0.6830	0.9260	50.80	0.0054	-0.0022
		MEAN	0.0101	0.0046

NITRIC ACID(1) - WATER(2)

BOUBLIK T.,KUCHYNKA K.:COLL.CZECH.CHEM.COMM. 25,579(1960).

P =   200.00

ORDER	A12	B12	C12
1	-0.9669	2.5486	-
2	-3.2091	5.7536	1.1839

			DEV. IN VAPOR PHASE COMPN.	
X	Y	TEMP.	ORDER1	ORDER2
0.0850	0.0010	71.70	0.0033	0.0011
0.1320	0.0110	75.40	0.0049	-0.0011
0.1440	0.0150	77.20	0.0057	-0.0013
0.1920	0.0400	79.50	0.0113	0.0004
0.2840	0.1710	84.80	0.0066	-0.0009
0.3560	0.3410	86.50	0.0008	0.0066
0.3970	0.4700	86.30	-0.0211	-0.0090
0.4320	0.5500	85.10	-0.0104	0.0040
0.5580	0.7990	73.40	-0.0027	0.0027
0.6530	0.8930	64.60	0.0090	0.0059
0.6860	0.9310	61.90	-0.0057	-0.0103
0.7160	0.9400	58.20	0.0021	-0.0033
0.9330	0.9860	49.80	0.0093	0.0070
		MEAN	0.0071	0.0041

478

NITRIC ACID(1) - WATER(2)

BOURLIK T.,KUCHYNKA K.:COLL.CZECH.CHEM.COMM. 25,579(1960).

P =   400.00

ORDER	A12	B12	C12
1	-0.8573	2.3530	-
2	-1.3573	3.0643	0.2700

			DEV. IN VAPOR PHASE COMPN.	
X	Y	TEMP.	ORDER1	ORDER2
0.1350	0.0160	91.90	0.0025	0.0009
0.1810	0.0450	96.00	0.0000	-0.0024
0.2960	0.2050	102.40	-0.0023	-0.0034
0.3570	0.3210	103.20	0.0196	0.0208
0.3980	0.4550	102.80	-0.0113	-0.0091
0.4350	0.5440	101.60	-0.0080	-0.0056
0.5330	0.7390	94.10	0.0032	0.0038
0.5370	0.7490	94.00	0.0000	0.0005
0.5460	0.7700	92.70	-0.0063	-0.0060
0.7180	0.9310	78.50	0.0048	0.0027
0.8800	0.9860	68.50	-0.0008	-0.0020
0.8890	0.9880	67.80	0.0006	-0.0005
		MEAN	0.0050	0.0048

NITRIC ACID(1) - WATER(2)

BOUBLIK T.,KUCHYNKA K.:COLL.CZECH.CHEM.COMM. 25,579(1960).

P = 760.00

ORDER	A12	B12	C12
1	-0.7252	2.1221	-
2	-0.4227	1.6902	-0.1616

			DEV. IN VAPOR PHASE COMPN.	
X	Y	TEMP.	ORDER1	ORDER2
0.0880	0.0080	106.50	-0.0018	-0.0012
0.0900	0.0110	106.50	-0.0044	-0.0037
0.1170	0.0170	108.50	-0.0033	-0.0023
0.1450	0.0260	110.70	-0.0007	0.0007
0.2580	0.1380	117.80	0.0004	0.0018
0.3550	0.3240	120.60	0.0089	0.0081
0.3610	0.3400	120.60	0.0070	0.0060
0.3980	0.4400	120.30	-0.0040	-0.0056
0.4030	0.4450	120.10	0.0031	0.0014
0.5160	0.7050	113.50	-0.0094	-0.0106
0.6680	0.8930	98.20	-0.0013	-0.0005
0.6980	0.9120	96.00	0.0018	0.0028
0.8580	0.9780	85.80	0.0018	0.0026
		MEAN	0.0037	0.0036

480

NITRIC ACID(1) - WATER(2)

POTIER J.: ALGER, SCIENCES PHYSIQUES 4,99 (1958).

P = 450.00

ORDER	A12	B12	C12
1	-0.9974	2.5078	-
2	-10.5939	16.2929	4.8895

			DEV. IN VAPOR PHASE COMPN.	
X	Y	TEMP.	ORDER1	ORDER2
0.0520	0.0000	88.60	0.0010	0.0000
0.0690	0.0001	89.70	0.0021	-0.0001
0.0890	0.0022	92.00	0.0023	-0.0020
0.0960	0.0024	92.40	0.0032	-0.0022
0.1810	0.0320	100.00	0.0062	-0.0224
0.2410	0.0810	102.00	0.0145	-0.0303
0.3000	0.1140	103.80	0.0764	0.0464
0.3020	0.0870	104.40	0.1073	0.0786
0.3220	0.2380	104.70	-0.0024	-0.0160
0.3160	0.2160	105.00	0.0068	-0.0118
0.3220	0.2490	105.20	-0.0134	-0.0270
0.3410	0.2710	105.30	0.0073	0.0111
0.3680	0.3670	106.00	-0.0235	0.0066
0.4570	0.6070	104.00	-0.0387	0.0396
0.4980	0.7260	101.00	-0.0647	0.0084
0.5370	0.8230	97.50	-0.0858	-0.0270
0.6300	0.8860	91.50	-0.0179	-0.0013
0.7250	0.9630	81.90	-0.0246	-0.0330
0.7590	0.9610	79.50	-0.0066	-0.0202
0.8240	0.9630	74.60	0.0127	-0.0055
0.8750	0.9700	73.10	0.0165	-0.0011
0.9010	0.9680	72.20	0.0225	0.0066
0.9140	0.9790	71.40	0.0132	-0.0014
0.9380	0.9675	70.60	0.0275	0.0157
0.9610	0.9740	70.00	0.0231	0.0149
		MEAN	0.0249	0.0172

POTIER J.: ALGER, SCIENCES PHYSIQUES 4,99 (1958).

P =   600.00

ORDER	A12	B12	C12
1	-1.4979	3.4070	-
2	2.0206	-1.6024	-1.9668

			DEV. IN VAPOR PHASE COMPN.	
X	Y	TEMP.	ORDER1	ORDER2
0.0170	0.0001	94.20	-0.0001	0.0001
0.0290	0.0008	96.00	-0.0007	-0.0001
0.0510	0.0010	96.90	-0.0005	0.0010
0.0420	0.0007	97.00	-0.0004	0.0007
0.0430	0.0008	97.10	-0.0005	0.0006
0.0480	0.0008	97.30	-0.0004	0.0010
0.0790	0.0029	98.60	-0.0011	0.0022
0.0720	0.0020	99.20	-0.0006	0.0022
0.0840	0.0030	100.00	-0.0008	0.0029
0.1080	0.0105	102.00	-0.0056	0.0000
0.1390	0.0180	103.50	-0.0065	0.0015
0.1640	0.0290	105.00	-0.0084	0.0011
0.1870	0.0400	106.20	-0.0069	0.0033
0.1870	0.0590	108.20	-0.0259	-0.0157
0.2590	0.0950	110.80	0.0152	0.0200
0.3380	0.2670	113.50	0.0147	0.0004
0.3690	0.3680	113.70	0.0023	-0.0188
0.3920	0.4230	113.60	0.0164	-0.0076
0.4390	0.5630	112.80	0.0154	-0.0075
0.5330	0.7270	108.00	0.0678	0.0644
0.5280	0.7420	107.90	0.0440	0.0395
0.5460	0.8230	105.00	-0.0067	-0.0074
0.5890	0.9130	102.10	-0.0388	-0.0328
0.6000	0.9450	99.10	-0.0588	-0.0517
0.6330	0.9290	97.00	-0.0129	-0.0035
0.7060	0.9800	93.30	-0.0216	-0.0119
0.6910	0.9600	91.70	-0.0082	0.0018
0.7280	0.9800	99.20	-0.0134	-0.0044
0.7220	0.9840	87.80	-0.0195	-0.0103
0.8910	0.9840	79.50	0.0111	0.0137
0.9610	0.9980	76.80	0.0008	0.0015
		MEAN	0.0137	0.0106

NITRIC ACID(1) - WATER(2)

POTIER J.: ALGER, SCIENCES PHYSIQUES 4,98 (1958).

P = 760.00

ORDER	A12	B12	C12
1	-1.3383	3.1677	-
2	2.5149	-2.3211	-2.1925

			DEV. IN VAPOR PHASE COMPN.	
X	Y	TEMP.	ORDER1	ORDER2
0.0670	0.0025	104.00	-0.0011	0.0021
0.0720	0.0025	104.50	-0.0008	0.0029
0.1020	0.0103	106.50	-0.0054	0.0008
0.1020	0.0105	106.70	-0.0056	0.0006
0.1100	0.0120	107.00	-0.0057	0.0011
0.1350	0.0200	108.50	-0.0077	0.0009
0.1410	0.0230	109.50	-0.0088	0.0002
0.1620	0.0350	110.50	-0.0122	-0.0023
0.1810	0.0420	111.50	-0.0085	0.0016
0.1810	0.0420	112.00	-0.0085	0.0016
0.2170	0.0820	114.50	-0.0185	-0.0105
0.2330	0.0960	115.50	-0.0142	-0.0083
0.2820	0.1650	117.50	-0.0048	-0.0102
0.3000	0.1910	118.00	0.0068	-0.0042
0.0337	0.2110	119.00	0.0776	0.0555
0.3480	0.2970	119.20	0.0215	-0.0036
0.4500	0.5640	118.50	0.0444	0.0177
0.4740	0.6510	117.00	0.0175	-0.0031
0.5150	0.7620	115.00	-0.0054	-0.0140
0.5150	0.7360	114.90	0.0206	0.0120
0.5300	0.7640	113.00	0.0199	0.0155
0.5400	0.7680	112.60	0.0327	0.0309
0.5570	0.8570	111.50	-0.0302	-0.0280
0.5740	0.8640	108.80	-0.0141	-0.0085
0.6060	0.9360	106.00	-0.0498	-0.0397
0.6490	0.9450	102.00	-0.0225	-0.0097
0.7000	0.9600	97.50	-0.0083	0.0044
0.7230	0.9840	95.80	-0.0237	-0.0109
0.7380	0.9860	95.50	-0.0195	-0.0083
0.7550	0.9830	92.00	-0.0112	-0.0009
0.8020	0.9830	91.00	-0.0003	0.0075
0.8530	0.9840	87.20	0.0064	0.0115
0.8780	0.9880	86.90	0.0051	0.0090
0.9910	0.9970	82.80	0.0027	0.0029
		MEAN	0.0159	0.0100

NITRIC ACID(1)-WATER(2)

PROSEK J.: THESIS, UTZCHT-CSAV, PRAGUE, 1965.

P = 760.00

ORDER	A12	B12	C12
1	-0.9462	2.4621	-
2	1.3723	-0.7936	-1.3705

			DEV. IN VAPOR PHASE COMPN.	
X	Y	TEMP.	ORDER1	ORDER2
0.1785	0.0439	111.15	-0.0046	0.0011
0.2391	0.1027	114.80	-0.0038	-0.0014
0.3484	0.2985	119.85	0.0090	-0.0001
0.3995	0.4300	119.35	0.0062	-0.0022
0.4183	0.4746	119.05	0.0097	0.0033
0.4708	0.6119	117.20	-0.0001	0.0024
0.4761	0.6261	117.20	-0.0023	0.0013
0.4951	0.6760	116.10	-0.0110	-0.0038
		MEAN	0.0058	0.0019

SULPHURIC ACID(1) - WATER(2)

LUCHINSKII G.P.: ZH. FIZ. KHIM. 30,6,1207(1956).

P = 760.00

ORDER	A12	B12	C12
1	-17.4082	10.6623	-
2	82.0884	-68.1304	-37.5232

			DEV. IN VAPOR PHASE COMPN.	
X	Y	TEMP.	ORDER1	ORDER2
0.7420	0.0380	308.10	-0.0066	0.0013
0.7760	0.0580	321.30	-0.0031	-0.0050
0.8150	0.2120	331.10	0.0250	-0.0021
0.8550	0.4750	336.80	0.0473	0.0166
0.9160	0.9160	340.20	-0.0220	-0.0020
		MEAN	0.0208	0.0054

485

THREE-COMPONENT SYSTEMS

3.1. Nonelectrolytes

ACETONE(1)-ACETONITRILE(2)-WATER(3)

PRATT N.R.C.:TRANS. INST.CHEM.ENGRS.(LONDON) 25,43(1947).

ANTOINE VAPOR PRESSURE CONSTANTS

	A	B	C
(1)	7.23967	1279.870	237.500
(2)	7.24299	1397.929	238.894
(3)	7.96681	1668.210	228.000

P = 760.00

EQUATION	ORDER	BINARY CONSTANTS				TERNARY CONSTANTS	
		I-J	1-2	2-3	3-1		
MARGULES	3	AIJ	0.0482	0.8145	0.6867	C123	0.3462
		AJI	-0.0447	0.7375	0.8484		
MARGULES	4	AIJ	0.0317	0.9344	0.7450	C1123	1.0992
		AJI	-0.0664	0.8793	0.8806	C1223	0.6870
		DIJ	-0.0611	0.4697	0.1722	C1233	0.4629
ALPHA	2	AIJ	1.4209	13.5373	4.1332		
		AJI	-0.5385	5.3282	34.0904		
ALPHA	3	AIJ	9.9438	9.9341	4.9334	A123	3.4167
		AJI	3.2436	8.6222	2.4809	A231	-4.8283
		AIJJ	-8.3533	8.6752	-5.4350	A312	2.0635
		AJII	-3.8033	-5.3311	37.5587		

------------EXPERIMENTAL---------------					---MARGULES 3---		---MARGULES 4---		-----ALPHA 2----		-----ALPHA 3----	
X1	X2	Y1	Y2	TEMP.	ΔY1	ΔY2	ΔY1	ΔY2	ΔY1	ΔY2	ΔY1	ΔY2
0.0025	0.0168	0.0636	0.1960	92.20	-0.0123	-0.0450	-0.0113	-0.0135	-0.0032	-0.0251	-0.0004	0.0105
0.0043	0.0507	0.0805	0.3955	84.80	-0.0157	-0.0658	-0.0167	-0.0316	-0.0105	-0.0562	-0.0138	-0.0204
0.0084	0.1050	0.0751	0.4820	78.70	0.0076	-0.0232	0.0065	-0.0089	0.0115	-0.0333	0.0027	-0.0122
0.0136	0.0241	0.2330	0.1970	84.00	-0.0113	-0.0303	-0.0120	-0.0020	0.0045	-0.0227	0.0060	0.0063
0.0137	0.1960	0.0755	0.5440	76.80	0.0066	0.0004	0.0074	-0.0072	0.0136	-0.0158	0.0010	-0.0094
0.0192	0.2860	0.0756	0.5620	76.60	0.0051	0.0145	0.0066	0.0005	0.0159	0.0028	0.0012	0.0024
0.0198	0.6810	0.0354	0.6680	76.80	0.0051	-0.0002	0.0022	0.0149	0.0120	0.0073	0.0035	0.0086
0.0200	0.0284	0.3140	0.2010	80.00	-0.0252	-0.0283	-0.0285	-0.0018	-0.0148	-0.0250	-0.0170	0.0017
0.0258	0.4190	0.0704	0.5700	76.20	0.0047	0.0306	0.0050	0.0243	0.0187	0.0291	0.0028	0.0249
0.0364	0.6610	0.0721	0.6360	75.40	0.0018	0.0063	-0.0034	0.0219	0.0134	0.0074	-0.0021	0.0089
0.0429	0.8834	0.0801	0.7910	77.60	0.0077	-0.0109	0.0001	-0.0157	0.0057	-0.0589	0.0001	-0.0330
0.0608	0.0582	0.4450	0.2190	73.20	0.0257	-0.0302	0.0111	-0.0130	0.0259	-0.0322	0.0014	-0.0114
0.0810	0.6420	0.1430	0.5750	74.20	0.0124	0.0107	0.0013	0.0283	0.0319	-0.0004	0.0007	0.0044
0.0890	0.7430	0.1560	0.6130	75.00	0.0082	0.0166	-0.0050	0.0319	0.0178	-0.0114	-0.0073	0.0025
0.0961	0.8190	0.1720	0.6850	75.50	0.0099	0.0006	-0.0023	0.0074	0.0085	-0.0471	-0.0057	-0.0222
0.1110	0.6110	0.1870	0.5420	73.00	0.0210	0.0019	0.0071	0.0198	0.0443	-0.0141	0.0031	-0.0076
0.1390	0.7270	0.2380	0.5880	73.00	0.0081	-0.0010	-0.0074	0.0154	0.0141	-0.0408	-0.0166	-0.0214
0.1440	0.0455	0.6280	0.1060	67.90	0.0463	-0.0202	0.0286	-0.0111	0.0425	-0.0182	0.0155	-0.0056
0.1620	0.0540	0.6240	0.1048	66.00	0.0571	-0.0136	0.0387	-0.0052	0.0522	-0.0114	0.0204	0.0018

--------------EXPERIMENTAL--------------					---MARGULES 3---		---MARGULES 4---		-----ALPHA 2----		-----ALPHA 3----	
X1	X2	Y1	Y2	TEMP.	ΔY1	ΔY2	ΔY1	ΔY2	ΔY1	ΔY2	ΔY1	ΔY2
0.1830	0.1410	0.5170	0.2090	68.40	0.0579	-0.0202	0.0367	-0.0164	0.0707	-0.0212	0.0128	-0.0033
0.1960	0.5420	0.3160	0.4330	71.20	0.0233	0.0092	0.0048	0.0267	0.0497	-0.0133	-0.0127	-0.0007
0.2000	0.6400	0.3340	0.4920	71.20	0.0002	0.0032	-0.0168	0.0224	0.0111	-0.0334	-0.0355	-0.0146
0.2680	0.5010	0.4170	0.3680	68.60	0.0168	0.0092	-0.0013	0.0250	0.0363	-0.0160	-0.0350	0.0018
0.2860	0.5640	0.4360	0.4070	69.60	0.0091	-0.0005	-0.0060	0.0161	0.0157	-0.0364	-0.0392	-0.0138
0.3210	0.1870	0.6070	0.1740	65.40	0.0294	-0.0162	0.0140	-0.0160	0.0491	-0.0153	-0.0277	0.0037
0.3380	0.4520	0.5000	0.3210	67.60	0.0149	-0.0041	-0.0001	0.0081	0.0276	-0.0275	-0.0479	-0.0061
0.3420	0.4690	0.5080	0.3290	67.50	0.0078	-0.0033	-0.0064	0.0094	0.0168	-0.0295	-0.0542	-0.0072
0.3580	0.0918	0.7110	0.0890	63.20	0.0259	-0.0136	0.0194	-0.0136	0.0376	-0.0093	-0.0161	0.0049
0.4480	0.3420	0.6150	0.2260	65.00	0.0187	-0.0099	0.0110	-0.0061	0.0213	-0.0236	-0.0588	0.0005
0.4900	0.2640	0.6660	0.1750	64.30	0.0183	-0.0148	0.0149	-0.0155	0.0225	-0.0205	-0.0611	0.0031
				MEAN	0.0171	0.0152	0.0111	0.0150	0.0240	0.0235	0.0174	0.0092

ACETONE(1)-METHYL ALCOHOL(2)-ETHYL ALCOHOL(3)

AMER H.H.,PAXTON R.R.,VAN WINKLE M.:IND.ENG.CHEM.48,142(1956).

ANTOINE VAPOR PRESSURE CONSTANTS
```
         A          B          C
(1) 7.32967    1279.870    237.500
(2) 8.07246    1574.990    238.860
(3) 8.16290    1623.220    228.980
```

P = 760.00

EQUATION	ORDER	BINARY CONSTANTS					TERNARY CONSTANTS	
		I-J	1-2	2-3	3-1			
MARGULES	3	AIJ	0.2885	0.0019	0.0828		C123	0.2454
		AJI	0.2501	-0.0242	0.3197			
MARGULES	4	AIJ	0.2802	0.0448	0.0083		C1123	-0.1991
		AJI	0.2411	0.0224	0.2656		C1223	0.3930
		DIJ	-0.0275	0.1326	-0.2119		C1233	0.2992
ALPHA	2	AIJ	1.3848	0.5932	-0.0272			
		AJI	0.4605	-0.4172	2.2137			
ALPHA	3	AIJ	1.4753	-5.5924	1.0105		A123	-7.3980
		AJI	0.6220	-4.1890	3.2885		A231	2.8961
		AIJJ	-0.0259	6.2213	-1.2058		A312	0.5736
		AJII	-0.1861	3.7983	-0.5695			

--------------EXPERIMENTAL---------------					---MARGULES 3---		---MARGULES 4---		-----ALPHA 2----		-----ALPHA 3----	
X1	X2	Y1	Y2	TEMP.	ΔY1	ΔY2	ΔY1	ΔY2	ΔY1	ΔY2	ΔY1	ΔY2
0.0190	0.0460	0.0600	0.0700	76.50	0.0241	-0.0005	0.0157	0.0044	-0.0032	-0.0014	0.0036	-0.0022
0.0190	0.1000	0.0630	0.1460	75.90	0.0163	0.0005	0.0084	0.0065	-0.0074	-0.0009	-0.0014	-0.0034
0.0220	0.1600	0.0740	0.2220	74.80	0.0117	0.0027	0.0038	0.0079	-0.0116	0.0018	-0.0058	-0.0042
0.0180	0.2200	0.0520	0.3130	74.00	0.0159	-0.0097	0.0096	-0.0064	-0.0015	-0.0118	0.0031	-0.0207
0.0170	0.2810	0.0570	0.3700	73.00	0.0046	0.0068	-0.0008	0.0090	-0.0102	0.0042	-0.0056	-0.0121
0.0170	0.3300	0.0480	0.4340	72.30	0.0116	-0.0020	0.0066	0.0002	-0.0020	-0.0047	0.0052	-0.0433
0.0220	0.3800	0.0530	0.4890	71.30	0.0205	-0.0094	0.0150	-0.0057	0.0048	-0.0106	0.0049	0.0909
0.0200	0.4290	0.0480	0.5520	70.70	0.0175	-0.0201	0.0126	-0.0160	0.0040	-0.0220	0.0019	-0.0107
0.0210	0.4850	0.0530	0.5910	69.80	0.0138	-0.0063	0.0090	-0.0009	0.0006	-0.0080	-0.0015	-0.0088
0.0190	0.5350	0.0450	0.6450	69.30	0.0146	-0.0117	0.0101	-0.0059	0.0030	-0.0138	-0.0001	-0.0190
0.0130	0.5830	0.0350	0.6800	69.00	0.0059	0.0046	0.0025	0.0097	-0.0021	0.0007	-0.0058	-0.0044
0.0170	0.6490	0.0450	0.7300	68.00	0.0068	0.0027	0.0027	0.0082	-0.0031	0.0008	-0.0125	-0.0146
0.0170	0.7030	0.0450	0.7700	67.30	0.0062	0.0043	0.0022	0.0090	-0.0037	0.0034	-0.0397	-0.0438
0.0200	0.6900	0.0570	0.7580	67.20	0.0029	0.0015	-0.0016	-0.0067	-0.0086	0.0013	-0.0376	-0.0429
0.0180	0.7960	0.0450	0.8370	66.00	0.0084	0.0011	0.0047	0.0034	-0.0023	0.0032	0.0070	0.0087
0.0110	0.8500	0.0340	0.8820	65.60	-0.0007	0.0050	-0.0029	0.0051	-0.0078	0.0057	-0.0041	0.0051
0.0170	0.8960	0.0410	0.9110	65.00	0.0094	-0.0088	0.0069	-0.0089	-0.0014	-0.0027	0.0015	-0.0037
0.0200	0.9370	0.0400	0.9310	64.40	0.0189	-0.0124	0.0169	-0.0124	0.0059	-0.0019	0.0078	-0.0032
0.0490	0.0450	0.1550	0.0690	74.80	0.0332	-0.0076	0.0202	-0.0031	-0.0194	-0.0057	-0.0081	-0.0088

		EXPERIMENTAL			MARGULES 3		MARGULES 4		ALPHA 2		ALPHA 3	
X1	X2	Y1	Y2	TEMP.	$\Delta Y1$	$\Delta Y2$	$\Delta Y1$	$\Delta Y2$	$\Delta Y1$	$\Delta Y2$	$\Delta Y1$	$\Delta Y2$
0.0620	0.1050	0.1700	0.1500	73.40	0.0444	-0.0152	0.0327	-0.0089	-0.0075	-0.0098	0.0014	-0.0189
0.0600	0.1680	0.1570	0.2330	72.40	0.0428	-0.0213	0.0329	-0.0148	-0.0025	-0.0146	0.0031	-0.0289
0.0600	0.2400	0.1520	0.3130	71.40	0.0385	-0.0190	0.0304	-0.0125	-0.0014	-0.0113	-0.0008	-0.0318
0.0620	0.3030	0.1550	0.3750	70.50	0.0333	-0.0144	0.0266	-0.0076	-0.0033	-0.0062	-0.0113	-0.0314
0.0580	0.3790	0.1390	0.4590	69.70	0.0325	-0.0178	0.0263	-0.0104	0.0008	-0.0104	-0.0466	-0.0402
0.0640	0.4700	0.1520	0.5360	68.30	0.0267	-0.0151	0.0209	-0.0064	-0.0039	-0.0065	0.0480	-0.0951
0.0600	0.5310	0.1430	0.5850	67.70	0.0229	-0.0070	0.0166	0.0019	-0.0053	0.0012	0.0133	-0.0851
0.0620	0.6910	0.1330	0.7160	65.40	0.0312	-0.0181	0.0239	-0.0115	0.0029	-0.0053	0.0148	0.0768
0.0580	0.7360	0.1300	0.7490	65.00	0.0243	-0.0157	0.0171	-0.0103	-0.0031	-0.0017	0.0073	0.0347
0.0590	0.7930	0.1330	0.7850	64.30	0.0226	-0.0177	0.0159	-0.0141	-0.0059	0.0002	0.0011	0.0145
0.0600	0.8400	0.1310	0.8150	63.80	0.0263	-0.0223	0.0206	-0.0201	-0.0034	-0.0004	0.0018	0.0054
0.0580	0.8900	0.1260	0.8400	63.20	0.0269	-0.0180	0.0228	-0.0166	-0.0033	0.0080	0.0002	0.0087
0.1180	0.0510	0.2800	0.0830	71.40	0.0675	-0.0243	0.0570	-0.0209	-0.0015	-0.0190	0.0067	-0.0258
0.1230	0.1200	0.2860	0.1620	70.30	0.0561	-0.0273	0.0505	-0.0222	-0.0056	-0.0171	-0.0038	-0.0317
0.1130	0.1880	0.2570	0.2360	69.90	0.0547	-0.0250	0.0513	-0.0192	-0.0001	-0.0127	-0.0042	-0.0320
0.1140	0.2530	0.2510	0.3020	69.10	0.0521	-0.0239	0.0510	-0.0177	0.0021	-0.0105	-0.0123	-0.0300
0.1220	0.3170	0.2690	0.3560	67.90	0.0394	-0.0187	0.0405	-0.0125	-0.0078	-0.0044	-0.0409	-0.0116
0.1270	0.4040	0.2600	0.4400	66.60	0.0465	-0.0257	0.0479	-0.0194	0.0022	-0.0108	-0.0712	0.0474
0.1160	0.6670	0.2340	0.6520	63.70	0.0345	-0.0296	0.0292	-0.0255	-0.0060	-0.0074	-0.0158	0.0624
0.1070	0.7350	0.2210	0.6990	63.10	0.0298	-0.0264	0.0237	-0.0232	-0.0103	-0.0004	-0.0104	0.0333
0.1060	0.7870	0.2190	0.7340	62.60	0.0287	-0.0314	0.0231	-0.0294	-0.0124	-0.0006	-0.0104	0.0159
0.1150	0.8380	0.2200	0.7610	61.80	0.0420	-0.0434	0.0385	-0.0425	-0.0024	-0.0041	-0.0003	0.0009
0.1850	0.0630	0.3890	0.0860	68.80	0.0569	-0.0208	0.0509	-0.0183	-0.0074	-0.0137	-0.0072	-0.0233
0.1920	0.1280	0.3860	0.1630	67.70	0.0559	-0.0328	0.0570	-0.0295	-0.0034	-0.0212	-0.0109	-0.0372
0.1900	0.2010	0.3780	0.2290	67.00	0.0481	-0.0268	0.0541	-0.0233	-0.0070	-0.0125	-0.0249	-0.0290
0.1990	0.2680	0.3730	0.2950	65.90	0.0535	-0.0320	0.0627	-0.0290	0.0015	-0.0167	-0.0307	-0.0212
0.1930	0.3370	0.3600	0.3530	65.20	0.0493	-0.0267	0.0583	-0.0239	-0.0006	-0.0106	-0.0494	0.0124
0.1920	0.4010	0.3570	0.3980	64.30	0.0433	-0.0176	0.0509	-0.0151	-0.0055	-0.0005	-0.0681	0.0649
0.1860	0.4650	0.3430	0.4570	63.70	0.0426	-0.0236	0.0477	-0.0214	-0.0051	-0.0051	-0.0698	0.0994
0.1830	0.5900	0.3300	0.5560	62.30	0.0406	-0.0333	0.0407	-0.0326	-0.0069	-0.0085	-0.0362	0.0775
0.1710	0.6540	0.3150	0.5980	62.00	0.0358	-0.0282	0.0334	-0.0276	-0.0115	0.0006	-0.0260	0.0585
0.1760	0.7090	0.3150	0.6300	61.20	0.0394	-0.0334	0.0365	-0.0337	-0.0096	0.0022	-0.0158	0.0344
0.1620	0.7850	0.2950	0.6730	60.70	0.0383	-0.0279	0.0356	-0.0280	-0.0117	0.0157	-0.0121	0.0261
0.2690	0.0540	0.4860	0.0750	66.30	0.0528	-0.0234	0.0450	-0.0219	-0.0024	-0.0175	-0.0083	-0.0262
0.2650	0.1440	0.4640	0.1730	65.40	0.0547	-0.0360	0.0596	-0.0339	0.0014	-0.0245	-0.0135	-0.0393
0.2650	0.2170	0.4540	0.2360	64.40	0.0529	-0.0322	0.0632	-0.0305	0.0008	-0.0185	-0.0246	-0.0264
0.2830	0.2790	0.4610	0.2980	63.20	0.0534	-0.0429	0.0655	-0.0422	0.0026	-0.0283	-0.0326	-0.0158
0.2760	0.3500	0.4400	0.3500	62.50	0.0577	-0.0354	0.0692	-0.0357	0.0073	-0.0191	-0.0377	0.0249
0.2680	0.4270	0.4370	0.3930	61.70	0.0432	-0.0179	0.0520	-0.0193	-0.0071	0.0012	-0.0544	0.0770
0.2630	0.4940	0.4180	0.4570	61.00	0.0496	-0.0338	0.0554	-0.0359	-0.0008	-0.0107	-0.0395	0.0717
0.2550	0.5630	0.4080	0.5050	60.50	0.0447	-0.0350	0.0472	-0.0375	-0.0060	-0.0067	-0.0316	0.0612
0.2560	0.6270	0.4030	0.5430	59.80	0.0449	-0.0373	0.0453	-0.0399	-0.0068	-0.0014	-0.0203	0.0408
0.2530	0.6840	0.3920	0.5800	59.30	0.0487	-0.0443	0.0482	-0.0463	-0.0043	-0.0004	-0.0102	0.0206
0.3790	0.0740	0.5690	0.0930	63.30	0.0576	-0.0257	0.0477	-0.0239	0.0108	-0.0196	0.0012	-0.0299
0.3680	0.1570	0.5520	0.1830	62.60	0.0527	-0.0412	0.0561	-0.0392	0.0039	-0.0313	-0.0118	-0.0408
0.3680	0.2370	0.5310	0.2550	61.70	0.0606	-0.0447	0.0694	-0.0437	0.0111	-0.0325	-0.0122	-0.0251
0.3640	0.3090	0.5170	0.3080	60.80	0.0604	-0.0396	0.0701	-0.0400	0.0102	-0.0248	-0.0193	0.0080
0.3570	0.3750	0.5130	0.3500	60.20	0.0490	-0.0314	0.0575	-0.0332	-0.0013	-0.0133	-0.0332	0.0410
0.3570	0.4490	0.5010	0.4060	59.40	0.0503	-0.0371	0.0557	-0.0400	-0.0003	-0.0132	-0.0267	0.0475
0.3520	0.5180	0.4860	0.4570	58.70	0.0525	-0.0453	0.0553	-0.0485	0.0013	-0.0138	-0.0166	0.0347
0.3470	0.5850	0.4750	0.4970	58.10	0.0519	-0.0481	0.0528	-0.0507	-0.0003	-0.0071	-0.0094	0.0197
0.4910	0.0740	0.6520	0.0950	61.20	0.0518	-0.0285	0.0357	-0.0253	0.0066	-0.0237	0.0017	-0.0324

489

		EXPERIMENTAL			MARGULES 3		MARGULES 4		ALPHA 2		ALPHA 3	
X1	X2	Y1	Y2	TEMP.	ΔY1	ΔY2	ΔY1	ΔY2	ΔY1	ΔY2	ΔY1	ΔY2
0.4880	0.1600	0.6300	0.1900	60.10	0.0569	-0.0487	0.0543	-0.0448	0.0083	-0.0400	0.0019	-0.0430
0.4750	0.2530	0.6050	0.2600	59.30	0.0564	-0.0429	0.0595	-0.0407	0.0066	-0.0297	-0.0051	-0.0106
0.4690	0.3310	0.5910	0.3160	58.60	0.0519	-0.0410	0.0553	-0.0407	0.0021	-0.0221	-0.0121	0.0150
0.4740	0.3840	0.5810	0.3540	57.80	0.0551	-0.0441	0.0572	-0.0448	0.0052	-0.0189	-0.0071	0.0199
0.4710	0.4570	0.5690	0.3980	57.10	0.0522	-0.0441	0.0528	-0.0452	0.0025	-0.0083	-0.0055	0.0182
0.6380	0.0750	0.7390	0.0970	58.90	0.0540	-0.0285	0.0377	-0.0231	0.0045	-0.0234	0.0108	-0.0297
0.6220	0.1580	0.7080	0.1820	58.10	0.0560	-0.0424	0.0495	-0.0361	0.0057	-0.0314	0.0121	-0.0296
0.6190	0.2310	0.6810	0.2460	57.20	0.0629	-0.0492	0.0600	-0.0444	0.0132	-0.0305	0.0167	-0.0164
0.6350	0.2660	0.6930	0.2630	56.60	0.0488	-0.0412	0.0465	-0.0376	0.0004	-0.0158	0.0028	-0.0001
0.7680	0.0820	0.8150	0.1030	57.20	0.0445	-0.0266	0.0368	-0.0208	-0.0048	-0.0167	0.0092	-0.0201
0.7670	0.1660	0.7650	0.1830	56.10	0.0636	-0.0363	0.0607	-0.0321	0.0184	-0.0116	0.0275	-0.0087
0.8990	0.0680	0.8930	0.0800	56.00	0.0294	-0.0148	0.0292	-0.0135	-0.0023	0.0027	0.0052	0.0011
				MEAN	0.0385	0.0245	0.0365	0.0228	0.0051	0.0112	0.0163	0.0300

ACETONE(1)-METHYL ALCOHOL(2)-WATER(3)

CHU J.C.,GETTY R.J.,BRENNECKE L.F.,PAUL R.:DISTILLATION
EQUILIBRIUM DATA,REINHOLD,NEW YORK 1950.

ANTOINE VAPOR PRESSURE CONSTANTS

	A	B	C
(1)	7.23967	1279.870	237.500
(2)	8.07246	1574.990	238.860
(3)	7.96681	1668.210	228.000

P = 760.00

EQUATION	ORDER	BINARY	CONSTANTS			TERNARY	CONSTANTS
		I-J	1-2	2-3	3-1		
MARGULES	3	AIJ	0.2434	0.3292	0.6867	C123	0.3588
		AJI	0.3320	0.1542	0.8484		
MARGULES	4	AIJ	0.1973	0.3840	0.7450	C1123	0.7708
		AJI	0.2807	0.3916	0.8806	C1223	0.7651
		DIJ	-0.1650	0.4978	0.1722	C1233	0.3566
ALPHA	2	AIJ	1.5257	5.1719	4.1338		
		AJI	0.3644	-0.1490	34.0904		
ALPHA	3	AIJ	-0.7745	16.6605	4.9334	A123	12.7667
		AJI	-1.1155	3.9462	2.4809	A231	3.6840
		AIJJ	2.0965	-9.2108	-5.4350	A312	-6.7133
		AJII	1.2779	-4.2185	37.5587		

--------------EXPERIMENTAL---------------					---MARGULES 3---		---MARGULES 4---		-----ALPHA 2----		-----ALPHA 3----	
X1	X2	Y1	Y2	TEMP.	ΔY1	ΔY2	ΔY1	ΔY2	ΔY1	ΔY2	ΔY1	ΔY2
0.0124	0.0613	0.2090	0.2257	82.80	-0.0251	0.0266	-0.0172	0.0302	0.0258	-0.0198	0.0023	0.0176
0.0188	0.3778	0.0912	0.6450	72.80	0.0054	0.0015	0.0118	0.0036	0.0694	-0.0434	0.0011	0.0091
0.0253	0.6722	0.0773	0.7869	67.50	0.0006	0.0169	-0.0090	0.0360	0.0558	-0.0495	-0.0135	0.0334
0.0331	0.5564	0.0953	0.7191	69.20	0.0203	0.0025	0.0144	0.0225	0.0988	-0.0734	0.0050	0.0250
0.0429	0.8710	0.0945	0.8710	64.20	0.0055	-0.0011	-0.0078	0.0049	0.0477	-0.0699	-0.0143	0.0139
0.0449	0.8439	0.1018	0.8496	64.40	0.0051	0.0041	-0.0099	0.0125	0.0556	-0.0750	-0.0172	0.0225
0.0628	0.3042	0.2731	0.4420	70.00	0.0205	0.0274	0.0325	0.0196	0.1377	-0.0710	0.0073	0.0461
0.0639	0.1166	0.4383	0.2690	72.80	0.0124	-0.0278	0.0179	-0.0329	0.0939	-0.0935	0.0180	-0.0302
0.0724	0.4985	0.2176	0.6139	67.50	0.0179	-0.0084	0.0112	0.0084	0.1387	-0.1281	-0.0109	0.0312
0.0757	0.7150	0.1830	0.7230	64.70	0.0049	0.0143	-0.0170	0.0361	0.1010	-0.1096	-0.0329	0.0561
0.0840	0.0401	0.6037	0.0858	69.70	0.0187	-0.0030	0.0106	0.0004	0.0464	-0.0300	0.0209	-0.0043
0.0905	0.3940	0.3042	0.4955	67.60	0.0142	0.0020	0.0163	0.0067	0.1441	-0.1190	-0.0133	0.0406
0.0910	0.1828	0.4154	0.3280	69.70	0.0453	-0.0369	0.0522	-0.0449	0.1455	-0.1208	0.0368	-0.0254
0.0952	0.2668	0.3817	0.3746	68.60	0.0211	0.0022	0.0300	-0.0051	0.1401	-0.1009	0.0036	0.0266
0.0971	0.6271	0.2352	0.6615	64.40	0.0134	-0.0078	-0.0076	0.0171	0.1315	-0.1482	-0.0314	0.0459
0.1120	0.7580	0.2312	0.7180	62.50	0.0099	-0.0029	-0.0127	0.0150	0.1024	-0.1309	-0.0392	0.0500
0.1314	0.3729	0.3633	0.4565	65.60	0.0430	-0.0294	0.0422	-0.0241	0.1747	-0.1579	0.0065	0.0202
0.1515	0.4050	0.3827	0.4400	65.10	0.0354	-0.0063	0.0299	0.0038	0.1702	-0.1443	-0.0079	0.0525
0.1665	0.2395	0.4771	0.2904	65.60	0.0617	-0.0135	0.0614	-0.0146	0.1690	-0.1127	0.0349	0.0236

		EXPERIMENTAL			MARGULES 3		MARGULES 4		ALPHA 2		ALPHA 3	
X1	X2	Y1	Y2	TEMP.	$\Delta Y1$	$\Delta Y2$	$\Delta Y1$	$\Delta Y2$	$\Delta Y1$	$\Delta Y2$	$\Delta Y1$	$\Delta Y2$
0.1696	0.7026	0.2960	0.6407	61.10	0.0354	-0.0136	0.0140	0.0038	0.1393	-0.1575	-0.0291	0.0596
0.1697	0.3776	0.4226	0.4126	64.70	0.0329	-0.0164	0.0281	-0.0081	0.1645	-0.1512	-0.0110	0.0436
0.1758	0.4412	0.3955	0.4543	63.90	0.0360	-0.0126	0.0253	0.0020	0.1712	-0.1603	-0.0168	0.0570
0.1954	0.3476	0.4190	0.3970	63.90	0.0816	-0.0438	0.0768	-0.0368	0.2071	-0.1734	0.0364	0.0179
0.2191	0.1954	0.5742	0.2206	64.20	0.0466	-0.0141	0.0411	-0.0125	0.1359	-0.0982	0.0200	0.0231
0.2238	0.6497	0.3800	0.5694	60.30	0.0218	-0.0116	0.0032	0.0037	0.1293	-0.1622	-0.0527	0.0754
0.2365	0.0795	0.6920	0.1055	63.60	0.0382	-0.0191	0.0241	-0.0154	0.0795	-0.0564	0.0247	-0.0042
0.2708	0.4974	0.4574	0.4505	60.90	0.0357	-0.0188	0.0191	-0.0023	0.1606	-0.1805	-0.0419	0.0782
0.2712	0.2395	0.5813	0.2482	62.80	0.0421	-0.0201	0.0366	-0.0161	0.1401	-0.1201	0.0042	0.0341
0.2712	0.5115	0.4612	0.4500	60.60	0.0270	-0.0085	0.0099	0.0081	0.1506	-0.1709	-0.0523	0.0897
0.2950	0.3475	0.5210	0.3543	61.50	0.0551	-0.0443	0.0465	-0.0347	0.1709	-0.1787	-0.0053	0.0368
0.3155	0.4865	0.4865	0.4265	60.30	0.0389	-0.0162	0.0238	-0.0021	0.1573	-0.1765	-0.0441	0.0872
0.3314	0.2956	0.5614	0.2563	61.70	0.0589	0.0044	0.0522	0.0114	0.1649	-0.1168	0.0045	0.0802
0.3330	0.1383	0.7007	0.1461	62.10	0.0167	-0.0185	0.0093	-0.0153	0.0819	-0.0809	-0.0067	0.0177
0.3343	0.5475	0.5000	0.4550	58.70	0.0140	-0.0067	0.0010	0.0038	0.1163	-0.1567	-0.0725	0.0978
0.3510	0.4905	0.4935	0.4700	58.60	0.0511	-0.0658	0.0376	-0.0540	0.1598	-0.2214	-0.0379	0.0433
0.3842	0.5793	0.4972	0.4871	57.50	0.0296	-0.0254	0.0258	-0.0234	0.0854	-0.1140	-0.0290	0.0452
0.3875	0.3073	0.5952	0.2795	60.60	0.0423	-0.0175	0.0354	-0.0103	0.1449	-0.1451	-0.0225	0.0701
0.4085	0.1134	0.7150	0.1039	60.90	0.0382	-0.0040	0.0341	-0.0017	0.0937	-0.0574	0.0166	0.0314
0.4242	0.2615	0.6447	0.2145	59.80	0.0287	0.0077	0.0241	0.0126	0.1201	-0.1063	-0.0317	0.0899
0.4315	0.0940	0.7437	0.0966	61.10	0.0254	-0.0140	0.0225	-0.0122	0.0751	-0.0594	0.0085	0.0166
0.4715	0.4745	0.5726	0.4050	56.50	0.0258	-0.0213	0.0194	-0.0169	0.0931	-0.1323	-0.0507	0.0720
0.4722	0.1043	0.7366	0.1088	60.00	0.0341	-0.0183	0.0340	-0.0173	0.0854	-0.0697	0.0114	0.0192
0.4778	0.1572	0.7239	0.1185	59.50	0.0179	0.0165	0.0177	0.0179	0.0836	-0.0591	-0.0210	0.0736
0.5290	0.0865	0.7910	0.0605	59.40	-0.0007	0.0149	0.0028	0.0147	0.0433	-0.0300	-0.0218	0.0502
0.5590	0.3807	0.6334	0.3396	56.50	0.0248	-0.0195	0.0183	-0.0152	0.0910	-0.1340	-0.0546	0.0785
0.5702	0.2035	0.7211	0.1687	58.10	0.0165	0.0082	0.0167	0.0083	0.0851	-0.0929	-0.0488	0.0942
0.6140	0.2002	0.7315	0.1770	57.50	0.0171	0.0008	0.0175	-0.0002	0.0800	-0.1013	-0.0535	0.0914
0.6209	0.1633	0.7545	0.1388	57.80	0.0127	0.0075	0.0156	0.0055	0.0688	-0.0799	-0.0467	0.0854
0.6750	0.1025	0.7865	0.1080	57.60	0.0217	-0.0121	0.0286	-0.0158	0.0581	-0.0728	-0.0218	0.0450
0.7444	0.1880	0.7832	0.1805	56.00	0.0001	0.0049	-0.0003	0.0026	0.0389	-0.0868	-0.0645	0.0854
0.7529	0.1708	0.7852	0.1764	56.10	0.0084	-0.0000	0.0090	-0.0093	0.0445	-0.0951	-0.0558	0.0739
0.8100	0.0871	0.8370	0.0933	56.50	0.0141	0.0009	0.0187	-0.0049	0.0243	-0.0568	-0.0313	0.0566
0.8220	0.0406	0.8686	0.0633	57.00	0.0106	-0.0184	0.0171	-0.0226	0.0047	-0.0482	-0.0160	0.0125
0.8536	0.0497	0.8846	0.0569	56.20	0.0021	0.0004	0.0066	-0.0046	-0.0084	-0.0358	-0.0272	0.0355
0.8993	0.0370	0.9391	0.0434	56.50	-0.0258	0.0028	-0.0231	-0.0017	-0.0498	-0.0246	-0.0418	0.0246
				MEAN	0.0256	0.0144	0.0227	0.0143	0.1047	0.1047	0.0264	0.0467

ACETONE(1)-METHYL ALCOHOL(2)-WATER(3)

GRISWOLD J.,BUFORD C.B.:IND.ENG.CHEM. 41,2347(1949).

ANTOINE VAPOR PRESSURE CONSTANTS

	A	B	C
(1)	7.23967	1279.870	237.500
(2)	8.07246	1574.990	238.860
(3)	7.96681	1668.210	228.000

P = 760.00

EQUATION	ORDER	BINARY CONSTANTS				TERNARY CONSTANTS	
		I-J	1-2	2-3	3-1		
MARGULES	3	AIJ	0.1523	0.3292	0.6867	C123	0.5806
		AJI	0.2911	0.1542	0.8484		
MARGULES	4	AIJ	0.1616	0.3840	0.7450	C1123	-0.0181
		AJI	0.3029	0.3916	0.8806	C1223	0.9123
		DIJ	0.0344	0.4978	0.1722	C1233	2.0096
ALPHA	2	AIJ	1.2380	5.1719	4.1338		
		AJI	0.2393	-0.1490	34.0904		
ALPHA	3	AIJ	2.3944	16.6605	4.9334	A123	18.8483
		AJI	0.1628	3.9462	2.4809	A231	-9.4031
		AIJJ	-1.5978	-9.2108	-5.4350	A312	-3.1863
		AJII	0.4278	-4.2185	37.5587		

EXPERIMENTAL					MARGULES 3		MARGULES 4		ALPHA 2		ALPHA 3	
X1	X2	Y1	Y2	TEMP.	ΔY1	ΔY2	ΔY1	ΔY2	ΔY1	ΔY2	ΔY1	ΔY2
0.1000	0.1000	0.6100	0.1300	70.00	-0.0481	0.0384	-0.0425	0.0377	0.0246	-0.0146	-0.0170	0.0062
0.1000	0.2000	0.5200	0.2400	70.00	-0.0722	0.0632	-0.0844	0.0733	0.0496	-0.0293	-0.0237	0.0135
0.1000	0.3000	0.4300	0.3550	69.40	-0.0641	0.0579	-0.0843	0.0808	0.0845	-0.0631	-0.0082	0.0009
0.1000	0.4000	0.3700	0.4500	68.80	-0.0639	0.0554	-0.0868	0.0873	0.0964	-0.0867	-0.0070	-0.0037
0.1000	0.5000	0.3200	0.5400	68.00	-0.0578	0.0459	-0.0804	0.0795	0.1029	-0.1113	-0.0045	-0.0130
0.1000	0.6000	0.2600	0.6300	66.50	-0.0296	0.0272	-0.0502	0.0542	0.1213	-0.1378	0.0162	-0.0298
0.1000	0.7000	0.2200	0.7100	65.00	-0.0138	0.0111	-0.0304	0.0254	0.1178	-0.1496	0.0226	-0.0419
0.1000	0.8000	0.1850	0.7800	63.50	0.0025	-0.0020	-0.0075	-0.0010	0.0988	-0.1319	0.0264	-0.0444
0.2000	0.1000	0.7200	0.0700	65.00	-0.0318	0.0413	-0.0108	0.0336	0.0223	-0.0007	-0.0091	0.0124
0.2000	0.2000	0.6350	0.1800	65.50	-0.0411	0.0365	-0.0257	0.0330	0.0596	-0.0471	0.0032	-0.0133
0.2000	0.3000	0.5600	0.2800	65.50	-0.0425	0.0339	-0.0340	0.0392	0.0902	-0.0875	0.0137	-0.0296
0.2000	0.4000	0.5000	0.3700	64.50	-0.0420	0.0329	-0.0407	0.0452	0.1077	-0.1197	0.0164	-0.0381
0.2000	0.5000	0.4400	0.4700	63.50	-0.0298	0.0148	-0.0345	0.0282	0.1248	-0.1601	0.0253	-0.0590
0.2000	0.6000	0.3850	0.5550	62.50	-0.0133	0.0095	-0.0216	0.0179	0.1324	-0.1724	0.0337	-0.0613
0.2000	0.7000	0.3400	0.6250	61.00	0.0012	0.0010	-0.0065	0.0019	0.1142	-0.1534	0.0334	-0.0553
0.3000	0.1000	0.7600	0.0650	62.80	-0.0219	0.0243	-0.0038	0.0164	0.0269	-0.0142	0.0024	-0.0048
0.3000	0.2000	0.7000	0.1500	62.70	-0.0364	0.0292	-0.0148	0.0218	0.0492	-0.0501	0.0063	-0.0222
0.3000	0.3000	0.6250	0.2550	62.30	-0.0250	0.0122	-0.0062	0.0092	0.0871	-0.1060	0.0273	-0.0548
0.3000	0.4000	0.5700	0.3450	61.50	-0.0233	0.0067	-0.0116	0.0070	0.1037	-0.1440	0.0306	-0.0692

		EXPERIMENTAL			MARGULES 3		MARGULES 4		ALPHA 2		ALPHA 3	
X1	X2	Y1	Y2	TEMP.	ΔY1	ΔY2	ΔY1	ΔY2	ΔY1	ΔY2	ΔY1	ΔY2
0.3000	0.5000	0.5200	0.4150	60.50	-0.0181	0.0161	-0.0145	0.0165	0.1101	-0.1529	0.0307	-0.0605
0.3000	0.6000	0.4650	0.5000	59.50	-0.0014	0.0042	-0.0033	0.0027	0.1045	-0.1487	0.0344	-0.0588
0.4000	0.1000	0.7900	0.0600	61.50	-0.0251	0.0200	-0.0180	0.0143	0.0216	-0.0189	0.0023	-0.0110
0.4000	0.2000	0.7250	0.1550	60.80	-0.0201	0.0078	-0.0034	0.0020	0.0547	-0.0721	0.0220	-0.0467
0.4000	0.3000	0.6700	0.2420	60.00	-0.0183	0.0041	-0.0018	0.0002	0.0763	-0.1142	0.0302	-0.0660
0.4000	0.4000	0.6150	0.3220	59.40	-0.0109	0.0060	-0.0008	0.0031	0.0931	-0.1406	0.0369	-0.0703
0.4000	0.5000	0.5650	0.4020	58.40	-0.0029	0.0032	-0.0002	0.0004	0.0889	-0.1398	0.0342	-0.0619
0.5000	0.1000	0.8100	0.0650	60.00	-0.0250	0.0121	-0.0310	0.0097	0.0174	-0.0293	0.0031	-0.0213
0.5000	0.2000	0.7500	0.1600	59.40	-0.0158	-0.0016	-0.0077	-0.0034	0.0488	-0.0856	0.0247	-0.0592
0.5000	0.3000	0.7000	0.2320	58.60	-0.0126	0.0091	-0.0033	0.0078	0.0660	-0.1111	0.0312	-0.0612
0.5000	0.4000	0.6500	0.3170	57.60	-0.0062	0.0048	-0.0018	0.0030	0.0696	-0.1249	0.0299	-0.0585
0.6000	0.1000	0.8200	0.0800	59.10	-0.0167	-0.0010	-0.0326	0.0005	0.0188	-0.0468	0.0101	-0.0365
0.6000	0.2000	0.7700	0.1630	58.30	-0.0114	-0.0002	-0.0107	0.0025	0.0413	-0.0900	0.0245	-0.0576
0.6000	0.3000	0.7200	0.2450	57.30	-0.0053	0.0025	-0.0025	0.0038	0.0531	-0.1107	0.0274	-0.0545
0.7000	0.1000	0.8270	0.0950	58.20	-0.0023	-0.0098	-0.0204	-0.0046	0.0211	-0.0614	0.0195	-0.0447
0.7000	0.2000	0.7780	0.1840	57.20	0.0037	-0.0090	0.0016	-0.0040	0.0404	-0.0994	0.0288	-0.0535
0.8000	0.1000	0.8600	0.1000	57.20	-0.0082	-0.0037	-0.0191	0.0028	-0.0020	-0.0596	0.0047	-0.0284
				MEAN	0.0240	0.0183	0.0236	0.0215	0.0706	0.0940	0.0200	0.0396

494

CARBON TETRACHLORIDE(1)-BENZENE(2)-ISO-PROPYL ALCOHOL(3)

NAGATA I.:J.CHEM.ENG.DATA 10(2),106(1965).

ANTOINE VAPOR PRESSURE CONSTANTS
```
        A          B          C
(1) 6.93390   1242.430   230.000
(2) 6.90565   1211.033   220.790
(3) 7.75634   1366.142   197.970
```

P = 760.00

EQUATION	ORDER	BINARY CONSTANTS				TERNARY CONSTANTS	
		I-J	1-2	2-3	3-1		
MARGULES	3	AIJ	0.0543	0.4956	0.6957	C123	-0.4093
		AJI	0.0356	0.6909	0.4436		
MARGULES	4	AIJ	0.0556	0.5880	0.7785	C1123	-0.6490
		AJI	0.0370	0.7910	0.5216	C1223	-0.0381
		DIJ	0.0039	0.3126	0.2802	C1233	0.4036
ALPHA	2	AIJ	0.2312	4.2926	2.1167		
		AJI	0.0025	2.3459	4.0891		
ALPHA	3	AIJ	0.2878	9.8206	2.4773	A123	0.9360
		AJI	0.0631	2.7880	11.5822	A231	-1.9004
		AIJJ	-0.0510	-6.3771	2.5976	A312	2.5700
		AJII	-0.0636	1.9198	-8.9691		

--------------EXPERIMENTAL---------------					---MARGULES 3---		---MARGULES 4---		-----ALPHA 2----		-----ALPHA 3----	
X1	X2	Y1	Y2	TEMP.	ΔY1	ΔY2	ΔY1	ΔY2	ΔY1	ΔY2	ΔY1	ΔY2
0.0320	0.3620	0.0540	0.4700	73.50	-0.0030	0.0131	-0.0090	0.0119	-0.0098	0.0354	-0.0027	0.0330
0.0340	0.4560	0.0480	0.5120	71.90	0.0002	0.0130	-0.0056	0.0169	-0.0071	0.0358	0.0008	0.0345
0.0710	0.4300	0.1000	0.4720	72.00	-0.0033	0.0132	-0.0107	0.0190	-0.0159	0.0376	-0.0003	0.0269
0.3710	0.0520	0.5080	0.0660	70.20	0.0026	0.0065	-0.0071	0.0030	-0.0140	0.0054	0.0091	-0.0091
0.3010	0.0540	0.4710	0.0710	71.20	-0.0074	0.0130	-0.0102	0.0094	-0.0207	0.0124	-0.0025	-0.0020
0.3130	0.0830	0.4520	0.1110	71.30	-0.0027	0.0076	-0.0029	0.0028	-0.0173	0.0077	0.0081	-0.0127
0.4290	0.5280	0.4160	0.4370	73.80	-0.0026	0.0159	-0.0020	0.0116	0.0064	0.0379	0.0098	-0.0095
0.4680	0.5060	0.4580	0.4490	75.10	0.0031	-0.0010	0.0035	-0.0039	0.0120	0.0132	0.0110	-0.0229
0.4240	0.5550	0.4200	0.4870	75.50	0.0067	0.0099	0.0058	0.0062	0.0135	0.0229	0.0128	-0.0093
0.0920	0.8680	0.1000	0.7480	75.90	0.0006	0.0197	-0.0035	0.0015	-0.0013	0.0517	0.0010	0.0105
0.0920	0.7340	0.0960	0.6270	72.80	-0.0019	-0.0265	-0.0077	-0.0276	-0.0078	0.0263	0.0041	-0.0067
0.0780	0.6210	0.0880	0.5630	71.60	-0.0014	-0.0118	-0.0075	-0.0020	-0.0108	0.0306	0.0038	0.0124
0.0580	0.2240	0.1120	0.3480	73.50	-0.0045	0.0229	-0.0113	0.0224	-0.0127	0.0446	-0.0041	0.0318
0.0360	0.1040	0.0930	0.2110	75.90	-0.0062	0.0271	-0.0073	0.0356	-0.0015	0.0621	-0.0057	0.0461
0.1900	0.6510	0.1900	0.5450	71.40	-0.0045	-0.0178	-0.0082	-0.0198	-0.0103	0.0306	0.0099	-0.0165
0.1690	0.5100	0.1860	0.4680	70.90	-0.0041	-0.0089	-0.0054	-0.0011	-0.0179	0.0284	0.0108	-0.0054
0.1510	0.3730	0.1940	0.4040	71.60	-0.0024	0.0034	-0.0045	0.0079	-0.0204	0.0264	0.0086	-0.0003
0.1320	0.2380	0.2080	0.3220	71.70	-0.0008	0.0147	-0.0042	0.0139	-0.0177	0.0281	0.0044	0.0056
0.1060	0.1430	0.2030	0.2200	74.60	-0.0014	0.0360	-0.0035	0.0365	-0.0062	0.0533	0.0026	0.0348

	EXPERIMENTAL				MARGULES 3		MARGULES 4		ALPHA 2		ALPHA 3	
X1	X2	Y1	Y2	TEMP.	ΔY1	ΔY2	ΔY1	ΔY2	ΔY1	ΔY2	ΔY1	ΔY2
0.0950	0.0690	0.2310	0.1320	75.20	-0.0172	0.0203	-0.0120	0.0243	-0.0066	0.0367	-0.0140	0.0227
0.2700	0.6640	0.2690	0.5500	73.80	-0.0049	0.0038	-0.0091	-0.0063	-0.0027	0.0407	0.0068	-0.0127
0.2750	0.5360	0.2700	0.4520	71.20	-0.0092	-0.0159	-0.0051	-0.0152	-0.0130	0.0266	0.0161	-0.0259
0.2560	0.4250	0.2760	0.3840	71.50	-0.0119	-0.0024	-0.0038	0.0012	-0.0231	0.0301	0.0143	-0.0140
0.2370	0.2800	0.3000	0.3040	71.30	-0.0074	0.0087	-0.0000	0.0074	-0.0254	0.0240	0.0120	-0.0116
0.2030	0.1720	0.3150	0.2230	71.30	-0.0067	0.0226	-0.0034	0.0188	-0.0241	0.0295	0.0028	0.0026
0.1640	0.0800	0.3200	0.1310	74.10	-0.0097	0.0179	-0.0053	0.0168	-0.0100	0.0252	-0.0038	0.0087
0.3860	0.5520	0.3690	0.4560	73.10	-0.0027	0.0053	-0.0021	-0.0001	0.0050	0.0347	0.0137	-0.0195
0.3580	0.3060	0.3800	0.2810	70.10	-0.0148	0.0039	0.0014	0.0024	0.0188	0.0236	0.0188	-0.0236
0.3020	0.1820	0.3910	0.2030	70.30	-0.0075	0.0161	0.0013	0.0109	-0.0253	0.0218	0.0119	-0.0118
0.2620	0.0900	0.4110	0.1260	72.10	-0.0057	0.0116	-0.0031	0.0071	-0.0179	0.0132	0.0040	-0.0075
0.4900	0.4480	0.4590	0.3730	73.50	-0.0057	0.0020	-0.0005	-0.0002	0.0079	0.0257	0.0140	-0.0268
0.4700	0.3600	0.4400	0.2840	70.30	-0.0184	0.0107	-0.0030	0.0100	-0.0095	0.0369	0.0192	-0.0195
0.3920	0.2130	0.4330	0.2090	70.60	-0.0126	0.0069	0.0006	0.0027	-0.0216	0.0172	0.0199	-0.0235
0.6010	0.3510	0.5520	0.2970	73.00	0.0051	0.0041	0.0113	0.0036	0.0209	0.0193	0.0181	-0.0241
0.5940	0.2270	0.5420	0.1880	70.80	-0.0187	0.0018	-0.0075	0.0008	-0.0015	0.0169	0.0202	-0.0290
0.5510	0.1010	0.5670	0.1030	69.60	-0.0130	-0.0024	-0.0224	-0.0055	-0.0137	0.0004	0.0151	-0.0255
0.7130	0.2360	0.6410	0.2010	72.80	0.0094	0.0027	0.0144	0.0029	0.0286	0.0132	0.0174	-0.0221
0.7030	0.1180	0.6330	0.1000	70.00	-0.0158	0.0011	-0.0245	0.0002	0.0043	0.0076	0.0118	-0.0209
0.8300	0.1250	0.7410	0.1060	73.20	0.0175	0.0040	0.0138	0.0042	0.0374	0.0095	0.0123	-0.0111
				MEAN	0.0070	0.0114	0.0068	0.0101	0.0140	0.0267	0.0097	0.0178

CHLOROFORM(1)-METHYL ALCOHOL(2)-ETHYL ACETATE(3)

NAGATA I.:J.CHEM.ENG.DATA 7(3),367(1962).

ANTOINE VAPOR PRESSURE CONSTANTS

	A	B	C
(1)	6.93708	1171.200	227.000
(2)	8.07246	1574.990	238.860
(3)	7.10232	1245.239	217.911

P = 760.00

EQUATION	ORDER	BINARY CONSTANTS				TERNARY CONSTANTS	
		I-J	1-2	2-3	3-1		
MARGULES	3	AIJ	0.3418	0.4238	-0.4370	C123	0.6950
		AJI	0.7369	0.4180	-0.2566		
MARGULES	4	AIJ	0.3812	0.4401	-0.4547	C1123	0.8208
		AJI	0.7784	0.4363	-0.2725	C1223	0.7497
		DIJ	0.1321	0.0536	-0.0500	C1233	0.7659
ALPHA	2	AIJ	3.5023	2.9112	-0.7966		
		AJI	2.0056	1.1469	-0.2196		
ALPHA	3	AIJ	10.6348	2.4444	-0.9900	A123	-5.9399
		AJI	1.5322	1.5405	-0.6720	A231	-1.4669
		AIJJ	-9.1939	0.9027	0.1810	A312	-0.6327
		AJII	3.7869	-0.6736	0.5412		

-----------EXPERIMENTAL---------------					---MARGULES 3---		---MARGULES 4---		-----ALPHA 2----		-----ALPHA 3----	
X1	X2	Y1	Y2	TEMP.	ΔY1	ΔY2	ΔY1	ΔY2	ΔY1	ΔY2	ΔY1	ΔY2
0.0750	0.0770	0.0570	0.2380	72.00	0.0019	-0.0066	0.0005	-0.0033	0.0089	-0.0118	-0.0073	-0.0063
0.0820	0.2460	0.0600	0.4700	65.40	-0.0001	-0.0142	-0.0013	-0.0134	0.0214	-0.0252	-0.0158	-0.0165
0.1470	0.3030	0.1180	0.4940	64.10	0.0016	0.0070	-0.0001	0.0084	0.0371	-0.0170	-0.0120	-0.0015
0.1220	0.5390	0.1220	0.6180	61.90	0.0031	0.0032	0.0022	0.0062	0.0445	-0.0346	-0.0053	-0.0041
0.1230	0.4450	0.1080	0.5740	62.80	0.0026	0.0066	0.0014	0.0089	0.0415	-0.0244	-0.0095	-0.0006
0.0790	0.3470	0.0590	0.5130	63.80	0.0013	0.0150	0.0004	0.0157	0.0273	0.0002	-0.0145	0.0125
0.0600	0.4700	0.0500	0.5850	63.10	-0.0001	0.0090	-0.0004	0.0103	0.0254	-0.0069	-0.0134	0.0080
0.0580	0.6340	0.0640	0.6670	62.10	-0.0021	0.0048	-0.0016	0.0067	0.0285	-0.0214	-0.0128	0.0048
0.0480	0.5690	0.0450	0.6400	62.30	-0.0001	0.0009	0.0000	0.0026	0.0243	-0.0159	-0.0113	0.0015
0.0460	0.6510	0.0510	0.6810	62.20	-0.0016	0.0001	-0.0011	0.0018	0.0248	-0.0223	-0.0114	0.0008
0.0400	0.8610	0.0760	0.8140	62.40	-0.0072	-0.0008	-0.0046	-0.0034	0.0283	-0.0447	-0.0107	-0.0037
0.1040	0.6450	0.1260	0.6650	61.40	-0.0015	0.0029	-0.0015	0.0053	0.0406	-0.0396	-0.0083	-0.0023
0.1010	0.6570	0.1180	0.6730	61.40	0.0050	0.0008	0.0052	0.0031	0.0472	-0.0421	-0.0016	-0.0041
0.0630	0.7250	0.0830	0.7130	62.00	-0.0013	0.0040	-0.0001	0.0049	0.0352	-0.0340	-0.0093	0.0025
0.0860	0.7380	0.1180	0.7180	61.30	0.0027	-0.0028	0.0039	-0.0020	0.0458	-0.0500	-0.0023	-0.0068
0.0870	0.5750	0.0870	0.6440	62.00	0.0019	-0.0024	0.0017	-0.0000	0.0387	-0.0339	-0.0098	-0.0050
0.0700	0.7920	0.1110	0.7500	62.10	-0.0036	0.0016	-0.0015	0.0008	0.0397	-0.0477	-0.0075	-0.0024
0.2500	0.2410	0.2180	0.4420	64.50	0.0013	0.0110	-0.0009	0.0127	0.0367	-0.0192	0.0076	-0.0095
0.2190	0.3600	0.1970	0.5480	62.80	0.0061	-0.0186	0.0036	-0.0158	0.0458	-0.0574	0.0123	-0.0400

497

		EXPERIMENTAL			MARGULES 3		MARGULES 4		ALPHA 2		ALPHA 3	
X1	X2	Y1	Y2	TEMP.	ΔY1	ΔY2	ΔY1	ΔY2	ΔY1	ΔY2	ΔY1	ΔY2
0.1900	0.4320	0.1800	0.5800	62.50	0.0044	-0.0136	0.0022	-0.0104	0.0461	-0.0550	0.0071	-0.0320
0.1360	0.6780	0.1780	0.6630	61.10	0.0050	0.0044	0.0046	0.0069	0.0494	-0.0461	0.0074	-0.0067
0.1200	0.6370	0.1440	0.6550	61.30	0.0014	0.0047	0.0011	0.0074	0.0448	-0.0405	-0.0025	-0.0030
0.3140	0.1220	0.2810	0.3160	69.00	0.0049	0.0075	0.0041	0.0090	0.0314	-0.0114	0.0177	-0.0096
0.2220	0.3830	0.2040	0.5440	62.40	0.0075	-0.0046	0.0049	-0.0015	0.0471	-0.0453	0.0162	-0.0272
0.2020	0.5970	0.2420	0.6190	60.50	0.0131	-0.0088	0.0106	-0.0046	0.0497	-0.0559	0.0276	-0.0299
0.1800	0.6510	0.2400	0.6390	60.30	0.0045	-0.0051	0.0026	-0.0016	0.0432	-0.0537	0.0156	-0.0225
0.0970	0.8220	0.1780	0.7480	61.10	-0.0088	0.0012	-0.0062	-0.0005	0.0421	-0.0569	-0.0058	-0.0067
0.4160	0.0710	0.4090	0.2350	69.80	0.0065	-0.0038	0.0075	-0.0027	0.0231	-0.0190	0.0272	-0.0160
0.3640	0.2070	0.3290	0.4140	63.60	0.0118	0.0018	0.0102	0.0033	0.0382	-0.0326	0.0454	-0.0306
0.2890	0.4820	0.3220	0.5530	60.00	0.0084	-0.0081	0.0047	-0.0035	0.0329	-0.0489	0.0387	-0.0412
0.3470	0.3000	0.3290	0.4890	62.00	0.0084	-0.0132	0.0059	-0.0106	0.0355	-0.0541	0.0467	-0.0520
0.2230	0.6590	0.3190	0.6030	59.00	0.0026	-0.0008	-0.0004	0.0033	0.0298	-0.0406	0.0202	-0.0209
0.3840	0.3140	0.3760	0.4750	60.90	0.0078	-0.0044	0.0054	-0.0018	0.0283	-0.0442	0.0530	-0.0469
0.3450	0.4650	0.3880	0.5260	59.00	0.0094	-0.0134	0.0058	-0.0091	0.0222	-0.0461	0.0457	-0.0491
0.3380	0.5040	0.3980	0.5310	58.30	0.0108	-0.0145	0.0070	-0.0100	0.0201	-0.0428	0.0428	-0.0461
0.3270	0.5790	0.4330	0.5090	57.30	0.0047	0.0090	0.0007	0.0136	0.0075	-0.0086	0.0269	-0.0129
0.2650	0.6600	0.4010	0.5580	57.50	-0.0069	0.0053	-0.0105	0.0096	0.0063	-0.0193	0.0100	-0.0121
0.4860	0.2110	0.4630	0.4120	61.40	0.0066	-0.0141	0.0058	-0.0128	0.0224	-0.0520	0.0576	-0.0554
0.5070	0.1910	0.4850	0.3800	61.60	0.0037	0.0008	0.0031	0.0020	0.0191	-0.0368	0.0540	-0.0386
0.4700	0.2850	0.4670	0.4260	59.80	0.0011	0.0074	-0.0002	0.0091	0.0129	-0.0285	0.0519	-0.0369
0.4190	0.4070	0.4700	0.4780	57.90	-0.0092	-0.0083	-0.0117	-0.0053	-0.0053	-0.0343	0.0309	-0.0455
0.5970	0.0770	0.6030	0.2420	65.70	0.0011	-0.0026	0.0012	-0.0008	0.0150	-0.0327	0.0314	-0.0171
0.5270	0.2190	0.5090	0.3930	60.00	0.0011	-0.0014	0.0006	-0.0002	0.0137	-0.0381	0.0517	-0.0426
0.4990	0.3380	0.5150	0.4370	57.50	0.0005	-0.0080	-0.0003	-0.0067	0.0028	-0.0328	0.0430	-0.0460
0.4920	0.3630	0.5120	0.4490	57.30	0.0065	-0.0162	0.0056	-0.0147	0.0066	-0.0375	0.0459	-0.0518
0.4340	0.4860	0.5290	0.4480	56.00	-0.0091	0.0041	-0.0112	0.0066	-0.0189	-0.0005	0.0129	-0.0165
0.3970	0.5180	0.4880	0.4750	56.70	0.0059	-0.0017	0.0030	0.0017	-0.0006	-0.0101	0.0288	-0.0235
0.5880	0.1830	0.5700	0.3610	59.70	-0.0084	-0.0035	-0.0086	-0.0025	0.0052	-0.0407	0.0378	-0.0397
0.6530	0.0820	0.6520	0.2470	63.80	-0.0038	-0.0031	-0.0046	-0.0007	0.0128	-0.0377	0.0242	-0.0174
0.6080	0.2340	0.5840	0.3710	57.40	-0.0033	0.0020	-0.0030	0.0024	0.0070	-0.0295	0.0383	-0.0334
0.5990	0.3510	0.5800	0.4020	56.00	0.0251	-0.0189	0.0269	-0.0204	0.0243	-0.0274	0.0472	-0.0395
0.6800	0.2000	0.6340	0.3470	57.10	-0.0068	-0.0045	-0.0065	-0.0040	0.0115	-0.0393	0.0278	-0.0341
0.8240	0.1020	0.7360	0.2480	57.20	-0.0027	0.0027	-0.0066	0.0072	0.0358	-0.0437	0.0100	-0.0067
0.1950	0.1740	0.1580	0.3860	67.10	0.0025	0.0029	0.0006	0.0046	0.0323	-0.0134	-0.0071	-0.0066
0.2080	0.7450	0.3530	0.6220	58.10	-0.0036	-0.0039	-0.0054	-0.0016	0.0240	-0.0390	0.0093	-0.0178
0.1150	0.1790	0.0880	0.3820	67.50	-0.0008	0.0098	-0.0025	0.0114	0.0216	-0.0010	-0.0168	0.0063
0.1700	0.0720	0.1440	0.2410	72.30	0.0012	-0.0172	-0.0001	-0.0143	0.0154	-0.0204	-0.0087	-0.0197
0.4820	0.0670	0.4810	0.2350	68.80	0.0123	-0.0124	0.0133	-0.0114	0.0260	-0.0315	0.0378	-0.0245
0.2900	0.6440	0.4260	0.5400	57.20	-0.0020	0.0015	-0.0058	0.0058	0.0048	-0.0164	0.0145	-0.0151
0.2060	0.7400	0.3450	0.6170	58.50	-0.0033	0.0038	-0.0052	0.0062	0.0255	-0.0332	0.0104	-0.0111
0.4700	0.0710	0.4680	0.2350	68.50	0.0097	-0.0038	0.0107	-0.0028	0.0239	-0.0228	0.0353	-0.0167
0.5400	0.0750	0.5420	0.2440	66.40	0.0092	-0.0065	0.0098	-0.0052	0.0215	-0.0312	0.0386	-0.0203
0.4540	0.3250	0.4640	0.4400	59.10	0.0005	0.0077	-0.0011	0.0098	0.0098	-0.0254	0.0488	-0.0354
0.7310	0.0770	0.7040	0.2300	60.60	0.0061	-0.0035	0.0039	0.0001	0.0271	-0.0402	0.0231	-0.0108
0.5660	0.3420	0.5840	0.3910	55.90	-0.0066	0.0068	-0.0058	0.0065	-0.0083	-0.0071	0.0234	-0.0205
0.2460	0.1970	0.2060	0.4200	65.80	0.0062	-0.0059	0.0042	-0.0042	0.0392	-0.0299	0.0082	-0.0226
0.0330	0.8610	0.0580	0.8090	62.40	-0.0022	0.0074	0.0001	0.0049	0.0285	-0.0320	-0.0060	0.0048
0.0300	0.8980	0.0570	0.8380	62.60	-0.0010	0.0128	0.0017	0.0094	0.0309	-0.0285	-0.0033	0.0096
0.0760	0.3280	0.0590	0.5060	64.40	-0.0021	0.0103	-0.0031	0.0109	0.0224	-0.0034	-0.0176	0.0076
0.3290	0.6390	0.4680	0.5170	56.60	0.0123	-0.0119	0.0087	-0.0080	0.0086	-0.0147	0.0226	-0.0220
				MEAN	0.0050	0.0067	0.0044	0.0065	0.0262	0.0313	0.0223	0.0203

498

ETHYL ALCOHOL(1)-BENZENE(2)-HEPTANE(3)

OAKESON G.O.,WEBER J.H.:J.CHEM.ENG.DATA 5(3),279(1960).

ANTOINE VAPOR PRESSURE CONSTANTS

	A	B	C
(1)	8.16290	1623.220	228.980
(2)	6.90565	1211.033	220.790
(3)	6.90240	1268.115	216.900

P = 180.00

EQUATION	ORDER	BINARY	CONSTANTS			TERNARY	CONSTANTS
		I-J	1-2	2-3	3-1		
MARGULES	3	AIJ	0.8475	0.1338	0.8727	C123	-0.8615
		AJI	0.5880	0.1559	1.0611		
MARGULES	4	AIJ	0.9672	0.1293	1.0782	C1123	0.8355
		AJI	0.6976	0.1507	1.2325	C1223	-0.9031
		DIJ	0.3891	-0.0152	0.7257	C1233	-1.0936
ALPHA	2	AIJ	4.7128	1.3493	21.4143		
		AJI	7.5827	-0.1209	29.0415		
ALPHA	3	AIJ	3.6303	1.7870	43.2640	A123	24.6817
		AJI	18.0065	0.0310	7.0653	A231	48.3157
		AIJJ	6.6773	-0.4778	-37.0847	A312	-4.7670
		AJII	-13.3299	-0.1007	48.7230		

-----EXPERIMENTAL-----					---MARGULES 3---		---MARGULES 4---		-----ALPHA 2----		-----ALPHA 3----	
X1	X2	Y1	Y2	TEMP.	ΔY1	ΔY2	ΔY1	ΔY2	ΔY1	ΔY2	ΔY1	ΔY2
0.1180	0.7710	0.2900	0.6360	33.30	0.0066	0.0000	0.0192	-0.0022	0.0314	-0.0895	-0.0024	-0.0183
0.1940	0.7100	0.3420	0.5880	32.80	-0.0032	0.0087	-0.0082	0.0245	0.0303	-0.0913	-0.0290	0.0089
0.3360	0.5780	0.3410	0.5840	32.70	0.0218	-0.0211	0.0059	0.0023	0.0850	-0.1524	-0.0015	-0.0176
0.3910	0.5200	0.3500	0.5630	32.80	0.0174	-0.0149	0.0024	0.0040	0.0942	-0.1632	-0.0025	-0.0124
0.4590	0.4570	0.3630	0.5500	32.90	0.0086	-0.0130	-0.0017	-0.0035	0.0978	-0.1732	-0.0042	-0.0141
0.8260	0.0880	0.5590	0.2290	36.30	-0.0087	0.0178	0.0062	-0.0017	0.0299	-0.1107	-0.0472	0.0482
0.7810	0.0830	0.5340	0.1980	36.20	-0.0150	0.0176	0.0276	-0.0016	0.0477	-0.1157	-0.0404	0.0448
0.7240	0.0830	0.5230	0.1810	36.20	-0.0280	0.0165	0.0345	0.0003	0.0520	-0.1174	-0.0393	0.0367
0.6650	0.0730	0.5210	0.1500	36.20	-0.0304	0.0142	0.0483	0.0044	0.0508	-0.1052	-0.0315	0.0259
0.5070	0.0740	0.5110	0.1270	36.30	-0.0124	0.0107	0.0375	0.0131	0.0499	-0.0942	-0.0196	0.0124
0.3160	0.0670	0.5080	0.1020	36.90	0.0314	-0.0016	0.0240	0.0065	0.0372	-0.0766	-0.0092	-0.0007
0.2270	0.0790	0.4970	0.1130	37.00	0.0427	-0.0070	0.0207	-0.0015	0.0321	-0.0817	-0.0083	-0.0047
0.1060	0.0800	0.4740	0.1170	38.00	0.0134	-0.0108	0.0245	-0.0158	0.0049	-0.0751	-0.0056	-0.0154
0.0280	0.1850	0.3580	0.2410	39.20	-0.1062	0.0318	-0.0455	0.0050	-0.0639	-0.0623	-0.0287	-0.0112
0.0300	0.2880	0.3160	0.3460	37.40	-0.0742	0.0268	-0.0167	-0.0080	-0.0366	-0.0861	-0.0161	-0.0201
0.1150	0.3680	0.3510	0.3900	34.70	0.0560	-0.0374	0.0567	-0.0539	0.0784	-0.1911	0.0049	-0.0233
0.1550	0.4430	0.3480	0.4360	33.80	0.0581	-0.0385	0.0425	-0.0429	0.0914	-0.2048	-0.0053	-0.0123
0.1650	0.4950	0.3380	0.4780	33.60	0.0562	-0.0457	0.0407	-0.0450	0.0942	-0.2119	-0.0060	-0.0187
0.1440	0.5660	0.3200	0.5200	33.40	0.0478	-0.0392	0.0440	-0.0398	0.0838	-0.1940	-0.0046	-0.0207

499

		EXPERIMENTAL			MARGULES 3		MARGULES 4		ALPHA 2		ALPHA 3	
X1	X2	Y1	Y2	TEMP.	ΔY1	ΔY2	ΔY1	ΔY2	ΔY1	ΔY2	ΔY1	ΔY2
0.1030	0.6050	0.3020	0.5390	33.60	0.0294	-0.0210	0.0435	-0.0335	0.0608	-0.1631	-0.0039	-0.0199
0.0700	0.5900	0.2870	0.5370	34.30	0.0067	-0.0049	0.0365	-0.0307	0.0365	-0.1398	-0.0038	-0.0222
0.0760	0.6910	0.2710	0.5970	33.80	0.0054	0.0001	0.0317	-0.0185	0.0351	-0.1197	0.0005	-0.0244
0.0460	0.7910	0.2340	0.6660	34.30	-0.0404	0.0445	-0.0081	0.0207	-0.0146	-0.0406	-0.0062	-0.0147
0.0720	0.6640	0.2740	0.5790	33.80	0.0031	0.0030	0.0313	-0.0184	0.0335	-0.1217	-0.0016	-0.0210
0.2740	0.5190	0.3420	0.5180	32.90	0.0433	-0.0396	0.0162	-0.0201	0.1093	-0.2146	-0.0078	-0.0146
0.3540	0.4400	0.3700	0.4710	33.00	0.0204	-0.0163	-0.0099	-0.0020	0.1081	-0.2110	-0.0213	0.0066
0.5120	0.3260	0.2900	0.4480	33.30	0.1016	-0.0109	0.0896	-0.0179	0.2177	-0.2203	0.0832	0.0024
0.6020	0.2500	0.4220	0.4000	33.80	-0.0142	0.0024	-0.0115	-0.0178	0.1055	-0.2050	-0.0267	0.0159
0.7010	0.2010	0.4530	0.3850	34.40	-0.0143	0.0091	-0.0037	-0.0190	0.0870	-0.1767	-0.0295	0.0229
0.6380	0.2630	0.4180	0.4390	33.60	-0.0090	0.0041	-0.0042	-0.0185	0.1014	-0.1890	-0.0189	0.0108
0.3060	0.4190	0.3650	0.4500	33.30	0.0412	-0.0335	0.0034	-0.0214	0.1166	-0.2279	-0.0122	-0.0043
0.4820	0.2940	0.4000	0.3990	33.60	0.0038	-0.0098	-0.0148	-0.0155	0.1181	-0.2228	-0.0192	0.0081
0.5340	0.2170	0.4330	0.3280	34.30	-0.0116	0.0031	-0.0143	-0.0085	0.1041	-0.2010	-0.0267	0.0192
0.5200	0.3940	0.4000	0.4820	33.10	-0.0210	0.0343	-0.0264	0.0330	0.0786	-0.1393	-0.0303	0.0331
0.4690	0.1930	0.4430	0.2800	34.70	-0.0011	0.0016	-0.0130	-0.0013	0.0966	-0.1851	-0.0221	0.0154
0.4350	0.1500	0.4640	0.2220	35.40	0.0052	0.0021	-0.0039	0.0048	0.0807	-0.1557	-0.0187	0.0110
0.3230	0.1520	0.4670	0.2030	35.60	0.0272	-0.0030	-0.0067	0.0043	0.0671	-0.1414	-0.0178	0.0069
0.2560	0.1460	0.4650	0.1940	36.00	0.0431	-0.0128	0.0057	-0.0065	0.0600	-0.1351	-0.0119	-0.0030
0.0620	0.8820	0.2400	0.7170	33.60	-0.0364	0.0439	-0.0104	0.0239	-0.0196	-0.0026	0.0046	-0.0146
0.2310	0.1660	0.4460	0.2120	35.80	0.0555	-0.0154	0.0169	-0.0110	0.0709	-0.1434	-0.0031	-0.0034
0.1190	0.1550	0.4340	0.2030	36.70	0.0394	-0.0199	0.0384	-0.0284	0.0403	-0.1250	0.0017	-0.0190
0.2270	0.2060	0.4300	0.2540	35.40	0.0566	-0.0226	0.0151	-0.0195	0.0796	-0.1666	-0.0053	-0.0057
0.7180	0.1370	0.4870	0.2780	35.10	-0.0227	0.0185	0.0059	-0.0065	0.0747	-0.1565	-0.0393	0.0409
0.1990	0.2950	0.3970	0.3320	34.60	0.0596	-0.0344	0.0225	-0.0349	0.0897	-0.1972	-0.0076	-0.0094
0.2930	0.3030	0.3950	0.3600	34.10	0.0485	-0.0354	0.0015	-0.0300	0.1101	-0.2211	-0.0089	-0.0086
0.3570	0.2930	0.4010	0.3580	33.60	0.0299	-0.0199	-0.0119	-0.0158	0.1128	-0.2175	-0.0142	0.0033
0.0620	0.0660	0.4540	0.0960	38.90	-0.0402	0.0023	0.0009	-0.0065	-0.0249	-0.0506	-0.0059	-0.0105
0.9180	0.0450	0.6900	0.1520	39.50	0.0191	0.0134	-0.0036	0.0134	-0.0389	-0.0402	-0.0356	0.0429
0.2800	0.5290	0.3410	0.5250	32.80	0.0407	-0.0361	0.0150	-0.0153	0.1075	-0.2081	-0.0077	-0.0131
0.2140	0.6070	0.3290	0.5530	32.80	0.0380	-0.0316	0.0230	-0.0133	0.0867	-0.1802	-0.0095	-0.0134
0.3960	0.5720	0.3440	0.6210	32.40	0.0142	-0.0132	0.0042	0.0024	0.0679	-0.1006	0.0171	-0.0281
0.0850	0.4910	0.3170	0.4760	34.40	0.0256	-0.0247	0.0470	-0.0488	0.0533	-0.1726	-0.0041	-0.0253
0.3310	0.6460	0.3320	0.6450	32.40	0.0182	-0.0172	0.0067	-0.0003	0.0549	-0.0776	0.0217	-0.0308
0.4430	0.5210	0.3530	0.6050	32.50	0.0103	-0.0100	0.0030	0.0017	0.0744	-0.1128	0.0145	-0.0261
0.3240	0.6310	0.3340	0.6230	32.40	0.0192	-0.0170	0.0066	0.0038	0.0654	-0.1049	0.0109	-0.0243
				MEAN	0.0301	0.0183	0.0203	0.0157	0.0708	0.1433	0.0160	0.0175

ETHYL ALCOHOL(1)-BENZENE(2)-METHYLCYCLOPENTANE(3)

SINOR J.E.,WEBER J.H.:J.CHEM.ENG.DATA 5,244(1960).

ANTOINE VAPOR PRESSURE CONSTANTS

	A	B	C
(1)	8.16290	1623.220	228.980
(2)	6.90565	1211.033	220.790
(3)	6.86283	1186.059	226.042

P = 750.00

EQUATION	ORDER	BINARY	CONSTANTS			TERNARY	CONSTANTS
		I-J	1-2	2-3	3-1		
MARGULES	3	AIJ	0.8475	0.1364	0.7074	C123	-0.3707
		AJI	0.5880	0.1439	1.0018		
MARGULES	4	AIJ	0.9672	0.1416	0.8360	C1123	0.6646
		AJI	0.6976	0.1494	1.1397	C1223	0.0417
		DIJ	0.3891	0.0182	0.4825	C1233	-0.7064
ALPHA	2	AIJ	4.7128	0.0930	12.4437		
		AJI	7.5827	0.7214	6.8185		
ALPHA	3	AIJ	3.6303	0.3923	32.0315	A123	14.5869
		AJI	18.0065	0.9546	2.4289	A231	5.3667
		AIJJ	6.6773	-0.3257	-26.3753	A312	0.8045
		AJII	-13.3299	-0.1519	5.1803		

---------------EXPERIMENTAL---------------					---MARGULES 3---		---MARGULES 4---		-----ALPHA 2----		-----ALPHA 3----	
X1	X2	Y1	Y2	TEMP.	ΔY1	ΔY2	ΔY1	ΔY2	ΔY1	ΔY2	ΔY1	ΔY2
0.0520	0.5950	0.2250	0.4190	67.00	-0.0195	0.0088	0.0007	-0.0012	-0.0555	0.0267	0.0112	-0.0069
0.3950	0.0370	0.3960	0.0330	60.20	-0.0326	-0.0004	-0.0364	0.0028	-0.0347	-0.0059	-0.0411	0.0024
0.6900	0.1280	0.4320	0.1650	62.70	-0.0139	0.0141	0.0127	0.0097	0.0018	0.0118	-0.0124	0.0415
0.5040	0.1850	0.3800	0.1740	61.60	-0.0111	0.0082	-0.0051	0.0072	0.0060	-0.0018	-0.0069	0.0257
0.0960	0.0570	0.3080	0.0460	61.30	-0.0171	-0.0031	-0.0119	-0.0056	-0.0550	-0.0075	-0.0087	-0.0052
0.8780	0.0530	0.5940	0.1100	67.70	0.0169	0.0127	0.0140	0.0144	-0.0432	0.0381	-0.0157	0.0516
0.1740	0.1040	0.3280	0.0770	61.00	0.0157	-0.0024	-0.0019	-0.0049	-0.0267	-0.0100	-0.0026	-0.0011
0.6900	0.1530	0.4420	0.2000	63.40	-0.0157	0.0155	0.0023	0.0064	-0.0047	0.0170	0.0014	0.0168
0.1990	0.5650	0.3270	0.3950	64.50	0.0295	-0.0086	0.0227	-0.0001	-0.0180	0.0261	-0.0117	0.0331
0.4770	0.3160	0.3870	0.2920	63.10	-0.0046	0.0083	-0.0093	0.0015	-0.0001	0.0127	-0.0087	0.0254
0.4020	0.3810	0.3730	0.3240	63.40	0.0066	-0.0005	-0.0039	-0.0019	-0.0019	0.0107	-0.0013	0.0191
0.3890	0.2860	0.3600	0.2360	62.10	0.0071	0.0005	-0.0053	-0.0019	0.0039	-0.0027	-0.0029	0.0139
0.3990	0.2060	0.3610	0.1740	61.30	0.0018	-0.0001	-0.0088	-0.0005	0.0030	-0.0103	-0.0029	0.0302
0.3240	0.4940	0.3960	0.3840	64.40	-0.0128	-0.0017	-0.0247	0.0066	-0.0416	0.0268	-0.0418	0.0302
0.3230	0.2890	0.3470	0.2270	61.90	0.0161	-0.0071	-0.0013	-0.0093	0.0006	-0.0105	0.0022	0.0089
0.2180	0.2960	0.3280	0.2130	61.80	0.0222	-0.0082	0.0076	-0.0136	-0.0130	-0.0109	0.0042	0.0021
0.2240	0.1920	0.3300	0.1380	61.30	0.0233	-0.0016	0.0037	-0.0053	-0.0115	-0.0103	0.0040	0.0047
0.1060	0.2050	0.2920	0.1450	61.80	0.0017	-0.0007	0.0131	-0.0106	-0.0387	-0.0077	0.0073	-0.0041
0.1750	0.4980	0.3170	0.3390	63.60	0.0236	-0.0056	0.0189	-0.0051	-0.0218	0.0161	0.0042	0.0108

| X1 | X2 | Y1 | Y2 | TEMP. | ---MARGULES 3--- | | ---MARGULES 4--- | | -----ALPHA 2---- | | -----ALPHA 3---- | |
					ΔY1	ΔY2	ΔY1	ΔY2	ΔY1	ΔY2	ΔY1	ΔY2
0.0940	0.3140	0.2840	0.2180	62.80	-0.0069	-0.0011	0.0107	-0.0146	-0.0472	-0.0034	0.0049	-0.0061
0.0910	0.4140	0.2780	0.2790	63.60	-0.0060	0.0033	0.0108	-0.0096	-0.0475	0.0087	0.0061	-0.0013
0.5700	0.3280	0.4260	0.3590	64.60	-0.0083	0.0181	-0.0081	0.0029	-0.0104	0.0343	-0.0242	0.0501
0.1240	0.5050	0.2900	0.3450	64.00	0.0180	-0.0087	0.0235	-0.0138	-0.0291	0.0101	0.0119	-0.0022
0.4900	0.4180	0.4160	0.4080	65.10	-0.0039	0.0143	-0.0078	0.0092	-0.0170	0.0390	-0.0289	0.0491
0.1010	0.6550	0.2790	0.4500	65.50	0.0111	-0.0021	0.0203	-0.0016	-0.0388	0.0344	0.0081	0.0039
0.3660	0.5360	0.3950	0.4450	65.40	0.0075	0.0056	-0.0020	0.0164	-0.0247	0.0443	-0.0298	0.0460
0.1270	0.6870	0.3020	0.4790	66.10	0.0196	-0.0093	0.0242	-0.0029	-0.0363	0.0379	0.0004	0.0092
0.2620	0.6170	0.3630	0.4680	65.50	0.0239	-0.0113	0.0146	0.0053	-0.0241	0.0375	-0.0178	0.0293
0.1200	0.7540	0.3040	0.5380	67.40	0.0185	-0.0110	0.0258	-0.0055	-0.0422	0.0460	-0.0051	0.0115
0.2810	0.2340	0.3440	0.1730	61.30	0.0149	-0.0010	-0.0052	-0.0037	-0.0081	-0.0088	-0.0013	0.0099
0.0510	0.8720	0.2170	0.6740	71.00	0.0012	0.0037	0.0239	-0.0099	-0.0457	0.0507	0.0128	-0.0051
0.4970	0.1920	0.3710	0.1790	62.50	-0.0002	0.0076	0.0042	0.0062	0.0138	-0.0018	0.0012	0.0256
0.2270	0.5780	0.3350	0.4200	64.20	0.0324	-0.0134	0.0238	-0.0014	-0.0126	0.0251	0.0008	0.0168
0.6130	0.2190	0.4130	0.2470	63.10	-0.0095	0.0178	-0.0010	0.0055	0.0046	0.0189	-0.0107	0.0455
0.2920	0.4400	0.3380	0.3350	62.70	0.0304	-0.0109	0.0164	-0.0075	0.0041	0.0057	0.0087	0.0130
0.6690	0.2320	0.4540	0.3020	64.70	-0.0118	0.0195	-0.0079	0.0000	-0.0133	0.0337	-0.0258	0.0544
0.3240	0.3550	0.3470	0.2790	62.00	0.0186	-0.0088	0.0029	-0.0098	0.0022	-0.0046	0.0034	0.0109
0.7460	0.1780	0.4970	0.2710	65.70	-0.0088	0.0235	-0.0076	0.0057	-0.0263	0.0465	-0.0320	0.0652
0.1790	0.3800	0.3370	0.2500	62.10	0.0008	0.0055	-0.0059	-0.0005	-0.0399	0.0109	-0.0142	0.0158
0.3520	0.1090	0.3560	0.0900	60.50	0.0059	-0.0009	-0.0106	0.0016	-0.0030	-0.0117	-0.0050	0.0057
0.7630	0.0670	0.4570	0.0990	63.00	-0.0052	0.0112	0.0388	0.0128	-0.0005	0.0112	-0.0083	0.0361
0.6500	0.0790	0.4070	0.0920	61.50	-0.0174	0.0091	0.0312	0.0136	0.0081	0.0013	-0.0082	0.0257
0.5690	0.0710	0.3860	0.0760	60.90	-0.0166	0.0028	0.0237	0.0087	0.0087	-0.0065	-0.0071	0.0134
0.4950	0.0830	0.3710	0.0780	60.70	-0.0083	0.0040	0.0110	0.0095	0.0092	-0.0064	-0.0034	0.0129
0.4220	0.0790	0.3630	0.0720	60.50	-0.0014	-0.0012	-0.0010	0.0034	0.0035	-0.0110	-0.0046	0.0052
0.2320	0.0990	0.3400	0.0750	60.70	0.0180	-0.0021	-0.0080	-0.0026	-0.0167	-0.0104	-0.0039	0.0010
0.0470	0.1070	0.2580	0.0840	63.00	-0.0570	0.0017	-0.0254	-0.0044	-0.0782	-0.0032	-0.0089	-0.0060
				MEAN	0.0143	0.0072	0.0128	0.0063	0.0212	0.0176	0.0107	0.0197

ETHYL ALCOHOL(1)-METHYLCYCLOPENTANE(2)-HEXANE(3)

KAES G.L.,WEBER J.H.:J.CHEM.ENG.DATA 7(3),344(1962).

ANTOINE VAPOR PRESSURE CONSTANTS
	A	B	C
(1)	8.16290	1623.220	228.980
(2)	6.86283	1186.059	226.042
(3)	6.87776	1171.530	224.366

P = 760.00

EQUATION	ORDER	BINARY CONSTANTS				TERNARY CONSTANTS	
		I-J	1-2	2-3	3-1		
MARGULES	3	AIJ	1.0018	0.0000	0.7817	C123	-0.7155
		AJI	0.7074	0.0000	0.9889		
MARGULES	4	AIJ	1.1397	0.0000	0.9361	C1123	0.8407
		AJI	0.8360	0.0000	1.1487	C1223	-0.3219
		DIJ	0.4825	0.0000	0.5483	C1233	-0.8477
ALPHA	2	AIJ	6.8185	0.0000	17.1742		
		AJI	12.4437	0.0000	8.8090		
ALPHA	3	AIJ	2.4289	0.0000	31.3435	A123	18.4402
		AJI	32.0315	0.0000	3.7045	A231	1.1908
		AIJJ	15.1803	0.0000	-21.5236	A312	-2.2390
		AJII	-26.3753	0.0000	13.0333		

EXPERIMENTAL					MARGULES 3		MARGULES 4		ALPHA 2		ALPHA 3	
X1	X2	Y1	Y2	TEMP.	ΔY1	ΔY2	ΔY1	ΔY2	ΔY1	ΔY2	ΔY1	ΔY2
0.0500	0.8390	0.2690	0.6330	62.50	-0.0424	0.0392	-0.0184	0.0260	-0.0718	0.0641	0.0010	0.0184
0.1560	0.7270	0.3340	0.5710	60.50	0.0181	-0.0285	0.0139	-0.0099	-0.0361	0.0134	0.0014	0.0095
0.2920	0.6060	0.3540	0.5470	60.20	0.0145	-0.0275	-0.0021	0.0011	-0.0142	-0.0052	0.0022	0.0105
0.4160	0.4850	0.3640	0.5190	60.10	-0.0067	-0.0123	-0.0166	0.0017	-0.0023	-0.0161	0.0045	0.0095
0.5220	0.3820	0.3830	0.4830	60.20	-0.0323	0.0043	-0.0316	-0.0021	-0.0048	-0.0169	-0.0031	0.0130
0.6260	0.2800	0.3820	0.4400	60.30	-0.0260	0.0074	-0.0161	-0.0163	0.0142	-0.0232	0.0139	0.0066
0.7310	0.1840	0.4150	0.3710	61.10	-0.0280	0.0135	-0.0143	-0.0146	0.0068	-0.0141	0.0087	0.0079
0.7980	0.1050	0.4520	0.2580	62.20	-0.0257	0.0107	-0.0064	-0.0064	-0.0085	-0.0099	0.0004	0.0018
0.7520	0.0890	0.4070	0.1840	60.50	-0.0195	0.0171	0.0255	0.0061	0.0121	-0.0132	0.0155	0.0043
0.6160	0.1910	0.3970	0.2870	60.00	-0.0508	0.0133	-0.0322	-0.0066	-0.0091	-0.0266	-0.0034	0.0065
0.5370	0.2750	0.3830	0.3550	59.80	-0.0401	0.0058	-0.0383	-0.0108	-0.0068	-0.0284	0.0021	0.0085
0.4090	0.4000	0.3560	0.4250	59.70	-0.0025	-0.0166	-0.0183	-0.0137	0.0020	-0.0309	0.0176	0.0044
0.1510	0.6340	0.3200	0.4960	60.30	0.0345	-0.0339	0.0257	-0.0182	-0.0245	-0.0010	0.0207	0.0084
0.0450	0.7360	0.2590	0.5570	62.30	-0.0386	0.0272	-0.0183	0.0179	-0.0702	0.0492	0.0090	0.0177
0.0400	0.6310	0.2430	0.4900	62.30	-0.0334	0.0102	-0.0129	-0.0017	-0.0636	0.0270	0.0190	0.0074
0.2150	0.4750	0.3330	0.3820	59.80	0.0371	-0.0211	0.0113	-0.0128	-0.0141	-0.0105	0.0241	0.0196
0.2880	0.3980	0.3420	0.3600	59.40	0.0248	-0.0260	-0.0059	-0.0230	-0.0062	-0.0317	0.0236	0.0048
0.4190	0.2930	0.3620	0.3090	59.40	-0.0124	-0.0006	-0.0302	-0.0090	-0.0052	-0.0289	0.0131	0.0104
0.5440	0.1890	0.3690	0.2470	59.30	-0.0289	0.0110	-0.0173	-0.0029	0.0048	-0.0296	0.0139	0.0053

503

EXPERIMENTAL					MARGULES 3		MARGULES 4		ALPHA 2		ALPHA 3	
X1	X2	Y1	Y2	TEMP.	ΔY1	ΔY2	ΔY1	ΔY2	ΔY1	ΔY2	ΔY1	ΔY2
0.6230	0.1000	0.3680	0.1440	59.60	-0.0211	0.0232	0.0281	0.0170	0.0162	-0.0131	0.0177	0.0111
0.5200	0.0970	0.3560	0.1180	59.00	-0.0135	0.0159	0.0188	0.0134	0.0104	-0.0166	0.0143	0.0070
0.4050	0.2000	0.3490	0.1980	59.20	0.0016	0.0145	-0.0146	0.0063	0.0053	-0.0182	0.0205	0.0172
0.2760	0.3090	0.3430	0.2680	59.30	0.0235	-0.0091	-0.0114	-0.0145	-0.0108	-0.0241	0.0201	0.0133
0.1650	0.4780	0.3190	0.3670	59.90	0.0427	-0.0198	0.0236	-0.0173	-0.0173	-0.0066	0.0296	0.0197
0.0980	0.4800	0.3000	0.3460	60.30	0.0196	-0.0010	0.0205	-0.0093	-0.0373	0.0114	0.0256	0.0274
0.1370	0.3790	0.3110	0.2810	59.60	0.0362	-0.0074	0.0222	-0.0164	-0.0222	-0.0057	0.0295	0.0219
0.1590	0.3250	0.3060	0.2590	59.30	0.0488	-0.0198	0.0276	-0.0293	-0.0070	-0.0244	0.0391	0.0061
0.3000	0.2080	0.3390	0.1760	58.90	0.0226	0.0107	-0.0097	0.0035	-0.0035	-0.0129	0.0213	0.0205
0.4270	0.1030	0.3470	0.1070	58.90	0.0016	0.0132	0.0072	0.0111	0.0063	-0.0149	0.0147	0.0085
0.2760	0.1100	0.3260	0.0970	58.70	0.0330	0.0031	0.0074	-0.0005	0.0040	-0.0155	0.0226	0.0058
0.1650	0.2220	0.3150	0.1720	59.30	0.0362	-0.0038	0.0144	-0.0137	-0.0137	-0.0156	0.0261	0.0122
0.0490	0.3270	0.2660	0.2450	60.90	-0.0382	0.0067	-0.0120	-0.0104	-0.0627	0.0053	0.0106	0.0147
0.0970	0.1750	0.3000	0.1280	59.80	-0.0010	0.0032	0.0055	-0.0075	-0.0360	-0.0058	0.0133	0.0126
0.1700	0.1180	0.3220	0.0850	59.10	0.0221	0.0084	0.0032	0.0025	-0.0190	-0.0039	0.0114	0.0149
0.0440	0.1140	0.2440	0.0860	61.00	-0.0495	0.0048	-0.0170	-0.0025	-0.0475	-0.0003	0.0114	0.0072
0.0120	0.0570	0.1550	0.0460	64.00	-0.0850	0.0035	-0.0608	0.0016	-0.0667	0.0043	-0.0183	0.0051
0.0160	0.1630	0.1560	0.1310	64.00	-0.0588	0.0082	-0.0291	0.0014	-0.0486	0.0095	0.0122	0.0109
0.0150	0.2710	0.1560	0.2210	64.50	-0.0575	0.0114	-0.0296	0.0010	-0.0550	0.0161	0.0107	0.0143
0.0190	0.5290	0.1760	0.4170	64.60	-0.0485	0.0325	-0.0258	0.0187	-0.0608	0.0447	0.0180	0.0273
0.0150	0.8150	0.1650	0.6820	65.40	-0.0644	0.0483	-0.0460	0.0364	-0.0722	0.0609	0.0009	0.0138
0.0580	0.7690	0.2770	0.5710	61.80	-0.0271	0.0263	-0.0081	0.0197	-0.0645	0.0536	0.0079	0.0224
0.4310	0.4680	0.3640	0.5050	59.90	-0.0088	-0.0027	-0.0174	0.0083	-0.0000	-0.0100	0.0062	0.0167
0.3050	0.6450	0.3580	0.5910	60.30	0.0103	-0.0181	-0.0047	0.0079	-0.0146	0.0055	-0.0058	0.0137
0.6350	0.3190	0.4100	0.5070	60.70	-0.0432	0.0208	-0.0326	0.0008	-0.0084	-0.0054	-0.0156	0.0189
0.8260	0.1320	0.4960	0.3690	63.30	-0.0272	0.0048	-0.0287	-0.0133	-0.0274	0.0030	-0.0147	0.0040
0.9190	0.0390	0.5780	0.1870	67.10	0.0452	-0.0275	0.0245	-0.0228	-0.0198	-0.0069	0.0301	-0.0300
0.8250	0.0450	0.4510	0.1250	62.10	-0.0050	0.0032	0.0443	0.0001	-0.0016	-0.0124	0.0077	-0.0067
0.5520	0.0270	0.3660	0.0650	59.00	-0.0188	-0.0230	0.0601	-0.0211	0.0027	-0.0360	-0.0039	-0.0280
0.1640	0.0630	0.3030	0.0460	58.90	0.0324	0.0049	0.0185	0.0020	-0.0023	-0.0034	0.0230	0.0076
0.3230	0.6000	0.3520	0.5650	60.30	0.0151	-0.0232	-0.0001	0.0037	-0.0056	-0.0051	0.0051	0.0097
0.3220	0.5570	0.3580	0.5100	60.10	0.0081	-0.0136	-0.0089	0.0118	-0.0126	0.0001	0.0030	0.0217
0.3360	0.5000	0.3450	0.4800	59.90	0.0187	-0.0280	-0.0005	-0.0091	0.0020	-0.0229	0.0205	0.0054
0.2660	0.5180	0.3370	0.4510	59.80	0.0326	-0.0305	0.0092	-0.0115	-0.0044	-0.0172	0.0239	0.0104
0.3570	0.3600	0.3510	0.3460	59.40	0.0073	-0.0086	-0.0186	-0.0096	-0.0030	-0.0248	0.0201	0.0136
0.3280	0.5880	0.3490	0.5650	60.20	0.0173	-0.0310	0.0021	-0.0045	-0.0018	-0.0148	0.0093	0.0016
0.3730	0.2950	0.3520	0.2860	59.40	0.0035	0.0013	-0.0219	-0.0054	-0.0026	-0.0234	0.0195	0.0159
0.2150	0.3110	0.3340	0.2570	59.30	0.0326	-0.0158	0.0002	-0.0228	-0.0162	-0.0254	0.0214	0.0092
				MEAN	0.0280	0.0154	0.0191	0.0107	0.0208	0.0182	0.0144	0.0123

504

METHYL ACETATE(1)-CHLOROFORM(2)-BENZENE(3)

NAGATA I.:J.CHEM.ENG.DATA 7(3),360(1962).

ANTOINE VAPOR PRESSURE CONSTANTS

	A	B	C
(1)	7.06525	1157.622	219.724
(2)	6.93708	1171.200	227.000
(3)	6.90565	1211.033	220.790

P = 760.00

EQUATION	ORDER	BINARY CONSTANTS				TERNARY CONSTANTS	
		I-J	1-2	2-3	3-1		
MARGULES	3	AIJ	-0.3192	-0.0566	0.0249	C_{123}	0.0343
		AJI	-0.2542	-0.1221	0.1592		
MARGULES	4	AIJ	-0.3137	-0.0555	-0.0149	C_{1123}	-0.1695
		AJI	-0.2487	-0.1207	0.1295	C_{1223}	0.1677
		DIJ	0.0160	0.0039	-0.1146	C_{1233}	-0.0308
ALPHA	2	AIJ	-0.3531	0.4667	-0.2628		
		AJI	-0.6047	-0.5621	1.4993		
ALPHA	3	AIJ	-0.3368	-0.9700	-0.1683	A_{123}	-1.2470
		AJI	-0.9582	-1.4310	1.0273	A_{231}	0.7131
		AIJJ	-0.1432	1.4631	-0.2280	A_{312}	0.1449
		AJII	0.5058	0.9113	0.6835		

EXPERIMENTAL					MARGULES 3		MARGULES 4		ALPHA 2		ALPHA 3	
X1	X2	Y1	Y2	TEMP.	$\Delta Y1$	$\Delta Y2$	$\Delta Y1$	$\Delta Y2$	$\Delta Y1$	$\Delta Y2$	$\Delta Y1$	$\Delta Y2$
0.0780	0.8440	0.0560	0.8990	63.90	0.0039	0.0000	0.0034	-0.0001	0.0121	-0.0091	-0.0047	0.0068
0.0900	0.7560	0.0780	0.8290	65.10	0.0024	0.0008	0.0013	0.0015	0.0142	-0.0115	-0.0114	0.0124
0.0930	0.6550	0.0920	0.7290	67.20	0.0054	0.0110	0.0042	0.0132	0.0197	-0.0010	-0.0103	0.0244
0.0730	0.6470	0.0710	0.7440	67.30	0.0070	-0.0035	0.0056	-0.0010	0.0193	-0.0138	-0.0066	0.0075
0.0600	0.5950	0.0680	0.6930	68.50	0.0019	0.0033	0.0005	0.0063	0.0127	-0.0044	-0.0092	0.0118
0.0690	0.5460	0.0810	0.6440	69.10	0.0050	-0.0012	0.0038	0.0024	0.0168	-0.0079	-0.0060	0.0068
0.0570	0.4500	0.0790	0.5420	70.90	0.0033	0.0041	0.0022	0.0078	0.0121	0.0020	-0.0034	0.0078
0.0750	0.3810	0.1080	0.4610	71.80	0.0085	-0.0020	0.0081	0.0019	0.0177	-0.0021	0.0031	0.0000
0.0690	0.3120	0.1080	0.3850	72.90	0.0099	-0.0030	0.0094	0.0002	0.0157	-0.0012	0.0065	-0.0029
0.0560	0.2640	0.1020	0.3320	73.80	0.0016	-0.0006	0.0005	0.0018	0.0042	0.0016	-0.0010	-0.0013
0.0520	0.2060	0.0950	0.2630	74.90	0.0092	-0.0002	0.0077	0.0014	0.0087	0.0018	0.0067	-0.0013
0.0450	0.1360	0.1030	0.1750	75.70	-0.0030	0.0033	-0.0055	0.0040	-0.0077	0.0041	-0.0065	0.0020
0.0450	0.0630	0.1030	0.0810	76.60	0.0069	0.0031	0.0027	0.0033	-0.0030	0.0030	0.0010	0.0022
0.1540	0.7600	0.1330	0.8150	64.80	0.0027	-0.0012	0.0021	-0.0017	0.0146	-0.0136	-0.0107	0.0118
0.1250	0.6920	0.1190	0.7600	66.30	0.0037	0.0027	0.0028	0.0037	0.0189	-0.0113	-0.0137	0.0186
0.1480	0.6460	0.1490	0.7120	66.70	0.0053	-0.0009	0.0049	0.0003	0.0224	-0.0153	-0.0131	0.0162
0.1530	0.5760	0.1700	0.6380	67.40	0.0049	0.0006	0.0054	0.0027	0.0233	-0.0124	-0.0125	0.0155
0.1320	0.5320	0.1630	0.6020	68.70	-0.0016	-0.0018	-0.0011	0.0015	0.0160	-0.0116	-0.0166	0.0099
0.1210	0.4660	0.1560	0.5350	69.90	0.0059	-0.0020	0.0068	0.0021	0.0219	-0.0077	-0.0052	0.0052

EXPERIMENTAL					MARGULES 3		MARGULES 4		ALPHA 2		ALPHA 3	
X1	X2	Y1	Y2	TEMP.	ΔY1	ΔY2	ΔY1	ΔY2	ΔY1	ΔY2	ΔY1	ΔY2
0.1360	0.3790	0.1860	0.4430	70.60	0.0130	-0.0138	0.0157	-0.0100	0.0276	-0.0156	0.0065	-0.0114
0.1390	0.3310	0.2050	0.3770	70.90	0.0085	-0.0034	0.0118	-0.0000	0.0214	-0.0033	0.0044	-0.0032
0.1050	0.2140	0.1800	0.2530	73.30	0.0109	-0.0015	0.0114	0.0005	0.0140	0.0019	0.0087	-0.0027
0.1060	0.1510	0.1970	0.1780	73.60	0.0088	-0.0004	0.0077	0.0006	0.0069	0.0028	0.0058	-0.0016
0.1080	0.0700	0.2240	0.0640	74.20	0.0033	0.0183	-0.0018	0.0184	-0.0067	0.0201	-0.0029	0.0178
0.2260	0.7030	0.2130	0.7400	65.30	0.0017	0.0008	0.0016	-0.0001	0.0118	-0.0097	-0.0104	0.0138
0.2080	0.6700	0.1970	0.7230	65.70	0.0100	-0.0084	0.0100	-0.0091	0.0241	-0.0221	-0.0059	0.0081
0.2000	0.6000	0.2150	0.6450	66.70	0.0030	0.0016	0.0039	0.0018	0.0207	-0.0129	-0.0146	0.0185
0.1960	0.5420	0.2250	0.5850	67.60	0.0043	0.0016	0.0063	0.0028	0.0233	-0.0112	-0.0118	0.0158
0.1860	0.4760	0.2320	0.5180	68.40	0.0035	0.0003	0.0069	0.0027	0.0227	-0.0090	-0.0090	0.0095
0.1900	0.5310	0.2180	0.5780	67.90	0.0077	-0.0012	0.0099	0.0005	0.0269	-0.0133	-0.0079	0.0122
0.1900	0.4680	0.2370	0.5110	68.40	0.0051	-0.0030	0.0088	-0.0006	0.0243	-0.0120	-0.0068	0.0056
0.1820	0.2850	0.2720	0.3110	70.50	0.0076	-0.0045	0.0134	-0.0022	0.0208	-0.0036	0.0060	-0.0060
0.1620	0.2130	0.2630	0.2360	71.40	0.0091	-0.0036	0.0126	-0.0020	0.0164	-0.0004	0.0084	-0.0056
0.1720	0.1360	0.2890	0.1480	71.80	0.0159	-0.0027	0.0164	-0.0023	0.0177	0.0006	0.0148	-0.0044
0.1360	0.0800	0.2550	0.0900	73.10	0.0138	-0.0003	0.0095	-0.0002	0.0068	0.0021	0.0091	-0.0010
0.2530	0.6650	0.2470	0.6950	65.50	0.0045	0.0005	0.0047	-0.0008	0.0145	-0.0099	-0.0076	0.0140
0.2730	0.5910	0.2900	0.6120	66.00	0.0045	0.0011	0.0059	-0.0004	0.0175	-0.0112	-0.0086	0.0156
0.2730	0.5350	0.3090	0.5510	66.80	0.0032	0.0020	0.0062	0.0010	0.0188	-0.0105	-0.0094	0.0152
0.2660	0.4760	0.3230	0.4880	67.30	-0.0000	0.0030	0.0049	0.0028	0.0176	-0.0082	-0.0105	0.0123
0.2460	0.4050	0.3160	0.4190	68.20	0.0061	0.0001	0.0129	0.0013	0.0245	-0.0071	-0.0006	0.0039
0.2010	0.3300	0.2860	0.3520	69.70	0.0051	-0.0015	0.0117	0.0008	0.0212	-0.0032	0.0019	-0.0019
0.2360	0.2920	0.3380	0.2980	69.20	0.0042	0.0003	0.0123	0.0018	0.0197	-0.0008	0.0035	-0.0018
0.2240	0.2250	0.3390	0.2320	69.80	0.0073	-0.0026	0.0139	-0.0015	0.0191	-0.0010	0.0082	-0.0056
0.2440	0.1470	0.3810	0.1450	69.80	0.0093	-0.0010	0.0123	-0.0006	0.0171	0.0015	0.0109	-0.0036
0.2260	0.0760	0.3820	0.0760	70.50	0.0065	-0.0011	0.0021	-0.0012	0.0069	0.0010	0.0054	-0.0023
0.3350	0.5960	0.3560	0.5950	65.50	-0.0015	0.0040	-0.0008	0.0023	0.0029	-0.0017	-0.0087	0.0133
0.3260	0.5110	0.3750	0.5080	66.10	-0.0006	0.0027	0.0028	0.0009	0.0112	-0.0080	-0.0092	0.0129
0.3060	0.4870	0.3600	0.4870	66.60	0.0016	0.0036	0.0060	0.0024	0.0162	-0.0076	-0.0074	0.0131
0.3030	0.4240	0.3780	0.4210	67.10	0.0000	0.0018	0.0067	0.0014	0.0162	-0.0075	-0.0060	0.0067
0.3060	0.3600	0.3970	0.3510	67.40	0.0032	0.0017	0.0116	0.0019	0.0194	-0.0048	0.0008	0.0021
0.3040	0.2900	0.4180	0.2770	67.70	-0.0001	0.0021	0.0091	0.0027	0.0153	-0.0011	0.0006	-0.0009
0.2930	0.2150	0.4220	0.2030	68.30	0.0043	0.0012	0.0120	0.0019	0.0175	0.0013	0.0069	-0.0027
0.2890	0.1590	0.4380	0.1490	68.40	-0.0012	0.0000	0.0031	0.0005	0.0092	0.0014	0.0014	-0.0034
0.3000	0.0760	0.4660	0.0680	68.60	0.0036	0.0006	-0.0015	-0.0006	0.0092	0.0020	0.0048	-0.0010
0.3850	0.5250	0.4300	0.5040	65.40	-0.0007	0.0043	0.0010	0.0022	0.0028	-0.0011	-0.0053	0.0116
0.3870	0.4820	0.4450	0.4580	65.70	0.0011	0.0037	0.0041	0.0015	0.0073	-0.0039	-0.0033	0.0104
0.3910	0.4180	0.4710	0.3890	65.80	-0.0003	0.0035	0.0049	0.0018	0.0087	-0.0050	-0.0027	0.0070
0.3600	0.3850	0.4470	0.3620	66.70	0.0022	0.0033	0.0093	0.0024	0.0149	-0.0048	0.0001	0.0052
0.3250	0.3580	0.4190	0.3440	67.50	0.0019	0.0016	0.0102	0.0014	0.0171	-0.0051	0.0000	0.0018
0.3770	0.2160	0.5110	0.1880	66.60	0.0012	0.0026	-0.0005	0.0032	0.0131	0.0003	0.0044	-0.0018
0.3720	0.1410	0.5280	0.1210	66.60	-0.0027	0.0003	-0.0054	0.0010	0.0075	0.0001	0.0003	-0.0033
0.3670	0.0720	0.5350	0.0610	66.80	0.0014	-0.0004	0.0052	-0.0001	0.0081	0.0001	0.0030	-0.0023
0.4480	0.4910	0.5180	0.4380	64.80	-0.0143	0.0162	-0.0133	0.0145	-0.0166	0.0155	-0.0156	0.0199
0.4390	0.4270	0.5140	0.3870	65.20	0.0018	0.0026	0.0052	0.0007	0.0053	-0.0036	0.0008	0.0059
0.4420	0.3580	0.5500	0.2950	65.20	-0.0108	0.0232	-0.0054	0.0220	-0.0046	0.0160	-0.0100	0.0232
0.4300	0.2810	0.5490	0.2400	65.50	-0.0011	0.0043	0.0059	0.0043	0.0078	-0.0014	0.0012	0.0010
0.4310	0.2240	0.5640	0.1870	65.60	-0.0014	0.0029	0.0048	0.0036	0.0078	-0.0009	0.0015	-0.0015
0.4390	0.1410	0.5860	0.1150	65.30	0.0021	-0.0005	0.0031	0.0006	0.0103	-0.0022	0.0044	-0.0043
0.4070	0.0910	0.5750	0.0710	65.80	-0.0057	0.0033	-0.0102	0.0040	0.0024	0.0031	-0.0039	0.0008
0.5030	0.4250	0.5760	0.3710	64.30	0.0027	0.0006	0.0041	-0.0008	-0.0004	-0.0007	0.0030	0.0026
0.4930	0.3550	0.5900	0.2990	64.40	-0.0007	0.0052	0.0028	0.0042	0.0009	-0.0008	0.0006	0.0053

--------------EXPERIMENTAL--------------					---MARGULES 3---		---MARGULES 4---		-----ALPHA 2----		-----ALPHA 3----	
X1	X2	Y1	Y2	TEMP.	ΔY1	ΔY2	ΔY1	ΔY2	ΔY1	ΔY2	ΔY1	ΔY2
0.4940	0.3030	0.6030	0.2500	64.50	0.0005	0.0038	0.0051	0.0036	0.0038	-0.0026	0.0026	0.0019
0.4730	0.2850	0.5890	0.2340	64.70	-0.0009	0.0065	0.0046	0.0065	0.0046	0.0003	0.0013	0.0037
0.4960	0.1550	0.6340	0.1200	64.30	0.0025	0.0010	0.0036	0.0025	0.0076	-0.0022	0.0039	-0.0028
0.4940	0.0810	0.6490	0.0590	64.20	-0.0014	0.0021	-0.0084	0.0033	0.0034	0.0008	-0.0017	-0.0002
0.5690	0.3570	0.6550	0.2900	63.30	0.0046	0.0002	0.0057	-0.0005	0.0005	-0.0017	0.0054	0.0010
0.5610	0.3050	0.6590	0.2410	63.40	0.0058	0.0026	0.0080	0.0026	0.0039	-0.0025	0.0072	0.0020
0.5710	0.2230	0.6870	0.1680	63.30	0.0032	0.0020	0.0053	0.0033	0.0022	-0.0037	0.0040	-0.0005
0.5540	0.1570	0.6870	0.1140	63.40	-0.0005	0.0033	-0.0003	0.0052	0.0005	-0.0009	-0.0006	0.0001
0.5570	0.0820	0.6980	0.0570	63.20	0.0019	0.0018	-0.0050	0.0035	0.0028	-0.0003	-0.0004	-0.0002
0.6380	0.2940	0.7300	0.2200	62.50	0.0048	0.0005	0.0053	0.0007	0.0002	-0.0022	0.0044	0.0017
0.6370	0.2210	0.7380	0.1610	62.40	0.0087	-0.0010	0.0094	0.0004	0.0042	-0.0065	0.0080	-0.0017
0.6550	0.1480	0.7630	0.1030	61.80	0.0080	-0.0012	0.0069	0.0011	0.0023	-0.0062	0.0050	-0.0025
0.6460	0.0760	0.7650	0.0500	61.80	0.0061	0.0007	-0.0001	0.0027	0.0005	-0.0021	0.0006	-0.0005
0.7110	0.2220	0.8000	0.1530	61.20	0.0062	-0.0012	0.0061	-0.0002	0.0018	-0.0057	0.0035	0.0007
0.7220	0.1460	0.8120	0.0980	61.00	0.0109	-0.0029	0.0101	-0.0008	0.0034	-0.0081	0.0065	-0.0023
0.7150	0.0840	0.8140	0.0520	60.90	0.0084	0.0013	0.0048	0.0035	-0.0011	-0.0023	0.0016	0.0010
0.7890	0.1510	0.8610	0.0980	60.10	0.0101	-0.0047	0.0098	-0.0033	0.0058	-0.0098	0.0051	-0.0017
0.7840	0.0660	0.8600	0.0400	59.90	0.0122	-0.0008	0.0100	0.0010	0.0004	-0.0041	0.0040	-0.0001
0.8620	0.0720	0.9140	0.0430	58.90	0.0091	-0.0028	0.0090	-0.0017	0.0018	-0.0067	0.0025	-0.0005
0.2070	0.3460	0.2910	0.3670	69.50	0.0030	-0.0007	0.0096	0.0015	0.0198	-0.0034	-0.0010	-0.0004
				MEAN	0.0049	0.0029	0.0065	0.0028	0.0118	0.0055	0.0057	0.0062

METHYL ALCOHOL(1)-CARBON TETRACHLORIDE(2)-BENZENE(3)

SCATCHARD G.,TICKNOR L.B.:J.AM.CHEM.SOC.74,15,3724(1952).

ANTOINE VAPOR PRESSURE CONSTANTS

	A	B	C
(1)	8.07246	1574.990	238.860
(2)	6.93390	1242.430	230.000
(3)	6.90565	1211.033	220.790

T = 34.68

EQUATION	ORDER	BINARY CONSTANTS				TERNARY CONSTANTS	
		I-J	1-2	2-3	3-1		
MARGULES	3	AIJ	1.0304	0.0623	0.7218	C123	-0.0783
		AJI	0.7465	0.0381	0.9463		
MARGULES	4	AIJ	1.2187	0.0619	0.8612	C1123	1.0235
		AJI	0.9347	0.0377	1.0952	C1223	0.3477
		DIJ	0.6502	-0.0013	0.5157	C1233	-0.4202
ALPHA	2	AIJ	15.2691	0.2957	9.5751		
		AJI	14.1510	-0.0214	13.1416		
ALPHA	3	AIJ	8.6735	0.2332	20.4934	A123	32.4880
		AJI	59.6431	-0.0633	5.2724	A231	-15.4754
		AIJJ	48.2036	0.0669	-17.1916	A312	13.7877
		AJII	-55.3299	0.0361	20.2503		

--------------EXPERIMENTAL--------------					---MARGULES 3---		---MARGULES 4---		-----ALPHA 2----		-----ALPHA 3---	
X1	X2	Y1	Y2	PRESS.	ΔY1	ΔY2	ΔY1	ΔY2	ΔY1	ΔY2	ΔY1	ΔY2
0.2075	0.1900	0.4920	0.1472	291.11	0.0174	-0.0099	0.0025	-0.0115	-0.0046	0.0082	0.0020	0.0003
0.2110	0.3879	0.4804	0.2774	302.13	0.0195	-0.0068	0.0127	-0.0086	-0.0016	0.0189	0.0049	0.0059
0.1987	0.5876	0.4733	0.3999	308.63	0.0226	-0.0118	0.0171	-0.0035	-0.0060	0.0191	0.0048	0.0036
0.3781	0.3122	0.5043	0.2747	307.23	0.0187	-0.0091	0.0068	-0.0082	0.0152	0.0062	0.0022	0.0023
0.5543	0.2078	0.5308	0.2501	308.13	0.0020	-0.0055	0.0123	-0.0119	0.0161	0.0003	-0.0001	-0.0030
0.7599	0.1076	0.5903	0.2080	298.80	0.0040	-0.0027	0.0202	-0.0046	-0.0055	0.0147	-0.0035	0.0017
				MEAN	0.0140	0.0076	0.0119	0.0081	0.0081	0.0112	0.0029	0.0028

METHYL ALCOHOL(1)-CARBON TETRACHLORIDE(2)-BENZENE(3)

SCATCHARD G.,TICKNOR L.B.:J.AM.CHEM.SOC.74,3724(1952).

ANTOINE VAPOR PRESSURE CONSTANTS

	A	B	C
(1)	8.07246	1574.990	238.860
(2)	6.93390	1242.430	230.000
(3)	6.90565	1211.033	220.790

T = 55.00

EQUATION	ORDER	BINARY CONSTANTS				TERNARY CONSTANTS	
		I-J	1-2	2-3	3-1		
MARGULES	3	AIJ	0.9867	0.0543	0.7083	C123	-0.1430
		AJI	0.7371	0.0356	0.9298		
MARGULES	4	AIJ	1.0964	0.0556	0.8528	C1123	0.3170
		AJI	0.8521	0.0370	1.0566	C1223	-0.2529
		DIJ	0.4132	0.0039	0.5264	C1233	0.2787
ALPHA	2	AIJ	14.6800	0.2312	8.4506		
		AJI	11.3592	0.0025	13.5071		
ALPHA	3	AIJ	4.5480	0.2878	17.5444	A123	22.3477
		AJI	20.6092	0.0631	6.4434	A231	1.9442
		AIJJ	20.7948	-0.0510	-14.6006	A312	-0.9281
		AJII	-16.9666	-0.0636	17.7464		

---------------EXPERIMENTAL---------------					---MARGULES 3---		---MARGULES 4---		-----ALPHA 2----		-----ALPHA 3----	
X1	X2	Y1	Y2	PRESS.	ΔY1	ΔY2	ΔY1	ΔY2	ΔY1	ΔY2	ΔY1	ΔY2
0.1880	0.1960	0.5152	0.1387	665.26	0.0144	-0.0088	0.0077	-0.0084	-0.0090	0.0031	0.0055	0.0024
0.1983	0.3961	0.5130	0.2624	690.29	0.0132	-0.0100	0.0048	-0.0106	-0.0103	0.0110	0.0062	0.0050
0.1982	0.3963	0.5109	0.2628	689.67	0.0153	-0.0103	0.0069	-0.0109	-0.0082	0.0107	0.0083	0.0047
0.1945	0.5922	0.5068	0.3733	706.51	0.0180	-0.0115	0.0190	-0.0130	-0.0117	0.0174	0.0045	0.0063
0.3590	0.3230	0.5397	0.2539	711.17	0.0142	-0.0099	0.0022	-0.0085	0.0104	0.0001	0.0105	0.0014
0.5557	0.2134	0.5672	0.2316	717.20	-0.0024	-0.0021	-0.0009	-0.0057	0.0154	-0.0044	0.0118	-0.0027
0.7515	0.1115	0.6209	0.1925	703.34	0.0017	-0.0047	0.0036	-0.0064	-0.0017	0.0034	0.0098	-0.0049
0.8433	0.0814	0.6736	0.1890	680.86	0.0157	-0.0120	0.0041	-0.0082	-0.0261	0.0172	0.0077	-0.0062
				MEAN	0.0119	0.0087	0.0062	0.0090	0.0116	0.0084	0.0081	0.0042

METHYL ALCOHOL(1)-HEPTANE(2)-TOLUENE(3)

BENEDICT M.,JOHNSON C.A.,SOLOMON E.,RUBIN L.C.:TRANS.AM.INST.
     CHEM.ENGRS. 41,371,(1945).

ANTOINE VAPOR PRESSURE CONSTANTS
        A   B   C
(1) 8.07246   1574.990   238.860
(2) 6.90240   1268.115   216.900
(3) 6.95334   1343.943   219.377

P =   760.00

EQUATION	ORDER	BINARY CONSTANTS				TERNARY CONSTANTS	
		I-J	1-2	2-3	3-1		
MARGULES	3	AIJ	0.9819	0.1451	0.7965	C123	-0.2911
		AJI	1.0624	0.0913	0.8674		
MARGULES	4	AIJ	1.2818	0.1203	0.9123	C1123	0.6053
		AJI	1.2753	0.0611	1.0240	C1223	-0.0836
		DIJ	0.9362	-0.0862	0.4724	C1233	-0.2384
ALPHA	2	AIJ	109.1542	0.7964	4.6598		
		AJI	36.0505	0.0116	28.8824		
ALPHA	3	AIJ	19.3278	-0.8584	12.4602	A123	65.6203
		AJI	40.5350	-1.0498	10.8896	A231	-5.3505
		AIJJ	107.1288	1.6389	-11.6283	A312	6.1957
		AJII	-33.0575	0.9496	47.8002		

--------------EXPERIMENTAL--------------					---MARGULES 3---		---MARGULES 4---		-----ALPHA 2----		-----ALPHA 3----	
x1	x2	Y1	Y2	TEMP.	ΔY1	ΔY2	ΔY1	ΔY2	ΔY1	ΔY2	ΔY1	ΔY2
0.3172	0.2533	0.7458	0.1326	62.96	0.0275	-0.0147	0.0156	-0.0106	0.0059	0.0542	0.0060	0.0008
0.3711	0.0962	0.7797	0.0635	63.34	0.0082	-0.0024	0.0041	0.0010	-0.0036	0.0473	-0.0050	-0.0006
0.5409	0.3412	0.7520	0.2048	59.97	-0.0007	0.0008	0.0007	-0.0043	-0.0003	0.0297	-0.0055	0.0107
0.6038	0.1717	0.7697	0.1412	61.35	-0.0120	0.0077	-0.0042	0.0031	-0.0022	0.0498	-0.0062	0.0069
0.6114	0.0911	0.7864	0.0911	62.51	-0.0092	0.0023	0.0048	0.0038	-0.0011	0.0483	-0.0033	-0.0007
0.7494	0.0787	0.7927	0.1123	61.59	-0.0096	0.0018	-0.0003	0.0044	-0.0077	0.0471	-0.0044	0.0015
0.7733	0.1510	0.7563	0.1969	59.96	-0.0009	0.0012	0.0044	-0.0057	0.0069	0.0224	0.0049	0.0004
0.8486	0.0684	0.7815	0.1517	60.77	0.0234	-0.0210	0.0142	-0.0101	0.0000	0.0319	0.0125	-0.0080
				MEAN	0.0114	0.0065	0.0060	0.0054	0.0035	0.0413	0.0060	0.0037

510

3.2. Electrolytes

HYDROGEN CHLORIDE(1)-SULPHURIC ACID(2)-WATER(3)
STORONKIN A.V.,SUSAREV M.P:VESTNIK LENINGR.UNIV.
6,119(1952).

T = 25.00

ORDER	A13	B13	C13	D13	A23	B23	C23	D23
1	-5.5923	17.5465	-	-31.1108	-	-	-	-
2	9.7138	-0.4355	-11.9442	13.4784	-	-	-	-

| --------------EXPERIMENTAL-------------- | | | | | DEVIATION IN CALCD. VAPOR PHASE COMPN. | | | |
| | | | | | ----ORDER 1---- | | ----ORDER 2---- | |
X1	X2	Y1	Y2	PRESS.	ΔY1	ΔY2	ΔY1	ΔY2
0.0100	0.0400	0.0001	0.0000	20.47	0.0002	0.0000	0.0000	0.0000
0.0100	0.0600	0.0003	0.0000	18.25	0.0016	0.0000	0.0000	0.0000
0.0100	0.0800	0.0009	0.0000	15.88	0.0069	0.0000	0.0002	0.0000
0.0100	0.1000	0.0039	0.0000	13.37	0.0183	0.0000	0.0001	0.0000
0.0100	0.1200	0.0146	0.0000	11.20	0.0341	0.0000	-0.0005	0.0000
0.0100	0.1400	0.0511	0.0000	9.29	0.0362	0.0000	-0.0029	0.0000
0.0100	0.1600	0.1549	0.0000	8.27	-0.0213	0.0000	-0.0048	0.0000
0.0100	0.1700	0.2505	0.0000	8.18	-0.0930	0.0000	-0.0034	0.0000
0.0100	0.1800	0.3810	0.0000	8.60	-0.2003	0.0000	-0.0025	0.0000
0.0100	0.1900	0.5210	0.0000	9.59	-0.3185	0.0000	0.0090	0.0000
0.0100	0.2000	0.6610	0.0000	12.03	-0.4388	0.0000	0.0147	0.0000
0.0100	0.2100	0.7650	0.0000	15.02	-0.5257	0.0000	0.0286	0.0000
0.0200	0.0400	0.0003	0.0000	19.54	0.0007	0.0000	-0.0001	0.0000
0.0200	0.0800	0.0039	0.0000	14.80	0.0204	0.0000	0.0000	0.0000
0.0200	0.1200	0.0515	0.0000	10.59	0.0779	0.0000	-0.0026	0.0000
0.0200	0.1600	0.3919	0.0000	10.28	-0.0934	0.0000	-0.0058	0.0000
0.0300	0.0200	0.0001	0.0000	20.40	0.0001	0.0000	0.0001	0.0000
0.0300	0.0400	0.0008	0.0000	18.57	0.0021	0.0000	-0.0001	0.0000
0.0300	0.0600	0.0025	0.0000	16.08	0.0137	0.0000	0.0003	0.0000
0.0300	0.0800	0.0103	0.0000	13.83	0.0450	0.0000	0.0003	0.0000
0.0300	0.1000	0.0377	0.0000	11.71	0.0953	0.0000	-0.0004	0.0000
0.0300	0.1100	0.0701	0.0000	10.94	0.1153	0.0000	-0.0018	0.0000
0.0300	0.1200	0.1252	0.0000	10.43	0.1182	0.0000	-0.0038	0.0000
0.0300	0.1250	0.1629	0.0000	10.29	0.1105	0.0000	-0.0037	0.0000
0.0300	0.1300	0.2091	0.0000	10.35	0.0943	0.0000	-0.0033	0.0000
0.0300	0.1400	0.3280	0.0000	10.85	0.0345	0.0000	-0.0016	0.0000
0.0300	0.1500	0.4792	0.0000	11.96	-0.0615	0.0000	-0.0045	0.0000
0.0300	0.1600	0.6308	0.0000	14.69	-0.1035	0.0000	-0.0040	0.0000
0.0300	0.1800	0.8386	0.0000	26.27	-0.2912	0.0000	0.0137	0.0000
0.0400	0.0400	0.0016	0.0000	17.44	0.0054	0.0000	0.0001	0.0000
0.0400	0.0800	0.0245	0.0000	12.89	0.0835	0.0000	0.0004	0.0000
0.0400	0.1200	0.2503	0.0000	11.02	0.1309	0.0000	-0.0023	0.0000
0.0400	0.1600	0.8013	0.0000	23.39	-0.1856	0.0000	0.0001	0.0000
0.0500	0.0400	0.0039	0.0000	16.39	0.0107	0.0000	0.0001	0.0000
0.0500	0.0800	0.0544	0.0000	12.16	0.1344	0.0000	0.0003	0.0000
0.0500	0.1200	0.4256	0.0000	12.42	0.0982	0.0000	-0.0012	0.0000
0.0500	0.1600	0.9091	0.0000	37.95	-0.1769	0.0000	-0.0105	0.0000

		EXPERIMENTAL			DEVIATION IN CALCD. VAPOR PHASE COMPN.			
					ORDER 1		ORDER 2	
X1	X2	Y1	Y2	PRESS.	ΔY1	ΔY2	ΔY1	ΔY2
0.0600	0.0200	0.0021	0.0000	17.59	0.0022	0.0000	0.0001	0.0000
0.0600	0.0400	0.0083	0.0000	15.39	0.0238	0.0000	0.0005	0.0000
0.0600	0.0600	0.0307	0.0000	13.15	0.0941	0.0000	0.0022	0.0000
0.0600	0.0800	0.1110	0.0000	11.72	0.1876	0.0000	0.0003	0.0000
0.0600	0.0850	0.1492	0.0000	11.36	0.1998	0.0000	-0.0019	0.0000
0.0600	0.0900	0.1923	0.0000	11.55	0.2072	0.0000	-0.0001	0.0000
0.0600	0.1000	0.3119	0.0000	12.08	0.1841	0.0000	-0.0020	0.0000
0.0600	0.1100	0.4606	0.0000	13.56	0.1206	0.0000	-0.0028	0.0000
0.0600	0.1200	0.6101	0.0000	16.31	0.0423	0.0000	0.0025	0.0000
0.0700	0.0400	0.0179	0.0000	14.38	0.0444	0.0000	0.0008	0.0000
0.0700	0.0800	0.2093	0.0000	11.83	0.2197	0.0000	-0.0008	0.0000
0.0800	0.0400	0.0371	0.0000	13.46	0.0767	0.0000	0.0013	0.0000
0.0800	0.0800	0.3558	0.0000	12.94	0.2073	0.0000	-0.0044	0.0000
0.0900	0.0400	0.0738	0.0000	12.85	0.1201	0.0000	0.0020	0.0000
0.0900	0.0800	0.5311	0.0000	15.47	0.1526	0.0000	-0.0084	0.0000
0.1000	0.0100	0.0205	0.0000	14.65	0.0104	0.0000	0.0007	0.0000
0.1000	0.0200	0.0398	0.0000	13.67	0.0418	0.0000	0.0018	0.0000
0.1000	0.0300	0.0764	0.0000	12.95	0.0974	0.0000	0.0023	0.0000
0.1000	0.0400	0.1399	0.0000	13.99	0.1651	0.0000	0.0024	0.0000
0.1000	0.0500	0.2411	0.0000	24.11	0.2118	0.0000	0.0010	0.0000
0.1000	0.0600	0.3828	0.0000	38.28	0.2069	0.0000	-0.0040	0.0000
0.1000	0.0700	0.5216	0.0000	52.16	0.1780	0.0000	0.0148	0.0000
0.1000	0.0800	0.6790	0.0000	67.90	0.1015	0.0000	0.0069	0.0000
0.1000	0.0800	0.6840	0.0000	68.40	0.0965	0.0000	0.0019	0.0000
0.1100	0.0400	0.2423	0.0000	13.03	0.1967	0.0000	0.0067	0.0000
0.1150	0.0400	0.3138	0.0000	13.54	0.1953	0.0000	0.0044	0.0000
0.1200	0.0400	0.3865	0.0000	14.38	0.1912	0.0000	0.0094	0.0000
0.1300	0.0400	0.5564	0.0000	17.86	0.1450	0.0000	0.0058	0.0000
				MEAN	0.1161	0.0000	0.0034	0.0000

HYDROCHLORIC ACID(1)-FERRIC CHLORIDE(2)-WATER(3)

SUSAREV M.P.,PROKOF'EVA R.V.: ZHUR. FIZ. KHIM. 37,2408(1963).

T =   25.00

ORDER	A13	B13	C13	D13	A23	B23	C23	D23
1	-5.7367	17.8810	-	-88.3945	-	-	-	-
2	13.7188	-4.8674	-15.2124	44.8540	-	-	-	-

| --------------EXPERIMENTAL--------------- | | | | | DEVIATION IN CALCD. VAPOR PHASE COMPN. | | | |
| | | | | | ----ORDER 1---- | | ----ORDER 2---- | |
X1	X2	Y1	Y2	PRESS.	ΔY1	ΔY2	ΔY1	ΔY2
0.0100	0.0400	0.0004	0.0000	18.05	0.0002	0.0000	-0.0003	0.0000
0.0100	0.0600	0.0031	0.0000	14.81	-0.0019	0.0000	-0.0026	0.0000
0.0200	0.0200	0.0005	0.0000	20.39	-0.0003	0.0000	-0.0005	0.0000
0.0200	0.0400	0.0011	0.0000	17.17	0.0007	0.0000	-0.0008	0.0000
0.0200	0.0600	0.0069	0.0000	13.96	-0.0034	0.0000	-0.0050	0.0000
0.0300	0.0400	0.0030	0.0000	16.49	0.0014	0.0000	-0.0022	0.0000
0.0400	0.0100	0.0005	0.0000	20.13	-0.0002	0.0000	-0.0003	0.0000
0.0400	0.0200	0.0011	0.0000	18.59	0.0006	0.0000	-0.0007	0.0000
0.0400	0.0300	0.0025	0.0000	17.01	0.0024	0.0000	-0.0016	0.0000
0.0400	0.0400	0.0068	0.0000	15.49	0.0024	0.0000	-0.0047	0.0000
0.0400	0.0500	0.0137	0.0000	13.87	-0.0008	0.0000	-0.0083	0.0000
0.0400	0.0600	0.0286	0.0000	12.43	-0.0139	0.0000	-0.0146	0.0000
0.0500	0.0400	0.0124	0.0000	14.80	0.0054	0.0000	-0.0073	0.0000
0.0500	0.0600	0.0480	0.0000	11.80	-0.0218	0.0000	-0.0148	0.0000
0.0600	0.0200	0.0043	0.0000	16.91	0.0041	0.0000	-0.0023	0.0000
0.0600	0.0400	0.0226	0.0000	13.96	0.0101	0.0000	-0.0108	0.0000
0.0600	0.0600	0.0849	0.0000	11.48	-0.0404	0.0000	-0.0110	0.0000
0.0700	0.0400	0.0411	0.0000	13.09	0.0160	0.0000	-0.0150	0.0000
0.0800	0.0100	0.0067	0.0000	16.31	0.0034	0.0000	-0.0022	0.0000
0.0800	0.0200	0.0151	0.0000	14.99	0.0177	0.0000	-0.0051	0.0000
0.0800	0.0300	0.0381	0.0000	13.67	0.0273	0.0000	-0.0149	0.0000
0.0800	0.0400	0.0710	0.0000	12.70	0.0244	0.0000	-0.0154	0.0000
0.0800	0.0500	0.1369	0.0000	12.05	-0.0247	0.0000	-0.0058	0.0000
0.0900	0.0400	0.1163	0.0000	12.27	0.0357	0.0000	-0.0033	0.0000
0.0900	0.0600	0.2601	0.0000	11.88	-0.0915	0.0000	0.2050	0.0000
0.1000	0.0200	0.0520	0.0000	13.29	0.0548	0.0000	-0.0061	0.0000
0.1000	0.0400	0.1793	0.0000	12.27	0.0508	0.0000	0.0347	0.0000
0.1100	0.0200	0.0978	0.0000	12.95	0.0800	0.0000	-0.0098	0.0000
0.1200	0.0100	0.0957	0.0000	13.16	0.0468	0.0000	-0.0142	0.0000
0.1200	0.0200	0.1680	0.0000	12.87	0.1090	0.0000	-0.0069	0.0000
0.1200	0.0300	0.2740	0.0000	13.23	0.1099	0.0000	0.0510	0.0000
0.1200	0.0400	0.3690	0.0000	13.60	0.0724	0.0000	0.1776	0.0000
0.1300	0.0100	0.1652	0.0000	13.19	0.0720	0.0000	-0.0113	0.0000
0.1300	0.0200	0.2670	0.0000	13.40	0.1329	0.0000	0.0258	0.0000
0.1300	0.0400	0.4710	0.0000	15.30	0.0863	0.0000	0.2440	0.0000
0.1340	0.0400	0.5080	0.0000	16.40	0.0938	0.0000	0.2625	0.0000
0.1400	0.0100	0.2520	0.0000	13.45	0.1106	0.0000	0.0183	0.0000
0.1400	0.0200	0.3990	0.0000	15.08	0.1336	0.0000	0.0602	0.0000
0.1500	0.0100	0.4140	0.0000	15.54	0.0910	0.0000	0.0148	0.0000
0.1500	0.0200	0.5360	0.0000	18.10	0.1214	0.0000	0.0961	0.0000
0.1600	0.0100	0.5670	0.0000	19.17	0.0755	0.0000	0.0353	0.0000
0.1600	0.0200	0.6640	0.0000	23.20	0.0972	0.0000	0.1147	0.0000
0.1700	0.0100	0.6980	0.0000	25.69	0.0589	0.0000	0.0547	0.0000
0.1800	0.0100	0.8040	0.0000	37.20	0.0376	0.0000	0.0551	0.0000
				MEAN	0.0451	0.0000	0.0374	0.0000

HYDROCHLORIC ACID(1)-POTASSIUM CHLORIDE(2)-WATER(3)

STORONKIN A.V.,MARKUZIN N.P.: ZHUR. FIZ. KHIM. 29,111(1955).

T = 25.00

ORDER	A13	B13	C13	D13	A23	B23	C23	D23
1	-5.5923	17.5465	-	-12.5529	-	-	-	-
2	9.7138	-0.4355	-11.9442	-2.1946	-	-	-	-

| | | | | | DEVIATION IN CALCD. VAPOR PHASE COMPN. | | | |
| --------EXPERIMENTAL-------- | | | | | ----ORDER 1---- | | ----ORDER 2---- | |
X1	X2	Y1	Y2	PRESS.	ΔY1	ΔY2	ΔY1	ΔY2
0.0093	0.0708	0.0000	0.0000	19.71	0.0000	0.0000	0.0000	0.0000
0.0187	0.0622	0.0001	0.0000	19.56	-0.0001	0.0000	-0.0001	0.0000
0.0284	0.0539	0.0003	0.0000	19.21	-0.0001	0.0000	-0.0001	0.0000
0.0382	0.0458	0.0005	0.0000	18.75	-0.0002	0.0000	-0.0002	0.0000
0.0481	0.0384	0.0009	0.0000	18.45	-0.0004	0.0000	-0.0003	0.0000
0.0581	0.0318	0.0015	0.0000	17.87	-0.0005	0.0000	-0.0004	0.0000
0.0682	0.0259	0.0028	0.0000	17.41	-0.0009	0.0000	-0.0007	0.0000
0.0783	0.0209	0.0045	0.0000	16.75	-0.0008	0.0000	-0.0009	0.0000
0.0885	0.0167	0.0081	0.0000	15.89	-0.0009	0.0000	-0.0005	0.0000
0.0987	0.0133	0.0148	0.0000	15.07	-0.0005	0.0000	-0.0006	0.0000
0.1086	0.0107	0.0274	0.0000	14.37	-0.0006	0.0000	-0.0004	0.0000
0.1189	0.0086	0.0503	0.0000	13.64	0.0036	0.0000	-0.0005	0.0000
0.1290	0.0074	0.0844	0.0000	13.21	0.0162	0.0000	0.0096	0.0000
0.1321	0.0068	0.1130	0.0000	13.05	0.0068	0.0000	-0.0009	0.0000
0.1351	0.0065	0.1325	0.0000	13.12	0.0096	0.0000	0.0008	0.0000
0.1391	0.0061	0.1651	0.0000	13.11	0.0116	0.0000	0.0014	0.0000
0.1492	0.0052	0.2780	0.0000	13.80	0.0115	0.0000	0.0003	0.0000
0.1593	0.0047	0.4156	0.0000	15.21	0.3177	0.0000	0.0121	0.0000
0.1693	0.0042	0.5864	0.0000	19.45	-0.0037	0.0000	0.0017	0.0000
0.1793	0.0040	0.7299	0.0000	25.99	-0.0140	0.0000	0.0023	0.0000
				MEAN	0.0050	0.0000	0.0017	0.0000

514

HYDROCHLORIC ACID(1)-POTASSIUM CHLORIDE(2)-WATER(3)
    HALA F.,BOUBLIK T.:COLL.CZECH.CHEM.COMM. 29,2412 (1964).

T =   75.90

ORDER	A13	B13	C13	D13	A23	B23	C23
1	-3.5016	12.7060	-	-10.5331	-	-	-
2	-80.2251	95.7581	63.0566	-71.9030	-	-	-

| --------------EXPERIMENTAL-------------- | | | | | DEVIATION IN CALCD. VAPOR PHASE COMPN. | | | |
| | | | | | ----ORDER 1---- | | ----ORDER 2---- | |
X1	X2	Y1	Y2	PRESS.	ΔY1	ΔY2	ΔY1	ΔY2
0.0899	0.0150	0.0276	0.0000	212.80	-0.0011	0.0000	0.0006	0.0000
0.1038	0.0081	0.0511	0.0000	206.20	0.0003	0.0000	0.0022	0.0000
0.1053	0.0108	0.0604	0.0000	203.70	-0.0001	0.0000	0.0010	0.0000
0.1054	0.0077	0.0549	0.0000	205.40	0.0010	0.0000	0.0026	0.0000
0.1084	0.0066	0.0626	0.0000	204.10	0.0020	0.0000	0.0028	0.0000
0.1088	0.0091	0.0703	0.0000	202.70	0.0001	0.0000	-0.0003	0.0000
0.1106	0.0096	0.0793	0.0000	201.40	-0.0005	0.0000	-0.0030	0.0000
0.1109	0.0088	0.0771	0.0000	202.10	0.0015	0.0000	-0.0008	0.0000
0.1122	0.0058	0.0791	0.0000	202.30	-0.0007	0.0000	-0.0022	0.0000
0.1132	0.0096	0.0906	0.0000	200.70	0.0002	0.0000	-0.0060	0.0000
				MEAN	0.0007	0.0000	0.0021	0.0000

NITRIC ACID(1)-LITHIUM NITRATE(2)-WATER(3)
PROSEK J.: THESIS, UTZCHT-CSAV, PRAGUE, 1965.

P = 760.00

ORDER	A13	B13	C13	D13	A23	B23	C23	D23
1	-0.9462	2.4621	-	2.0182	-	-	-	-
2	1.0712	-0.3730	-1.1867	3.5648	-	-	-	-

| | | | | | DEVIATION IN CALCD. VAPOR PHASE COMPN. | | | |
| | | | | | ----ORDER 1---- | | ----ORDER 2---- | |
X1	X2	Y1	Y2	TEMP.	ΔY1	ΔY2	ΔY1	ΔY2
				EXPERIMENTAL				
0.3724	0.0431	0.5065	0.0000	120.65	-0.0067	0.0000	-0.0039	0.0000
0.3724	0.0431	0.5037	0.0000	120.65	-0.0039	0.0000	-0.0011	0.0000
0.3842	0.0200	0.4515	0.0000	120.20	0.0072	0.0000	0.0042	0.0000
0.3842	0.0200	0.4515	0.0000	120.20	0.0072	0.0000	0.0042	0.0000
0.3588	0.0100	0.3563	0.0000	120.00	0.0059	0.0000	-0.0005	0.0000
0.3605	0.0102	0.3563	0.0000	120.00	0.0108	0.0000	0.0045	0.0000
				MEAN	0.0070	0.0000	0.0031	0.0000

NITRIC ACID(1)-MAGNESIUM NITRATE(2)-WATER(3)
PROSEK J.: THESIS, UTZCHT-CSAV, PRAGUE, 1965.

P = 760.00

ORDER	A13	B13	C13	D13	A23	B23	C23	D23
1	-0.9462	2.4621	-	4.1905	- -	- -	- -	- -
2	1.0712	-0.3730	-1.1867	10.3382	- -	- -	- -	- -

DEVIATION IN CALCD. VAPOR PHASE COMPN.

----------EXPERIMENTAL----------					----ORDER 1----		----ORDER 2----	
$X1$	$X2$	$Y1$	$Y2$	TEMP.	$\Delta Y1$	$\Delta Y2$	$\Delta Y1$	$\Delta Y2$
0.1352	0.0597	0.1527	0.0000	120.70	-0.0113	0.0000	-0.0113	0.0000
0.1345	0.0587	0.1495	0.0000	120.60	-0.0134	0.0000	-0.0133	0.0000
0.2475	0.0392	0.3091	0.0000	121.60	-0.0069	0.0000	-0.0105	0.0000
0.2490	0.0393	0.3091	0.0000	121.50	-0.0023	0.0000	-0.0059	0.0000
0.3300	0.0336	0.4655	0.0000	121.10	0.0225	0.0000	0.0222	0.0000
0.3305	0.0332	0.4690	0.0000	121.10	0.0175	0.0000	0.0171	0.0000
0.1711	0.0315	0.0980	0.0000	116.50	0.0059	0.0000	0.0073	0.0000
0.1694	0.0301	0.0989	0.0000	116.50	-0.0018	0.0000	-0.0001	0.0000
0.3765	0.0160	0.4790	0.0000	120.20	0.0093	0.0000	0.0061	0.0000
0.3757	0.0158	0.4820	0.0000	120.20	0.0028	0.0000	-0.0006	0.0000
0.3220	0.0126	0.3070	0.0000	120.30	0.0164	0.0000	0.0097	0.0000
0.3205	0.0121	0.3048	0.0000	120.30	0.0116	0.0000	0.0050	0.0000
				MEAN	0.0101	0.0000	0.0091	0.0000

FOUR. AND MORE-THAN-FOUR-COMPONENT SYSTEMS

ETHYL ALCOHOL(1)-BENZENE(2)-METHYLCYCLOPENTANE(3)-HEXANE(4)

BELKNAP R.C.,WEBER J.H.:J.CHEM.ENG.DATA 6(4),485(1961).

ANTOINE VAPOR PRESSURE CONSTANTS

	A	B	C
(1)	8.16290	1623.220	228.980
(2)	6.90565	1211.033	220.790
(3)	6.86283	1186.059	226.042
(4)	6.87776	1171.530	224.366

P = 760.00

CONSTANTS	MARGULES 3	MARGULES 4	ALPHA 2	ALPHA 3
BINARY	A12 = 0.8475 A21 = 0.5880	A12 = 0.9672 A21 = 0.6976 D12 = 0.3891	A12 = 4.7128 A21 = 7.5827	A12 = 3.6303 A21 = 18.0065 A122= 6.6773 A211=-13.3299
	A23 = 0.1364 A32 = 0.1439	A23 = 0.1416 A32 = 0.1494 D23 = 0.0182	A23 = 0.0930 A32 = 0.7214	A23 = 0.3923 A32 = 0.9546 A233= -0.3257 A322= -0.1519
	A13 = 1.0018 A31 = 0.7074	A13 = 1.1397 A31 = 0.8360 D13 = 0.4825	A13 = 6.8185 A31 = 12.4437	A13 = 2.4289 A31 = 32.0315 A133= 15.1803 A311=-26.3753
	A24 = 0.2063 A42 = 0.1729	A24 = 0.2337 A42 = 0.1934 D24 = 0.0822	A24 = 0.1138 A42 = 1.1582	A24 = 0.3982 A42 = 1.3852 A244= -0.3125 A422= -0.1384
	A14 = 0.9889 A41 = 0.7817	A14 = 1.1487 A41 = 0.9361 D14 = 0.5483	A14 = 8.8090 A41 = 17.1742	A14 = 3.7045 A41 = 31.3435 A144= 13.0333 A411=-21.5236
TERNARY	C123=-0.3707	C1123= 0.6646 C1223= 0.0417 C1233=-0.7064		A123= 14.5869 A231= 5.3667 A312= 0.8045
	C134=-0.7155	C1134= 0.8407 C1334=-0.3219 C1344=-0.8477		A134= 18.4402 A341= 1.1908 A413= -2.2390

COMP.	EXPERIMENTAL X	EXPERIMENTAL Y	TEMP.	MARG 3 ΔY	MARG 4 ΔY	ALPHA2 ΔY	ALPHA3 ΔY
(1)	0.3790	0.3430	59.10	0.0036	-0.0056	0.0029	-0.0116
(2)	0.0790	0.0680		-0.0120	-0.0043	-0.0209	-0.0125
(3)	0.0600	0.0610		0.0088	0.0005	-0.0085	0.0050
(4)	0.4810	0.5290		-0.0014	0.0084	0.0254	0.0181

COMP.	X	Y	TEMP.	MARG 3 ΔY	MARG 4 ΔY	ALPHA2 ΔY	ALPHA3 ΔY
(1)	0.2490	0.3280	59.30	0.0213	0.0062	-0.0055	0.0069
(2)	0.0830	0.0640		-0.0102	-0.0067	-0.0182	-0.0142
(3)	0.1700	0.1420		0.0081	0.0013	-0.0102	0.0153
(4)	0.4990	0.4650		-0.0182	0.0002	0.0349	-0.0069
(1)	0.2400	0.3190	59.70	0.0135	0.0112	-0.0018	-0.0191
(2)	0.1770	0.1260		-0.0098	-0.0062	-0.0269	-0.0131
(3)	0.0820	0.0710		0.0074	-0.0002	-0.0055	0.0113
(4)	0.5010	0.4830		-0.0102	-0.0038	0.0352	0.0219
(1)	0.1170	0.3050	60.10	0.0038	0.0083	-0.0347	0.0067
(2)	0.0900	0.0630		-0.0024	-0.0024	-0.0108	-0.0116
(3)	0.2760	0.2070		0.0024	-0.0046	-0.0025	0.0191
(4)	0.5170	0.4250		-0.0038	-0.0012	0.0481	-0.0143
(1)	0.1180	0.2850	60.40	0.0012	0.0115	-0.0205	-0.0017
(2)	0.1880	0.1310		-0.0060	-0.0079	-0.0223	-0.0192
(3)	0.1810	0.1410		0.0056	-0.0017	-0.0027	0.0184
(4)	0.5130	0.4430		-0.0009	-0.0019	0.0454	0.0025
(1)	0.1120	0.2810	61.10	-0.0235	-0.0021	-0.0279	-0.0395
(2)	0.3210	0.2010		0.0127	0.0056	-0.0121	0.0028
(3)	0.0780	0.0670		0.0024	-0.0030	-0.0040	0.0091
(4)	0.4880	0.4520		0.0074	-0.0015	0.0429	0.0267
(1)	0.0190	0.1810	64.20	-0.0881	-0.0740	-0.0704	-0.0115
(2)	0.0680	0.0640		0.0001	0.0001	-0.0032	-0.0056
(3)	0.3700	0.3400		0.0082	0.0008	0.0048	-0.0001
(4)	0.4750	0.4790		0.0159	0.0091	0.0048	-0.0467
(1)	0.0310	0.1960	63.20	-0.0512	-0.0291	-0.0540	0.0000
(2)	0.1760	0.1290		0.0116	0.0083	-0.0001	-0.0057
(3)	0.2840	0.2190		0.0195	0.0112	0.0195	0.0240
(4)	0.5090	0.4570		0.0191	0.0086	0.0336	-0.0192
(1)	0.0400	0.2080	63.20	-0.0549	-0.0315	-0.0508	-0.0198
(2)	0.2900	0.2070		0.0094	0.0024	-0.0094	-0.0118
(3)	0.1670	0.1340		0.0108	0.0053	0.0072	0.0171
(4)	0.5030	0.4520		0.0337	0.0229	0.0519	0.0135
(1)	0.0330	0.1800	64.50	-0.0592	-0.0331	-0.0465	-0.0356
(2)	0.4070	0.2950		0.0071	-0.0034	-0.0150	-0.0112
(3)	0.0670	0.0580		0.0043	0.0021	0.0018	0.0074
(4)	0.4930	0.4680		0.0468	0.0334	0.0587	0.0385

				MARG 3	MARG 4	ALPHA2	ALPHA3
MEAN DEVIATIONS FOR COMPONENT:			(1)	0.0320	0.0213	0.0315	0.0152
			(2)	0.0081	0.0048	0.0139	0.0108
			(3)	0.0078	0.0031	0.0067	0.0127
			(4)	0.0157	0.0091	0.0381	0.0208
			MEAN	0.0159	0.0096	0.0225	0.0149

519

ETHYL ALCOHOL(1)-BENZENE(2)-METHYLCYCLOPENTANE(3)-HEXANE(4)

KAES G.L.,WEBER J.H.:J.CHEM.ENG.DATA 7(3),344(1962).

ANTOINE VAPOR PRESSURE CONSTANTS

	A	B	C
(1)	8.16290	1623.220	228.980
(2)	6.90565	1211.033	220.790
(3)	6.86283	1186.059	226.042
(4)	6.87776	1171.530	224.366

P = 760.00

CONSTANTS	MARGULES 3	MARGULES 4	ALPHA 2	ALPHA 3
BINARY	A12 = 0.8475	A12 = 0.9672	A12 = 4.7128	A12 = 3.6303
	A21 = 0.5880	A21 = 0.6976	A21 = 7.5827	A21 = 18.0065
		D12 = 0.3891		A122= 6.6773
				A211=-13.3299
	A23 = 0.1364	A23 = 0.1416	A23 = 0.0930	A23 = 0.3923
	A32 = 0.1439	A32 = 0.1494	A32 = 0.7214	A32 = 0.9546
		D23 = 0.0182		A233= -0.3257
				A322= -0.1519
	A13 = 1.0018	A13 = 1.1397	A13 = 6.8185	A13 = 2.4289
	A31 = 0.7074	A31 = 0.8360	A31 = 12.4437	A31 = 32.0315
		D13 = 0.4825		A133= 15.1803
				A311=-26.3753
	A24 = 0.2063	A24 = 0.2337	A24 = 0.1138	A24 = 0.3982
	A42 = 0.1729	A42 = 0.1934	A42 = 1.1582	A42 = 1.3852
		D24 = 0.0822		A244= -0.3125
				A422= -0.1384
	A14 = 0.9889	A14 = 1.1487	A14 = 8.8090	A14 = 3.7045
	A41 = 0.7817	A41 = 0.9361	A41 = 17.1742	A41 = 31.3435
		D14 = 0.5483		A144= 13.0333
				A411=-21.5236
TERNARY	C123=-0.3707	C1123= 0.6646		A123= 14.5869
		C1223= 0.0417		A231= 5.3667
		C1233=-0.7064		A312= 0.8045
	C134=-0.7155	C1134= 0.8407		A134= 18.4402
		C1334=-0.3219		A341= 1.1908
		C1344=-0.8477		A413= -2.2390

COMP.	-------EXPERIMENTAL-------			MARG 3	MARG 4	ALPHA2	ALPHA3
	X	Y	TEMP.	ΔY	ΔY	ΔY	ΔY
(1)	0.1570	0.3030	59.20	0.0154	0.0211	-0.0090	0.0035
(2)	0.0590	0.0500		-0.0098	-0.0086	-0.0189	-0.0168
(3)	0.0520	0.0410		0.0018	-0.0001	-0.0049	0.0028
(4)	0.7330	0.6060		-0.0074	-0.0124	0.0328	0.0104

520

COMP.	X	Y	TEMP.	MARG 3 ΔY	MARG 4 ΔY	ALPHA2 ΔY	ALPHA3 ΔY
(1)	0.1070	0.2940	59.80	-0.0031	0.0146	-0.0264	0.0063
(2)	0.0450	0.0350		-0.0028	-0.0029	-0.0097	-0.0095
(3)	0.1050	0.0810		0.0006	-0.0033	-0.0077	0.0035
(4)	0.7430	0.5900		0.0054	-0.0084	0.0438	-0.0002
(1)	0.0390	0.2230	62.10	-0.0461	-0.0121	-0.0465	0.0107
(2)	0.0600	0.0450		0.0050	0.0038	-0.0023	-0.0049
(3)	0.1770	0.1310		0.0111	0.0036	0.0069	0.0129
(4)	0.7250	0.6010		0.0300	0.0046	0.0419	-0.0188
(1)	0.0150	0.1290	65.20	-0.0372	-0.0113	-0.0324	0.0249
(2)	0.0460	0.0370		0.0064	0.0063	0.0029	0.0013
(3)	0.2340	0.2000		0.0006	-0.0064	0.0038	0.0019
(4)	0.7050	0.6350		0.0292	0.0104	0.0247	-0.0292
(1)	0.0100	0.1080	65.80	-0.0511	-0.0336	-0.0420	-0.0062
(2)	0.1180	0.1020		0.0068	0.0062	0.0001	-0.0018
(3)	0.1530	0.1340		-0.0001	-0.0033	0.0026	0.0030
(4)	0.7200	0.6570		0.0433	0.0297	0.0383	0.0040
(1)	0.0350	0.2210	62.20	-0.0724	-0.0406	-0.0609	-0.0192
(2)	0.1230	0.0960		0.0050	0.0020	-0.0089	-0.0125
(3)	0.1110	0.0860		0.0056	0.0011	0.0021	0.0077
(4)	0.7320	0.5970		0.0618	0.0376	0.0677	0.0240
(1)	0.0900	0.2770	60.10	-0.0255	0.0004	-0.0291	-0.0109
(2)	0.1110	0.0850		-0.0046	-0.0064	-0.0209	-0.0196
(3)	0.0570	0.0420		0.0039	0.0011	-0.0011	0.0059
(4)	0.7420	0.5960		0.0262	0.0049	0.0510	0.0247
(1)	0.0840	0.2250	61.60	0.0137	0.0417	0.0131	0.0254
(2)	0.1560	0.1160		-0.0039	-0.0077	-0.0245	-0.0220
(3)	0.0560	0.0430		0.0026	-0.0003	-0.0018	0.0053
(4)	0.7040	0.6160		-0.0125	-0.0337	0.0132	-0.0087
(1)	0.0150	0.0950	66.30	-0.0197	0.0018	-0.0072	0.0265
(2)	0.1630	0.1410		0.0015	-0.0013	-0.0102	-0.0137
(3)	0.1060	0.0920		-0.0001	-0.0028	0.0000	0.0023
(4)	0.7160	0.6720		0.0183	0.0023	0.0174	-0.0151
(1)	0.0160	0.1260	65.30	-0.0530	-0.0314	-0.0367	-0.0117
(2)	0.2100	0.1720		0.0057	0.0011	-0.0093	-0.0129
(3)	0.0580	0.0470		0.0037	0.0023	0.0032	0.0053
(4)	0.7170	0.6550		0.0436	0.0279	0.0428	0.0193
MEAN DEVIATIONS FOR COMPONENT: (1)				0.0337	0.0209	0.0303	0.0145
(2)				0.0052	0.0046	0.0108	0.0115
(3)				0.0030	0.0024	0.0034	0.0051
(4)				0.0278	0.0172	0.0374	0.0154
MEAN				0.0174	0.0113	0.0205	0.0116

HH

ETHYL ALCOHOL(1)-BENZENE(2)-METHYLCYCLOPENTANE(3)-HEXANE(4)

SINOR J.E.,WEBER J.H.:J.CHEM.ENG.DATA 5,244(1960).

ANTOINE VAPOR PRESSURE CONSTANTS

	A	B	C
(1)	8.16290	1623.220	228.980
(2)	6.90565	1211.033	220.790
(3)	6.86283	1186.059	226.042
(4)	6.87776	1171.530	224.366

P = 760.00

CONSTANTS	MARGULES 3	MARGULES 4	ALPHA 2	ALPHA 3
BINARY	A12 = 0.8475 A21 = 0.5880	A12 = 0.9672 A21 = 0.6976 D12 = 0.3891	A12 = 4.7128 A21 = 7.5827	A12 = 3.6303 A21 = 18.0065 A122= 6.6773 A211=-13.3299
	A23 = 0.1364 A32 = 0.1439	A23 = 0.1416 A32 = 0.1494 D23 = 0.0182	A23 = 0.0930 A32 = 0.7214	A23 = 0.3923 A32 = 0.9546 A233= -0.3257 A322= -0.1519
	A13 = 1.0018 A31 = 0.7074	A13 = 1.1397 A31 = 0.8360 D13 = 0.4825	A13 = 6.8185 A31 = 12.4437	A13 = 2.4289 A31 = 32.0315 A133= 15.1803 A311=-26.3753
	A24 = 0.2063 A42 = 0.1729	A24 = 0.2337 A42 = 0.1934 D24 = 0.0822	A24 = 0.1138 A42 = 1.1582	A24 = 0.3982 A42 = 1.3852 A244= -0.3125 A422= -0.1384
	A14 = 0.9889 A41 = 0.7817	A14 = 1.1487 A41 = 0.9361 D14 = 0.5483	A14 = 8.8090 A41 = 17.1742	A14 = 3.7045 A41 = 31.3435 A144= 13.0333 A411=-21.5236
TERNARY	C123=-0.3707	C1123= 0.6646 C1223= 0.0417 C1233=-0.7064		A123= 14.5869 A231= 5.3667 A312= 0.8045
	C134=-0.7155	C1134= 0.8407 C1334=-0.3219 C1344=-0.8477		A134= 18.4402 A341= 1.1908 A413= -2.2390

COMP.	EXPERIMENTAL			MARG 3	MARG 4	ALPHA2	ALPHA3
	X	Y	TEMP.	ΔY	ΔY	ΔY	ΔY
(1)	0.2860	0.3290	61.50	0.0275	0.0148	0.0035	0.0148
(2)	0.2980	0.2230		-0.0235	-0.0173	-0.0265	-0.0266
(3)	0.2840	0.2790		0.0012	0.0024	-0.0007	0.0240
(4)	0.1500	0.1690		-0.0052	0.0002	0.0237	-0.0122

522

COMP.	EXPERIMENTAL			MARG 3	MARG 4	ALPHA2	ALPHA3
	X	Y	TEMP.	ΔY	ΔY	ΔY	ΔY
(1)	0.3580	0.3430	61.90	0.0167	0.0123	0.0094	-0.0050
(2)	0.3460	0.2750		-0.0121	-0.0068	-0.0158	0.0001
(3)	0.1550	0.1830		0.0113	-0.0025	-0.0061	0.0195
(4)	0.1400	0.1990		-0.0158	-0.0030	0.0124	-0.0146
(1)	0.1280	0.2910	62.30	0.0107	0.0112	-0.0263	0.0109
(2)	0.3690	0.2520		-0.0121	-0.0129	-0.0140	-0.0286
(3)	0.3520	0.3040		0.0039	0.0046	0.0173	0.0301
(4)	0.1520	0.1530		-0.0026	-0.0029	0.0229	-0.0123
(1)	0.3590	0.3460	60.30	0.0067	-0.0050	0.0042	0.0138
(2)	0.1520	0.1250		-0.0122	-0.0054	-0.0202	-0.0192
(3)	0.3410	0.3540		0.0022	0.0071	-0.0063	0.0217
(4)	0.1470	0.1750		0.0033	0.0033	0.0223	-0.0164
(1)	0.0850	0.2610	65.40	-0.0124	0.0036	-0.0410	-0.0204
(2)	0.6590	0.4450		0.0092	-0.0001	0.0192	0.0137
(3)	0.1100	0.1170		0.0032	-0.0032	0.0004	0.0099
(4)	0.1460	0.1770		0.0000	-0.0003	0.0213	-0.0032
(1)	0.6770	0.4020	61.20	-0.0123	0.0018	0.0103	-0.0067
(2)	0.1120	0.1320		0.0021	0.0086	-0.0015	0.0165
(3)	0.0850	0.1720		0.0101	-0.0176	-0.0191	-0.0016
(4)	0.1260	0.2940		0.0001	0.0072	0.0103	-0.0081
(1)	0.0680	0.2740	61.70	-0.0199	-0.0031	-0.0565	0.0117
(2)	0.1140	0.0830		0.0010	0.0004	-0.0048	-0.0159
(3)	0.6560	0.5030		0.0127	0.0023	0.0411	0.0221
(4)	0.1620	0.1400		0.0062	0.0004	0.0202	-0.0180
(1)	0.1530	0.2980	63.20	0.0160	0.0172	-0.0183	-0.0053
(2)	0.5160	0.3500		-0.0118	-0.0115	-0.0045	-0.0065
(3)	0.1840	0.1830		0.0024	-0.0033	0.0010	0.0189
(4)	0.1480	0.1690		-0.0067	-0.0023	0.0218	-0.0071
(1)	0.5140	0.3700	60.90	-0.0118	-0.0043	0.0096	-0.0031
(2)	0.1800	0.1710		-0.0074	-0.0008	-0.0152	-0.0014
(3)	0.1660	0.2310		0.0191	-0.0028	-0.0115	0.0163
(4)	0.1390	0.2280		0.0001	0.0078	0.0171	-0.0118
(1)	0.1540	0.3100	60.80	0.0207	0.0146	-0.0225	0.0180
(2)	0.1800	0.1280		-0.0089	-0.0066	-0.0160	-0.0255
(3)	0.5100	0.4170		-0.0110	-0.0043	0.0159	0.0249
(4)	0.1560	0.1450		-0.0008	-0.0038	0.0225	-0.0174

MEAN DEVIATIONS FOR COMPONENT:

				MARG 3	MARG 4	ALPHA2	ALPHA3
			(1)	0.0155	0.0088	0.0202	0.0110
			(2)	0.0100	0.0071	0.0138	0.0154
			(3)	0.0077	0.0050	0.0119	0.0189
			(4)	0.0041	0.0031	0.0195	0.0121
			MEAN	0.0093	0.0060	0.0163	0.0143

2,2,4-TRIMETHYLPENTANE(1)-METHYLCYCLOHEXANE(2)-TOLUENE(3)-PHENOL(4)

DRICKAMER H.G.,BROWN G.G.,WHITE R.R.:TRANS.AMER.INST.
CHEM.ENGRS.41,555(1945).

ANTOINE VAPOR PRESSURE CONSTANTS

	A	B	C
(1)	6.81189	1257.840	220.735
(2)	6.82689	1272.864	221.630
(3)	6.95334	1343.943	219.377
(4)	7.57893	1817.000	205.000

P = 760.00

CONSTANTS	MARGULES 3	MARGULES 4	ALPHA 2	ALPHA 3
BINARY	A12 = 0.0358 A21 =-0.0003	A12 = 0.0945 A21 = 0.0819 D12 = 0.2263	A12 = 0.0943 A21 = -0.0379	A12 = 9.6638 A21 = 9.8963 A122= -9.2095 A211=-10.2285
	A23 = 0.1194 A32 = 0.0705	A23 = 0.1227 A32 = 0.0740 D23 = 0.0105	A23 = 0.5663 A32 = -0.0122	A23 = 1.1386 A32 = 0.5247 A233= -0.5031 A322= -0.5535
	A34 = 0.3583 A43 = 0.2832	A34 = 0.2684 A43 = 0.1305 D34 =-0.4342	A34 = 11.8396 A43 = -0.4648	A34 = -2.1323 A43 = -1.6652 A344= 13.8654 A433= 0.8505
	A13 = 0.1164 A31 = 0.0946	A13 = 0.1060 A31 = 0.0840 D13 =-0.0351	A13 = 0.6914 A31 = -0.0447	A13 = -0.1468 A31 = -0.5431 A133= 0.8675 A311= 0.4314
	A24 = 0.5242 A42 = 0.6949	A24 = 0.7918 A42 = 0.9245 D24 = 0.9393	A24 = 22.8658 A42 = 0.7070	A24 = -4.0598 A42 = -1.9051 A244= 21.4909 A422= 1.5171
	A14 = 1.0592 A41 = 0.9559	A14 = 1.7943 A41 = 1.2474 D14 = 1.5975	A14 = 48.0900 A41 = 2.1458	A14 = -5.0173 A41 = -0.6255 A144= 42.6663 A411= 0.1565

COMP.	X	Y	TEMP.	MARG 3 ΔY	MARG 4 ΔY	ALPHA2 ΔY	ALPHA3 ΔY
(1)	0.0196	0.1909	124.90	-0.0115	0.1693	-0.0390	-0.0558
(2)	0.0557	0.2800		-0.0678	-0.0822	-0.0675	-0.0949
(3)	0.2340	0.4937		-0.0061	-0.1428	-0.0098	-0.0949
(4)	0.6907	0.0354		0.0854	0.0557	0.1164	0.1051

COMP.	EXPERIMENTAL		TEMP.	MARG 3 ΔY	MARG 4 ΔY	ALPHA2 ΔY	ALPHA3 ΔY
	X	Y					
(1)	0.0300	0.1058	117.80	0.0297	0.1283	0.0369	0.0284
(2)	0.1288	0.3549		-0.0433	-0.0577	-0.0487	-0.0768
(3)	0.3510	0.4618		0.0181	-0.0608	-0.0043	0.0639
(4)	0.4902	0.0775		-0.0046	-0.0098	0.0160	-0.0154
(1)	0.0306	0.1283	118.30	0.0407	0.1872	0.0327	0.0297
(2)	0.1525	0.4600		-0.0522	-0.1066	-0.0629	-0.0922
(3)	0.2319	0.3340		0.0071	-0.0723	-0.0042	0.0567
(4)	0.5850	0.0777		0.0044	-0.0083	0.0344	0.0058
(1)	0.0479	0.3216	123.80	0.0500	0.2494	0.0094	-0.0319
(2)	0.0183	0.0831		-0.0228	-0.0287	-0.0208	-0.0269
(3)	0.2535	0.4932		-0.0326	-0.1933	-0.0253	0.0274
(4)	0.6803	0.1021		0.0054	-0.0275	0.0367	0.0314
(1)	0.0496	0.0785	106.10	0.0085	0.0367	0.0434	0.0694
(2)	0.4312	0.5835		-0.0030	-0.0076	-0.0188	-0.0179
(3)	0.3341	0.3070		-0.0029	-0.0285	-0.0500	-0.0162
(4)	0.1851	0.0310		-0.0026	-0.0006	0.0253	-0.0353
(1)	0.0772	0.0969	102.20	0.0104	0.0336	0.0530	0.0740
(2)	0.6709	0.7494		0.0144	-0.0045	-0.0365	-0.0170
(3)	0.1251	0.1148		-0.0095	-0.0180	-0.0314	-0.0316
(4)	0.1268	0.0389		-0.0153	-0.0112	0.0149	-0.0254
(1)	0.0797	0.1882	115.60	0.0670	0.1963	0.0885	0.0996
(2)	0.2627	0.5263		-0.0426	-0.1215	-0.0713	-0.0735
(3)	0.1888	0.2115		-0.0117	-0.0568	-0.0305	0.0017
(4)	0.4688	0.0740		-0.0127	-0.0180	0.0133	-0.0278
(1)	0.0854	0.2945	116.10	0.0371	0.1492	0.0617	0.0153
(2)	0.0416	0.1125		-0.0222	-0.0237	-0.0254	-0.0235
(3)	0.4120	0.5220		-0.0110	-0.1168	-0.0493	0.0123
(4)	0.4610	0.0710		-0.0039	-0.0086	0.0130	-0.0040
(1)	0.0892	0.4060	117.20	0.0814	0.2268	0.0438	0.0217
(2)	0.0784	0.2485		-0.0538	-0.0881	-0.0538	-0.0564
(3)	0.1756	0.2640		-0.0254	-0.1154	-0.0266	0.0152
(4)	0.6568	0.0815		-0.0022	-0.0233	0.0366	0.0195
(1)	0.0953	0.1316	104.90	0.0093	0.0313	0.0670	0.0821
(2)	0.3868	0.4816		-0.0029	-0.0017	-0.0235	0.0404
(3)	0.4108	0.3606		0.0018	-0.0244	-0.0618	-0.0887
(4)	0.1071	0.0262		-0.0083	-0.0052	0.0183	-0.0339
(1)	0.1048	0.3710	116.70	0.0345	0.1453	0.0525	0.0099
(2)	0.0513	0.1420		-0.0342	-0.0384	-0.0385	-0.0337
(3)	0.3481	0.4242		-0.0063	-0.1061	-0.0393	0.0174
(4)	0.4958	0.0628		0.0060	-0.0008	0.0253	0.0064
(1)	0.1209	0.3035	111.60	0.0769	0.1991	0.0940	0.0991
(2)	0.2385	0.4895		-0.0662	-0.1452	-0.0999	-0.0878
(3)	0.1370	0.1502		-0.0128	-0.0503	-0.0266	-0.0069
(4)	0.5036	0.0568		0.0022	-0.0036	0.0325	-0.0043

COMP.	EXPERIMENTAL			MARG 3	MARG 4	ALPHA2	ALPHA3
	X	Y	TEMP.	ΔY	ΔY	ΔY	ΔY
(1)	0.1350	0.3898	112.00	0.0717	0.1675	0.0842	0.0649
(2)	0.1270	0.2845		-0.0462	-0.0715	-0.0627	-0.0477
(3)	0.2240	0.2620		-0.0234	-0.0879	-0.0469	-0.0130
(4)	0.5140	0.0637		-0.0020	-0.0080	0.0254	-0.0042
(1)	0.2036	0.4368	110.00	0.0704	0.1223	0.1108	0.0901
(2)	0.1766	0.3133		-0.0414	-0.0572	-0.0756	-0.0346
(3)	0.1823	0.1844		-0.0186	-0.0552	-0.0481	-0.0324
(4)	0.4375	0.0655		-0.0105	-0.0099	0.0129	-0.0230
(1)	0.2556	0.6775	110.00	0.0489	0.0451	0.0490	0.0265
(2)	0.0559	0.1161		-0.0281	-0.0080	-0.0374	-0.0198
(3)	0.1400	0.1515		-0.0250	-0.0433	-0.0432	-0.0211
(4)	0.5485	0.0549		0.0042	0.0063	0.0316	0.0145
(1)	0.3996	0.5017	100.60	0.0340	0.0461	0.1407	0.0191
(2)	0.0713	0.0977		-0.0140	-0.0080	-0.0316	0.1000
(3)	0.4312	0.3760		-0.0148	-0.0394	-0.1293	-0.1012
(4)	0.0979	0.0246		-0.0051	0.0014	0.0202	-0.0179
(1)	0.4102	0.5836	107.20	0.0434	0.0500	0.1374	0.0398
(2)	0.0395	0.0607		-0.0126	-0.0054	-0.0240	0.0415
(3)	0.3440	0.3157		-0.0299	-0.0521	-0.1303	-0.0734
(4)	0.2063	0.0400		-0.0009	0.0075	0.0169	-0.0080
(1)	0.7215	0.7830	101.70	0.0415	0.0224	0.0829	0.0122
(2)	0.0555	0.0682		-0.0115	-0.0002	-0.0314	0.0436
(3)	0.1049	0.0973		-0.0126	-0.0148	-0.0532	-0.0357
(4)	0.1181	0.0515		-0.0174	-0.0075	0.0018	-0.0200

MEAN DEVIATIONS FOR COMPONENT:							
			(1)	0.0426	0.1226	0.0682	0.0483
			(2)	0.0322	0.0476	0.0461	0.0516
			(3)	0.0150	0.0710	0.0450	0.0367
			(4)	0.0107	0.0118	0.0273	0.0223
			MEAN	0.0251	0.0632	0.0466	0.0397

526

METHYL ALCOHOL(1)-2-METHYLPENTANE(2)-3-METHYLPENTANE(3)-HEXANE(4)

VILIM O.:COLL.CZECH.CHEM.COMMUN.26,2124(1961).

ANTOINE VAPOR PRESSURE CONSTANTS

	A	B	C
(1)	8.07246	1574.990	238.860
(2)	6.83910	1135.410	226.572
(3)	6.84887	1152.368	227.129
(4)	6.87776	1171.530	224.366

P = 745.00

CONSTANTS	MARGULES 3	MARGULES 4	ALPHA 2	ALPHA 3
BINARY	A12 = 0.9627 A21 = 0.9047	A12 = 1.1203 A21 = 1.0451 D12 = 0.6336	A12 = 13.6661 A21 = 20.8786	A12 = 18.4755 A21 = 20.7485 A122= -5.5471 A211= 5.8661
	A13 = 0.9280 A31 = 0.9039	A13 = 1.1610 A31 = 1.1850 D13 = 0.9962	A13 = 14.8781 A31 = 20.5066	A13 = 18.6866 A31 = 19.3754 A133= -4.9991 A311= 5.1310
	A14 = 0.9432 A41 = 0.9973	A14 = 1.0616 A41 = 1.1440 D14 = 0.6455	A14 = 16.4013 A41 = 15.5911	A14 = 15.7699 A41 = 10.3615 A144= -2.9642 A411= 6.4179

COMP.	-------EXPERIMENTAL-------			MARG 3 ΔY	MARG 4 ΔY	ALPHA2 ΔY	ALPHA3 ΔY
	X	Y	TEMP.				
(1)	0.2010	0.4240	48.20	0.0188	0.0483	-0.0151	0.0148
(2)	0.2960	0.2620		-0.0263	-0.0404	-0.0225	-0.0120
(3)	0.1730	0.1290		-0.0057	-0.0068	0.0090	0.0089
(4)	0.3320	0.1880		0.0102	-0.0041	0.0257	-0.0147
(1)	0.3530	0.4470	46.80	0.0216	0.0321	-0.0120	0.0120
(2)	0.2450	0.2580		-0.0296	-0.0380	-0.0226	-0.0108
(3)	0.1449	0.0990		0.0212	0.0288	0.0380	0.0379
(4)	0.2580	0.1920		-0.0092	-0.0190	0.0006	-0.0351
(1)	0.5102	0.4750	46.40	-0.0130	-0.0125	-0.0203	-0.0050
(2)	0.1810	0.2260		-0.0007	-0.0062	-0.0012	0.0135
(3)	0.1015	0.1205		-0.0083	0.0078	0.0035	0.0047
(4)	0.2060	0.1790		0.0215	0.0104	0.0176	-0.0136
(1)	0.8410	0.5130	48.70	-0.0007	-0.0422	0.0057	-0.0010
(2)	0.0581	0.2090		-0.0177	-0.0126	-0.0148	0.0038
(3)	0.0352	0.1060		-0.0022	0.0330	0.0097	0.0133
(4)	0.0675	0.1750		0.0175	0.0187	-0.0035	-0.0191

| COMP. | ------EXPERIMENTAL------ | | | MARG 3 | MARG 4 | ALPHA2 | ALPHA3 |
	X	Y	TEMP.	ΔY	ΔY	ΔY	ΔY
(1)	0.2450	0.4500	48.40	0.0130	0.0354	-0.0266	0.0022
(2)	0.2310	0.2010		-0.0111	-0.0196	-0.0055	0.0057
(3)	0.1849	0.1380		-0.0018	-0.0008	0.0162	0.0177
(4)	0.3400	0.2110		0.0000	-0.0150	0.0159	-0.0255
(1)	0.4330	0.4180	46.10	0.0357	0.0393	0.0141	0.0322
(2)	0.3120	0.3230		0.0063	-0.0115	0.0121	0.0226
(3)	0.0897	0.1150		-0.0317	-0.0184	-0.0202	-0.0222
(4)	0.1665	0.1440		-0.0103	-0.0094	-0.0060	-0.0327
(1)	0.9030	0.5460	52.20	0.0513	-0.0094	0.0219	0.0084
(2)	0.0282	0.1520		-0.0255	-0.0142	-0.0098	0.0076
(3)	0.0237	0.1090		-0.0132	0.0268	0.0085	0.0151
(4)	0.0450	0.1930		-0.0126	-0.0032	-0.0206	-0.0311
(1)	0.3000	0.4240	45.60	0.0260	0.0403	-0.0133	0.0118
(2)	0.3790	0.3350		-0.0131	-0.0314	-0.0018	0.0021
(3)	0.1700	0.1345		-0.0068	-0.0014	0.0127	0.0074
(4)	0.1585	0.1080		-0.0075	-0.0090	0.0009	-0.0228
(1)	0.4180	0.4440	45.80	0.0116	0.0173	-0.0120	0.0085
(2)	0.2860	0.3480		-0.0525	-0.0683	-0.0480	-0.0386
(3)	0.1245	0.1140		-0.0004	0.0113	0.0145	0.0119
(4)	0.1720	0.0940		0.0413	0.0396	0.0454	0.0182
(1)	0.6750	0.4530	46.20	-0.0112	-0.0203	0.0029	0.0112
(2)	0.1730	0.3360		-0.0176	-0.0398	-0.0302	-0.0192
(3)	0.0789	0.1230		0.0053	0.0312	0.0141	0.0112
(4)	0.0750	0.1030		0.0085	0.0139	-0.0018	-0.0182
(1)	0.8810	0.4860	49.80	0.0593	0.0075	0.0389	0.0306
(2)	0.0628	0.3090		-0.0601	-0.0576	-0.0474	-0.0298
(3)	0.0301	0.1090		-0.0023	0.0376	0.0143	0.0148
(4)	0.0285	0.0970		0.0020	0.0114	-0.0068	-0.0167
(1)	0.2920	0.4680	48.40	0.0215	0.0288	-0.0185	0.0048
(2)	0.1130	0.1140		-0.0163	-0.0138	-0.0121	-0.0003
(3)	0.1485	0.1280		-0.0124	-0.0029	0.0038	0.0122
(4)	0.4490	0.2910		0.0062	-0.0132	0.0257	-0.0177
(1)	0.4000	0.4700	47.80	0.0220	0.0238	-0.0073	0.0111
(2)	0.0954	0.0970		-0.0020	0.0036	0.0015	0.0130
(3)	0.1394	0.1420		-0.0171	-0.0017	-0.0004	0.0083
(4)	0.3720	0.2910		-0.0029	-0.0256	0.0062	-0.0324
(1)	0.8860	0.5480	50.80	0.0217	-0.0261	0.0189	0.0058
(2)	0.0173	0.0920		-0.0217	-0.0122	-0.0157	-0.0036
(3)	0.0243	0.1050		-0.0159	0.0228	0.0004	0.0098
(4)	0.0750	0.2580		0.0130	0.0125	-0.0066	-0.0150
MEAN DEVIATIONS FOR COMPONENT: (1)				0.0234	0.0274	0.0163	0.0114
(2)				0.0215	0.0264	0.0175	0.0130
(3)				0.0103	0.0165	0.0118	0.0139
(4)				0.0116	0.0147	0.0131	0.0223
MEAN				0.0167	0.0212	0.0147	0.0152

METHYL ALCOHOL(1)-ISOPRENE(2)-2-METHYLPENTANE(3)-
-3-METHYLPENTANE(4)-HEXANE(5)-METHYLCYCLOPENTANE(6)

VILIM O.:COLL.CZECH.CHEM.COMMUN.26,2124(1961).

ANTOINE VAPOR PRESSURE CONSTANTS
	A	B	C
(1)	8.07246	1574.990	238.860
(2)	6.90334	1080.996	234.670
(3)	6.83910	1135.410	226.572
(4)	6.84887	1152.368	227.129
(5)	6.87776	1171.530	224.366
(6)	6.86283	1186.059	226.042

P = 745.00

CONSTANTS	MARGULES 3	MARGULES 4	ALPHA 2	ALPHA 3
BINARY	A_{12} = 0.9664 A_{21} = 0.9030	A_{12} = 1.1725 A_{21} = 0.9682 D_{12} = 0.5139	A_{12} = 5.5223 A_{21} = 32.4236	A_{12} = 4.1819 A_{21} = 60.7063 A_{122}= 6.9817 A_{211}=-34.3272
	A_{13} = 0.9627 A_{31} = 0.9047	A_{13} = 1.1203 A_{31} = 1.0451 D_{13} = 0.6336	A_{13} = 13.6661 A_{31} = 20.8786	A_{13} = 18.4755 A_{31} = 20.7485 A_{133}= -5.5471 A_{311}= 5.8661
	A_{14} = 0.9280 A_{41} = 0.9039	A_{14} = 1.1610 A_{41} = 1.1850 D_{14} = 0.9962	A_{14} = 14.8781 A_{41} = 20.5066	A_{14} = 18.6866 A_{41} = 19.3754 A_{144}= -4.9991 A_{411}= 5.1310
	A_{15} = 0.9432 A_{51} = 0.9973	A_{15} = 1.0616 A_{51} = 1.1440 D_{15} = 0.6455	A_{15} = 16.4013 A_{51} = 15.5911	A_{15} = 15.7699 A_{51} = 10.3615 A_{155}= -2.9642 A_{511}= 6.4179
	A_{16} = 0.9058 A_{61} = 0.9054	A_{16} = 0.8585 A_{61} = 0.8301 D_{16} =-0.2129	A_{16} = 19.6342 A_{61} = 17.2622	A_{16} = -3.7103 A_{61} = 2.8196 A_{166}= 12.5798 A_{611}= -1.3463

	-------EXPERIMENTAL-------			MARG 3	MARG 4	ALPHA2	ALPHA3
COMP.	X	Y	TEMP.	ΔY	ΔY	ΔY	ΔY
(1)	0.1040	0.1840	41.40	0.0780	0.1136	0.1252	0.0853
(2)	0.2560	0.5570		-0.1534	-0.1732	-0.2929	-0.1678
(3)	0.0960	0.0420		0.0201	0.0174	0.0298	0.0257
(4)	0.1350	0.0760		0.0026	0.0009	0.0238	0.0148
(5)	0.3600	0.1210		0.0515	0.0418	0.1017	0.0483
(6)	0.0490	0.0200		0.0012	-0.0005	0.0123	-0.0063

II

| COMP. | EXPERIMENTAL | | | MARG 3 | MARG 4 | ALPHA2 | ALPHA3 |
	X	Y	TEMP.	ΔY	ΔY	ΔY	ΔY
(1)	0.0930	0.1950	43.80	0.0748	0.1065	0.1262	0.0696
(2)	0.1830	0.4530		-0.1433	-0.1567	-0.2675	-0.1461
(3)	0.2360	0.1620		0.0039	-0.0026	0.0132	0.0238
(4)	0.0820	0.0380		0.0138	0.0130	0.0222	0.0237
(5)	0.2310	0.1000		0.0206	0.0153	0.0429	0.0232
(6)	0.1750	0.0520		0.0303	0.0245	0.0631	0.0058
(1)	0.5830	0.3300	42.20	0.0247	0.0205	0.0653	0.0069
(2)	0.1250	0.4090		-0.0476	-0.0543	-0.1494	-0.0548
(3)	0.0640	0.0830		-0.0070	-0.0009	0.0050	0.0277
(4)	0.0320	0.0500		-0.0161	-0.0061	-0.0067	0.0015
(5)	0.1210	0.0700		0.0448	0.0459	0.0574	0.0579
(6)	0.0750	0.0580		0.0011	-0.0049	0.0286	-0.0392
(1)	0.4630	0.3420	43.60	0.0558	0.0540	0.0733	0.0193
(2)	0.1130	0.3600		-0.0926	-0.0903	-0.1715	-0.0754
(3)	0.0760	0.0730		0.0015	0.0067	0.0115	0.0314
(4)	0.0280	0.0190		0.0055	0.0116	0.0116	0.0169
(5)	0.2330	0.1550		0.0237	0.0161	0.0445	0.0385
(6)	0.0870	0.0510		0.0062	0.0020	0.0305	-0.0307
(1)	0.2060	0.4730	50.20	0.0073	0.0197	-0.0140	-0.0572
(2)	0.0000	0.0000		0.0000	0.0000	0.0000	0.0000
(3)	0.1130	0.1070		-0.0092	-0.0090	-0.0191	0.0637
(4)	0.0530	0.0510		-0.0097	-0.0068	-0.0104	0.0245
(5)	0.1480	0.0860		0.0127	0.0122	0.0055	0.0519
(6)	0.4800	0.2830		-0.0011	-0.0161	0.0379	-0.0830
(1)	0.6330	0.4510	48.20	0.0323	0.0176	0.0470	-0.0065
(2)	0.0000	0.0000		0.0000	0.0000	0.0000	0.0000
(3)	0.0840	0.1570		-0.0085	-0.0029	-0.0285	0.0931
(4)	0.0620	0.1090		-0.0109	0.0123	-0.0157	0.0626
(5)	0.0310	0.0460		-0.0006	0.0042	-0.0097	0.0107
(6)	0.1900	0.2370		-0.0123	-0.0312	0.0069	-0.1599
MEAN DEVIATIONS FOR COMPONENT:			(1)	0.0455	0.0553	0.0752	0.0408
			(2)	0.1092	0.1186	0.2203	0.1110
			(3)	0.0084	0.0066	0.0178	0.0442
			(4)	0.0097	0.0085	0.0151	0.0240
			(5)	0.0256	0.0226	0.0436	0.0384
			(6)	0.0087	0.0132	0.0299	0.0541
			MEAN	0.0345	0.0375	0.0670	0.0521

530

Formula	Name	Other components	Page
$C_2H_4Cl_2$	1,2-Dichloroethane	Benzene	302–305
		Carbon tetrachloride	30
		Ethyl alcohol	148–150
		*Cyclo*hexane	330
		Methyl alcohol	79–82
		Propyl alcohol	127–130
		Toluene	131–132
		1,1,1-Trichloroethane	114
C_2H_4O	Ethylene oxide	Water	133
$C_2H_4O_2$	Acetic acid	Water	445–446
C_2H_5Br	Ethyl bromide	Benzene	134
		Ethyl alcohol	135
		Heptane	136
C_2H_6O	Ethyl alcohol	Acetone	217–222
		Acetonitrile	137
		Benzene	138–144
		Butyl alcohol	145–146
		Carbon tetrachloride	31–33
		Chloroform	55–59
		Decane	147
		1,2-Dichloroethane	148–150
		Ethyl acetate	151–153
		Ethylbenzene	154
		Ethyl bromide	135
		Ethyl ether	274–281
		Heptane	155–158
		Hexane	159
		*Cyclo*hexane	160
		Methyl alcohol	89
		Methyl*cyclo*hexane	161–166
		Methyl*cyclo*pentane	167
		*Iso*octane	168–169
		Pentane	287–288
		Propyl alcohol	170–173
		*Iso*propyl alcohol	174
		Toluene	175–183
		Trichloroethylene	184
		Triethylamine	355–357
		Water	185–195
		Acetone and methyl alcohol	488–490
		Benzene and heptane	499–500
		Benzene and methyl*cyclo*pentane	501–502
		Hexane and methyl*cyclo*pentane	503–504
		Benzene, hexane, and methyl-*cyclo*pentane	518–523
C_3H_6O	Acetone	Acetonitrile	196
		Benzene	197–201
		Butyl alcohol	202–203
		Carbon disulphide	437–439
		Carbon tetrachloride	204–208
		Chlorobenzene	209
		Chloroform	210–216
		Ethyl alcohol	217–222
		Ethyl ether	270–271

Formula	Name	Other components	Page
C_4H_9Br	l-Bromobutane	Butyl alcohol 1-Chlorobutane Heptane	266 267 376
C_4H_9Cl	1-Chlorobutane	1-Bromobutane Heptane	267 268
$C_4H_{10}O$	Butyl alcohol	Acetone Benzene 1-Bromobutane Carbon tetrachloride Ethyl alcohol Heptane Methyl alcohol 1,2,3,4-Tetrahydronaphthalene Toluene	202–203 295–298 266 29 145–146 371 34 269 361
$C_4H_{10}O$	tert.-Butyl alcohol	Benzene	299
$C_4H_{10}O$	Ethyl ether	Acetone Carbon disulphide Carbon tetrachloride Ethyl alcohol	270–271 272 273 274–281
$C_4H_{11}N$	Diethylamine	Water	282–284
$C_5H_4O_2$	Furfuraldehyde	Toluene 2,2,4-Trimethylpentane	362 399
C_5H_8	Isoprene	Methyl alcohol Hexane, methyl alcohol, 2-methyl- pentane, 3-methylpentane, and methylcyclopentane	93 529–530
C_5H_{10}	Cyclopentane	Benzene	285
C_5H_{12}	Pentane	Acetone Ethyl alcohol	286 287–288
C_5H_{12}	Isopentane	Carbon disulphide	289
$C_6H_4Cl_2$	p-Dichlorobenzene	Benzene	301
C_6H_5Cl	Chlorobenzene	Acetone 1,2-Ethylene dibromide Hexane 1-Nitropropane	209 123–124 346–347 290–291
$C_6H_5O_2N$	Nitrobenzene	Cyclohexane	336
C_6H_6	Benzene	Acetone Acetonitrile Aniline Butyl alcohol tert.-Butyl alcohol Carbon disulphide Carbon tetrachloride	197–201 115 292–294 295–298 299 440–442 24–28

Formula	Name	Other components	Page
C_6H_{12} (*cont.*)	*Cyclo*hexane	Carbon tetrachloride	35–40
		1,2-Dichloroethane	330
		Ethyl alcohol	331–334
		Methyl acetate	236
		Methyl*cyclo*hexane	335
		Nitrobenzene	336
		*Iso*propyl alcohol	337–338
		Toluene	339
		2,2,3-Trimethylbutane	340
C_6H_{12}	Methyl*cyclo*pentane	Benzene	341–342
		Ethyl alcohol	167
		Methyl alcohol	94
		Toluene	343
		Benzene and ethyl alcohol	501–502
		Ethyl alcohol and hexane	159
		Benzene, ethyl alcohol, and hexane	518–523
		·Isoprene, hexane, methyl alcohol, 2-methylpentane and 3-methylpentane	529–530
C_6H_{14}	Hexane	Acetone	223
		Benzene	344–345
		Chlorobenzene	346–347
		Ethyl alcohol	348–353
		Methyl alcohol	91–92
		Propyl alcohol	354
		Ethyl alcohol and methyl*cyclo*pentane	503–504
		Benzene, ethyl alcohol and methyl*cyclo*pentane	501–502
		Methyl alcohol, 2-methylpentane, and 3-methylpentane	527–528
		Isoprene, methyl alcohol, 2-methylpentane, 3-methylpentane, and methyl*cyclo*pentane	529–530
C_6H_{14}	2-Methylpentane	Methyl alcohol	95
		Hexane, methyl alcohol, and 3-methylpentane	527–528
		Hexane, isoprene, methyl alcohol, 3-methylpentane, and methyl*cyclo*pentane	529–530
C_6H_{14}	3-Methylpentane	Methyl alcohol	96
		Hexane, methyl alcohol, and 2-methylpentane	527–528
		Hexane, isoprene, methyl alcohol, 2-methylpentane, and methyl*cyclo*pentane	529–530
$C_6H_{15}N$	Triethylamine	Ethyl alcohol	355–357
		Water	357–360
$C_7H_5F_3$	Trifluorotoluene	Bromine	22

Formula	Name	Other components	Page
C₇H₈	Toluene	Benzene	322–325
		Butyl alcohol	361
		1,2-Dichloroethane	131–132
		Ethyl acetate	264
		Ethyl alcohol	175–183
		Furfuraldehyde	362
		Heptane	379–386
		*Cyclo*hexane	339
		Methyl alcohol	101–102
		Methyl*cyclo*hexane	372–375
		Methyl*cyclo*pentane	343
		Octane	363–368
		Phenol	369
		Phosgene	112
		*Iso*propyl alcohol	370–371
		2,2,4-Trimethylpentane	403–406
		Methyl alcohol and heptane	510
		Methyl*cyclo*hexane, phenol, and 2,2,4-trimethylpentane	524–526
C₇H₈O	*m*-Cresol	Benzene	300
C₇H₁₄	Methyl*cyclo*hexane	Benzene	315
		Ethyl alcohol	161–166
		Heptane	378
		*Cyclo*hexane	335
		Phenol	371
		*Iso*propyl alcohol	250
		Toluene	372–375
		2,2,4-Trimethylpentane	400
		Phenol, toluene, and 2,2,4-trimethylpentane	524–526
C₇H₁₆	Heptane	Benzene	307–308
		1-Bromobutane	376
		Butyl alcohol	377
		Carbon tetrachloride	34
		1-Chlorobutane	268
		Ethyl alcohol	155–158
		Ethyl bromide	136
		Methyl alcohol	90
		Methyl*cyclo*hexane	378
		Toluene	379–383
		Benzene and ethyl alcohol	155–158
		Methyl alcohol and toluene	510
C₇H₁₆	2,4-Dimethylpentane	Benzene	385
C₇H₁₆	2,2,3-Trimethylbutane	*Cyclo*hexane	340
C₈H₈	Styrene	Ethylbenzene	389
		Propyl alcohol	240
C₈H₁₀	Ethylbenzene	Ethyl alcohol	154
		Octane	394–397
		1-Octene	393

Formula	Name	Other components	Page
C_8H_{10} (*cont.*)	Ethylbenzene	Propyl alcohol	238–239
		Styrene	389–390
		2,2,5-Trimethylhexane	408
C_8H_{10}	*m*-Xylene	Aniline	391
C_8H_{10}	*p*-Xylene	Aniline	392
		Ethyl acetate	265
C_8H_{16}	1-Octene	Ethylbenzene	393
C_8H_{18}	Octane	Ethylbenzene	394–397
		Propionic acid	396
		Toluene	363–368
		2,2,4-Trimethylpentane	401
C_8H_{18}	*Iso*octane	Ethyl alcohol	168–169
C_8H_{18}	2,2,4-Trimethylpentane	Furfuraldehyde	399
		Methyl*cyclo*hexane	400
		Octane	401
		Phenol	402
		Toluene	403–406
		Methyl*cyclo*hexane, phenol, and toluene	524–526
C_9H_{10}	α-Methylstyrene	Phenol	407
C_9H_{20}	2,2,5-Trimethylhexane	Ethylbenzene	408
$C_{10}H_8$	Naphthalene	1-Hexadecene	409
		Tetradecane	410–415
$C_{10}H_{12}$	1,2,3,4-Tetrahydronaphthalene	Butyl alcohol	269
$C_{10}H_{16}$	α-Pinene	β-Pinene	416–418
$C_{10}H_{16}$	β-Pinene	α-Pinene	416–418
$C_{10}H_{18}$	*trans*-Decaline	Decane	419–421
$C_{10}H_{22}$	Decane	*trans*-Decaline	419–421
		Ethyl alcohol	147
$C_{12}H_{26}$	Dodecane	1-Hexadecene	422–425
		1-Octadecene	426–428
$C_{14}H_{30}$	Tetradecane	1-Hexadecene	429–436
		Naphthalene	410–415
$C_{16}H_{32}$	1-Hexadecene	Dodecane	422–425
		Naphthalene	409
		Tetradecane	429–436
$C_{18}H_{36}$	1-Octadecene	Dodecane	426–428
CS_2	Carbon disulphide	Acetone	437–439
		Benzene	440–442
		Carbon tetrachloride	443
		Chloroform	444
		Ethyl ether	272
		*Iso*pentane	289
$FeCl_3$	Ferric chloride	Hydrogen chloride and water	513
HBr	Hydrogen bromide	Water	457–458
HCl	Hydrogen chloride	Water	459–464
		Ferric chloride and water	513

Formula	Name	Other components	Page
HCl (*cont.*)	Hydrogen chloride	Potassium chloride and water	514–515
		Sulphuric acid and water	511–512
HF	Hydrogen fluoride	Water	465–467
HNO₃	Nitric acid	Water	470–484
		Lithium nitrate and water	516
		Magnesium nitrate and water	517
H₂O	Water	Acetic acid	445–446
		Acetone	230–232
		Acetonitrile	119
		Allyl alcohol	233
		Aniline	447
		Diethylamine	282–284
		Dimethylformamide	448
		Ethyl alcohol	185–195
		Ethylene oxide	133
		Hydrazine	449–454
		Hydrogen bromide	457–458
		Hydrogen chloride	459–464
		Hydrogen fluoride	465–469
		Methyl alcohol	104–111
		Nitric acid	470–484
		Phenol	455–456
		Propyl alcohol	241–244
		*Iso*propyl alcohol	259–269
		Sulphuric acid	485
		Triethylamine	357–530
		Acetone and acetonitrile	468–487
		Acetone and methyl alcohol	491–494
		Ferric chloride and hydrogen chloride	513
		Hydrogen chloride and sulphuric acid	459–464
		Lithium nitrate and nitric acid	516
		Magnesium nitrate and nitric acid	517
		Potassium chloride and hydrogen chloride	514–515
H₂SO₄	Sulphuric acid	Water	485
		Hydrogen chloride and water	459–464
KCl	Potassium Chloride	Hydrogen chloride and water	514–515
LiNO₃	Lithium nitrate	Nitric acid and water	516
Mg(NO₃)₂	Magnesium nitrate	Nitric acid and water	517
N₂H₄	Hydrazine	Water	449–454

REFERENCES

1. ANTOINE C., *C.R. Acad. Sci.*, *Paris* **107**, 681, 836, 1143 (1888).
2. BLACK C., Am. Petrol. Inst., Carnegie Inst. Technol., Pittsburg, Research Project 44.
3. BROWN I. and SMITH F., *Australian J. Chem.* **12**, 407 (1959).
4. CLARK A. M., *Trans. Faraday Soc.* **41**, 718 (1945).
5. DEMING W. E., *Statistical Adjustment of Data*, Wiley, New York, 1944.
6. DREISBACH R. R., *Physical Properties of Chemical Compounds*, I–III, Adv. Chem. Series, **15** (1955), **22** (1959), **29** (1961).
7. GEFFNER J. and WORTHING A. G., *Treatment of Experimental Data*, Wiley, New York, 1946.
8. HÁLA E., *Coll. Czech. Chem. Commun.* **22**, 1727 (1957).
9. HÁLA E., *Coll. Czech. Chem. Commun.* **24**, 2453 (1959).
10. HÁLA E., *Coll. Czech. Chem. Commun.* **28**, 1780 (1963).
11. HÁLA E., *Coll. Czech. Chem. Commun.* **31**, 908 (1966).
12. HÁLA E. and BOUBLÍK T., *Coll. Czech. Chem. Commun.* **29**, 2412 (1964).
13. HÁLA E., PICK J., FRIED V. and VILÍM O., *Vapour–Liquid Equilibrium*, Pergamon Press, Oxford, 1967.
14. KRETSCHMER C. B. and WIEBE R., *J. Am. Chem. Soc.* **71**, 1793, 3176 (1949).
15. MANDEL J., *The Statistical Analysis of Experimental Data*, Wiley, New York, 1964.
16. MARGULES M., *S.B. Akad. Wiss. Wien*, Math. Naturw. Kl. II, **104**, 1243 (1895).
17. POLÁK J. and MERTL I., *Coll. Czech. Chem. Commun.* **30**, 3526 (1965).
18. PRAHL W. H., *Ind. Eng. Chem.* **43**, 1767 (1951).
19. SPINNER I. H., LU B. C. Y. and GRAYDON W. F., *Ind. Eng. Chem.* **48**, 147 (1956).
20. TAYLOR R. S. and SMITH L. B., *J. Am. Chem. Soc.* **44**, 2450 (1956).
21. VAN LAAR J. J., *Z. Phys. Chem.* **72**, 723 (1910).
22. VAN LAAR J. J., *Z. Phys. Chem.* **185**, 35 (1929).
23. WICHTERLE I., *Coll. Czech. Chem. Commun.* **31**, 3821 (1966).
24. WILLINGHAM C. B., TAYLOR W. J., PIGNOCCO J. M. and ROSSINI F. D., *J. Res. Natl. Bur. Standards* **35**, 219 (1945).
25. WOHL K., *Trans. Am. Inst. Chem. Engrs.* **42**, 215 (1946).
26. WOHL K., *Chem. Eng. Progr.* **49**, 218 (1953).